Future City

Volume 19

As of 2008, for the first time in human history, half of the world's population now live in cities. And with concerns about issues such as climate change, energy supply and environmental health receiving increasing political attention, interest in the sustainable development of our future cities has grown dramatically.

Yet despite a wealth of literature on green architecture, evidence-based design and sustainable planning, only a fraction of the current literature successfully integrates the necessary theory and practice from across the full range of relevant disciplines.

Springer's *Future City* series combines expertise from designers, and from natural and social scientists, to discuss the wide range of issues facing the architects, planners, developers and inhabitants of the world's future cities. Its aim is to encourage the integration of ecological theory into the aesthetic, social and practical realities of contemporary urban development.

Elisa Pozo Menéndez • Ester Higueras García
Editors

Urban Design and Planning for Age-Friendly Environments Across Europe: North and South

Developing Healthy and Therapeutic Living Spaces for Local Contexts

Springer

Editors
Elisa Pozo Menéndez
ABIO Research Group
Universidad Politécnica de Madrid
Madrid, Spain

Ester Higueras García
ABIO Research Group
Universidad Politécnica de Madrid
Madrid, Spain

ISSN 1876-0899 ISSN 1876-0880 (electronic)
Future City
ISBN 978-3-030-93874-1 ISBN 978-3-030-93875-8 (eBook)
https://doi.org/10.1007/978-3-030-93875-8

This Springer imprint is published by the registered company Springer Nature Switzerland AG
The registered company address is: Gewerbestrasse 11, 6330 Cham, Switzerland

This book is dedicated to the inspiring persons that put a light on the way.

Foreword

Cities That Care. The Dream of Life[1]

We are all fragile and have different types of needs and interests throughout life. To guarantee the right to equal opportunities and treatment, anyone should be able to function individually, autonomously and safely, regardless of their abilities, in any space, public or private. These objectives guide the development of the concepts of universal design and accessibility chain, aimed at promoting the transformation of spaces and different elements of the environment, so that they are useful and flexible for people with different abilities.

There is a deep social segregation between autonomous, independent people who are in an active and productive stage, as it is understood from the liberal economy, and those vulnerable, dependent people excluded from this activity. Active life is signified by individualism and depends on the need for care, until now with an insufficient response, poorly designed in both qualitative and quantitative terms.

This individualism, which explains the lack of recognition of the value of care, is confronted with the vulnerability or dependence that, with different manifestations, characterizes us throughout life. Some elements, such as early schooling, the regulation of the family conciliation, the full incorporation of women into working life or the rejection of any patriarchal or paternalistic behaviour, drive this contradiction. The current model of social care, including the residential centres for the elderly, whose functioning has been publicly revealed with particular harshness in our environment due to COVID-19, is pending reform.

The ability to address each disability situation in a specific way is understood as an indicator of the maturity of a social system, of the dignity of a society and of an advanced and caring democracy. In essence, it is about putting people at the centre of the system.

[1]This presentation is inspired by references of two books: *Building and Dwelling. Ethics for the City* by Richard Sennett (2019) and *Time of Care. Another Way of Being in the World* by Victoria Camps (2021).

Public space, especially in large cities, excludes the elderly and/or dependent people. Community involvement is necessary to identify and respond to the needs of all people. Small careful actions in the proximity of densified urban areas, such as the placement of benches, the existence of a vegetal shade, some chairs (not anchored to the ground) in a park, a fountain, a safe space in the open air, are highly effective in increasing the autonomy, health and safety of the elderly and/or dependent persons, and improve the life of the entire community.

Cities are open and complex systems, in which there is a strong difference between what is built and what is inhabited, what is designed and what is lived, precisely due to the complexity of the variables involved. Knowing how to manage (as well as knowing how to see) complexity in time is a continuous work of the citizen community. Investigating how a built space is transformed into a place lived by the inhabitants of a city is a pending task in any urban system.

Among the city model inhabited by a population that is mainly active, productive, conditioned by working hours and focused on individualistic functioning, and the complex residential model in consolidated areas, which integrates heterogeneous populations in age, interests or activities, there is a notable difference that is perceived in the way in which public space is inhabited. Some just occupy a space, and others inhabit a place.

The problem arises when that consolidated and heterogeneous city is taken as a commodity and is transformed, emptying itself of inhabitants (elderly, dependent, fragile), for the sake of a supposed progress. The management of the city requires an ethic of care, from the planning of the uses in the consolidated centres, such as the small intervention in a residential area on the edge. The design of open urban forms, defined through a compromise between builders and inhabitants, is still unprecedented in our environment.

It is good and timely to carry out analysis of good practices and comparative studies on the effectiveness of intervention in a public space that serves everyone, and it is urgent to generalize these interventions in the face of the unstoppable reality of a society in which the dimension of ageing represents an augmentation of the decline in the autonomy of an increasing number of people.

The contributions provided in this book are essential to understand how to intervene in our cities to respond to the profound social, demographic, economic, labour, social, health and environmental changes that are taking place, and to which it is necessary to have specific and integrated responses that are based on an ethic of care, which involves us and holds everyone responsible.

Inés Palanca Sánchez was born in Salamanca, Spain. She has a degree in Medicine and Surgery from Complutense University of Madrid, Spain. Her professional career is focused on planning and quality management, and it has been developed in recent years in health and social institutions linked to different public administrations (Ministry of Health and national health institutions and from the Community of Madrid). A relevant part of the activity that she has developed in her work has been focused on incorporating the perspective of users, both workers and people cared for in the centres, in the design of organizational or structural interventions, exchanging with professionals from different sectors, in a multiprofessional team and the transfer or access of knowledge and tools designed in different areas. She is responsible for the management of 20 centres of the Institute for the Elderly and Social Services (IMSERSO), linked to the Ministry of Social Rights and the 2030 Agenda, which care for people with disabilities, the elderly or those suffering from potentially disabling diseases and which are located in different autonomous communities.

IMSERSO (Institute for the Elderly and Social Services), Government of Spain
Madrid, Spain
inespalanca@imserso.es

Inés Palanca Sánchez

Acknowledgements

We want to thank all the contributors to this book for all the hard work during an extremely difficult period both for individuals and society. COVID-19 outbreak has challenged our health, both physical and mental, and overwork has not stopped. We express our sincere gratitude for the commitment of all the contributors to finish this book and also the meetings and exchange with all of them, as well as the persons interviewed and guided visits for the study cases in the beginning of 2020, just before the pandemic arrived in Europe.

Elisa Pozo Menéndez

Ester Higueras García

Introduction

Cities are nowadays presenting multiple challenges, where health inequities arise as one of the major concerns. Public health has been the main aim for urban planning since the industrial revolution by the end of the eighteenth century, where epidemics due to poor quality conditions and insalubrity were the principal problems. However, the urban planning of that period was based on an unprecedented development period, where massive production characterized many sectors, such as agriculture, construction, transports and energy. Since 1990, better environmental conditions, social services and health advances and technology made possible to life longer in a relatively good quality of life. However, the rapid expansion and development relegate our connection with the natural environment. Additionally, climate change and urgent answers for healthier lifestyles and environmental conditions make it urgent to take action to create healthier environments for people, in harmony with natural resources. This book offers a specific perspective on urban settings, as most of the population is urbanized, and it is expected to continue this progressive augmentation to develop healthier urban environments. Taking this tendency into account, sustainability of the planet and the resources are in crisis because of massive urbanization worldwide, which delivers a higher consumption of goods and services. Probably, the worldwide COVID-19 pandemic evidences the problematics of the cities. Extreme weather events have been also occurring recently in many European countries, challenging again the resiliency of cities and territories. Again, as in many other crises, the most vulnerable ones are the most affected. Among everything, COVID-19 has highlighted the frailty and vulnerability of the care system. We have suffered the death of many beloved ones and society was alarmed of the massive loss of older people living in care homes.

However, under these difficult circumstances, best practices and exceptional work from many professionals and researchers have made possible to start going out from the worst scenario. Equally, societies are more concerned about health, both physical and mental, and many have questioned the urban lifestyle of our present cities. Health has become again one of the major concerns in all policies, and here, regarding cities and territories, it is important to understand that humans' health is compulsorily intertwined with the natural environment and the development of a new culture of care, both for ourselves and urban ecosystems.

This book is the result of networking in different countries across Europe, sharing common aspirations to go forward healthier and towards more inclusive cities and neighbourhoods. Multidisciplinary and intergenerational profiles and different nationalities were selected in order to have a global perspective in terms of healthy ageing and designing urban spaces for "ageing in place". The result includes perspectives from European Southern countries as well as Northern ones, in order to learning from experiences, perspectives and innovations from all over Europe. The book is divided in three different parts, in order to introduce the challenges for cities in relation with the quality of life and healthy ageing; shows specific methodologies or perspectives applied to study cases in order to understand the impact of design and policies on people's health; and, finally, includes several study cases for inspiration in order to innovate in terms of healthier environments and cities to live (Fig. 1).

The first part of the book offers a general vision about health inequities in cities. It is necessary to consider healthy and active ageing at all stages of life to get older with a good quality of life, which is directly related with the meaningful role of older adults in society.

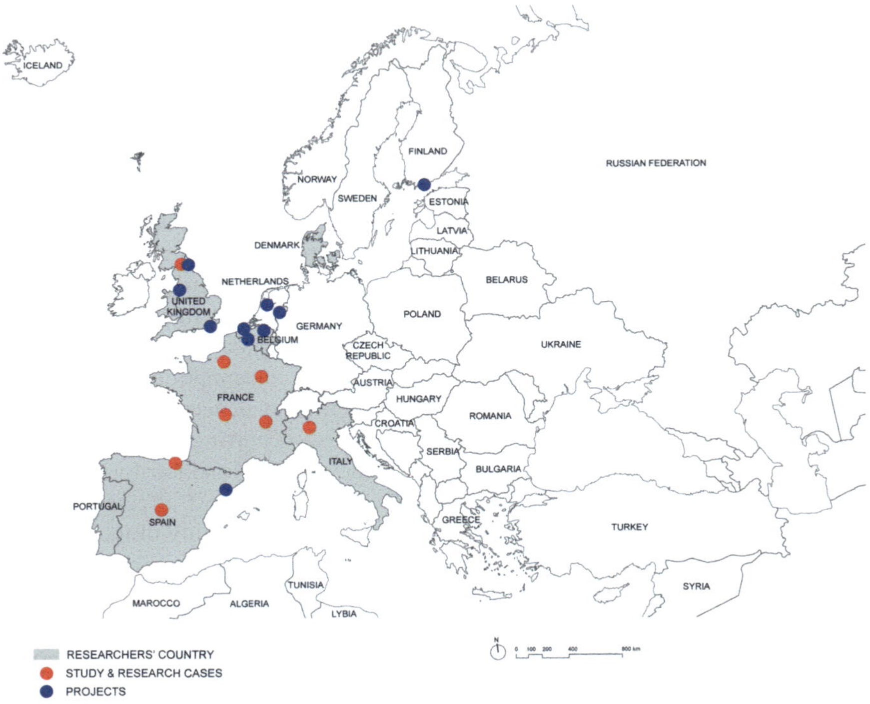

Fig. 1 Nationalities and cities that are represented along the contributions of this book. Light grey shows the researchers' nationalities, red dots indicate research and study cases, and blue dots are specific projects that have been visited and are included in the third part of the book. (Source: Pozo Menéndez, 2021)

Angela Freitas and Paula Santana start the book with a complete framework of the evolution of health policies and urban trends in European cities. The international framework for health promotion and main objectives and challenges in cities are presented to understand how they can be adapted in governments and local agendas. In this chapter the urgent need of a health-equity-oriented urban planning scheme and new urban models that can transform the current built environments into healthier and vibrant spaces is highlighted.

Camille Picard brings in the second chapter an interesting perspective from outside Europe for social innovation. The comparison with Japan, one of the countries with a higher expectancy of life, offers a very valuable perspective of the commitment of governments to implement new policies and innovative models for enabling environments for care and to maintain meaningful social bounds till the end of life. The findings show the importance of the social, cultural, economic and political context from which social innovation emerges. It is a key element to understand that innovation needs a fertile framework to propose different alternatives and create a positive outcome from a given problem. Social innovation led by elderly people recognizes their value in society and their human rights as part of the community.

The third chapter continues to explore innovation from the perspective of the so-called *silver economy*. Montserrat Calvo and José Antonio Ondiviela analyse the relationship between age-friendly and inclusive cities in relation with talent and generation of economic growth. If a city, an urban environment, is comfortable for older people, it will be also desirable for the rest of the population. Even more, considering the percentage of population that is older than 50 years, in the next decades an increasing need for new services and products to satisfy this specific group of the population is expected.

The fourth chapter goes into detail about the quality of life. Carmen Rodríguez-Blázquez and Maria João Forjaz examine which dimensions and elements matter to feel satisfied in later life and the different scales and tools that can be used for measurement and evaluation. Additionally, they explain the importance of an enabling and supportive residential environment and review the influence of the different living alternatives on older adults' quality of life.

This perspective is completed with the fifth chapter by Gloria Fernández-Mayoralas, Vicente Rodríguez-Rodríguez, Diego Sánchez-González and Fermina Rojo-Pérez, where they analyse the relationship between the self-image of older adults and the residential environment in different settings, such as nursing homes, family dwelling and cohousing. Results show a close relationship between the living environment and how older adults perceive themselves, where social participation and the availability of inclusive spaces are key elements that determine this self-image.

In direct relation with the quality of life, the sixth chapter by Álvaro García-Soler, Heitor García-Lantarón, Sara Marsillas Rascado, Elena del Barrio Truchado, Cristina Buiza Bueno and Pura Díaz Veiga presents the care system challenges of current societies and urban environments. Some projects developed by the Matia Foundation and Matia Institute are presented in order to illustrate some examples of the new model of the healthy city of care.

Finally, to close the first part of the book, the seventh chapter written by Miguel Padeiro, Ângela Freitas, Cláudia Costa, Adriana Loureiro and Paula Santana analyses the recent episode of the COVID-19 pandemic from an urban planning and design perspective. The authors show interesting interventions at the local level to explore mechanisms for enhancing resilience in cities in order to prevent and mitigate future pandemics or other emergencies.

The second part offers a closer perspective at the local scale to healthier environments from the urban, social, innovation and ethnographic perspective.

The eighth chapter written by Rose Gilroy brings a general framework for inclusive residential projects for all in the United Kingdom, with an innovative project that is being developed at Newcastle University. An intergenerational approach to solve current challenges of younger and older generations is discussed in this chapter, with a holistic view of a segregated society that is trying to find solutions to problems in specific and separated sectors, but the solution probably can satisfy both young and old citizens. Multigenerational homes and communities need physical and social architecture for care and respect and living in a more sustainable way.

Another vision from housing solutions for healthy ageing is explained by Heitor García Lantarón, Mayte Sancho and Maider Azurmendi. The ninth chapter explains the Danish paradigm for active ageing and the application of these principles to a specific project of lifetime homes in northern Spain. Architectural design conditions not only individual and collective relationships, but also the relation with the environment and urban surroundings. The lifetime homes project works from two complementary perspectives, home care and design, both necessary for meeting the principal objective: ageing in place.

Chapter 10 brings an evaluation of the development and management of cohousing initiatives in Spain. Fermina Rojo-Pérez, Diego Sánchez-González, Vicente Rodríguez-Rodríguez and Gloria Fernández-Mayoralas explore through deep interviews the features that make this residential option a healthy environment for ageing in place, reducing and preventing loneliness, at the same time that it provides an optimization of resources and socio-health services thanks to the management model by a cooperative. This emergent initiative is contextualized in the age-friendly communities' strategy for active and healthy ageing in place.

Health and sustainability meet in architectural and urban design, to adapt to the local climate and geographic conditions in order to use rationally the local resources, but also to provide wellbeing and quality of life. The design of a building or a public space not only needs to answer to specific programmes and ergonomics for addressing healthier lifestyles, but also needs to be articulated with the surrounding environment, with an appropriate scale and geometry, as well as to answer to specific climate conditions, to assure energy efficiency and also comfort and wellbeing. Neveen Hamza illustrates in Chap. 11 the education methodology for architecture students in Newcastle University not only to raise awareness about ageing and dementia, but also to make the future architects of society answer the future challenges with a sensitive approach at the human scale, taking into account the conditions of the most vulnerable ones.

In parallel with housing alternatives and construction of living environments, the public space also has to be adapted not only for the increasing demand on healthier environments, but also to meet the challenges of a climate change scenario that is also threatening health public systems across Europe. In the twelfth chapter, Emilia Román López, Rafael Córdoba Hernández and Teresa Eiroa Escalada show the intersection between sustainable indicators for a climate change scenario and the healthy city characteristics, explaining the multiple opportunities for adaptation and mitigation of the urban built environment in order to reduce the urban heat island effect and mitigate the risk of extreme weather conditions in cities.

Chapter 13 shows a specific study case from the point of view of older adults in public space, how they use it (permanence, frequency of use, activities) and how they perceive thermal and acoustic comfort. This analysis study is presented by María Teresa Baquero Larriva. She compares two public spaces in Madrid (Spain) and Newcastle-upon-Tyne (United Kingdom) to understand the similarities and disparities about comfort in public space by this group of the population in two different European contexts. Indeed, environmental factors that affect the subjective perception of comfort influence the permanence of older people in outdoor spaces.

For climate change mitigation and adaptation, as well as to promote both physical and mental health, integrating nature-based solutions in the built environment is needed nowadays. However, while green lawns and vegetation are easy to find in European northern countries, Mediterranean cities offer a different landscape and other climate condition that need to be understood in order to successfully integrate these designs and activities in nature. The fourteenth chapter written by Elisa Pozo Menéndez brings this perspective on mental health and green infrastructure in cities. A historical approach to understand the close relationship between gardens and health shows that this culture has been lost. A study case in a vulnerable city of Madrid shows the existing gap between environmental planning and health facilities.

Enabling environments are needed in outdoor public spaces, to make people remain as long as possible at home and in their communities. The fifteenth chapter by Veerle Baert, Elise Cornelis and Patricia DeVriendt identifies the different items and projects that have been implemented across Belgium to support the creation of dementia-friendly communities. Moreover, an interdisciplinary approach is needed to fully understand the process of ageing and the feeling of belonging with specific elements of the urban space and landscape. The sixteenth chapter by Caterina Gallerani shows from an ethnographic perspective the importance of remarkable places in the cities, such as markets. In many cities across Europe markets have been adapted and renovated, but not always maintaining the essential features that characterized them but accelerating a gentrification process. This contribution explores the immaterial heritage of a city, the social relations that are related and intertwined with the memory of a place.

On the other hand, although heritage plays an important role for a sense of belonging to a place and a community, new spaces for innovation and for fostering new solutions for care and wellbeing are also needed in cities. Living Labs and the dynamics in territories are presented in the seventeenth chapter by Luciana Castro Gonçalves. France offers this vibrant environment for innovation and economic activity related with ageing and the silver economy paradigm. Talent, innovation and co-design methodologies in relation to territorial dynamics are studied in three different contexts in France.

The second part of the book ends with the eighteenth chapter, a special contribution by Barbara Douglas, who shows the innovative approach of Voice organisation and Elders Council of Newcastle members in the United Kingdom. The beginning of this chapter acknowledges that the success of communities in becoming age-friendly depends on the inclusion of older people as principal characters of their own innovation process in a close relationship and engaging the co-creation process with researchers and innovation labs. She reminds us that older people are not a homogeneous group, and in this sense, a variety of alternatives and solutions shall be developed, in order to provide a wide range of options to be chosen to live life meaningfully.

Finally, to offer a practical perspective for further research, the third part of the book brings projects across Europe that illustrate the different realities in terms of innovation towards age- and dementia-friendliness. A variety of projects have been selected from bigger to smaller scale to show a variety of initiatives that are emerging to meet the evolving needs of the population. The healthy city is a sustainable one, based on proximity relations of trust and support within communities and with smart technologies and solutions to improve care and quality of life. Ten projects have been selected across Europe to illustrate innovation and different approaches to improve the quality of life and promote health and wellbeing in the built environment.

The Conclusion discusses three main questions that have been addressed by all the researchers and contributors across Europe:

- Which is the main challenge for addressing urban design from the perspective of ageing?
- How can societies transform into value, having inclusive environments for all ages?
- Regarding your work/research, what are the main strategies in cities that should be implemented in urban planning and design to encourage ageing in place?

The whole book is a multidisciplinary and international vision across Europe, facing similar challenges about ageing and quality of life. Each country, region and city has its own context, including territorial dynamics, climate, culture and socio-economic development, as well as multiple perspectives, to address the common challenges of demographic transition, climate change and healthy urban environments. This book offers this multiple vision across Europe, to better understand the challenges of ageing and care and contribute to the post-pandemic healthy and sustainable city.

ABIO Research Group
Universidad Politécnica de Madrid,
Madrid, Spain
e.pozo@upm.es

Elisa Pozo Menéndez
Ester Higueras García

Contents

About the Editors and Contributors

Editors

Elisa Pozo Menéndez is architect and urban planner, MsC in *Villes Durables* (Université Paris-Est Marne-la-Vallée - Institut Français d'Urbanisme, 2014) and MsC in Environment and Bioclimatic Architecture from the Universidad Politécnica de Madrid (MAYAB-UPM, 2019). She is researcher in the Bioclimatic Architecture in a Sustainable Environment (ABIOUPM) and PhD candidate on dementia, healthy ageing and urbanism. Since 2022 she is researcher in Matia Research Institute and co-founder and CEO of mita atelier sllp. She has experience as coordinator of European projects at Universidad Politécnica de Madrid on innovative education and urban health, and she collaborates as architect and consultant with different studios and engineering teams in urban studies, mobility plans, architecture projects and competitions. She also teaches workflows and digital tools at Istituto Europeo di Design since 2017. Since 2006 she has collaborated with multiple NGOs and associations, working with vulnerable collectives. In 2020 she receives the third prize of the Youth Talent Awards in the Solidarity category of the Region of Madrid.

Ester Higueras Garcia is doctor of Architecture and professor in the Master's programme "Environment and Bioclimatic Architecture and Territorial Planning" at the Universidad Politécnica de Madrid (UPM). She has participated in many courses of postgraduate degree and conferences. She is a team member of the research group ABIO-UPM (Bioclimatic Architecture in a Sustainable Environment-UPM). Since her PhD thesis in 1997, she has been combining both theory and practice in projects of environmental planning, bioclimatic urban design, passive strategies applied to urban scale, cultural landscape and environmental policies.

Contributors

Maider Azurmendi Ormazabal Community Co-operation, Matia Fundazioa, Donostia-San Sebastián, Spain

Veerle Baert Department of Social Work, Artevelde University of Applied Sciences, Ghent, Belgium

María Teresa Baquero Larriva Universidad Politécnica de Madrid, Madrid, Spain

Cristina Buiza Bueno Matia Fundazioa and Matia Research Institute, Donostia-San Sebastián, Spain

Montserrat Calvo Social and Cultural Anthropologist, Madrid, Spain

Luciana Castro Gonçalves ESIEE Paris, Université Gustave Eiffel, Paris, France

Rafael Córdoba Hernández Department of Urban and Regional Planning, Escuela Técnica Superior de Arquitectura, Universidad Politécnica de Madrid, Madrid, Spain

Elise Cornelis Department of Occupational Therapy, Research and Development Centres Health & Care, Artevelde University of Applied Sciences, Ghent, Belgium

Cláudia Costa Centre of Studies in Geography and Spatial Planning (CEGOT), Department of Geography and Tourism, University of Coimbra, Coimbra, Portugal

Patricia DeVriendt Faculty of Medicine and Health Sciences, Department of Rehabilitation Sciences, Ghent University, Ghent, Belgium
Department of Occupational Therapy, Artevelde University of Applied Sciences, Ghent, Belgium
Mental Health Research group, Frailty in Ageing Research Group, Vrije Universiteit Brussel, Belgium

Elena del Barrio Truchado Matia Fundazioa and Matia Research Institute, Donostia-San Sebastián, Spain

Pura Díaz-Veiga Matia Fundazioa and Matia Research Institute, Donostia-San Sebastián, Spain

Barbara Douglas Elders Council of Newcastle, Newcastle upon Tyne, UK

Teresa Eiroa Escalada Department of Urban and Regional Planning, Escuela Técnica Superior de Arquitectura, Universidad Politécnica de Madrid, Madrid, Spain

Colin Elkington Newcastle University, Newcastle upon Tyne, UK

Gloria Fernández-Mayoralas Institute of Economics, Geography and Demography (IEGD); Spanish National Research Council (CSIC), Research Group on Ageing (GIE-CSIC), Madrid, Spain

Maria João Forjaz National Centre of Epidemiology and REDISSEC, Carlos III Institute of Health, Madrid, Spain

Stuart Franklin JDDK Architects, Newcastle upon Tyne, UK

Ângela Freitas Centre of Studies in Geography and Spatial Planning (CEGOT), Department of Geography and Tourism, University of Coimbra, Coimbra, Portugal

Caterina Gallerani Università degli Studi di Milano-Bicocca, Milan, Italy

Heitor García-Lantarón Architectural Design Department, Matia Fundazioa, Donostia-San Sebastián, Spain

Álvaro García-Soler Matia Fundazioa and Matia Research Institute, Donostia-San Sebastián, Spain

Rose Gilroy Newcastle University, Newcastle upon Tyne, UK

Neveen Hamza Newcastle University, Newcastle upon Tyne, UK

Ester Higueras García ABIO Research Group, Universidad Politécnica de Madrid, Madrid, Spain

Adriana Loureiro Centre of Studies in Geography and Spatial Planning (CEGOT), Department of Geography and Tourism, University of Coimbra, Coimbra, Portugal

Sara Marsillas Rascado Matia Fundazioa and Matia Research Institute, Donostia-San Sebastián, Spain

José Antonio Ondiviela WorldWide Observatory for Attractive Cities, Universidad Francisco de Vitoria, Madrid, Spain

Miguel Padeiro Centre of Studies in Geography and Spatial Planning (CEGOT), Department of Geography and Tourism, University of Coimbra, Coimbra, Portugal

Inés Palanca-Sánchez IMSERSO (Institute for the Elderly and Social Services), Government of Spain, Madrid, Spain

Camille Picard Université Paris Est (Lab'Urba) and Kyoto Prefectural University, Paris, France

Elisa Pozo Menéndez ABIO Research Group, Universidad Politécnica de Madrid, Madrid, Spain

Carmen Rodríguez-Blázquez National Centre of Epidemiology and CIBERNED, Carlos III Institute of Health, Madrid, Spain

Vicente Rodríguez-Rodríguez Institute of Economics, Geography and Demography (IEGD); Spanish National Research Council (CSIC), Research Group on Ageing (GIE-CSIC), Ageing Network of the Latin American Population Association (ALAP), Madrid, Spain

Fermina Rojo-Pérez Institute of Economics, Geography and Demography (IEGD); Spanish National Research Council (CSIC), Research Group on Ageing (GIE-CSIC), Ageing Network of the Latin American Population Association (ALAP), Madrid, Spain

Emilia Román López Department of Urban and Regional Planning, Escuela Técnica Superior de Arquitectura, Universidad Politécnica de Madrid, Madrid, Spain

Diego Sánchez-González Department of Geography, Faculty of Geography and History, Universidad Nacional de Educación a Distancia (UNED), Madrid, Spain

Mayte Sancho Castiello Matia Research Institute, Donostia-San Sebastián, Spain

Paula Santana Centre of Studies in Geography and Spatial Planning (CEGOT), Department of Geography and Tourism, University of Coimbra, Coimbra, Portugal

Abbreviations

AA	Active Ageing
ADL	Activities of Daily Living
AFCC	Age-Friendly Cities and Communities
AIP	Ageing in Place
ALH	Ageing Like at Home
BPSD	Behavioural and Psychological Symptoms of Dementia
CCM	Chronic Care Model
DFC	Dementia-Friendly Community
DSDC	Dementia Services Development Centre
ECCM	Expanded Chronic Care Model
EHPAD	*Établissement d'Hébergement pour Personnes Âgées Dépendantes* [Residential and Medical Facility for Dependent Elderly People]
ESN	European Social Network
EU	European Union
GDP	Gross Domestic Product
GVA	Gross Value Added
HIA	Health Impact Assessments
IFIC	International Foundation for Integrated Care
IoT	Internet of Things
LLs	Living Labs
LTCF	Long-Term Care Facilities
MCI	Mild Cognitive Impairment
NCDs	Non-Communicable Diseases
OECD	Organisation for Economic Co-operation and Development
PAR	Participatory Action Research
PCC	Person-Centred Care
PHC	Primary Health Care
PwD	People with Dementia
QoL	Quality of Life
RCN	Resilient Cities Network
SSE	Social Solidarity Economy

SCN	Smart Cities Network
SDG	Sustainable Development Goals
UIA	Urban Innovative Actions
ULL	Urban Living Lab
UCLG	United Cities and Local Governments Network
WHO	World Health Organization
WZC	*Woonzorgcentrum* [Home Care Centre/Nursing Home]

Part I
The Post-pandemic City for Healthy Ageing and Wellbeing

Chapter 1
Urban Planning and Health Equity. Integrating Global and European Perspectives in Local

Ângela Freitas and Paula Santana

Abstract Promoting health equity is a place-based issue that goes way beyond the health sector alone. Urban planning plays a key role in addressing inequities in the social, economic, built, and physical determinants of health. The WHO and EU have been stressing the need of placing health equity at the core of urban policies to ensure that everyone has fair access to resources and opportunities to be healthy. Currently, there is more guidance than ever before, with a wealth of frameworks, tools, guides, and best practices. However, for most city governments, the integration of a health equity into urban planning processes remains a challenge. The chapter begins with an overview of the relationship between urban planning and health equity, how it evolved across time, and how it shaped emerging research and practice on healthy urban planning. Then, a set of guiding points are presented considering the added value of informing health equity-oriented urban planning: (i) conceptual frameworks addressing complexities of urban health, (ii) urban health assessment tools, (iii) inclusive and participatory urban planning, (iv) Health Equity in All Policies approach and participatory governance, (v) innovative urbanism and networking, and (vi) new urban models changing the urban design for health.

Keywords Urban health · Equity · Healthy urban planning · Frameworks

Â. Freitas (✉) · P. Santana
Centre of Studies in Geography and Spatial Planning (CEGOT), Department of Geography and Tourism, University of Coimbra, Coimbra, Portugal
e-mail: angela.freitas@uc.pt; paulasantana@uc.pt

E. Pozo Menéndez, E. Higueras García (eds.), *Urban Design and Planning for Age-Friendly Environments Across Europe: North and South*, Future City 19,
https://doi.org/10.1007/978-3-030-93875-8_1

1.1 Urban Planning and the Emergence of Health Equity Research: An Overview

Presently, we hear and talk more and more about the relationship between urban place and health. While this is not new in urban health research, the COVID-19 pandemic has brought health inequities to the forefront of public discussion, shedding light on existing social, structural, and place-based factors of inequities between and within cities (Bambra et al., 2020). Negative health impacts of COVID-19 are disproportionately being felt by people living in vulnerable circumstances, namely among elderly people (Lee, 2021). Long before, we had evidence that whether it is excessive mortality or morbidity or access to resources to be healthy, all were not equally distributed. Even in wealthier countries with good health systems, the pandemic has amplified inequities in the social, built, and physical environments, raising fundamental questions about social justice and health equity both in cities of developed and developing countries (Bambra et al., 2020; UN-Habitat, 2021). Despite health equity has always been an ideal, this pursuit has not translated into equitable societies and current pandemic times are a strong reminder to accelerate the development for better, healthier, and equitable urban environments (Capolongo et al., 2020; UN-Habitat, 2021).

Before diving into how urban planning and health equity views have evolved across time, it is important to first outline both concepts.

In the words of Susan Fainstein (n.d.) urban planning is the process of "design and regulation of the uses of space that focus on the physical form, economic functions, and social impacts of the urban environment and on the location of different activities within it". As a spatial management mechanism, urban planning influences the social, physical, and economic environments and how cities function affecting people's daily lives in systematic ways, thus being recognized as having a key role in human health and well-being (Barton & Tsourou, 2000; Barton et al., 2009).

Following the definition of WHO, equity "is the absence of avoidable or remediable differences among groups of people, whether those groups are defined socially, economically, demographically, or geographically. Health inequities therefore involve more than inequality with respect to health determinants, access to the resources needed to improve and maintain health or health outcomes" (WHO, 2021). Three features, when combined, turn a mere difference in health into an inequity in health. A difference in health that is systematic, socially produced (and, therefore, modifiable), and unfair is an inequity in health.

Braveman et al. (2017, p. 2) reinforce that "health equity means that everyone has a fair and just opportunity to be as healthy as possible. This requires removing obstacles to health such as poverty, discrimination, and their consequences, including powerlessness and lack of access to good jobs with fair pay, quality education and housing, safe environments, and health care". In the context of public policies, urban planning, and measurement, health equity should be assumed as a principle and commitment to reduce and eliminate disparities in health and in its determinants (Braveman, 2006, 2014).

The role of place, and the importance of how it is managed, in promoting or not promoting health equity is becoming increasingly recognized and understood (Rubin, 2015). Residents living in a neighbourhood with affordable and adequate housing, good air quality, safe access to green spaces and parks, supermarkets with fruits and vegetables, good schools, public transport, safe and walkable routes, healthcare services, cultural and social services, etc., are more likely to be in good health and well-being. On the opposite, residents in neighbourhoods lacking these amenities are more likely to suffer from chronic diseases, such as obesity, diabetes, asthma, heart disease, and depression. Evidence shows that unserved neighbourhoods are home to a large number of poor, low educated, deprived, or vulnerable populations (Corburn, 2017).

1.1.1 From the Sanitary to Urban Social Movements

As stated by Barton and colleagues "Health and urban planning are natural allies" (Barton et al., 2009, p. 91). Urban planning and public health perspectives have been intertwined across time interacting and intersecting in practice, being particularly evident during the nineteenth and early twentieth centuries. Both evolved together in response to unhealthy urban conditions (Barton & Tsourou, 2000).

The sanitary movement of urban planning was an approach to public health that first developed in the United Kingdom around 1830. The John Snow's map of the cholera outbreak of 1854, alongside being the pioneer of disease mapping and epidemiology, transformed the way public health practitioners looked at the city and contributed to change the organization of the city of London, namely its sanitary water supplies and waste removal systems (Fine et al., 2013).

Major urban planning reforms stem from efforts, during the sanitary movement, to decrease the spread of infectious diseases related to overcrowding, poor quality and unsanitary housing conditions, and social disorder caused by uncontrolled urbanization brought by the Industrial Revolution in the 1700s (Duhl et al., 1999). The Haussmann's renovation of Paris (1852–1870) is a well-known example of a modernization plan that transformed the city and inspired urban development plans throughout Europe. Another influential urban design approach was the Garden City of Howard Ebenezer (1902), which shaped the development of residential and suburban areas in the United States and the United Kingdom as a response to the need for improvement in the quality of life and as an attempt of reversing overpopulation in the city centres (Hall, 2014). The pressing need for urban renewal in the aftermath of the First World War, along with the need to address tuberculosis inspired the Modernist movement ideas focused on urban functionality (Campbell, 2005). Le Corbusier's "Contemporary City" (1922) plan and its "Radiant City" (1925) were a demonstration of the overall concern for a healthy urban environment, bringing circulation, open spaces, sunlight, and ventilation into the built environment. One of the main features was the strict separation of the societal functions in the city into zones for workplaces, residences, shopping and entertainment, and government buildings (Lopez, 2012).

This type of urban development focused on functionality, e.g., separating residential areas from industrial sites, was of key importance for public health at the time since people would no longer live exposed to industrial pollution. Nevertheless, this model also gave place to discriminatory zoning, urban fragmentation, suburbanization, and socio-spatial segregation, with differential distribution of health-promoting resources, as public facilities and services, and opportunities such as housing, medical care, education, employment, transportation, and accessible and varied public space, across neighbourhoods, with impacts on health outcomes such as obesity, diabetes, and mental health (Wilson et al., 2008). In most American cities, residential segregation is a fundamental cause of health disparities: people living in highly segregated and isolated and low-income neighbourhoods have lower housing quality, higher concentrations of poverty, and less access to good jobs and education. Therefore, they experience higher risks of illness and mortality (Schwarz, 2016).

In the 1960s and 70s, urban social movements sparked public debate and gave stimulus to the development of critical approaches to urban development models (Fainstein, 2010). Some of the most influential urban scholars that shaped the theory and practice on the topics of "right to the city", "urban justice", and "urbanization of justice" in these decades were Jane Jacobs with her book "Death and life of great American cities" (1961), Henri Lefebvre with "La droit à la ville" (1968), and David Harvey with "Social justice and the city" (1973). Under their readings, urban space starts to be understood as a social production, constituted by social relations rather only by its territorial and physical characteristics, reproducing and contributing to inequities and injustice (Fainstein, 2010; Soja, 2010).

In parallel, a "health inequities knowledge base" begin to emerge (1970s and early 1980s), with a growing research documenting health differences attributable to disparities in the distribution of resources and opportunities and unequal exposures to the social, economic, political, cultural, and environmental contexts in which people live (Collins & Hayes, 2010).

1.1.2 From Ottawa Charter to the Healthy Cities Movement: Paving the Way for Healthy Urban Planning

The Ottawa Charter for Health Promotion (1986) was seminal to the establishment of health equity research and coincided with the broadening scope of "health promotion" considering that reducing health inequities would need not only lifestyle-oriented changes but mostly environmental and societal changes (Collins & Hayes, 2010). This direction gave rise to the WHO "Healthy Cities Movement" which started to encourage policymakers and urban planners to embrace a different approach to urban development (Ashton et al., 1986).

In addition, the world was also becoming increasingly urbanized with more and more people living in urban settings. In the latter decade of the twentieth century, considerable attention starts to be devoted to the health effects of the rapid and often unplanned growth and urban sprawling resulting from the dramatic rise of urban population since the 1950s (Frumkin, 2002; Galea et al., 2005). The growing burden of urban-associated diseases (e.g., allergies, asthma, cardiovascular disease, obesity, diabetes, depression, respiratory disease), are linked for example to higher exposure to air pollution, sedentary lifestyles, and unhealthy diet, risk factors that can be modified by urban planning (Flies et al., 2019).

The study of how characteristics of the urban environment affect health and related inequities was established as a distinct field of research in the 1990s. The urban health field emerged to understand the factors affecting health and health inequities in cities within four broad dimensions that frame questions on how city living can affect health and can be viewed as target areas that urban policies seek to address: the demographic characteristics, the physical environment, the social environment, and the access to health and social services (Galea & Vlahov, 2005; Vlahov & Galea, 2003; Vlahov et al., 2007). Following a growing body of research, the International Society for Urban Health (ISUH) was founded in 2002 at The New York Academy of Medicine and since then has been serving academia and practitioners as a platform for advancing urban health and health equity worldwide.

As the WHO Healthy Cities project was expanding, the concept of healthy urban planning, its principles and practices begin to evolve in the 2000s, focusing on the positive aspects of urban planning and its potential to improve both the quality of built environments and of individuals and communities. The WHO defines healthy urban planning as planning for people, meaning putting the needs of people and communities at the heart of the urban planning process and considering the implications of policy decisions for health and well-being. The book "Healthy Urban Planning. A WHO Guide to planning for people" (2000) provided the first guidelines for healthy urban planning, stimulating, and informing debate on the relevance of bringing the consideration of health impacts into the core of urban planning decisions (Barton & Tsourou, 2000). Targeted to those involved in the design and development of cities, this framework sets out equity as one of its crucial principles along with health promotion (Barton et al., 2009).

For three decades, the Healthy Cities movement has been a driver of change in the creation of healthier urban settings and instrumental in providing frameworks and guides to local policymakers and urban planners. This initiative engages local governments in health development, emphasizing the relevance of municipal level. Policies and urban planning plans implemented by municipal governments are critical components to address health inequities (Galea et al., 2005; Collins & Hayes, 2010; Corburn, 2013).

Today, the WHO European Healthy Cities Network brings together more than 1400 municipalities. Integrating health equity into urban development and planning is one of the principles of the pillar of Place in the current Healthy Cities vision (Phase VII: 2019–2025): People, Participation, Prosperity, Planet, Place and Peace (WHO Regional Office for Europe, 2018). This framework is inspired by and aligned with the Sustainable Development Goals (SDGs), since global problems are being dealt with at a local level and the Healthy Cities vision acknowledges this.

1.1.3 Advancing Urban Health Equity in the Sustainable Development Goals Era

The SDGs can be seen as an operational framework to consider urban health equity globally, providing locally an opportunity to implement policies with a "Health in All Policies" view (Ramirez-Rubio et al., 2019).

The recent work from Ramirez-Rubio et al. (2019) elucidates very well the interlinkages between health and the SDGs, making the case for putting "urban health" at the core of SDG implementation, as an intersectoral arena that links both the public health and the urban planning sectors. The authors created a comprehensive, conceptual framework by linking urban health dimensions to SDGs, targets, and indicators (Fig. 1.1). Health equity is an explicit condition in the context of implementing SDG3 (Ensure healthy lives and promote well-being for all at all ages) and SDG11 (Make cities and human settlements inclusive, safe, resilient, and sustainable) being indirectly embodied in many others. For example, by providing all people access to decent housing, clean air and water, nutritious food, safe transport and mobility, opportunities for physical activity, and protection from pollution, among others, cities are ensuring no one is left behind, thus, promoting health equity.

1.1.4 Urban Health Equity: A Cross-Cutting Issue and Public Concern

In the last few decades, a wealth of evidence on how urban planning shape neighbourhood environment and, ultimately, influence health, contributing to creating or reducing health inequities, has been published (Corburn, 2005, 2017; Santana et al., 2009; Barton et al., 2010, 2015; Diez Roux & Mair, 2010; Northridge & Freeman, 2011; Friel et al., 2011; Borrell et al., 2013, 2014; Santana et al., 2015; Giles-Corti et al., 2016). At the same time, a wide range of guidelines, tools, and resources providing entry points seeking policy and planning change had been released by international organizations (WHO and UN-Habitat, 2010, 2016; WHO, 2012; WHO Regional Office for the Western Pacific, 2015).

In the European arena, many initiatives and joint actions have been taking place, with European Union funding several projects aiming to advance health equity at regional and urban levels (e.g., EURO-HEALTHY, INEQ-Cities, SOPHIE, Joint Action Health Equity Europe, Health Equity Pilot Project, Urban Health cluster). In 2019, the WHO Europe stressed the need of placing health equity as a central pillar of every policy through the release of "The European Health Equity Status Report"

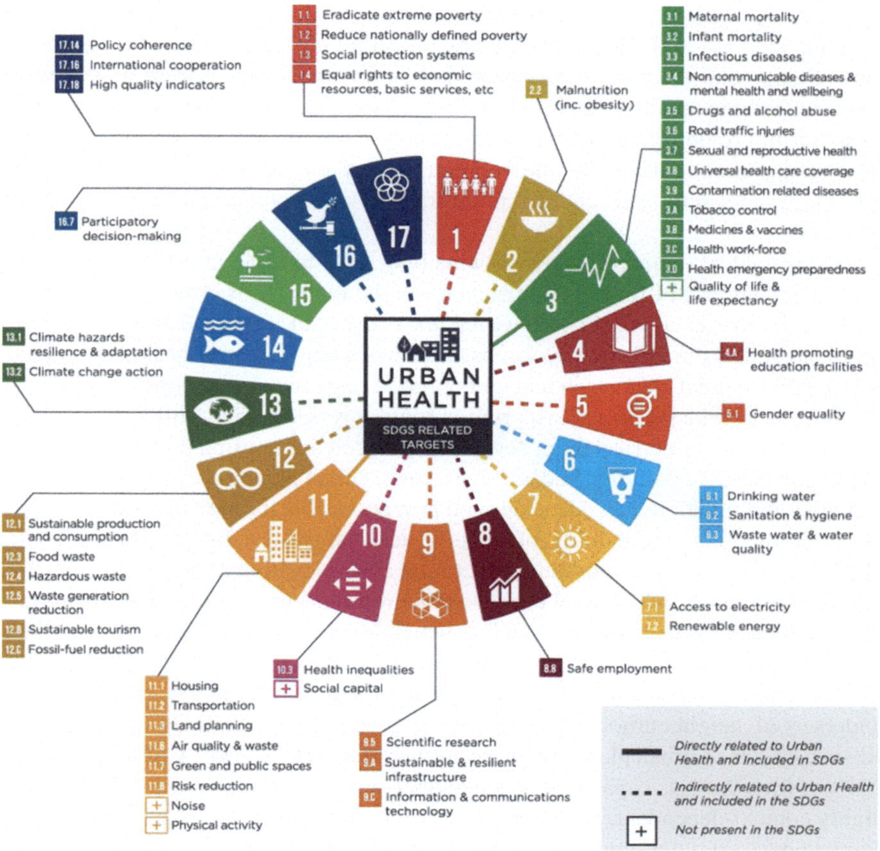

Fig. 1.1 Urban health within the SDGs: a conceptual framework. (Source: Ramirez-Rubio et al., 2019, Fig. 1)

(WHO Regional Office for Europe, 2019). The most recent WHO and UN-Habitat sourcebook on integrating health into urban and territorial planning starts with the question "If the purpose of planning is not for human and planetary health, then what is it for?" (UN-Habitat and WHO, 2020).

The current and fast-growing research along with increased public advocacy to creating healthier cities is obvious across the globe. Achieving health equity is moving from a niche concern of urban health researchers, practitioners, and global health organizations into the broad public sphere – the so-called health social movements (Brown et al., 2004), along with grassroots movements for environmental justice and social equity (Rubin, 2015). The current health social movement paradigm demands from political leaders and urban planners to integrate health considerations into all aspects of urban governance, with an emphasis on guaranteeing values such as health equity and citizen empowerment (Kim, 2020).

Currently, there are numerous public campaigns demanding policy and urban planning changes at the local level, many reflecting global concerns, being the climate change and associated environmental justice movement, strongly related to environmental inequality, the most notorious (Santana et al., 2021). Other examples focus on improving many of the urban health determinants, as, for example, the built environment conditions: the creation of new public parks, the provision of housing, road safety interventions, more transportation choices, access to medical services, etc.

In this context, municipal governments play a pivotal role in promoting overall health, making use of formal competencies in the majority of urban health dimensions (Loureiro et al., 2015). However, they also can, unintendedly, exacerbate health inequities through their decisions. Even when implementing health-promoting built environmental changes in a neighbourhood (e.g., new home construction, new green parks or bike lanes or transportation routes), residents may still face discrimination or encounter barriers (e.g., new parks can increase housing prices, new traffic routes can increase pollution and accident rates). Urban planning can influence what activities and functions can take place and how accessible they are to different population groups (e.g., availability of parks to take physical activity) (Barton et al., 2009). Another example is urban regeneration and renovation of houses in gentrifying old city centres, which could expose older residents to displacement and loss of well-established social resources, all of which affect health and well-being (Ellen et al., 2020).

Municipal governments can act as "game-changers" and target investments to underserved neighbourhoods and prioritize resources for mitigating potentially adverse effects of urban planning interventions (Galea et al., 2005; Collins & Hayes, 2010). Nevertheless, capacities of municipal governments at the local level are highly context-dependent, and dependent on the form, function, legal framework of competencies, statutory powers, and political priorities, which vary across countries and in some cases within countries and regions (Collins & Hayes, 2010).

1.2 Key Perspectives Informing Health Equity-Oriented Urban Planning

Making the case for health equity requires a comprehensive overview of the urban context, local actors, and resources. Each one of the points below is rooted in current and emerging trends in research and practice in the field of health equity and healthy urban planning. The description is not exhaustive and is only intended to provide a general snapshot of relevant insights and directions towards change in the city planning process.

1.2.1 Frameworks Addressing the Systemic Nature of Urban Health Equity

The relations between urban environment and health are complex (Siri, 2016) posing challenges to health and equitable urban planning (Pineo et al., 2019). Cities are complex systems, and as such they need a systemic approach encompassing the physical, social, economic, and political urban environment (Glouberman et al., 2006; Rydin et al., 2012). This complexity makes it difficult to establish direct cause-and-effect relationships between features in the built environment and health outcomes (Pineo et al., 2019).

Many frameworks can be used to help understand the systemic nature of health and the complexity inherent to associated policy responses. All share the common assumption that there is no single factor explaining health inequities and no single solution or policy response. The European Health Equity Status Report identifies five essential conditions for health equity: (i) good quality and accessible health services, (ii) income security and social protection, (iii) decent living conditions, (iv) social and human capital, and (v) decent work and employment conditions (WHO Regional Office for Europe, 2019).

One of the best-known conceptual frameworks to serve as a point of reference for analysing health equity is the "Health Map" developed by Barton and Grant (2006), as a tool of local investigation of the relationship between health and place. This tool is informed by the Dahlgren and Whitehead social determinants of health model, showing how various spheres of living conditions and the social, economic, built, and physical environmental determinants of health are interconnected and affected by planning decisions.

The EURO-HEALTHY project developed a multidimensional framework to analyse population health inequities at the European regional and metropolitan level, across nine areas of concern: (i) health outcomes, (ii) economic conditions, social protection, and security, (iii) education, (iv) demographic change, (v) lifestyle and health behaviours, (vi) physical environment, (vii) built environment, (viii) road safety, (ix) healthcare resources and expenditure, and (xi) healthcare performance (Santana et al., 2017, 2020; Freitas et al., 2018). This multidimensional framework was applied and adapted by (Freitas et al., 2020) in the assessment of urban health inequities at the neighbourhood level in the city of Lisbon.

A relational view of inequities in health is considered needed to understand the complexities of relations between the individual and socio-demographic characteristics and the place of residence to which is exposed: (a) people with lower incomes, lower education levels or unemployed are more likely to have poor health in childhood, as adults and older adults (life-course inequity) and, (b) those living in low income and deprived neighbourhoods with no available green spaces and community amenities, exposed to air pollution and urban decay, living in poor-quality or insecure housing are more likely to report poor health and to suffer from a variety of health problems (place-based inequity). Both are intrinsically connected and intertwined, since people with low socioeconomic status are more likely to live in low-income neighbourhoods, reinforcing inequity (Cummins et al., 2007).

Corburn (2017) conceptualized this approach in the "Relational Framework of Urban Place and Heath Equity", emphasizing the interaction between six intersecting aspects of urban places: people (valuing their culture, norms, knowledge, and history), physical (human built and natural environments), process (how a place is felt and perceived by people and institutions), politics (how different political organizations and stakeholders influence decisions), policies (decisions that shape the distribution and access to resources), and power (networks of support and social connections) (Fig. 1.2). This approach, applied to healthy urban planning, recognizes that these intersecting aspects vary in time and space, thus being essential to apply an "adaptive management" where monitoring of policy impacts and public participation is essential to integrate all aspects.

A recent work by Pineo (2020), proposing a new framework named as "Towards Healthy uRbanism: InclusiVe Equitable Sustainable (THRIVES)," extends previous conceptual frameworks and reorients focus towards the pressing and current global challenges as climate crisis and widening inequalities and social injustice. The THRIVES framework raises attention towards the urgency of considering health impacts in urban development and overcoming structural barriers to health that are caused by urban design processes that exclude vulnerable groups (low income, children, women, disabled, elderly, and migrants). The framework is driven by three

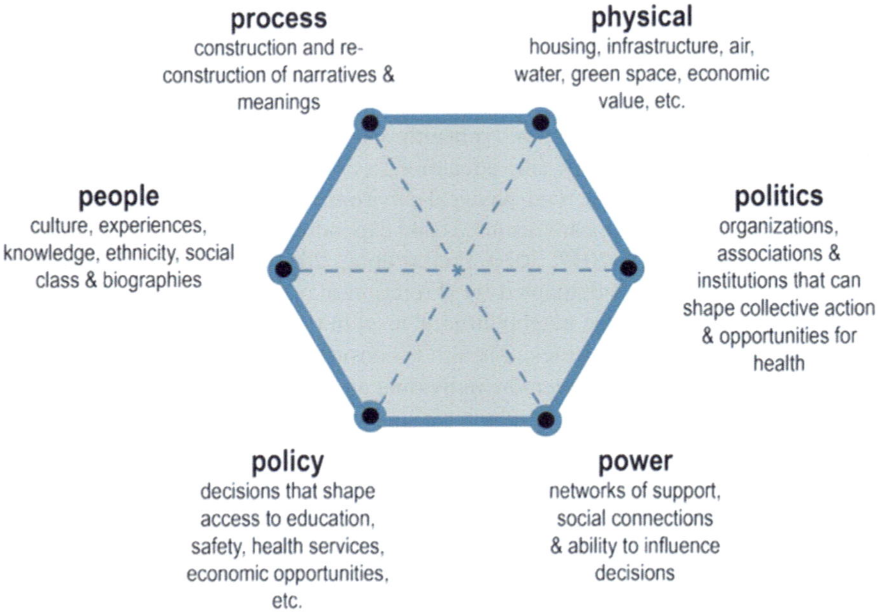

Fig. 1.2 Relational framework of urban place and health equity. (Source: Corburn, 2017, Fig. 1)

core principles that should inform design and policy decisions towards healthy urbanism: inclusive, equitable, and sustainable (Fig. 1.3).

Following the examples of conceptual frameworks presented above, a health equity lens analysis is needed to inform policy and planning strategies. Without a proper understanding of the local context, city planners may conduct unfavorable actions exacerbating existent inequities. Thus, it is important to consider a place-based and data-driven approach. This means applying a population health approach (Kindig & Stoddart, 2003) with a spatial lens: an overview of how the population's health outcomes and health determinants are geographically distributed constitutes the starting point to identify local health equity challenges.

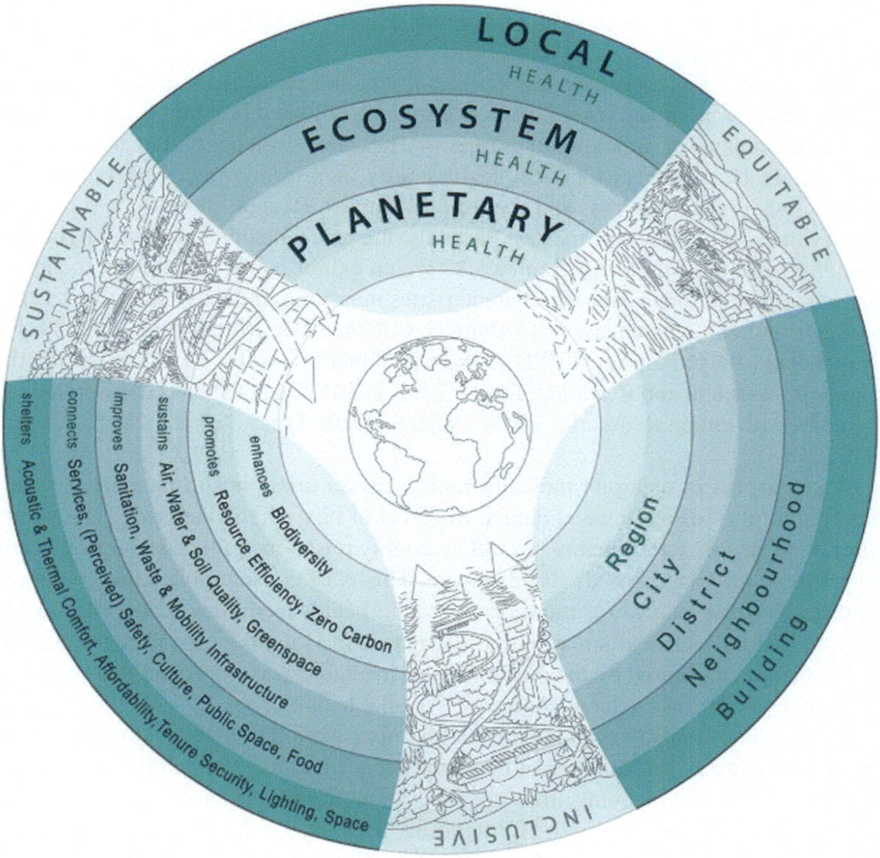

Fig. 1.3 THRIVES Framework (Towards Healthy uRbanism: InclusiVe Equitable Sustainable). The illustrated spatial scales align to the three scales of health impact: planetary, ecosystem, and local. Concentric rings in the image demonstrate interconnections across each scale and help to visually align the scales of health impact with scales of urban decision-making related to the built environment. (Source: Pineo, 2020, Fig. 1)

1.2.2 Urban Health Indicator Tools Informing Assessment and Place-Based Interventions

Each municipality should analyse the underlying conditions of health inequities and then determine what and where priorities should be placed. This should be specific and tailored to each locality and local context, first by identifying indicators that can be acted upon, second by considering the differences that cause local inequities (what indicators, in which neighbourhoods), and third, designing interventions prioritizing those who face disadvantage and vulnerability rooted in those identified differences (whose groups are more likely to be affected). In this context, the adoption of place-based approaches allows for the analysis of health inequities through a lens that incorporates data on local indicators (social, physical, and built environment) and the perspectives from stakeholders and community on what the main health problems are.

Health impact assessments and community health need assessments sought to understand how populations are disproportionately affected or could be affected by new developments or policies and programs. These tools can be used by urban planners and decision-makers for local health equity monitoring processes, helping to identify, understand, access, and measure inequities. Recent literature reviews on the development and use of metrics for analysing urban health equity highlight the existent plethora of indicator frameworks which exhibit great diversity in the aim, spatial scale, domains analysed, characteristics measured, indicator selection methodology, and data visualization (Corburn & Cohen, 2012; Rothenberg et al., 2015; Prasad et al., 2016; Pineo et al., 2018). Worthy of mention is the WHO Urban Health Equity Assessment and Response Tool – Urban HEART, since it has been applied in a number of cities across the globe (WHO, 2010; Prasad et al., 2015; Novoa et al., 2018).

A key aspect of using an indicator tool to assess urban health inequities is the incorporation of disaggregated data at the level of city neighbourhood. The use of "area-level health determinants" would invariably reveal inequities, stemming from a locational or "place-based" disadvantage. This is of considerable value to urban planners because allows for monitoring related health inequities across place and time and address critical situations. Although there is a similarity in the domains and indicators measured by the different tools available, it is important that they are adapted, considering the local context and available data.

In the field of tools to evaluate the health impacts of plans and development proposals, there are good examples of how to approach and mainstream health considerations into urban planning, through a set of policy requirements and standard assessments. Here, is worthy to highlight the "Healthy Urban Planning Checklist" in London, UK (NHS London Healthy Urban Development Unit, 2014) and the "Healthy Built Environment Checklist" in New South Wales, Australia (NSW Ministry of Health, 2020). Both checklists can be used to ensure that health issues are embedded into local plans, master plans, and major planning applications and can be customized for local use to reflect local circumstances and priorities.

Current urban ageing and the associated need for developing supportive urban places also boosted the emergence of assessment frameworks and tools capable of measuring the age-friendliness of a place. The goal is to inform and guide local policymakers and planners in developing and implementing interventions that promote ageing in place (e.g., people to age actively and healthily while living in the community, with some level of independence, rather than in residential care) (van Hoof et al., 2018). The WHO provide a toolbox of eight policy domains for age-friendly action nested in three dimensions of supportive and equitable environments: (i) physical environment (outdoor amenities like public spaces, transport and mobility, and housing), (ii) social environment (social participation, social inclusion, and non-discrimination and civic engagement and employment) and (iii) municipal services (communication and information across sectors and the provision of community and health services) (WHO Regional Office for Europe, 2016).

Making sound decisions to determine the priority health inequities to address is both a technical and a social process. It is important to consider the locality's needs, assets, and points of view of community and local stakeholders (assets-based approach). In the field of place-based assessments using local knowledge and citizen's perceptions, it is worthy to mention the "Place Diagram Tool", developed by Project for Public Spaces (Project for Public Spaces, n.d., 2016), and "Place Standard Tool", jointly developed by the NHS Scottish Government and the public body of Architecture and Design of Scotland (NHS Health Scotland, n.d.). Both tools are being used to prompt local discussions and inform urban planning, allowing to assess the assets as well as the needs of a place, looking at its social, physical, and built features.

1.2.3 *Inclusive and Participatory Urban Planning: Engaging All Concerned Parties and Citizens Through Collaboration and Co-creation*

The process of promoting health equity through urban planning must be itself an equitable and participatory process, involving local people, community organizations, and decision-makers. Participation and engagement, particularly for those with the least voice, lead to more effective public policies for health equity (Boyce & Brown, 2007). Local organizations can play a key role in driving positive changes in the community and, thus, addressing existent inequities. Establishing local collaborations that engage diverse cross-sector concerned partners is a way of accounting for the wide range, and sometimes competitive, interests of local stakeholders and thus assure collective responsibility (Impact on Urban Health, 2021).

Furthermore, public deliberation is key to ensuring that local priorities and views are integrated into healthy urban planning. The involvement of citizens, including those experiencing health inequities, contributes to understanding and identifying possible negative unintended consequences that may limit intervention effectiveness. In recent years, new collaborative governance models have been boosting the direct involvement of citizens in urban planning processes moving from a simple public deliberation to a co-creation process where citizens are considered key partners in co-creating solutions of urban design addressing place-based inequities in health (Boyce & Brown, 2007).

In this context, the placemaking movement is becoming a mainstream trend across the globe through established networks such as the European Placemaking Network and PlacemakingX. By design, placemaking initiatives create opportunities and spaces for gathering and socialization. As a participatory process, it offers opportunities for people to work together on a common urban project, transforming public spaces in the local community and creating new social connections. The main aim is to create public places that are accessible to everyone in a community, including streets, markets, squares, parks, and publicly owned or accessible lots. By engaging and bringing diverse populations together, including different ages, ethnicities, socioeconomic status, and cultures, placemaking processes, and the public places that result, are equitable by nature and promote health equity in many ways (Project for Public Spaces, 2016).

1.2.4 Health Equity in All Policies: Breaking the Silos Through Policy Alignment, Accountability, and Participatory Governance

Presently, many city leaders are being challenged to change their governance models shifting from a "siloed" planning (city departments operating as silos) to a more integrated, holistic, and collaborative one. The "Health Equity in All Policies" is an inter-sectoral and multi-level governance approach, a "silo-breaking" model that incorporates health equity considerations in the design and evaluation of all policies, strategies, and plans being developed in a city. In this approach, health equity means promoting an equitable distribution of city resources, namely, directing them to reduce inequalities (targeting those to people and neighbourhoods most in need) and ensuring openness and collaboration in the decision-making process. This stands not only by increasing coordination across city departments but also through joint participation of community organizations. Both can contribute to effectively address health inequities shaped by urban planning decisions (Corburn et al., 2014a, b).

A catalyst for adopting this approach is a formal health-in-all-policies ordinance and strategy and the development of municipal health strategies oriented to the promotion of health and equity. Although the health impact is difficult to quantify or measure, this type of strategy can set the direction for many of the plans, projects, and interventions that follow. For example, masterplans that include broad health and equity goals can guide changes to urban development processes. Examples from some cities/good practices in embedding health equity into their urban policies and strategies come from Richmond, California (USA) (Corburn et al., 2014a, b) and Utrecht (The Netherlands) (Weber, 2019).

1.2.5 Innovative Urbanism: Urban Experimentation and Networking

Cities have always been associated with transformative ideas and novel social initiatives. The geographical proximity of stakeholders and multidisciplinary interaction enable innovation dynamics (Florida et al., 2017). Innovative urban planning is defined "as an activity on innovations creation and their implementation to obtain a socio-economic, political, environmental or other effect" (Kolyasnikov, 2017, p. 1). By the end of the 2000s, an evolving need for new public policy and urban planning approaches started to emerge both from policymakers' side as from citizens' side. In the aftermath of the economic and social crisis, visible changes in public policies appeared in the delivery of concrete public services at municipal level. It became evident that the traditional top-down urban planning interventions were not possible anymore. Innovative approaches to urban planning started to be developed mostly regarding the transformation and use of public spaces, with citizens much more involved and active over urban development (Tosics, 2019).

The European Union has played a key role in promoting pilot projects, initiatives, and city's best practices exchange through its programmes "Urban Innovative Actions (UIA)" and "URBACT". These programmes contribute to build capacity as well as promote the development of knowledge and skills in many urban dimensions that are fundamental to promote urban health equity. Cities are provided with resources to test new and innovative solutions to address urban challenges related to social inclusion, employment, pollution, mobility, etc., which are recognized drivers of many urban health inequities.

In this context, urban experimentation is an increasing asset and celebrates practice in urban governance and an important vehicle not only to understand the urban health challenges but also to transforming the city into a healthier and more equitable place. Scholl and Kraker (2021, p. 161) define urban planning by experiment "as an approach that uses experimentation to innovate and improve urban planning instruments, approaches, and outcomes". Urban experiments are increasingly taking place in the context of Urban Living Labs (ULL), spaces that facilitate experimentation based on citizens' participation and their direct involvement in finding solutions to current urban challenges such as climate adaptation, urban mobility and health, and social inclusion. The Dutch city labs are well-known examples in Europe (Scholl & Kraker, 2021). ULLs have also the potential to leverage the creation of urban health observatories, serving the purpose of producing, linking, and translating data, developing locally relevant health knowledge, and influencing urban planning policies.

From the perspective of an emerging new urban planning paradigm, it is worthy to mention the "soft" urban planning approaches linked to tactical urbanism, that is, urban space transformations through short-term, low-cost, temporary, and community-based projects, intended to improve urban environments (Lydon & Garcia, 2015). Examples of tactical urbanism are bike lanes, parklets, and pop-up parks. Although implemented to be temporary (and more recently as a COVID-19 mitigation strategy), tactical urbanism interventions could be designed with a long-term vision to support a more permanent healthy urban design and long-term health equity vision, by improving conditions in disadvantaged and unserved neighbourhoods (Rojas-Rueda & Morales-Zamora, 2021).

All current urban innovations occur in a time when networking is key in the transformation of cities into equitable and healthier places. City networks provide unique programs and opportunities for cities to exchange, share, create, and learn from each other. They also provide new ways of collaboration, enhancing public participation in decision-making processes. Urban policymakers and planners should understand the role of city networks in leveraging urban experiments and urban planning innovations benefiting urban health and equity. Beyond the most obvious in the field of urban health – WHO European Healthy Cities Network –, examples of city networks contributing to boost healthy urban planning include the United Cities and Local Governments Network (UCLG), Smart Cities Network, WHO Global Network for Age-friendly Cities and Communities, C40 Cities, Resilient Cities Network, among others.

1.2.6 New Urban Models: Changing the Urban Design for Health

New urban models have been emerging in the last decade driven by the current need for remedy the negative effects of car-driven cities and associated air and noise pollution with well-known impacts on health and health equity. COVID-19 pandemic has been reinforcing and accelerating the adoption or introduction of these models, in cities across Europe and beyond, as a strategy to mitigate the impacts of successive lockdowns. The most trendy and well-known urban models calling for a more compact, walkable, and proximity to the city, where putting people at the centre is the main assumption (human-scale oriented planning), are the "closer-together city" of Stahle (2016), the "15-minute city" of Carlos Moreno in Paris (Moreno et al., 2021), and the Superblock model of Barcelona. One of the characteristics of these models is the so-called "demobility", that is, reducing travel time, which contributes to respect the zero-carbon objective and to increase sociability and urban liveability. One practical example on the incorporation of these principles into city planning is the "20-minute Neighbourhood Pilot Program" developed within the city of Melbourne Plan for 2017–2050. This plan aims to transform the city into 20-minute neighbourhoods, giving residents the ability to meet most of their daily needs within a small distance walk from home, with safe cycling and local transport options (living locally) (Fig. 1.4).

Fig. 1.4 Features of 20-Minute Neighbourhood. (Source: State of Victoria, 2019)

These models, rather than applied to a whole city, can be introduced in specific neighbourhoods or districts, in the view of promoting healthier conditions. For instance, Mueller et al. (2020) estimated that almost 700 premature deaths could be prevented annually with the Barcelona Superblocks, along with further gains in life expectancy linked with green space development and uptake of public and active transport.

1.3 Final Considerations and Challenges in Creating a Healthy City for All

There are a wide range of health determinants that are controlled by urban planning, being context-dependent and resulting from the local distribution and allocation of resources. Most studied drivers of urban health inequities are housing, transport, mobility, air pollution, access to services and goods, public space, and social inclusion.

Promoting health equity is a place-based issue: to develop concrete actions, it is important to integrate evidence, indicators' data, and the perspectives of local stakeholders and community on the nature of systematic, structural, and unfair inequalities in the urban determinants of health. Acknowledged the mechanisms through which individual and community health are affected by urban planning decisions across time, its role ought to be recognized as a form of "spatial medicine" (Rice, 2020). The COVID-19 pandemic, and the associated need to address risk factors of contagion and mitigate unintended impacts, can work as an opportunity for the integration of health considerations more broadly into urban planning processes.

Now, even more than in the past, a multidisciplinary and integrated approach is essential. The relationship between urban planning and health equity is complex and multidimensional by nature: there is no single indicator to reflect it nor a single "right" solution or single policy addressing it. The integration of global perspectives and solutions into local levels needs to be undertaken through a systemic lens, which means shifting from fragmented or single policy interventions to a comprehensive basket of policies reflecting the multiple domains in which urban health inequities are produced or exacerbated. A clear place-based orientation to address urban health inequities is needed: all city plans should be informed by evidence on health inequalities and then orient investment and interventions to priority neighbourhoods. A collaborative approach to health and planning is key: local governments should work with public health leads and citizens to understand and take account of the health needs of local communities and apply innovative actions capable to overcome existent barriers to improving health and well-being of all population groups and neighbourhoods.

With growing ageing population in cities, the main challenges for addressing a health equity-oriented urban planning are to redesign or adapt urban infrastructure to the needs of people, from a life-course approach. This means planning the city for all ages and preparing built environment for a safe and healthier ageing in place. First, urban planning should recognize that the three components of successful ageing – avoidance of disease, maintenance of functionality, and active engagement in public life – can all be impacted positively or negatively by the urban environment. The "city of proximity" urban model appears to be the one that provides more benefits to promote an inclusive and healthier environment: it provides an even, equitable and accessible distribution of urban services, including shops, social, and health centres, which do not require large distances to travel.

The following considerations should be incorporated in urban planning and design to promote a healthier and equitable place to grow old: (i) intersectoral work and participatory planning, overcoming barriers in social participation and promoting the active involvement of all citizens of all ages in the process of (re)designing and evaluating urban places; (ii) urban innovation, addressing the adequacy of community services (integrated care), housing (adaptation, accessibility, maintenance), functioning of public transport, embracing the way people move and adjusting to the interests and needs of ageing population; (iii) inclusive design, promoting a barrier-free environment and socially inclusive public spaces (intergenerational and multicultural).

As Soja argues in "Seeking Spatial Justice", justice has a geography and the equitable distribution of resources, services, and access is a basic human right (Soja, 2010). When the city is creating environments where all people, regardless of age, gender, social position, health status, or disability, are respected and have opportunities to participate and contribute, is also planning for health equity.

Acknowledgments This work received support from the Centre of Studies in Geography and Spatial Planning (CEGOT), funded by national funds through the Foundation for Science and Technology (FCT) under the reference UID/GEO/04084/2020. Angela Freitas is a recipient of an Individual Doctoral Fellowship funded by national funds through the FCT, under the reference SFRH/BD/123091/2016.

References

Ashton, J., Grey, P., & Barnard, K. (1986). Healthy cities – WHO's new public health initiative. *Health Promotion International, 1*(3), 319–324.

Bambra, C., Riordan, R., Ford, J., & Matthews, F. (2020). The COVID-19 pandemic and health inequalities. *Journal of Epidemiology and Community Health, 74*(11), 964–968.

Barton, H., & Grant, M. (2006). A health map for the local human habitat. *The Journal of the Royal Society for the Promotion of Health, 126*(6), 252–253.

Barton, H., & Tsourou, C. (2000). *Healthy urban planning: A WHO guide to planning for people*. WHO Regional Office for Europe.

Barton, H., Grant, M., Mitcham, C., & Tsourou, C. (2009). Healthy urban planning in European cities. *Health Promotion International, 24*(S1), 91–99.

Barton, H., Grant, M., & Guise, R. (2010). *Shaping neighbourhoods for local health and global sustainability*. Routledge.

Barton, H., Thompson, S., Burgess, S., & Grant, M. (Eds.). (2015). *The Routledge handbook of planning for health and well-being. Shaping a sustainable and healthy future* (1st ed.). Routledge.

Borrell, C., Pons-Vigués, M., Morrison, J., & Díez, E. (2013). Factors and processes influencing health inequalities in urban areas. *Journal of Epidemiology and Community Health, 67*(5), 389–391.

Borrell, C., Rodríguez Sanz, M., Pasarín, M. I., Marí Dell'olmo, M., Palència, L., Gotsens, M., … Morrison, J. (2014). Socioeconomic inequalities in mortality in 16 European cities. *Scandinavian Journal of Public Health, 42*(3), 245–254.

Boyce, T., & Brown, C. (2007). *Engagement and participation for health equity*. WHO Regional Office for Europe.

Braveman, P. (2006). Health disparities and health equity: Concepts and measurement. *Annual Review of Public Health, 27*, 167–194.

Braveman, P. (2014). What are health disparities and health equity? We need to be clear. *Public Health Reports, 129*(1_suppl2), 5–8.

Braveman, P., Arkin, E., Orleans, T., Proctor, D., & Plough, A. (2017). *What is health equity? And what difference does a definition make?* Princeton. Retrieved from https://www.rwjf.org/en/library/research/2017/05/what-is-health-equity-.html

Brown, P., Zavestoski, S., McCormick, S., Mayer, B., Morello-Frosch, R., & Altman, R. G. (2004). Embodied health movements: New approaches to social movements in health. *Sociology of Health & Illness, 26*(1), 50–80.

Campbell, M. (2005). What tuberculosis did for modernism: The influence of a curative environment on modernist design and architecture. *Medical History, 49*(4), 463–488.

Capolongo, S., Rebecchi, A., Buffoli, M., Appolloni, L., Signorelli, C., Fara, G. M., & D'Alessandro, D. (2020). COVID-19 and cities: From urban health strategies to the pandemic challenge. A decalogue of public health opportunities. *Acta Biomed, 91*(2), 13–22.

Collins, P. A., & Hayes, M. V. (2010). The role of urban municipal governments in reducing health inequities: A meta-narrative mapping analysis. *International Journal for Equity in Health, 9*(13), 1–20.
Corburn, J. (2005). Urban planning and health disparities: Implications for research and practice. *Planning Practice and Research, 20*(2), 111–126.
Corburn, J. (2013). *Healthy city planning: From neighbourhood to national health equity*. Routledge.
Corburn, J. (2017). Urban place and health equity: Critical issues and practices. *International Journal of Environmental Research and Public Health, 14*(2), 117.
Corburn, J., & Cohen, A. (2012). Why we need urban health equity indicators: Integrating science, policy, and community. *PLoS Medicine, 9*(8), e1001285.
Corburn, J., Curl, S., & Arredondo, G. (2014a). A health-in-all-policies approach addresses many of Richmond, California's place-based hazards, stressors. *Health Affairs, 33*(11), 1905–1913.
Corburn, J., Curl, S., Arredondo, G., & Malagon, J. (2014b). Health in all urban policy: City services through the prism of health. *Journal of Urban Health, 91*(4), 623–636.
Cummins, S., Curtis, S., Diez-Roux, A. V., & Macintyre, S. (2007). Understanding and representing 'place' in health research: A relational approach. *Social Science & Medicine, 65*(9), 1825–1838.
Diez Roux, A., & Mair, C. (2010). Neighborhoods and health. *Annals of the New York Academy of Sciences, 1186*, 125–145.
Duhl, L., Sanchez, A., & WHO. (1999). *Healthy cities and the city planning process: A background document on links between health and urban planning*. WHO Regional Office for Europe.
Ellen, I., Goddard, P., & Captanian, A. (2020). *Gentrification and the health of legacy residents, health affairs health policy brief.*
Fainstein, S. (2010). *The just city*. Cornell University Press.
Fainstein, S. (n.d.). *Urban planning*. Retrieved from https://www.britannica.com/topic/urban-planning. Last access 30 Apr 2021.
Fine, P., Victora, C., Rothman, K. J., Moore, P. S., Chang, Y., Curtis, V., … Deaton, A. (2013). John Snow's legacy: Epidemiology without borders. *Lancet, 381*(9874), 1302–1311.
Flies, E. J., Mavoa, S., Zosky, G. R., Mantzioris, E., Williams, C., Eri, R., … Buettel, J. C. (2019). Urban-associated diseases: Candidate diseases, environmental risk factors, and a path forward. *Environment International, 133*(Pt A), 105187.
Florida, R., Adler, P., & Mellander, C. (2017). The city as innovation machine. *Regional Studies, 51*(1), 86–96.
Freitas, Â., Santana, P., Oliveira, M. D., Almendra, R., Bana e Costa, J. C, & Bana e Costa, C. A. (2018). Indicators for evaluating European population health: A Delphi selection process. *BMC Public Health, 18*(1), 557.
Freitas, Â., Rodrigues, T. C., & Santana, P. (2020). Assessing urban health inequities through a multidimensional and participatory framework: Evidence from the EURO-HEALTHY project. *Journal of Urban Health, 97*(6), 857–875.
Friel, S., Akerman, M., Hancock, T., Kumaresan, J., Marmot, M., Melin, T., & Vlahov, D. (2011). Addressing the social and environmental determinants of urban health equity: Evidence for action and a research agenda. *Journal of Urban Health, 88*(5), 860–874.
Frumkin, H. (2002). Urban sprawl and public health. *Public Health Reports, 117*(3), 201–217.
Galea, S., & Vlahov, D. (2005). Urban health: Evidence, challenges, and directions. *Annual Review of Public Health, 26*, 341–365.
Galea, S., Freudenberg, N., & Vlahov, D. (2005). Cities and population health. *Social Science & Medicine, 60*, 1017–1033.
Giles-Corti, B., Vernez-Moudon, A., Reis, R., Turrell, G., Dannenberg, A. L., Badland, H., … Owen, N. (2016). City planning and population health: A global challenge. *Lancet, 388*(10062), 2912–2924.

Glouberman, S., Gemar, M., Campsie, P., Miller, G., Armstrong, J., Newman, C., ... Groff, P. (2006). A framework for improving health in cities: A discussion paper. *Journal of Urban Health, 83*(2), 325–338.

Hall, P. (2014). *Cities of tomorrow: An intellectual history of urban planning and design since 1880* (4th ed.). Wiley-Blackwell.

Impact on Urban Health. (2021). *Global perspectives on urban health.*

Kim, J. (2020). *What is a "healthy city"? The answer lies in the paradigm of the beholder.* CHETRE. Retrieved from https://chetre.org/2020/11/urban-health-paradigms/

Kindig, D., & Stoddart, G. (2003). What is population health? *American Journal of Public Health, 93*(3), 380–383.

Kolyasnikov, V. A. (2017). Significant and basic innovations in urban planning. In *IOP conference series: Materials science and engineering* (Vol. 262, p. 012117). Institute of Physics Publishing.

Lee, S. (2021). COVID-19 amplifiers on health inequity among the older populations. *Frontiers in Public Health, 8*, 609695.

Lopez, R. P. (2012). *The built environment and public health*. Jossey-Bass.

Loureiro, A., Freitas, A., Barros, C., & Santana, P. (2015). O papel dos municípios na promoção da saúde na Amadora, Lisboa, Mafra e Oeiras. In Santana (Ed.), *Território e Saúde Mental em tempos de crise*. Imprensa da Universidade de Coimbra.

Lydon, M., & Garcia, A. (2015). *Tactical urbanism: Short-term action for long-term change.* Island Press-Center for Resource Economics.

Moreno, C., Allam, Z., Chabaud, D., Gall, C., & Pratlong, F. (2021). Introducing the "15-Minute City": Sustainability, resilience and place identity in future post-pandemic cities. *Smart Cities, 4*(1), 93–111.

Mueller, N., Rojas-Rueda, D., Khreis, H., Cirach, M., Andrés, D., Ballester, J., ... Nieuwenhuijsen, M. (2020). Changing the urban design of cities for health: The superblock model. *Environment International, 134*, 105132.

NHS Health Scotland. (n.d.). *Place standard. How good is your place*? Retrieved from https://www.placestandard.scot/. Last access 20 Apr 2021.

NHS London Healthy Urban Development Unit. (2014). *Healthy urban planning checklist.*

Northridge, M. E., & Freeman, L. (2011). Urban planning and health equity. *Journal of Urban Health, 88*(3), 582–597.

Novoa, A. M., Pérez, G., Espelt, A., Echave, C., et al. (2018). The experience of implementing urban HEART Barcelona: A tool for action. *Journal of Urban Health, 95*(5), 647–661.

NSW Ministry of Health. (2020). *Healthy built environment checklist. A guide for considering health in development policies, plans and proposals*. St. Leonards, NSW.

Pineo, H. (2020). Towards healthy urbanism: Inclusive, equitable and sustainable (THRIVES) – An urban design and planning framework from theory to praxis. *Cities & Health*, 1–15.

Pineo, H., Glonti, K., Rutter, H., Zimmermann, N., Wilkinson, P., & Davies, M. (2018). Urban health indicator tools of the physical environment: A systematic review. *Journal of Urban Health, 95*(5), 613–646.

Pineo, H., Zimmermann, N., & Davies, M. (2019). Urban planning. Leveraging the urban planning system to shape healthy cities. In S. Galea, C. K. Ettman, & D. Vlahov (Eds.), *Urban health*. Oxford University Press.

Prasad, A., Kano, M., Dagg, K., Mori, H., et al. (2015). Prioritizing action on health inequities in cities: An evaluation of Urban Health Equity Assessment and Response Tool (Urban HEART) in 15 cities from Asia and Africa. *Social Science & Medicine, 145*, 237–242.

Prasad, A., Gray, C. B., Ross, A., & Kano, M. (2016). Metrics in urban health: Current developments and future prospects. *Annual Review of Public Health, 37*(1), 113–133.

Project for Public Spaces. (2016). *The case for healthy places. Improving health outcomes through placemaking.*

Project for Public Spaces. (n.d.). *The place diagram tool. What makes a great place?* Retrieved from https://www.pps.org/article/grplacefeat. Last access 20 Apr 2021.

Ramirez-Rubio, O., Daher, C., Fanjul, G., Gascon, M., Mueller, N., Pajín, L., … Nieuwenhuijsen, M. J. (2019). Urban health: An example of a "health in all policies" approach in the context of SDGs implementation. *Globalization and Health, 15*(1), 1–21.
Rice, L. (2020). After Covid-19: Urban design as spatial medicine. *Urban Design International*, 1–6.
Rojas-Rueda, D., & Morales-Zamora, E. (2021). Built environment, transport, and COVID-19: A review. *Current Environmental Health Reports, 5*, 1–8.
Rothenberg, R., Stauber, C., Weaver, S., Dai, D., Prasad, A., & Kano, M. (2015). Urban health indicators and indices-current status. *BMC Public Health, 15*(1), 494.
Rubin, V. (2015). *Regional planning for health equity* (Sustainable communities series). Oakland. Retrieved from https://www.policylink.org/sites/default/files/Regional-Planning-for-Health-Equity_FINAL.pdf
Rydin, Y., Bleahu, A., Davies, M., Dávila, J. D., Friel, S., De Grandis, G., … Wilson, J. (2012). Shaping cities for health: Complexity and the planning of urban environments in the 21st century. *Lancet, 379*(9831), 2079–2108.
Santana, P., Santos, R., & Nogueira, H. (2009). The link between local environment and obesity: A multilevel analysis in the Lisbon Metropolitan Area, Portugal. *Social Science & Medicine, 68*, 601–609.
Santana, P., Costa, C., Marí-Dell'Olmo, M., Gotsens, M., & Borrell, C. (2015). Mortality, material deprivation and urbanization: Exploring the social patterns of a metropolitan area. *International Journal for Equity in Health, 14*(1), 1–13.
Santana, P., Costa, C., Freitas, Â., Stefanik, I., Quintal, C., Bana e Costa, C, … Vieira, A. (2017). *Atlas of population health in European Union regions*. Imprensa da Universidade de Coimbra.
Santana, P., Freitas, Â., Stefanik, I., Costa, C., Oliveira, M., Rodrigues, T. C., et al. (2020). Advancing tools to promote health equity across European Union regions: The EURO-HEALTHY project. *Health Research Policy and Systems, 18*(1), 18.
Santana, P., Almendra, R., Pilot, E., Doreleijers, S., & Krafft, T. (2021). Environmental inequalities in global health. In R. Haring, I. Kickbusch, D. Ganten, & M. Moeti (Eds.), *Handbook of global health*. Springer.
Scholl, C., & Kraker, J. (2021). The practice of urban experimentation in Dutch City Labs. *Urban Planning, 6*(1), 161–170.
Schwarz, D. (2016). *Residential segregation and health?* Retrieved from https://www.rwjf.org/en/blog/2016/03/whats-the-connection-between-residential-segregation-and-health.html. Last access 22 May 2021.
Siri, J. (2016). Sustainable, healthy cities: Making the most of the urban transition. *Public Health Reviews, 37*, 22.
Soja, E. (2010). *Seeking spatial justice* (6th ed.). University of Minnesota Press.
Stahle, A. (2016). *Closer together: This is the future of cities*. Dokument Press.
State of Victoria. (2019). *20-minute neighbourhoods. Creating a more liveable Melbourne*. The State of Victoria Department of Environment, Land, Water and Planning. Retrieved from https://www.planning.vic.gov.au/__data/assets/pdf_file/0031/428908/Creating-a-more-liveable-Melbourne.pdf
Tosics, I. (2019). *New urban planning: Long-lasting innovation or just a temporary illusion?* Retrieved from https://www.blog.urbact.eu/2019/01/new-urban-planning-part-1/. Last access 26 May 2021.
UN-Habitat. (2021). *Cities and pandemics: Towards a more just, green and healthy future*. UN-Habitat.
UN-Habitat and WHO. (2020). *Integrating health in urban and territorial planning: A sourcebook*. UN-Habitat and World Health Organization.
van Hoof, J., Kazak, J. K., Perek-Białas, J. M., & Peek, S. T. M. (2018). The challenges of urban ageing: Making cities age-friendly in Europe. *International Journal of Environmental Research and Public Health, 15*(11), 2473.
Vlahov, D., & Galea, S. (2003). Urban health: A new discipline. *Lancet, 362*(9390), 1091–1092.

Vlahov, D., Freudenberg, N., Proietti, F., Ompad, D., Quinn, A., Nandi, V., & Galea, S. (2007). Urban as a determinant of health. *Journal of Urban Health, 84*(3 Suppl), 16–26.
Weber, M. (2019). How the city of Utrecht develops a health and equity in all policies approach. *European Journal of Public Health, 29*(Supplement_4), ckz185.733.
WHO. (2010). *Urban Health Equity Assessment and Response Tool (Urban HEART)*. Retrieved from http://www.who.int/kobe_centre/measuring/urbanheart/en/. Last access 25 July 2018.
WHO. (2012). *Addressing the social determinants of health: The urban dimension and the role of local government*. WHO Regional Office for Europe.
WHO. (2021). *Equity*. Retrieved from https://www.who.int/healthsystems/topics/equity/en/. Last access 22 May 2021.
WHO and UN-Habitat. (2010). *Hidden cities: Unmasking and overcoming health inequities in urban settings*. WHO and UN-Habitat.
WHO and UN-Habitat. (2016). *Global report on urban health: Equitable, healthier cities for sustainable development*. WHO and UN-Habitat.
WHO Regional Office for Europe. (2016). *Creating age-friendly environments in Europe: A tool for local policy-makers and planners*. WHO Regional Office for Europe.
WHO Regional Office for Europe. (2018). *Copenhagen Consensus of Mayors. Healthier and happier cities for all. A transformative approach for safe, inclusive, sustainable and resilient societies*. WHO Regional Office for Europe.
WHO Regional Office for Europe. (2019). *Healthy, prosperous lives for all: The European health equity status report*. WHO Regional Office for Europe.
WHO Regional Office for the Western Pacific. (2015). *Healthy cities: Good health is good politics: Toolkit for local governments to support healthy urban development*. WHO Regional Office for the Western Pacific.
Wilson, S., Hutson, M., & Mujahid, M. (2008). How planning and zoning contribute to inequitable development, neighborhood health, and environmental injustice. *Environmental Justice, 1*(4), 211–216.

Ângela Freitas is a Doctoral Researcher in Human Geography at the University of Coimbra and member of the Centre of Studies in Geography and Spatial Planning (CEGOT). Her dissertation explores the impact of urban policies on health equity and the role that local government can play in promoting population health and reducing inequities in health determinants. Her research interests lie in the relationship between local environment and health, urban planning, policy health impact assessment, and participatory research.

Paula Santana is a Senior Researcher and Full Professor of Health Geography in the Department of Geography and Tourism, at University of Coimbra and member of the Centre of Studies in Geography and Spatial Planning (CEGOT). She has more than 30 years of experience in studying the connection between place and health, focusing on the geographical disparities in health, urban health, healthy urban planning, and health-promoting policies, at European, national, and local levels.

Chapter 2
Social Innovation and Adaptability for Ageing in Place in Cities: A Comparison Between France and Japan

Camille Picard

Abstract Facing an ageing population, governments must respond to new challenges. Socio-demographic evolutions have highlighted the diversity of appropriations of our living environments and the difficulties faced by the most vulnerable. The need for adaptation is now known and new solutions must be found. In Japan, with a "community-based integrated care system", the government is trying to promote new practices of solidarity in neighbourhood communities in order to enable older people to age in place. This scheme is based on the idea of having inclusive environments where all the care needed by the population can be provided within the neighbourhood. The system has been developed in such a way that it gives a lot of autonomy to local governments and stakeholders. It also enables social innovation which will be a key component to apprehend how support for the elderly is provided according to the localities' own resources. In France, the notion of community support has yet to develop despite an ageing population. Support is mostly provided by public authorities. Nevertheless, citizens are taking action and socially innovative initiatives are being developed. The demonstration will be illustrated by a comparison between France and Japan around the question: how can enable social innovation help the creation of inclusive neighbourhood communities?

The findings show the importance of the social, cultural, economic and political context from which social innovation emerges. In Japan and France, social innovation appears as a response to the challenges of an ageing population. From the perspective of ageing, one of the main challenges in urban design is to include elderly people in the process. Social innovation, led by elderly people helps recognise their value and their human rights. Fostering the emergence of social innovations appears as a solution to develop inclusivity and ageing in place.

Keywords Social innovation · Ageing · France · Japan · Cities

C. Picard (✉)
Université Paris Est (Lab'Urba) and Kyoto Prefectural University, Paris, France

E. Pozo Menéndez, E. Higueras García (eds.), *Urban Design and Planning for Age-Friendly Environments Across Europe: North and South*, Future City 19,
https://doi.org/10.1007/978-3-030-93875-8_2

2.1 Introduction and Methodology

Demographic changes and the increase in the number of people over 65 in the world require us to rethink our living environments. Japan is the first country to be confronted with these issues today. By comparison, in 2015, while Japan had 47 people over 65 for every 100 people of working age, the OECD[1] average was 28 (Germany 35, France 33 and the United States 25) (OECD, n.d.). The ageing of the population leads to rethinking our living environments. Many questions and even issues might emerge if we conceive our cities as we used to. Faced with the rise in the number of elderly people, ageing in place remains a priority for governments and individuals, particularly in France and Japan. Indeed, the "home" represents many implications such as intimacy, identity and total control of one's environment. The "home" is closely related to the living environment (or "habitat" in French), which covers a wider space than the dwelling, both in material and immaterial ways.

In order to respond to the challenges and needs of a growing part of the population, new forms of care, mutual help and, more broadly, types of housing are being created. In that context, we will try to understand how countries, with their own social, cultural and political specificities, have created or not a favourable context from which socially innovative initiative can arise.

In order to provide some answers, the first part is devoted to the definition and the different theories of social innovation. In a second part, the research is positioned in this state of the art according to several selected elements. Finally, two case studies from Japan and France are presented and compared.

The elements presented from the Japanese example are based on two field investigations carried out in 2019 and 2020 for a dissertation on housing policies promoting home care in the context of the ageing population. About forty Japanese elderly people were met during these visits. Some of these meetings were organised in groups and provided an opportunity to discuss a questionnaire about living conditions. Six individuals were met several times in their homes, where semi-directive interviews were carried out. Finally, other exchanges took place in a more informal way. Most of the interviewees were members of a local association. Two districts of the city of Tokyo were observed in particular, Bunkyō and Adachi. This choice was based on a statistical analysis of the city's boroughs, which made it possible to identify two localities with different characteristics (the Adachi Borough being more accurately described as a "popular" district).

The elements presented for France come from the involvement in a Participatory Action Research started in 2019 that the researcher joined in 2020. Six fields throughout France are being investigated and I have joined the scientific coordination of the field located in the city of Rouen. Several collective (2) and semi-individual (5) interviews were conducted in addition to the collective work (collective bibliography etc.).

[1]Organisation for Economic Co-operation and Development. Retrieved from: https://www.oecd.org/

2.2 What Social Innovation Means in France and Japan? Positioning the Research

2.2.1 Social Innovation as a Part of Economic, Social and Political Systems

A number of authors, mainly Anglo-Saxon, have contributed to theorising social innovation. From their perspective, social innovation is a tool to improve the socio-economic system rather than disrupting a pre-existing one. It therefore exists in accordance with and in the context from which it arises.

Japanese literature shares many similarities. Tanimoto appears to be one of the most published researchers on this subject; many Japanese authors refer to his papers. He has drawn heavily on Anglo-Saxon theories in his approach to social innovation, particularly through his study of the city of San Francisco and social entrepreneurship (Tanimoto, 2008a, b; Tanimoto et al., 2007). He defines social innovation as "developing new social goods and services, and unique systems to address social problems" such as health, education, the environment and aid to developing countries etc. The creation of new organisational patterns to provide these goods and services is also a feature (Ibid.). This relatively broad interpretation does not situate social innovation as a disruptive movement towards the current economic system, nor as an empowerment tool.

Aoo, on the other hand, points out the evolving use of social innovation theories, notably European theories that differ from "traditional" approaches and focus more on the macro-level implications of innovation, its social effects and systemic changes (Aoo, 2019). In Japan, the literature on the subject is mainly developed by researchers trained in economic and business approaches. There is little interaction with European research. Conversely, since the 2000s, North American theories have moved from purely economic and entrepreneurial concerns to a vision that integrates more diverse elements, including those from European literature (Ibid.).

According to Richez-Battesti and echoed by the Jean-Baptiste Godin Institute, two currents complement each other in the Anglophone literature (Godin Institute Contribution, 2013; Richez-Battesti et al., 2012). First, the market-profit school of thought views social enterprises as a form of organisation that solves the finance-related issues of non-profit organisations or organisations developing profit-making activities (Ibid.). Today, this term refers to any organisation, whether profit-making or not, which carries out an economic activity for a social purpose (Ibid.). Second, the social innovation school of thought emphasises innovations driven by an entrepreneur whose capacity and willingness are decisive in the achievement of a social goal (Ibid.). Social entrepreneurship appeared in the 1980s and 1990s in the Anglo-Saxon world (in 1993, Harvard University launched the Social Enterprise Initiative) (Besançon et al., 2014). It then spread through foundations and associations.

The Ashoka association, one of the first to work on these issues and whose creation dates back to 1980, defines the social innovator as "someone who puts his or her entrepreneurial qualities at the service of solving a large-scale social and/or environmental issue" (Ibid.). Thus, social innovation depends on the capabilities of the individual entrepreneur.

These Anglo-Saxon approaches are more in line with a market-driven logic and do not disrupt current models (economic, social etc.). Besançon et al. summarise this vision as "providing a solution to a given problem or need" which "will be the result of the will of one or more individuals motivated by altruism. This altruism will be embodied in two figures, the entrepreneur and the philanthropist, for whom profit is not an objective; the philanthropist will be represented by wealthy individuals, foundations or even large companies or economic groups" (Besançon et al., 2014).

2.2.2 Social Innovation as a Disruptive Tool

In a number of publications, social innovation is presented and/or defined as a means of disrupting a pre-existing system (political, economic, social or cultural). Among other things, it is seen as an instrument for democratisation and as a tool for citizens to appropriate their environment. This point of view is shared by a majority of francophone authors. According to them, what pushes an innovation beyond novelty is the break with usual practices (Besançon et al., 2014).

Researchers at the Godin Institute have classified social innovation according to the different approaches on the subject. First is the social enterprise school, which emerges largely from the EMES, a network of researchers on social enterprises (Godin Institute Contribution, 2013). Second is the institutionalist approach put forward by the CRISES in Quebec (Ibid.). Although the first approach considers the enterprise as a producer of social innovation, it does not primarily seek profit and its social action is reflected both in its purpose and its process. Innovation is not the sole work of an individual, but the result of democratic and collective work. The economic and social aspects of the social enterprise are intertwined. Defourny (2004) has defined the social enterprise under two categories that were summed up by Besançon et al. (2014) (e.g., Table 2.1).

Furthermore, they comprehend practices as being socially innovative if they can be qualified as "solidarity practices" (Ibid.). They define solidarity practices as all

Table 2.1 Table of economic, entrepreneurial and social dimensions of the social enterprise

Economic and entrepreneurial dimension	Social dimension
Continuous activity in the production of goods and/or services Autonomy in management High economic risk-taking Limited paid employment	Explicit objective of service to the community Initiative from a group of citizens Decision-making power decoupled from capital Participatory and democratic dynamics Limited profit distribution

Source: Besançon et al. (2014)

actions related to justice, democracy, reciprocity, proximity etc. (Ibid.). According to them, this additional specification makes the theorisation of social innovation more complete. In France, for example, social innovation is widely used at a regional level in the context of the Social Solidarity Economy (SSE).

Therefore, the dynamics of disruption also involve citizen empowerment. Indeed, social innovation thus appears as a means to increase individuals' capabilities to act in and on society. In the International Handbook on Social Innovation: Collective action, social learning and transdisciplinary research, the authors approach this aspect of the definition as "progressive solutions to a whole range of problems of exclusion, deprivation, alienation, lack of wellbeing, and also to those actions that contribute positively to significant human progress and development" (Moulaert, 2013).

In a similar perspective, Hillier et al. characterise social innovation under three dimensions: the satisfaction of unmet needs – adding that they are not or no longer perceived as important by the market or the state; changes in social dynamics; and, above all, an "increase in socio-political capacity and access to the resources necessary for the materialisation of rights, the satisfaction of human needs and participation" (empowerment dimension) (Hillier et al., 2004). In their argument, the authors thus set out a political vision of social innovation.

Furthermore, they distinguish in their work four movements (the first three correspond to our definition's second category) in which social innovation has been "either mobilised or analysed" (Ibid.):

- Firstly, in the 1990s, the management sciences emphasised the positive consequences of the improvement of social capital.
- Secondly, social innovation is referred to in theories that cross management and scientific research, notably in the analysis of the positive both social and environmental externalities of an economic activity.
- Thirdly, the authors have identified the use of social innovation in "intellectual and social creation" within the arts and creative studies (defined by Mumford in 1992).
- Finally, social innovation is found in territorial development, especially highlighted by Moulaert et al. in Globalisation and Integrated Area Development in European Cities in 2002. This approach proposes to counteract the issues of local development by organising development at a neighbourhood scale (Integrated Area Development). This concept gives more importance to democratic and citizen participation aspects of local governance.

Moulaert et al. suggest rethinking the ALMOLIN (ALternativeMOdels for Local INnovation) model developed as part of the research funded by the European Commission called SINGOCOM in the early 2000s (Moulaert et al., 2017). Through this model, the authors examine the role of social innovation in the development of neighbourhood communities. They draw several key conclusions from this analysis:

- Any movement originates from a broad philosophical or normative vision (e.g., social solidarity), adapted to the political, social and cultural contexts of the locality.
- The emphasis is on the community, democratic governance and the notion of local identity.
- A re-emergence of social economy and mutual help.
- Tensions between local initiatives and central government due to the resurgence of local democracy.
- Tensions between social innovation and institutionalisation: In order to last, innovations need to be appropriated and spread. However, the institutionalisation can divert the innovation from its primary objective.
- Tensions between communities and society as a whole.
- Social innovation by purpose as opposed to social innovation by process: social innovation is often put forward as a process, but the authors defend a vision of social innovation as concrete solutions to the issues raised.

Furthermore, social innovation responds to human needs and leads to transformations in social dynamics – notably through governance, empowerment and easier access to resources (Ibid.). Therefore, the authors see it as synonymous with changes in institutions contributing to greater "social inclusion". The institutions are in reference to laws, regulations, organisations, mores, mechanisms and processes of socialisation (Ibid.).

This first part sets out the author's view of social innovation but in order to position the French and Japanese case studies, we identified the limits of the theories presented above.

2.2.3 Limitations of Social Innovation Theories

Researchers at the Godin Institute consider that no theoretical definition allows for a sufficiently precise identification of social innovation projects (Godin Institute Contribution, 2013). The authors of the TEPSIE report identified a "reductionist" definition of social innovation as opposed to "holistic" ones (TEPSIE, 2014). The limitations of both will be presented here.

First of all, social innovation is used by major international institutions and by States as a tool for modernising public policies, particularly in the social and welfare fields (Besançon et al., 2014). This translates into a belief in the private sector and citizens to compensate for the "shortcomings" of public authorities in terms of social support. Taken to an extreme, this view of innovation can be problematic. Too much reliance on the private sector and on citizens could lead to government withdrawal from social issues.

Secondly, limits can be identified in the economic behaviour of the entrepreneur (mainly referred to in the Anglo-Saxon literature). Indeed, social innovation should not be reduced to a philanthropic activity that depends solely on the goodwill and investment capacity of individuals. In addition, a hierarchical and power relationship between the innovation's bearer and its beneficiaries might be established. From this perspective, the TEPSIE report warns against approaches centred around the social enterprise and the social entrepreneur (TEPSIE, 2014). They omit the public sphere's capacities and forget the inherent limits of the private sector (Ibid.).

Hillier et al. warn against a vision of "innovative behaviour" limited to an "optimal behaviour" in a context of economic rationality and forgetting about social aspects (Hillier et al., 2004).

The excessive belief in the local institutions and organisations, especially local governments, is perceived as a limit (Moulaert, 2013). The local level is often described as the only relevant level for meeting human needs, omitting the national and international contexts that influence innovation (Ibid.). Furthermore, the delegation of social competences by the central government to local authorities can create an imbalance between localities in terms of services provided. To address these difficulties, Moulaert et al. propose to consider the articulation of the different scales (local, national, international) as well as the dynamics of innovation through institutions (Moulaert et al., 2017).

In addition, the question of measuring "novelty" and the perception of "needs" arises. These two notions, which are at the centre of the debate on social innovation, do not seem to have clear definitions. First, it is difficult to assess novelty, especially when it comes to the social field. Second, the notion of "need" is subject to the context in which the innovation takes place. Needs "are neither universal nor timeless, they correspond to a given economic, social and cultural context, they are sometimes imposed by this context, and finally they can be a source of alienation for individuals who have become dependent" (Godin Institute Contribution, 2013). The TEPSIE report comes to a similar conclusion by arguing the difficulty of defining what "social" is (social in purpose, in the process, in expressed needs etc.). The authors sum it up as follows: "what may appear 'social' (beneficial) to one group, at a given time, in a certain social strata or region, may prove irrelevant or even detrimental to others" (TEPSIE, 2014).

Finally, Francophone approaches to social innovation are more political, which makes them difficult to use in the Japanese context, where projects are not necessarily political. Market logic is not particularly important in the initiatives we observed either. It is therefore complex to use a single theory to qualify the studied projects. Several elements can be borrowed from these different schools of thought in order to obtain the most comprehensive vision of social innovation in the context of a France-Japan comparison.

2.3 Selected Elements of Definition in a Comparison Context

Faced with this multitude of interpretations, we choose to highlight the common elements that seem relevant to the demonstration. The purpose of this paper is not to develop a new definition of social innovation but to rely on a rich and already existing literature in order to build a comparative analysis grid. Typologies cannot be drawn from the case studies. Nevertheless, it is possible to measure their adequacy with the elements of definition mentioned above.

Making a positioning diagram, one of the most used methods in the comparative study does not sufficiently capture the complexity of the initiatives. For example, the SI-DRIVE working group distinguishes several potential entry points for analysing social innovation: the scope of the innovation (global vs. local); the capacity of the innovation to exist on its own or being dependent on other stakeholders; whether the innovation comes from a bottom-up or a top-down initiative; the network of stakeholders and the organisation of the innovation; the extent to which the innovation is embedded in the social context etc (SI-DRIVE, 2017).

In order to best capture the criteria that qualify social innovation from a comparative perspective, we have chosen to use a Radar chart (also called a radar, star or spider web diagram). The Radar chart diagram makes it possible to take into account a sufficient number of elements and to assign them "levels" of applications.

More concretely, we have chosen 5 criteria and set 4 levels for each of these criteria. Firstly, the extent of the initiatives in the territories will be assessed; whether they are international, national, regional or local. Secondly, the network of stakeholders mobilised around the projects studied and the extent of the network will be assessed; from the social entrepreneur as the sole bearer to a highly developed network of actors. Thirdly, the ways in which the initiatives disrupt pre-existing norms, regulations and the public sector will be observed. The level of disruption will be assessed according to the degree of involvement of the public sector (led by the public sector, in cooperation with the public sector, initiated outside the public sphere, but in line with norms and regulations, or with the intention of being disruptive). The fourth criterion is the mobilisation of volunteer citizens in the initiatives; ranging from non-mobilisation to projects entirely carried out by volunteers. Finally, the last criterion is the economic logic of the initiatives by measuring whether or not they assimilate a market and profit rationale (Table 2.2).

2.4 Enabling Social Innovation in France and Japan and the Consequences on the Creation of an Inclusive Neighbourhood

Investigations revealed that the two countries give room to develop socially innovative initiatives. Nonetheless, they both offer this possibility in different ways. In Japan, the initiatives emerge more often from the public sphere or in line with public authorities. In France, the initiatives emerge from citizens' initiatives.

Table 2.2 Table of selected criteria

Criteria	Levels 0	1	2	3
Area	International	National	Regional	Local/ neighbourhood
Stakeholders' network	Social entrepreneur on his own	Social organisation on its own	Stakeholders' network is partially developed	Stakeholders' network is well developed (social innovation *cluster*)
Disruption with norms and regulations	Public initiative	Initiative in cooperation with the public sector	Initiative outside the public framework but in line with norms and regulations	Disruption, political aspiration
Mobilisation of voluntary citizens	Initiatives that do not mobilise volunteers	Initiatives carried out partly by volunteers	Initiatives mostly carried out by volunteers	Initiatives carried out solely by volunteers
Economic issues	Initiatives that do not follow a market or profit logic	Initiatives that do not follow a market logic, but have economic needs	Initiatives that do not follow a market logic, but make limited profits	Initiatives that follow a market and profit logic

Source: Author's own

2.4.1 *Japanese Policies Favourable to Social Innovation*

In order to illustrate the Japanese case, the community-based integrated care system will be presented. Research has shown a shift in responsibility for the care of the elderly since the 1990s that led to a community support system.

In June 2016, the Prime Minister's Office decided to implement a plan for a "dynamic engagement of all citizens" (Prime Minister's Office, n.d.; Tobita, n.d.). The goal of the plan was to promote citizens' participation in society, in the workplace and in communities. More broadly, the plan is intended to create a social dynamic that would halt the decline in the number of births (Ibid.). These objectives are included in the Annual Report on the Ageing Society. The government's commitment is also supposed to allow individuals to age in good health and not be a cost, both human and financial, for carers and for the state (Ibid.). The public authorities' goals go hand in hand with the objective of developing communities "by considering all generations" (Ibid.). This effort is further encouraged in shrinking areas to facilitate access to care. The government's report estimates that in 2017, 245 local governments stated their intention to promote these communities and 79 non-profit organisations have already put actions in place (Cabinet Office, 2018).

In addition, government publications are filled with representations of elderly people playing an active role in society. In the February 2021 communication published by the government's public relations office, the valorisation of active elderly people was the subject of several articles (Public Relations Office – Government of

Japan, 2021). For example, it mentions "the potential of a super-ageing society" thanks to "active seniors" (Ibid.). Several examples of active elderly Japanese were highlighted: a 97-year-old pharmacist, a 92-year-old photographer, an 89-year-old cheerleader etc. (Ibid). It is not uncommon in Japan to encounter elderly workers: gardeners, station workers etc. There also exists a job fair solely aimed at people aged 55 years or older to find work after retirement (Tokyo Work Center, n.d.). These representations of an "ideal elderly person" in Japan do not reflect the individuality of the elderly. Shortcuts and stereotypes are sometimes being made and conveyed in Japanese society.

Nevertheless, whether it is in the central government publications, in local government initiatives or in the discourse of elderly people, there is a new emphasis on the local level to address the challenges of an ageing population. With the reaffirmed position to involve elderly people in local activities and communities, the Cabinet Office expects the elderly people to age well (Cabinet Office, 2018). As the history of Japan reveals, this idea is not a new one and is widely embraced by the population through their discourses. Indeed, for many people, the "home" is not limited to the physical structure of the dwelling but is understood through relationships and neighbourhood communities. This includes direct neighbours, neighbourhood associations, and also the place occupied by people who are no longer there (the room of deceased parents, children's bedrooms etc.).

According to Tsutsui, home support for the elderly and support for the most dependent has evolved several times before becoming a model of "community support" (Tsutsui, 2013):

- Until the late 1990s, informal support was carried out by relatives and the community. It is the "support by family" era.
- During the first part of the 2000s, elderly support was characterised by solidarity under common goals called "support by society."
- Since 2006, the support has been qualified as "community support" with the systematisation of community support networks.

As part of the implementation of the long-term care insurance system (long-term care insurance system or *kaigo hoken* in Japanese) in the 2000s, the Ministry of Health initiated a research group that characterised the community-based integrated care system (*chiiki hōkatsu kea shisutemu*) (Morikawa, 2014; Tsutsui, 2014). This system is defined in 2008 as "the system of the community that, having provided housing in accordance with needs as the basic function, can appropriately offer each elderly person not only health care and long-term care but also various forms of support services, in order to ensure safety, security and health in their everyday life" (Ibid.).

The goal of this measure is to allow each locality to develop its own system in accordance with its specificities (social and demographic characteristics, system of actors etc.). The government has set the year 2025 as an objective for the system to be implemented in all municipalities (Japan Association of Elderly Housing Advisers, 2018). The system gives enough room to develop new ways of tackling the issues related to the ageing population, thus creating social innovation. The system appears as a guideline allowing innovation, and not as a strict regulation.

The system aims to provide the opportunity for older people to remain in the area they have always lived in, called "the area/sphere of daily life" (*nichijōseikatsu keniki*) (Japan Association of Elderly Housing Advisers, 2018). This area or sphere is an area that can be covered in about 30 min of walking, which is equivalent to the middle school zone in Japan (Ibid.). The terms "community" and "neighbourhood" (*chiiki*) are used equivalently in this context. Thus, the system should make it possible for elderly people to access a range of nursing, medical and social services within a 30-min walking distance from their homes. In addition, there are mutual-help and community watch networks as well as various types of associative networks.

Socially innovative initiatives can be observed throughout the archipelago. For example, the association *komaji no uchi* located in the Bunkyo ward of Tokyo was founded in order to tackle issues related to the isolation and home vacancy (both issues related to the ageing population). In 2013, the founder, an elderly man, decided to open a vacant house he owned to the community. He calls it "the place where everybody is welcome". The association is now working with social workers from the public sector to provide information and care services to the most vulnerable. This arrangement and the activities developed in this context correspond to a "new" way to meet the "needs" of its community. Moreover, the association is part of the community-based integrated care system of the locality.

According to the description of the Japanese case, a portrait of social innovation can be drawn up according to the criteria outlined above (e.g., Table 2.3).

This policy emphasises the injunction for individuals to be involved in society. People are required to be active in their community in order to create an ideal environment for living and ageing well. These measures are further reinforced by the fact that the government's discourse is oriented towards valuing active elderly people (Martine, 2012).

The next few years will be decisive in order to observe the concrete socially innovative actions of solidarity, mutual aid and vigilance implemented within the neighbourhood communities. The sustainability of these innovations also needs to be observed.

Table 2.3 Table of selected criteria in the case of Japan

Criteria	Level	Level number
Area	Local/neighbourhood.	3
Stakeholders' network	Stakeholders' network is well developed (social innovation *cluster*).	3
Rupture with norms and regulations	Public initiative.	0
Mobilisation of voluntary citizens	Initiatives mostly carried out by volunteers.	2
Economic issues	Initiatives that do not follow a market logic but have economic needs.	1

Source: Author's own

2.4.2 The Experience of a Participatory Action Research in France

In France, socially innovative initiatives often come from citizens' movements that defend a political vision of ageing. Unlike Japan, initiatives are not necessarily government-led. In the following section, the case of a Participatory Action Research (PAR) taking place in France will be presented. The PAR is called RAPSoDIÂ which is an acronym for Participative Research-Action, Solidarity, Home, Innovation in Ageing.

The PAR has been initiated by a national association called Hal'âge. The association's goal is to develop and support social innovation, at the intersection of housing and ageing. With the help of a sponsor (New Agency of Active Solidarity – ANSA) and a funder (Housing Foundation – Fondation du Domicile), the project was started in 2019. In addition to these three main stakeholders, more participants are involved in the research. Indeed, there are three types of "researchers" involved in the project. Since they are not paid to participate in the project, the researchers have no real status but can be considered as volunteers. Nonetheless, the participants are encouraged to benefit from it partly on a personal level (publication, communication, new knowledge acquisition etc.). The three types of "researchers" involved are as followed:

- Academic researchers that come from various universities and fields throughout France.
- Researchers from the Hal'âge association referred to as co-researchers.
- Researchers from the fields that are being investigated, also called co-researchers.

In total, around 65 people are involved in the project.

The research started from observing the housing possibilities existing in France for elderly people and focusing on the alternatives. Currently, there are two main categories of housing specifically for the elderly. The first one, evidently the least medicalised option, is one's home. The second option is to go to a specialised residence for the elderly with services or a retirement home. The best-known type of specialised residence is the EHPAD, which is an acronym for Residential and Medical Facility for Dependent Elderly People. In recent years, alternatives to these two options have been flourishing: cohousing, intergenerational cohabitation, intergenerational residence etc. Some of these alternatives were not adequately designed for elderly people. On the other hand, these types of housing are attractive because they participate in creating a sense of community. For that reason, the research focuses on how people can age and stay autonomous in these types of housing. More precisely, the goal is to answer the following question: how can we stay autonomous through mutual help? This raises even more questions such as the meaning of being autonomous. Indeed, during the research process, the participants had to deconstruct their idea of autonomy. At the beginning of the process, the majority of the field's co-researchers perceived autonomy as being able to be completely independent and not having to rely on anybody. Through collective work, they rethought the concept of autonomy as being something you can acquire through a group and through a community. After laying these foundations, the research then focused on the definition of mutual help: With whom? How? To what extent?

Three different phases have been identified to conduct the project. The first phase was about framing and problematising the research. During that time the research fields were identified and chosen. It lasted between September 2019 and January 2020. The second phase, which is currently happening, and which is expected to end in August 2021, is about learning, analysing and experimenting. Data is being collected on-site and practical experimentation will be soon implemented. Finally, the third phase will be the opportunity to share and promote the results of the research at a larger scale through international seminars and publications.

Six different research fields are being investigated throughout France:

- Two of them are located in cities with more than 400,000 inhabitants (including the city of Paris).
- Two of them have more than 100,000 inhabitants.
- The last two have fewer than 60,000 inhabitants.

All of them were chosen because a project of alternative housing was under construction or already constructed. One of them, located in the city of Rouen, differs from the rest of them. The project is focusing on creating mutual help not only at the residence level (called the La Grand Mare or GM neighbourhood) but also at the neighbourhood level (Fig. 2.1).

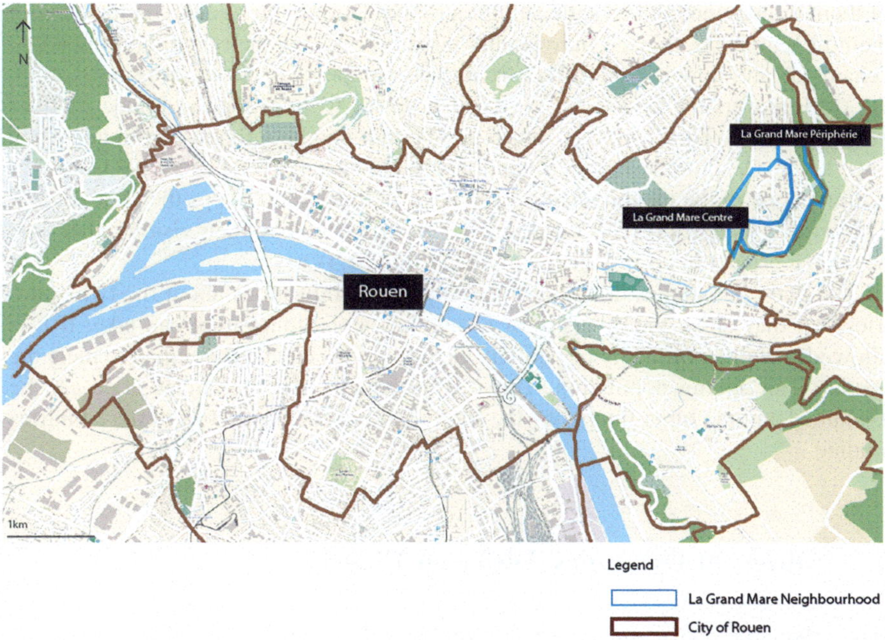

Fig. 2.1 Situation map of La Grand Mare neighbourhood. (Source: Author's own edited from Google Maps)

The neighbourhood has several interesting characteristics. First, the area is separated from the rest of the city by abundant green spaces and by topography (the neighbourhood is high up in relation to the rest of the city). Second, the neighbourhood is perceived as divided by the inhabitants. Half is composed of collective housing and is classified as a priority neighbourhood. The other half is composed of single home units, constructed during the 1970s. Also, the population of these two parts come from different social backgrounds. The inhabitants involved in the PAR, who live in the single home units, also perceived their neighbourhood as two distinct areas.

For this field, qualitative and collaborative methods have been put in place to help identify levers for mutual-help actions. More precisely, mental maps have been drawn and interviews are being conducted with a larger group of inhabitants. Moreover, a collaborative bibliography has been constructed in order to develop a common vocabulary between academic researchers and field researchers.

For now, more questions have been raised working on the La Grand Mare neighbourhood than answers provided. Some concerns have been pointed out regarding the capacity of creating a sense of mutual help between very different populations. Do people need to share the same social background, share common activities or is it possible to create "spontaneous" mutual help? Indeed, the inhabitants involved in the PAR come from the same social background: they were all participating in some sort of social movements in the 1960s, were working in alternative education for children and were able to socially ascend. This ascension was, among other things, characterised by the transition from collective housing to single-family units. Most of them transitioned in the same neighbourhood. During interviews conducted with them in June 2020, they all individually talked about how their neighbourhood had "changed" and how it now suffers from "bad reputation" (dirtier, drug trafficking, etc.). Moreover, the interviews revealed a desire to improve the image of their neighbourhood. They aspire to return to a certain neighbourhood community they once knew. Despite these good intentions, being nostalgic can be an obstacle. For example, one of the inhabitants wants to have a project that "keeps the youth busy". This simple sentence reveals the desire to conform people to their vision rather than develop a project in accordance with them (Table 2.4).

The two case studies do not allow to make generalisations about the two countries, but they show the diversity of forms that social innovation can take. If it is allowed socially, politically or publicly, new solutions can be developed to meet the needs of an ageing population. The way in which these initiatives arise can be explained, among other things, by the socio-cultural characteristics of the countries studied.

2.5 Different Dynamics, Different Elderly

Using the table above, we were able to draw a Radar chart that helped situate the case studies (e.g., Fig. 2.2).

Table 2.4 Table of selected criteria in the case of France

Criteria	Level	Level number
Area	Local/neighbourhood	3
Stakeholders' network	Stakeholders' network partially developed	2
Rupture with norms and regulations	Initiative rupture, political aspiration	3
Mobilisation of voluntary citizens	Initiatives carried out solely by volunteers	3
Economic issues	Initiatives that do not follow a market or a profit logic	0

Source: Created by author

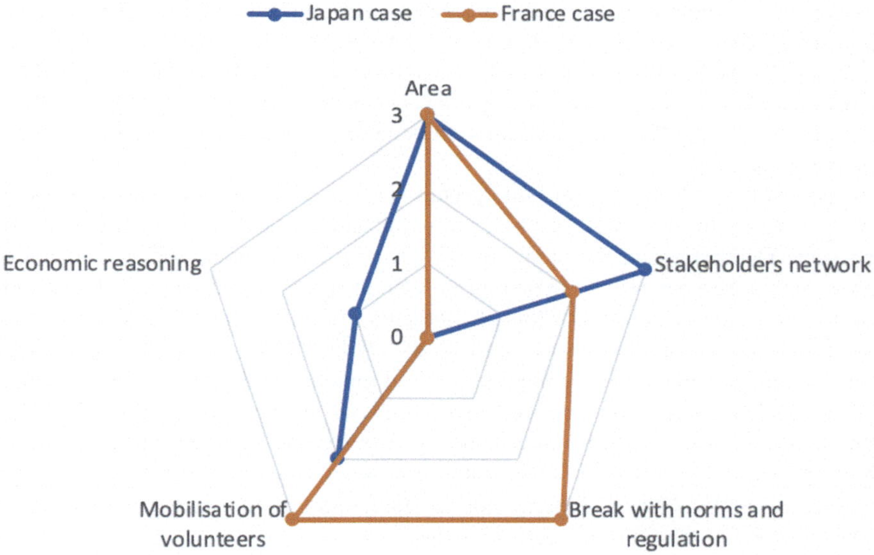

Fig. 2.2 Radar chart – a comparison between France and Japan on social innovation. (Source: Author's own)

The dynamics in the two countries are completely different although they might lead to similar ideas and projects. They both respond to the definition of social innovation by providing "new" solutions to identified needs that have not been met yet. The biggest difference between the two is "who" initiated the projects. These differences can be explained by socio-cultural differences that can be the subject of a separate paper. However, it is worth outlining the main explanations that give the keys to understand how social innovation can be implemented in different social and cultural contexts.

Preconceptions about Japan, the Japanese and individualism have been widely disseminated (Buisson, 2012). The idea that the Japanese are a homogeneous population in which individuality has little or no place has circulated for a very long time and continues to do so today. This image is not only shared in the West but has also been appropriated by the Japanese (Giraud and Galan, 2015). Furthermore, individuality has been associated with Western modernity and therefore does not fit in with a view of "traditional" Japanese society. Similarly, these issues are often approached in a binary way: individualistic societies vs. groupist societies. Galan, Giraud and Lozer and suggest analysing the "society-individual" relationship by means of the "We-I" balance used by Norbert Elias and by taking up Berque's remarks who emphasises that the Japanese would adapt their "self" to the position and context in which they find themselves, rather than "asserting it by imposing it on others" (Ibid.) There is indeed a romanticised vision of the collective in Japan that is not only embedded in discourse but also in history (Midgley et al., 1999).

In France, the dynamics are different. Social issues are more political, and the concept of community has a rather pejorative meaning. Indeed, the community is opposed, perhaps wrongly, to a democratic universalism. As a result, community is seen as a focus on differences, which is contrary to the values defended by public authorities. This is also the reason why the vocabulary used in these two case studies are different. In the PAR, the commitment is more political and expressed as so by the participants.

In conclusion, the diversity of interpretation of social innovation bears witness to the importance of the social, cultural, economic and political context from which it emerges. Depending on the status of the projects' instigators, social innovation emerges in coherence with or in disruption with the pre-existing systems. Japan and France are perfect examples of this phenomenon. In both countries, social innovation appears to be an obvious response to the challenges of an ageing population. These unprecedented socio-demographic developments, and the issues they create, call for the emergence of unprecedented solutions. In the two cases studied, the innovators are not of the same status and are therefore positioned in different innovative dynamics. Nevertheless, the "solutions" proposed are indeed "new" proposals that respond to "needs" that are not satisfied, either by the State (in the case of France) or the market (in the case of Japan).

Also, one of the main challenges for addressing urban design from the perspective of ageing, is to include elderly people in the process. Cities are often built from the perspective of young professionals, which imposes an ideal that does not correspond to the needs of everyone. In our two examples, the opportunity to develop social innovation is seized by the elderly people themselves. Thus, inclusion exists on two levels. First, the initiatives enable elderly people to remain independent in their living environment. Second, the innovations are developed for and by the elderly. Encouraging social innovation can be seen as an inclusive mechanism for all ages. To include older people in the development of cities is to recognise their value and their human rights. People should not be excluded because of their age. Regarding the results of this research, one of the strategies that can be put in place in order to encourage inclusivity and ageing in place is fostering the emergence of social innovations.

References

Advisers. (n.d.). *Revision of the 2018 law on long-term care insurancesystem*. BookWay.

Aoo, K. (2019). Social innovation in Japan: De facto social innovation and its limitation. *Atlas of Social Innovation – A Wordl of New Practises, 2*.

Besançon, E., Chochoy, N., & Guyon, T. (2014). *L'innovation sociale: Principes et fondements d'un concept*. L'Harmattan.

Buisson, A. (2012). Philippe Pelletier, La fascination du Japon: Idées reçues sur l'archipel japonais. *Géocarrefour, 87*(3–4), Article 3–4. Retrieved from: http://journals.openedition.org/geocarrefour/8566

Defourny, J. (2004). L'émergence du concept d'entreprise sociale. *Reflets et perspectives de la vie economique, Tome, XLIII*(3), 9–23.

Giraud, J. P., & Galan, C. (Eds.). (2015). *Individu-s et démocratie au Japon*. Presses Universitaires du Midi.

Godin Institute Contribution. (2013). *L'innovation sociale en pratiques solidaires Emergence, approches, caractérisation, définition, évaluation*. Retrieved from: http://www.institutgodin.fr/Is-InstitutGodin.pdf

Hillier, J., Moulaert, F., & Nussbaumer, J. (2004). Trois essais sur le rôle de l'innovation sociale dans le développement territorial. *Geographie, Economie, Societe, 6*(2), 129–152.

Japan Association of Elderly Housing Advisers. (2018). *Kōreisha sumaishurui to erabikata—Kōreisha sumaiadobaizākenteikōshikitekisuto [daini-ban] (kaigo hoken-hō 2018-nen kaiseitaiō) [Housing for Elderly: How to choose — Elderly Housing]*.

Martine, J. (2012). L'emploi des seniors au Japon: Un état des lieux du contexte et des pratiques. *Ebisu. Études Japonaises, 48*, 173–199. https://doi.org/10.4000/ebisu.671

Midgley, J., Goodman, R., White, G., & Kwon, H. (1999). The East Asian welfare model: Welfare orientalism and the state. *The Journal of Asian Studies, 58*. https://doi.org/10.2307/2659122

Morikawa, M. (2014). Towards community-based integrated care: Trends and issues in Japan's long-term care policy. *International Journal of Integrated Care, 14*. Retrieved from: https://www.ncbi.nlm.nih.gov/pmc/articles/PMC3943617/

Moulaert, F. (Ed.). (2013). *The international handbook on social innovation: Collective action, social learning and transdisciplinary research*. Elgar.

Moulaert, F., Mehmood, A., MacCallum, D., & Leubolt, B. (2017). Social innovation as a trigger for transformations—The role of research. https://doi.org/10.2777/68949

OECD. (n.d.). *Ageing and demographic change*. Retrieved from: https://www.oecd.org/economy/ageing-inclusive-growth/

Prime Minister's Office. (n.d.). *The Japan's plan for dynamic engagement of all citizens*. Retrieved from: https://japan.kantei.go.jp/content/jpnplnde_en.pdf

Public Relations Office – Government of Japan. (2021). *Active seniors: The potential of a "super-aged" society*. https://www.gov-online.go.jp/eng/publicity/book/hlj/20210201.html

Richez-Battesti, N., Petrella, F., & Vallade, D. (2012). L'innovation sociale, une notion aux usages pluriels : Quels enjeux et défis pour l'analyse? *Innovations, 38*(2), 15–36.

SI-DRIVE. (2017). *Towards a general theory and typology of social innovation*. Retrieved from: https://www.si-drive.eu/wp-content/uploads/2018/01/SI-DRIVE-Deliverable-D1_6-Theory-Report-2017-final-20180131.pdf

Tanimoto, K. (2008a). A conceptual framework of social entrepreneurship and social innovation cluster: A preliminary study. *Hitotsubashi Journal of Commerce and Management, 42*(1), 1–16. JSTOR.

Tanimoto, K. (2008b). The emergence of social entrepreneurship and its strategy in Japan. Corporate responsibility research conference.

Tanimoto, K., & Doi, M. (2007). Social innovation cluster in action: A case study of the San Francisco Bay area. *Hitotsubashi Journal of Commerce and Management, 41*, 1–17.

TEPSIE. (2014). *Social innovation theory and research a guide for researchers*. Transformative Social Innovation Theory. http://www.transitsocialinnovation.eu/resource-hub/social-innovation-theory-and-research%2D%2Da-guide-for-researchers

Tobita, E. (n.d.). Exploring the feasibility of "The Japan's plan for dynamic engagement of all citizens". *JRI Research Journal, 21*.

Tokyo Work Center. (n.d.). *Shiniashūgyō shien kyaraban (shiniaoshigotofea) [Senior employment support caravan (senior employment fair)]*. Tōkyō shigoto sentā. Retrieved from: https://www.tokyoshigoto.jp/senior/caravan/

Tsutsui, T. (2013). *2025-Nen nimuketaatarashī chiiki-dzukuri—Chiiki hōkatsu kea shisutemu no kōchiku o mezashite [New regional development for 2025 – Aiming to build an integrated community-based care system]*. Retrieved from: https://www.murc.jp/docs/care/care_04.pdf

Tsutsui, T. (2014). Implementation process and challenges for the community-based integrated care system in Japan. *International Journal of Integrated Care, 14*. Retrieved from: https://www.ncbi.nlm.nih.gov/pmc/articles/PMC3905786/

Camille Picard After obtaining a degree in Japanese and international relations, she obtained a Master's degree in urban planning at l'Ecole d'Urbanisme de Paris. Today, she is a PhD student in urban planning at the Paris Est University (Lab'Urba) and at the Kyoto Prefectural University. Her research focuses on the acknowledgment of the ageing population in housing policies in Japan and France.

Chapter 3
Silver Cities: Attractive Cities for Elderly

Montserrat Calvo and José Antonio Ondiviela

Abstract The demographic challenge posed by the increase in the population over 60 entails challenges in other orders such as social, health or urban. There is a negative perception of the capabilities of the elderly and a generalized mindshare that they are a social burden due to the increasing cost of retirement pensions and social/healthcare. However, those over 50 constitute a group with a great weight in the economy and that has given rise to the so-called silver economy. In addition, they hold a human/social capital and an immense capacity to help other generations. They have put this at the service of their families and society at all times, which has been especially evident during the last economic crises. According to the research carried out, the cities adapted to the elderly are also inclusive and attractive environments for the talented. Society in general, and urban environments in particular, must adapt so that the elderly feels comfortable and become attracted to them. This article addresses the reasons and proposes some measures to achieve that goal.

Keywords Silver cities · Smart cities · Elderly · Attractive cities · Silver economy · Ageing · Age-friendly

3.1 Silver Cities Challenges

> The term 'silver economy' was created by experts from the University of Oxford. It refers to the economy of people over 50 years old, including all their economic activities, products, demands and expenses. (Zsarnoczky, 2016, p. 105)

M. Calvo (✉)
Social and Cultural Anthropologist, Madrid, Spain
e-mail: mon@montsecalvo.com

J. A. Ondiviela
WorldWide Observatory for Attractive Cities, Universidad Francisco de Vitoria, Madrid, Spain

E. Pozo Menéndez, E. Higueras García (eds.), *Urban Design and Planning for Age-Friendly Environments Across Europe: North and South*, Future City 19,
https://doi.org/10.1007/978-3-030-93875-8_3

If we think about what cities should be like to better prepare and become more attractive to elder people, the “Silver-haired people”, we could think of the term “Silver Cities” to name them. In this chapter, we are going to study what characteristics they must accomplish to achieve that objective.

The elderly population is not homogeneous. Senior citizens are different in many ways. Age is the main distinguishing feature because there may be decades of difference between them. If we use two fundamental time milestones such as: the retirement age at 65 and the average life expectancy age at 80, then we obtain three groups of elder people (Adapted and extended from Zsarnoczky, 2016, p. 105):

- Young elder (active workers): aged 50–65.
- Adult elder (retired): 66–80.
- Oldest elderly: >80.

Demographic change will have important consequences for economic growth and return on investment.

3.1.1 Demography Trends

By 2050, 80% of elderly people in developed countries will live in urban areas (OECD, 2015). Many challenges must be faced to provide for everyone’s legitimate right to a long, happy and healthy life. Chronic and infectious diseases and microbial resistance due to the abuse of antibiotics are increasing. Pandemics’ spread along the world is underpinned by global mobility. Demographic data on ageing are increasing the demand for social and health services. Health services are usually offered by national or regional entities, but social services remain on the shoulders of cities. Unfortunately, cities are not prepared for this tsunami, neither financially nor infrastructurally, nor in terms of service provision capacity.

The current demographic curve in the advanced countries is shaped like a “baobab”, (See Fig. 3.1) with a large proportion of people in its central zone, threatened by a massive incorporation of elderly people into the retiree area, with a high demand for services and no contribution to Social Security to finance it. This situation will undoubtedly mean an exponential increase in service costs, which will lead to the collapse of the current system.

The direct effects of an ageing population should be addressed: The number of people in the EU (Eurostat, 2017) over 65 will increase by 70% by 2050, and the population over 60 is growing 3.5 times faster than the overall population. It is expected that the proportion of the population aged 65 or older in Europe will increase from 17% in 2010 to 30% by 2060. For the first time in history, this group will be the largest. The share of those aged 80 or over (the *oldest elderly*) is growing faster than any other segment of the population and is projected to triple by 2060. The old-age dependency ratio (population aged 65 years or over vs that 20–64 years old) is projected to more than double, from 28% in 2010 to 59% by 2060 (Fig. 3.2).

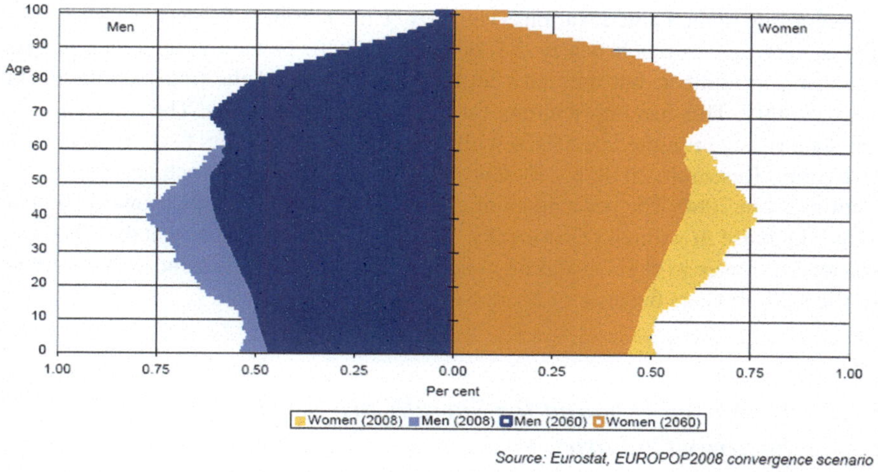

Fig. 3.1 Demographics evolution (2008–2060). (Source: Eurostat, 2011)

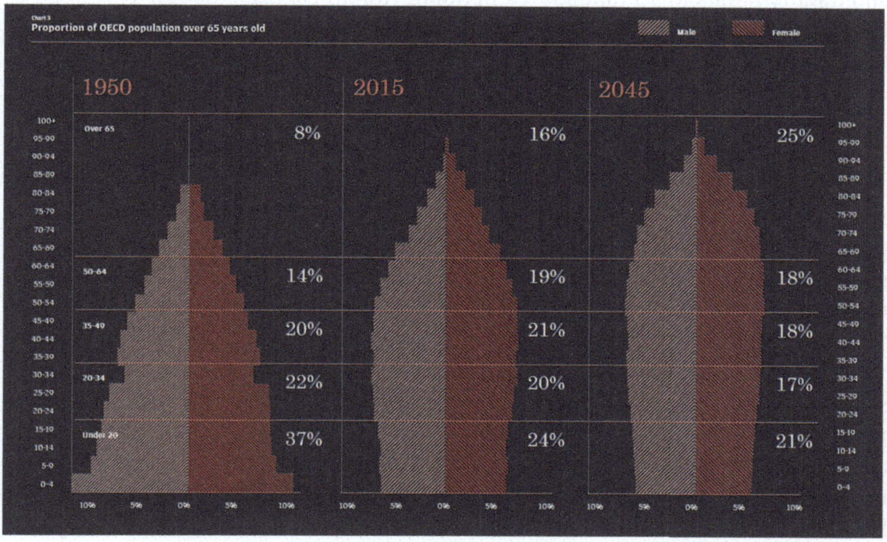

Fig. 3.2 Proportion of OECD population over 65 years old. (Source: Grosvenor Research, 2018, pp. 8–9, based on data from UN World Population Prospects 2017)

Japan leads global life expectancy (living more years from the age of 60) with 26 (UNDATA, 2014), which points to a total expectation of 86 years (data from 2012 with an upward trend). It is followed very closely by most modern countries at 85 and 84. If citizens reach retirement age at 65, this means that, on average, our population expects around 20 years of receiving a pension, social and health services and many others, without contributing to the maintenance of the system. This trend also contributes to increasing the total cost of the benefits guaranteed for retirees.

According to the United Nations (UN, 2017), the number of elderly people globally is expected to reach nearly 2.1 billion by 2050. Not only are older people increasing in number, but they take an increasing share of the total population in every country. This has implications for most sectors of society. The modern welfare society began in the late 1940s with a ratio of eight workers to each pensioner (Grosvenor Research, 2018, p. 19). Currently, this ratio is 4:1, which is projected to become 2:1 in 2060. The percentage of elder adults is continuously growing, with a greater increase in women (Zsarnoczky, 2016, p. 106). These new-age distributions will lead to fewer workers sustaining each elderly person. This trend will continue for the next 60 years because of decreasing fertility rates.

3.1.2 Increased Demand for Social/Care Services. Social Security Collapse

The number of face-to-face doctor visits per year in advanced countries exceeds 10 and rapidly points to 15 (already 17 in South Korea) (Statista, 2019). This trend is unaffordable. Most of these visits are made, obviously, by the elderly. With technological investments, mainly towards remote care and for devices to monitor patients at home, Nordic countries are reducing this to 4 visits or less (Sweden at 2.8).

There are only two ways to contain the cost of social and health services: simply cutting them, assuming the impact on the modern welfare society and the political cost; or ration their use, by using technology to get people to use the services strictly when truly essential. Going to the doctor to renew a chronic prescription or simply to socialize for a while is unsustainable, and unnecessary, as we have learnt from pandemic times. People need to socialize, and we need to avoid undesired loneliness, but with specialized services, not overloading regular medical resources. Some governments are considering setting up a basic minimal fee per visit, to prevent this behaviour. There are much cheaper ways to achieve those objectives. Our services continue to collapse every flu season. Technology can help us stay healthy without having to physically go to the doctor. First things first: prevent problems and maintain healthy habits with the advice of a personal monitoring system. Then, use remote care to deal with minor illnesses, so to prevent visiting the doctor if health indicators are under control and basic prescriptions are working. Finally, use remote specialized care for serious but chronic diseases, such as dementia, Alzheimer's, Parkinson's and so on.

Keep people at home for treatment unless really necessary to do otherwise.

A citizen with any disease at a hospital is considered a patient, and the responsibility falls upon the healthcare system, mostly provided by Regions or Central Government. The same person with the same non-urgent disease at home is a citizen with some dependencies, or who needs some care (mainly social). Then, the responsible authority is, basically, the city. This will imply an exponentially demanded service which cities are not prepared for. Anyway, it is widely proven that the best way to support a person with a chronic disease is by keeping them at home. It is cheaper for the system, and better for the citizen's health. Another alternative for elderly people is staying in a nursing home. As 90% of elderly people prefer to live at home, that means they link their happiness with that lifestyle, surrounded by family, neighbours and friends while maintaining access to community centres where they can connect with other seniors. When the elderly can no longer live in their own homes, they can be taken in by family members and get external assistance from caregivers while family members are at work. Many Nordic families build extra rooms at home to keep their loved ones closer. A person's home is more familiar, therefore if they stay in their homes as long as possible, they remain happier. At home, elderly people maintain some of their independence and confidence even though they may need assistance with certain things. They know where things are, routines are easier to maintain, everything is friendlier. Homes can be adapted to become age-friendly.

New technologies help avoid undesired loneliness by connecting people (Collier, 2019). Elderly people have different requirements, depending on their mental and physical condition. The plan is to always provide the most positive attitude, balancing:

- Socialization instead of isolation. Avoid undesired loneliness.
- Focus on health more than illness. Enjoy life, every day.
- Activity over inaction. Activity is always better (heart rate, blood pressure, glucose, cholesterol levels are very dependent on activity).
- Better to work on prevention than providing treatment.
- Freedom of choice instead of coercion. Always respect human dignity.
- Monitoring to prevent problems or to quickly react to alerts vs a Big Brother reality of continuous surveillance. The right to privacy is a must. Avoid cameras when possible.

Ageing population just adds to other social challenges at the centre of most political agendas like funding, social inclusion, high-quality healthcare, undesired loneliness, digital inclusion or access to quality housing.

With Bloomberg Philanthropies support, Barcelona's "Vincles" ("links") project provides a basic device as a communicator tool to keep the elderly always on, connected to family and friends, able to make videocalls, photos and socialization (Barcelona, 2020).

Physical Inclusiveness The UN Enable program published that 650 million people, approximately 10% of the world population, live with a disability, and more updated reports point to one billion (Ecosoc, 2008). Elderly people are at greater risk of cognitive deterioration, frailty and multiple chronic health conditions with implications for their independence, their quality of life (and that of their families), but also for the general sustainability of social assistance systems. The challenge now is to develop safe, intelligent and respectful digital solutions for integrated care which guarantee a truly personalized provision of health and social assistance, while consolidating advances in the efficiency of service delivery, and guaranteeing the trust of users and public officials on access, privacy and data exchange. Making provisions for disabled persons to find their way independently in urban areas, town centres in particular, should play a relevant role in all urban planning. Due to the greying of the population, it is expected that by 2050, 15% of the urban population will need some kind of support (UN, 2016). Some apps and tools are available to ease the accessibility of disabled persons (Walsh, 2019); however, they still are suboptimal applied (Hamraie, 2018).

3.1.3 Silver Economy Power

The silver economy is considered the general economy area that is relevant to the needs and demands of elder adults. It would include all the economic activity generated by the production, consumption and trade of goods and services relevant to this age group, both public and private, including their direct and indirect effects on the economy (Varnai et al., 2018, p. 6).

In 2015, in the European Union, there were 199 million people over 50, 39% of the total population (Ibid., p. 8). At that time, it accounted for around 4.2 trillion euros of Gross Domestic Product (GDP) and about 78 million jobs, of which 30 million were direct jobs. This represented 35.3% of employment in the entire EU (Ibid., p. 10). For the next decade, this group is expected to reach 222 million people (43% of the EU total). As an example of their purchase power, their household spending share is forecasted to reach 44% in 2025, around 5.0 trillion euros (Ibid., p. 9). *If it were a country, the silver economy would be the third-largest economy in the world, only behind the USA and China.* It generated profits for companies of almost €780 billion (Ibid., p. 9). Although the silver economy focuses on the elderly consumption patterns, there is an indirect effect in providing opportunities for all generations, both adults and young people. Challenges and opportunities are therefore interrelated (Fig. 3.3).

Currently, this age group has a higher educational level and purchasing power than previous generations, same as the use and access to new technologies. It is also expected that the level of well-being achieved by this generation will be even higher than that of the next. Real estate is the main investment held by elder people, which is favouring the proliferation of financial solutions such as reverse mortgages. This makes it possible to improve economic conditions in the retirement stage since a periodic amount that complements the pension received from the State is received (Grosvenor Research, 2018, p. 19). On the other hand, since they do not have to dedicate part of their income to expenses related to the purchase/rental of their living home, they allocate part of them to other purposes such as travel and entertainment. They take short trips more often and are more willing to buy expensive or luxury products. They are increasingly shopping online at times when others are working (Ibid., p. 24). Among their shopping habits, most of the people over 60 prefer to buy in stores near their residence, they buy fewer items on each occasion, but of greater value (Ibid., p. 25).

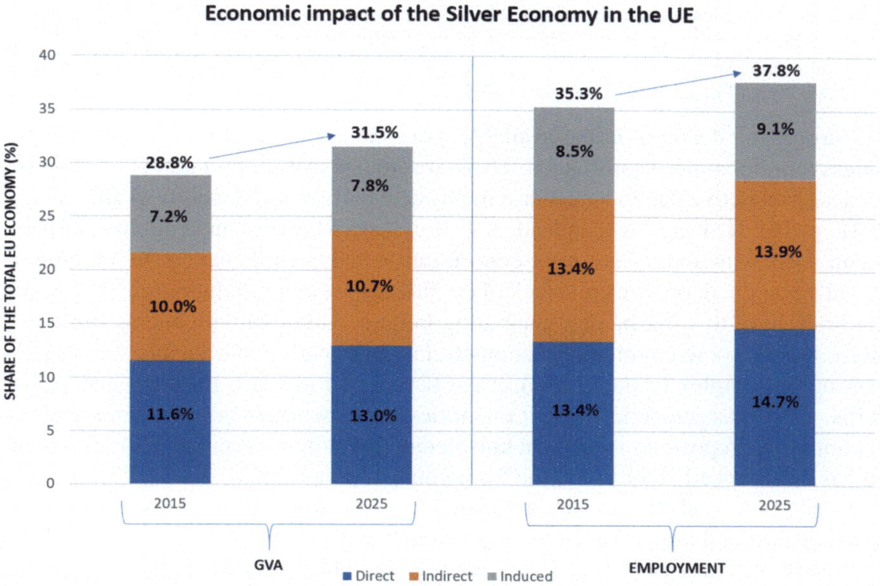

Fig. 3.3 Economic impact of the silver economy in the UE. The Silver Economy; GVA: gross value added. (Source: Authors' own based on Varnai et al., 2018, p. 11)

A whole market of products and services for this group is developing. New applications are also being created, such as the opportunity offered by the autonomous vehicle, especially targeting the elderly, when certain capabilities such as reflexes diminish. Another sector that has been growing for years is *travelling and hospitality*. In one of the researches we carried out, travelling was distinguished as the activity that most motivates this group.

3.2 Why Elderlies Are Important? Why Making Our Cities More Age-Friendly?

3.2.1 Why Are They So Relevant and Cannot Be Left Apart?

It is grossly unfair that the generation that won World War II, got our current welfare system, democracy and civil rights is despised in this way. But of course, the new generations who were born already at the time of social and economic prosperity do not know the effort that it did cost. They were born on the cusp of the mountain of well-being, and they are simply not willing to step down, not even for their own parents.

> By adapting cities and communities to elder people, we are turning them into inclusive and equitable places where no one is left behind, especially the most vulnerable senior people. In turn, equitable societies benefit everyone. Said by John Beard, Director of the Department of Ageing and Life Cycle of the WHO (2018, p. V)

In Europe, the decrease in the available workforce will lead to economic decline, unless retirement age is prolonged. There are studies that support that elder workers are less productive due to a decrease in physical and cognitive agility (Varnai et al., 2018, p. 16). Hearing loss can lead as well to loss of balance and cognitive impairment, such as memory loss, poor concentration and even dementia. In addition to social isolation, depression and lack of confidence are commonly found. They make the brain have to work harder, generating fatigue. Although with ageing there is a slight loss of some cognitive functions (related to hearing and vision, slower processing of complex or dual tasks), it is offset by an increase in social and verbal skills, as well as semantic/implicit memories. Elder workers bring benefits such as becoming the repository of tactical knowledge (know-how) accumulated over years and as experienced mentors for the team younger people (Ibid, pp. 26–27).

In addition to all the above, associated to a social nature, there is a huge economic aspect and impact on GDP, as previously explained.

In several Southern Europe countries or Asia, when physical distance allows it, they are also a fundamental support in the care of grandchildren: taking them to and from school, taking care of them after school classes end and parents have not yet finished their working day, or preparing a snack and babysitting when parents want to enjoy leisure. During the hardest years of the 2008–2013 crisis, pensions remained stable while salaries fell and unemployment increased, which caused many pensioners to have to help financially their children in what can be considered a *generational transfer*. Many paid for food or accommodate back their children at home. Many families endured the ravages of the crisis thanks to this generational transfer.

3.2.2 Elderly as a Social Burden

For this group, the majority of the public spending is focused on health and, secondly, on social protection services.

> An image of ageing as a social burden is disseminated, both by the media and, sometimes, by the scientific discourse. In addition, elder people tend to be spoken of as passive beings, anchored in tradition and lacking in attractiveness and social utility. This fatalistic vision of the phenomenon, however, usually ignores that, far from lacking initiative or stagnating in their attitudes and behaviours, they adapt to the new times like those of any other age. It is what they are doing at present, in effect, fitting into social modernity and contributing to its development like the most. (López & Díaz, 2013, p. 68)

It seems that the increase in the elderly population, as described above, leads to considering ageing as a social problem with global impact (Donizzetti, 2019, p. 1). The main concern about what the ageing process produces in people may be related to an internal desire to satisfy the social ideals of youth, typical from Western societies. This promotes an anti-ageing culture (Ibid., p. 7). In almost all societies there are established stereotypes and prejudices related to elder people. Education in this sense has proven to be the most valuable method for changing attitudes and anxiety produced by the anticipation of ageing itself (Ibid., p. 2). "Ageism" is a form of prejudice towards people for the mere fact of being older or being perceived as such. Gerontophobia and ageism are manifested in different social settings and productive sectors. A clear example is the high percentage of long-term unemployed among the elderly who are still active.

There is a big difference in the respect that society shows for the over-the-years acquired knowledge. In the Western model, the worker develops his professional career and, upon reaching a certain age, leaves their job permanently. In the Japanese model instead, after a few years of professional growth, the older worker gives way to a collaborator and remains in the background to continue contributing based on the accumulated experience over the years of work, without leaving the company.

3.2.3 Cities Must Transform Themselves: Adaptation to Elderly Friendly

We must prepare for the demographic challenge and reduce the perception of social burden, transforming challenges into opportunities for economic growth and social development.

As Felix (2018, p. 125) would say, human super-ageing is an unprecedented experience in the human history. Governments face a complex twofold demand: rising social costs as populations age vs cutting spending on education/unemployment with its impact on the younger generations. This could produce an intergenerational fracture.

One of the observed trends in urbanization is senior population migrating from rural to urban areas because they aspire to get access to better services, but most of the times cities are not prepared to welcome them (Zsarnoczky, 2016, p. 107).

By 2030, a third of households in OECD countries will include an elder person. In the next 30 years, due to divorces at middle age and widowhood later on, households with an elder owner living alone, mostly women, will largely increase. Although they have fewer needs for space in the homes, few of them change areas due to the roots generated in and with the environment. This produces an underutilization of space in larger houses. On the other hand, when they require assistance needs, they prefer to receive them at home and, only when there is no other alternative, move to nursing homes. Homes that are not well designed/adapted for the needs of the elderly can lead to greater home accidents (Grosvenor Research, 2018, p. 29).

Cities must also incorporate elderly friendly principles to make the urban environment more attractive, accessible, adaptable and appropriate for their needs. Currently, there are many initiatives that propose measures to promote active and healthy ageing. For example, the World Health Organization established the Decade of Healthy Ageing 2020–2030 and launched the Active and Healthy Ageing Plan. Previously, in 2010, it created the World Network of Elderly-Friendly Cities and

Communities (WHO, 2020). Its objective is to create inclusive, safe and accessible urban environments. It has a guide with indexes to measure the degree of friendliness of any city (WHO, 2015). The World Health Organization (WHO, 2007) defines an *Age-friendly City* as an inclusive and accessible community environment that optimizes opportunities for health, participation and security, in order that quality of life and dignity are ensured as people age. Although open to later improvements, it is worth highlighting a first guide of 16 indicators to establish the degree of age-friendliness of a community, developed for a pilot experience carried out by the WHO in 2007 in several cities of the world. The condition of an age-friendly community (not just cities) is not limited to urban services (Kano et al., 2018). In 2002, the United Nations defined "active ageing" as "the process of optimizing opportunities for health, participation and safety in order to improve the quality of life of the elderly". One of its determinants is the physical environment, since spaces adapted to the needs of the elderly can influence the level of dependency, social isolation, mobility problems and sedentary lifestyle. The first books on environmental gerontology emerged around 1970 (Baquero & Higueras, 2019, p. 110). Among the initiatives launched by the "Age-friendly World" Network, we can identify the city of Zaragoza (Spain), in which a group of elder people audits the streets to determine whether they have places to rest or public toilets, among other aspects, within the "Walk and discover a safe and accessible city" plan.

Another good ally to improve the lives of the elderly is technology, especially artificial intelligence, which is focusing on developing solutions to ageing problems such as hands tremors or hearing loss. Devices are being created, internal and external to the human body, to activate and pulse sensors connected to the brain/mind, to cognitively stimulate with neurofeedback or combat depression, among others. Also advances in pharmacology, preventive and curative medicine will ensure a better quality of life.

"The creation of environments adapted to the elderly requires action in many sectors (health, chronic care, transport, housing, work, social protection, information and communication) and by many actors (Government, service providers, civil society and older people and their organizations, families and friends). Working to create sustainable and accessible cities and communities for all involves a lifelong process in which the harmonization between the needs of people and the environments in which they live is progressively improved" (WHO, 2018, p. 1). The factors that influence healthy, pathologies-free ageing are intimately connected to the environment (Zsarnoczky, 2016, p. 106). Some studies show that the perception of a pleasant and clean environment is associated, to a certain extent, with the presence of vegetation. In addition, green areas, aesthetics, accessibility to walking areas, parks, children's play areas, security and maintenance are the elements that influence physical activity in public spaces (Baquero & Higueras, 2019, p. 120).

There must be greater private and public collaboration since many of the technological advances that will improve the lives of the elderly are in the hands of private companies. It is necessary to promote collaboration between transnational networks on smart, sustainable, resilient and inclusive cities. The WHO General Work Program highlights the need to follow the Plan of Action on Ageing and Health and the Decade of Healthy Ageing, including work on cities adapted to the elderly. (Ibid., p. 4). Based on the WHO pilot study plus some relevant additions (16 WHO Indicators extended to 21), Table 3.1 shows most important age-friendly indicators (initiatives) that facilitate the elderly living conditions.

Table 3.1 Main age-friendly indicators and initiatives that facilitate the elderly living conditions

Physical environment
1. Neighbourhood walkability Street lighting and adaptation for visually unpaired and reduced mobility, also hearing (replace sound signals with visual or tactile). Sidewalk curbs highlighted with illuminated physical reliefs and visual cues. In some cities, the height differences between sidewalks and road have been eliminated to avoid falls, as in Ottawa, Canada.
2. Accessibility of public spaces and buildings Wheel-chair accessible sidewalks. Easy to find indications. Visually unpaired aids (beep cons, 3D tactile maps, indoor maps navigation etc.).
3. Accessibility of public transportation vehicles Senior adapted buses with seats slightly inclined to make it easier to bend down and stand up or simply lean on; with space to place the walker in front or the cane to one side. The planning of cities must observe age-friendly principles, with adapted accessibility in all means of transport and spaces in the living quarters that allow their use by those with reduced mobility (Grosvenor Group, 2018, p. 32).
4. Accessibility of public transportation stops Tools to easily locate the right bus lane, ramps always better than stairs, information always available, easy to visualize (on the floor, front, voice/sound aided information systems etc.). 5G equipment, connectivity.
5. Affordability of housing Social housing should include special accommodations for elderly. We should keep in mind that they feel always better at home than at a nursing house.
6. Refurbishing of public spaces The exercise areas in the parks are outdoor on the ground. They cannot be used when the weather is not good (rain, sun heating machines, heat stress). Many studies agree that with age, thermal sensitivity decreases, this being one of the factors that make the elderly a vulnerable group facing thermal extremes (Baquero & Higueras, 2019, p. 110).
7. Refurbishing of commercial spaces Increased size of the labels/tags on the products in the stores is needed. For physical fatigue, allocate places to sit and rest while shopping is convenient.
8. Refurbishing workplaces If the retirement age is delayed, several generations will coincide in the work centres and this requires physical spaces adaptation. In particular, the lighting, ergonomics and acoustics needs will be different from each other.
Social environment
9. Positive social attitude towards older people Activities in schools between adults and children to foster relationships between generations (as in Loncoche, Chile). Creation of spaces for intergenerational relationships to avoid social fracture. Care networks, neighbourhood mutual support networks, social and community centres are more than needed because a project of biological and social reproduction is developed at home and district. Public space plays an important role in coping with isolation and loneliness. Create a sense of belonging through interaction and connectivity.

(continued)

Table 3.1 (continued)

10. Engagement in volunteer activity Many volunteerism activities and social giving have risen during COVID-19 times. Among them, helping elderly to do common activities are very frequent, like going shopping, walking the dog, paying taxes, using Internet for medical dates and many other.
11. Paid employment Paid works & tasks in addition to pension mean an incentive to put interest, dedication, active thought and are very healthy to mainly prevent mental disorders/dementia.
12. Engagement in socio-cultural activity Social participation and engagement, not only with family and friends, are crucial to stimulate the sense of being socially relevant, heard, fostering city co-creation. This prevents isolation, loneliness, sense of "abandoned-parked old junk".
13. Participation in local decision-making Experience is a great asset. Participation in political decisions is very empowerment builder. Sense of useful co-creation and future thinking makes life exciting.
14. Availability of information In this technology-driven, accelerated changing society, access to information is vital. Special formats, devices are needed to help elderly gain right access.
15. Availability of social and health services Crucial as explained. Social services are mainly provided by cities. Healthcare is mainly provided by regions or central systems. As age and demand increase, available, affordable, accessible, high quality care services are turning essential.
16. Economic security Economic stability is a must. Basically, the elderly cannot be dedicated to make money (work) but to care themselves, enjoy life and be as active as desired.
17. Intragenerational relationships Revitalization of associations for the elderly. Foster intragenerational relationships to avoid loneliness.
Impact
18. Quality of life Quality of life is the main and final indicator that measures the impact from all other actions / city transformation.
Equity
19. Population attributable risk Unbalanced societies (low GINI index or high concentration of wealth on few hands) make elderly lives miserable. Overpopulated societies are always treating the elderly as a burden, not as a human asset/capital.
20. Inequality between two reference groups Mentioned before, ruling generations decide over other, which must guarantee minimal ethical fairness. Social & Economic classes make a strong difference in the way the elderly live and spend their time. Access to high-quality care is also determined by economic power.

(continued)

Table 3.1 (continued)

Housing
21. Housing design Initiatives as co-living/co-housing, the "Grandma o Granny Pod" (homes for the elderly on the same parcel as their children to combine independence as well as close family contact). Take into account the needs of the elderly at their homes, with larger rooms to manoeuvre with wheelchairs and other reduced mobility aids, with accessibility from the street to home and a room for assistance staff (caregivers). Adapt the houses size and design, allowing a change to smaller houses and in the same area in which they already lived so as not to lose relationships and attachments with their neighbourhood because, as Ibáñez (2002, p. 20) refers, "the house is the privileged place for rooting the body into the world". With this, the largest housing stock will be free for young families who have the greatest need for space. Also, by being adapted to your needs, home accidents due to inadequate housing will be avoided (Grosvenor Group, 2018, p. 29). Establish a network of advisors who visit the elderly homes to ensure they are safe to age in. Although the different generations coexist for a longer time due to the increasing life expectancy, the living together time is shorter since residential independence has increased. The percentage of elder people in multi-generational households has decreased both in Europe, in the USA and even in Japan where the traditional family model included the coexistence of several generations in the same home. Elder people, after a more lasting marriage stage than ever, tend to model the family coexistence with "remote intimacy", with each one at their home although maintaining close ties between them (López & Díaz, 2013, p. 66, 72).

Source: Authors' own adapted and extended from WHO pilot study (WHO, 2018)

3.2.4 Why Attractive Cities Should Retain Elderly

Cities want to be attractive to elder people, not only because of an ethical question of equality between ages, but because talent also ages and cities are permanently competing for it, and in addition, the segment of elder people enjoys a superior economic position, generally due to their ability to generate wealth (savings and investments) due to their long working life.

There are cities and geographic areas specialized in attracting retired and affluent seniors. Places like Florida, the Canary Islands, the Balearic Islands, the southern coast of Europe, enjoy exceptional climatic and well-being conditions that make them very attractive to elder people. But we should not be confused: these touristic areas are also attractive for families, and in general, for anyone who can afford to live there. Therefore, cities want to be attractive to the elderly, despite the fact that this implies bearing a significant burden on social and health services, because it avoids generational and family fractures, retains its enormous talent and know-how and takes advantage of its economic contribution by helping youth to grow and the city's economy.

As previously stated, the silver economy represents a relevant percentage of the economy. Then, cities should be attractive to elder people to retain them and avoid them to move to another place when they enjoy greater purchasing power. The longevity economy is a cross-market with great job creation (Felix, 2018, p. 116). It is a challenge for municipal managers to retain this age group which contributes to their investments, purchases and taxes. They are direct consumers of goods and services and indirect generators of wealth for the city. Beyond that, they contribute to social change and should not be considered passive or resigned. By having greater purchasing power than the average, a whole area of market opportunity for premium products is being created in the tourism, insurance, automotive sector with autonomous cars, emotional well-being, culture or entertainment, among others.

3.3 Main Hypothesis

What is the ideal city for the elderly like?

We have seen many of the age-friendly features and initiatives that contribute to creating cities that are better prepared and adapted for the elderly. But which of these characteristics are the most important? To what extent will a city ready to retain and attract talented citizens also be an optimal city for elder people? Or from another way, an attractive city for talented citizens (those who are looking for the best cities to develop their full potential), will it also be suitable for the elderly? We are going to highlight a shortlist of main characteristics as hypothesis that we will try to corroborate with research activities.

1. Elderly = Human, most likely talented = > Attractive city for the talented MUST BE Attractive for the Elderly. You do not need another city.

   If we study the research (Ondiviela, 2020) on what elements make up an attractive city for talent (emotional magnetism: identity, dynamism, strategy; rational profitability: provided city services compared to cost of living), we will find that those elements are also necessary for the attractive city model for the elderly. What varies is the importance of some of these components compared to others more valued by the younger generations. And in the other way, an attractive city for the elderly must have a series of characteristics that make it attractive for any talented citizen as well. We will study it in Sect. 3.1.
2. Social/care services are the main drivers, as described and explained before.
3. Inclusive/accessible for disabled people. As these conditions are mainly impacting the elderly.
4. Adaptation. Physical (housing, transportation, spaces, mobility), social (inclusion, open, respectful, age equity, integrating, dignifying. Not a social burden but human capital). As explained before, as essential adaptations to keep in mind.
5. Safety. As fragility increases, the sense of safety (mental, physical, digital, social) is gaining relevance.
6. City identity is an important factor. The elders are more tied down, loyal to a particular city. They live more present than looking to the future. Except for a clear need or opportunity, elder people do not often think about moving cities; once established and accommodated in a city that meets their demands, they will hardly consider change. City Identity (history, customs, culture, gastronomy, climate, branding etc.) plays a preponderant role over other emotional magnets. Active elder people enjoy the city culture more as they enjoy more free time. The roots of a city are stronger. Fear of the unknown, laziness in the effort required to move and, in general, increased conformism make the emotional appreciation for a city and the sense of belonging superior to other emotional drivers.

3.3.1 *Attractive for the Elderly Means Attractive for the Talented*

An attractive city for the elderly must be attractive to talented citizens. We start from the hypothesis that the initiatives incorporated in the city for the elderly are valid for the rest of society, and also contribute to create an attractive city for any person at any age that is evaluating the alternatives to finally move there. That way, we can identify some of the most demanded areas to improve in the perfect city for elderly as very important drivers for making that city attractive for everybody. To mention the most relevant:

Identity

- Culturally attractive. A strong Identity builder, very much appreciated by elderly because they can better enjoy these attractions/services as they have the needed time for. Also, traditionally, the elderly better understands the value of history and past contributions. They do not need to work for the future as most are retired, but enjoy the present and remind the good memories from the past with joy.
- Climate. Enjoying a wonderful weather is very desired by the elderly, for general healthcare, and for mental health in particular. Exposure to the sun at places with a high ratio of sun hours has been proven as especially healthy. That's the reason why many decide to spend their retirement in Florida in United States or in the South in Europe (Balearic Islands, Canary Islands etc.). And these places are also attractive for everybody. The main reason why they are not attracting millions for living in the labour conditions, but telework is raising the new concept of smart working and many talented workers are deciding to move and work from these paradises (also the new group of tech nomads).
- Quality of food. Same quality of life and healthcare reasons.
- Natural wonderful places. Balnearies, touristic attractions.

Dynamism Social conditions to grow.

- Stable economy/government model. Same good economic conditions that make a city attractive for the talent, growing the wealth etc., are also guaranteeing the good provision of pension, and needed social services for the elderly. It is obvious that reaching the senior age in a rich city means different benefits than in other cities where social investments are not the top priority.
- Equity. Social inclusion. A city with high achievements in social equity, equality, social balance etc. is attractive for the elderly because it includes the respect, consideration and integration they need, but also same values are very important for attracting talent, as this approaching talent could come from different societies, and social tolerance will mean a door opened to settle in that city and contribute to its prosperity.

City Services Although with different weights, all city services especially designed for the elderly are also very much welcome by the rest. To mention the most likely appreciated by the elderly:

- Safety.
- Social services/health care. Special healthcare for the elderly? Most people use the same hospitals (with some exceptions).
- Environment (especially short-distance air quality, clean and drinkable water). Maybe the elderly better appreciate current conditions while younger citizens mainly think about the future (global warming, climate change), but everybody looks for a better place and take care of our planet.
- Excellent urban mobility. This seems to be one of the most demanded and critical city services, as a city is, in essence, a place where humans meet/encounter, so the way to make this happen, to connect them is considered the city's bloodstream. It is also very well studied that a high density is making a boosting effect on prosperity/creativity/competitiveness (Glaeser, 2011), reinforcing this concept of meeting catalyzer to make 1 + 1 + 1 = 111 (Arithmetic of Synergy) (Kotelnikov, 2010). It is easy to figure out that objectives may differ: elderly need an affordable, cheap, well-prepared urban mobility system, while young workers look for a well-connected fast one (concept of 15-minute city, where most important things – including job – are located at less than 15′ by an individual electric micro-vehicle) (Moreno, 2019). The elderly want to reduce use of cars because they dislike driving and its associated costs, while the younger are more convinced about reducing its use due to environmental and space availability conditions.

Affordable Cost of Living

- Affordable cities. This is again important for everybody. Same as the elderly cannot afford expensive areas and high cost of life due to their modest pensions, same think most of the millennials with their low wages. Both strongly consider the benefits in lower cost of living from alternative second cities (non-capitals), and also their better environment situations (always keeping a good connection with the main city if needed).

3.4 Research

Our object of study will be treated from different approaches, combining the quantitative and qualitative dimensions, as well as a review of the related literature, under a strategy of methodological "complementarity" (Bericat, 1998, p. 108).

The research analysis units have been: silver economy, active ageing, healthy ageing, gerontophobia, resilient cities, age-friendly cities, inclusive cities, smart cities, attractive cities.

3.4.1 Survey

We ran a survey at the largest SmartCity event in the world: the SmartCity Expo & WW Congress that Barcelona organizes every mid-November. At the 2018 event, we added our survey to the post-event feedback form, reaching more than 21,000 attendees. Around half of the attendees came from Western Europe and the other half from the rest of the world, (mainly Asia and the USA) which included people from 146 countries. Mean profile was a balance between technology decision-makers and government officials from cities, salespeople and technical staff from exhibitors. To become integrated with the event, we took the same main theme of "Cities to live IN", and we asked participants to think about city performance/services and cities magnetism (emotional magnets) from attractive cities they would consider to "Live IN". Find survey form at reference (Survey, 2018). We obtained 1550 answers, which for a total sample of N = 21,334 means a large, strong, solid, representative test, with Confidence = 95% and Error = 2% (SurveyMonkey, 2018). We can confirm that sample is a valid representative group of Cities & Technology Industry by 2018. Let us analyze the results (See results Data Summary at Fig. 3.4).

Our responding population was 42 on average, 25% with elderly people at their charge. It represents an unbalanced gender sample, with 67% males, which is unfortunately consistent with a very male-driven technology industry. Aged more than 50 from total answers was 400 or 26%, which leads to a good sample as well.

We consider three main layers (groups): Senior-Young-Active (50–59), Senior-Senior-Active (60–65) and Retired (+65). This retired group could be considered active as well, as they were participants at this global event and aged 67 on average. No older people (more than 80) were found in the survey/event. This has little impact or bias in the test, as this group is very unlikely motivated to look for a different city to live in. They simply ask for better conditions in the city they currently live.

By watching the results, first at Magnetism, there is a clear shift in preferences towards Identity-Past (history, climate, geo-position, gastronomy, branding) as main emotional component for those aged more than 50. Dynamism-Present (creativity, competitiveness, ethics) goes right after, then Strategy-Future (human capital, SmartCity plan, innovation). This confirms our hypothesis about a strong preference for City Identity as main emotional driver. Elderly people prefer cities with a strong identity, image and cultural assets, over new, recently founded, little branding and image cities. Roots start to be relevant, and they look at past and present more than future. People at this age want to enjoy the city's flavour, signs of historical identity, appreciate the people and happenings which made the city, built its assets and image. These old cities are basically more human, cosy, friendly if compared with new cold, distant, generic repetitive cities.

Regarding the city services, we asked the participants to sort these 10 groups (Table 3.2).

DATA AVERAGES

					CHILDREN		ELDER		AGE						GENDER		
CITY SERVICES - SCALE OF VALUES	RK	1-10	Over 100	INPUT	Yes	No	Yes	No	18-29	30-39	40-49	50-59	60-65	65+	Female	Male	N.A.
GOVERNANCE	8	2.85	6.95	3.74	3.71	3.77	3.76	3.73	3.67	3.84	3.71	3.70	3.72	4.11	3.80	3.71	3.92
EDUCATION	5	7.67	11.40	4.21	4.24	4.19	4.15	4.23	4.30	4.37	4.19	4.03	3.96	4.22	4.32	4.17	4.19
EMPLOYABILITY	6	7.11	10.87	4.16	4.10	4.22	4.07	4.19	4.23	4.31	4.10	4.11	3.60	4.00	4.32	4.08	4.12
CONNECTED CITY	9	1.83	6.01	3.64	3.63	3.65	3.68	3.63	3.66	3.55	3.61	3.76	3.81	3.72	3.61	3.66	3.42
SOCSERVICES / HEALTH	2	9.04	12.66	4.35	4.31	4.38	4.38	4.34	4.33	4.40	4.27	4.42	4.19	4.72	4.49	4.28	4.35
ENV. SUSTAINABILITY	3	8.95	12.57	4.34	4.32	4.35	4.33	4.34	4.38	4.32	4.35	4.33	4.26	4.44	4.40	4.32	4.00
CULTURAL SVS / TOURISM	10	1.00	5.24	3.56	3.53	3.58	3.60	3.54	3.42	3.56	3.55	3.64	3.61	3.78	3.60	3.54	3.42
URBAN MOBILITY / TRANSPORTATION	1	10.00	13.54	4.44	4.39	4.49	4.37	4.46	4.44	4.47	4.42	4.46	4.26	4.50	4.48	4.43	4.50
URBAN PLANNING	7	4.78	8.72	3.93	3.88	3.98	3.94	3.93	3.96	3.99	3.85	3.94	3.89	4.11	3.98	3.91	4.00
SAFETY (PHYSICAL/VIRTUAL)	4	8.37	12.04	4.28	4.29	4.28	4.36	4.26	4.23	4.21	4.28	4.40	4.30	4.44	4.35	4.26	3.81

					CHILDREN		ELDER		AGE						GENDER		
MAGNETISM PREFERENCES	RK	1-10	Over 100	INPUT	Yes	No	Yes	No	18-29	30-39	40-49	50-59	60-65	65+	Female	Male	N.A.
IDENTITY	2	8.43	35.33	4.14	4.12	4.15	4.12	4.14	4.08	4.08	4.14	4.23	4.07	4.44	4.18	4.12	3.96
DYNAMISM	1	10	36.93	4.19	4.18	4.19	4.19	4.19	4.13	4.27	4.17	4.16	4.02	4.44	4.21	4.18	4.23
STRATEGY	3	1	27.74	3.89	3.90	3.88	4.03	3.85	3.81	3.90	3.88	3.90	4.04	4.28	3.84	3.93	3.42

RANKINGS

					CHILDREN		ELDER		AGE						GENDER		
CITY SERVICES - SCALE OF VALUES	RK	1-10	Over 100	INPUT	Yes	No	Yes	No	18-29	30-39	40-49	50-59	60-65	65+	Female	Male	N.A.
GOVERNANCE	8	2.85	6.95	3.74	8	8	8	8	8	8	8	9	8	6	8	8	7
EDUCATION	5	7.67	11.40	4.21	5	6	5	5	4	3	5	6	5	5	6	5	3
EMPLOYABILITY	6	7.11	10.87	4.16	6	5	6	6	6	5	6	5	10	8	5	6	4
CONNECTED CITY	9	1.83	6.01	3.64	9	9	9	9	9	10	9	8	7	10	9	9	9
SOCSERVICES / HEALTH	2	9.04	12.66	4.35	3	2	1	3	3	2	4	2	4	1	1	3	2
ENV. SUSTAINABILITY	3	8.95	12.57	4.34	2	3	4	2	2	4	2	4	2	3	3	2	5
CULTURAL SVS / TOURISM	10	1.00	5.24	3.56	10	10	10	10	10	9	10	10	9	9	10	10	9
URBAN MOBILITY / TRANSPORTATION	1	10.00	13.54	4.44	1	1	2	1	1	1	1	1	2	2	2	1	1
URBAN PLANNING	7	4.78	8.72	3.93	7	7	7	7	7	7	7	7	6	6	7	7	5
SAFETY (PHYSICAL/VIRTUAL)	4	8.37	12.04	4.28	4	4	3	4	5	6	3	3	1	3	4	4	8
				n	786	764	383	1,167	230	424	496	325	57	18	482	1,040	26
					51%	49%	25%	75%	15%	27%	32%	21%	4%	1%	31%	67%	2%
									24	35	45	55	62	67			
										AVG AGE:	42.1						

Fig. 3.4 SmartCity Expo 2018 research summary. Colour codes: Blue is categories. In every category, dark green is best, red is worst with yellow in between. (Source: Authors' own elaboration)

Table 3.2 Grouped city services to evaluate

City services to evaluate
Governance
Education
Employability
Connected city
Social services/health
Sustainability/environment
Cultural services/tourism
Urban mobility/transportation
Urban planning
Safety (physical/virtual)

Source: Authors' own elaboration

Senior-Young-Active (50–59) answers are very aligned with the (40–49) group, being *Urban Mobility* the most important area, allowing people to connect, meet each other. The strongest change with younger people is the relevance of *Safety*, with moves from position 6 to 3 then top 1 for those >60.

As we move from 50 to >65 groups, the leading service is, as expected, *Social Services/Health* as the basic and most critical service, as retirement age is reached. *Urban mobility* keeps being very relevant from top 1 to the 2nd position as everybody needs it, as explained, the core component of city's essence. *Environmental sustainability* is also positioned at same places as younger, in 2,3,4 scores, meaning that it is important for everybody, although because of different motivations. *Employability* drops to the bottom position as retirement is reached, obviously. *Connectivity* is also little relevant as it is considered a city basic service more than a competitive advantage from one city to another. Only leading services need the latest technology, speed and latency and offer little value to retired (more targeted to younger). Lifelong *Education* keeps at position 5–6 although it is obviously less relevant than for younger people. This suggests that the elderly understand the value of continuous training, always learning concepts. *Urban planning* and *governance* bring little interest (same as young). And unfortunately, *Cultural services & Tourism* maintain lowest positions, although they experience a little increase. This means that culture as identity component is highly appreciated but culture as a service is the least valued service. So, elderly are proud of their roots, culture and city identity but prefer to spend time with family at home than attending cultural events. Travelling is a different option, considered as leisure and very much appreciated. As the audience was interested in technology, this low cultural preference is also easier to interpret, as both worlds are commonly (and wrongly in our opinion) considered as antagonist.

See in Appendix the list of 140 top world Cities ranked according to this survey preference.

3.4.2 Panel

In April 2020, a semi-structured anonymous survey was carried out (with closed and open questions), distributed through personal online social networks (Twitter, LinkedIn and Facebook) and snowball (people who recommended its completion to other contacts acquaintances and/or in their own online social networks). It was designed with a qualitative rather than quantitative approach, seeking the relevance of the responses rather than representativeness. The survey was closed when "discourse saturation" was reached with 118 responses. All of them in Spanish, and, due to the diffusion, it is very likely that they were all citizens living in Spain. Although the survey was aimed at people 50 years and older, 95% of the responses came from participants between the ages of 50 and 64, so they were, for the purposes of this article, among the young elders (active workers).

Regarding the profile of the participants and their gender, 45% were women and 55% were men, and the vast majority (80%) had a stable partner. Also, 55% lived with someone at their charge and car (elder or children) whereas 23% lived only with their partner. Finally, about their labour status, 12% were already retired, compared to 50% who worked full time and 15% worked part-time.

Regarding the more quantitative part of the survey, despite the number of participants, we can make some observations. On the question about being more or less happy than in the past, there was quite a bit of equality in every section, so we cannot highlight any conclusions in this regard. Regarding whether society values people over 50, there were most responses between levels 1 and 3 of a 5-level Likert scale. So, among this group, there was the perception of not being overly valued by society. Even being considered somewhat less valued at a professional level. All this, despite the fact that 77% considered that their age group was very well prepared to face the professional challenges required by the current labour market. This percentage rose to 86% when asked about their particular skills.

At the most qualitative part of the survey, the objective was to detect motivations, concerns and dreams of this age group. The main conclusions were:

- What motivates them the most in their day-to-day lives was job and relationships with friends and family.
- What brings the biggest excitement in your life, or would contribute if you could afford it, was travelling. About 42% gave this answer.
- The main "low" about ageing was focused on two large areas: health and professional worth. Both were considered in decline, although this perception was more associated with an emotional or even spiritual idea than with a clear fact.
- They considered that what society values the most from them was their experience, although, paradoxically, they did not consider that this was the case in real life. According to the participants, society did not value that the elderly was adaptable and was quite up to date with the skills and knowledge required by today's professional challenges. There were many preconceptions about them.
- In the self-declared unemployed group (19.7%), there was great anger against companies and recruiters for not valuing their knowledge and skills, as well as their human and social capital, as competitive advantages in front of other younger candidates (Table 3.3).

Table 3.3 Suggested ranking as Best Silver Cities

City	Country	SILVER RK	City	Country	SILVER RK
Stockholm	Sweden	1	Philadelphia	United States	71
Melbourne	Australia	2	Lisbon	Portugal	72
Berlin	Germany	3	Baltimore	United States	73
Zurich	Switzerland	4	Prague	Czech Republic	74
Vienna	Austria	5	Hong Kong	Hong Kong	75
Oslo	Norway	6	Jerusalem	Israel	76
Adelaide	Australia	7	Athens	Greece	77
Sydney	Australia	8	Ljubljana	Slovenia	78
Gothenburg	Sweden	9	Tallinn	Estonia	79
Bern	Switzerland	10	Tel Aviv	Israel	80
Montreal	Canada	11	Wroclaw	Poland	81
London	United Kingdom	12	Budapest	Hungary	82
Hamburg	Germany	13	Warsaw	Poland	83
Copenhagen	Denmark	14	Bratislava	Slovakia	84
Basel	Switzerland	15	Vilnius	Lithuania	85
Cologne	Germany	16	Zagreb	Croatia	86
Frankfurt	Germany	17	Taipei	Taiwan	87
Amsterdam	Netherlands	18	Buenos Aires	Argentina	88
Paris	France	19	Sofia	Bulgaria	89
Manchester	United Kingdom	20	Santiago	Chile	90
Barcelona	Spain	21	Córdoba	Argentina	91
Munich	Germany	22	Dubai	United Arab Emirates	92
Rotterdam	Netherlands	23	Shanghai	China	93
Geneva	Switzerland	24	Belgrade	Serbia	94
Canberra	Australia	25	Riga	Latvia	95
Valencia	Spain	26	Bucharest	Romania	96
Edinburgh	United Kingdom	27	Istanbul	Turkey	97
Madrid	Spain	28	Abu Dhabi	United Arab Emirates	98
Tokyo	Japan	29	Beijing	China	99
New York City	United States	30	Doha	Qatar	100
Lyon	France	31	Chengdu	China	101
Toronto	Canada	32	Moscow	Russia	102
Stuttgart	Germany	33	Chongqing	China	103
Seoul	South Korea	34	Shenyang	China	104
Luxembourg	Luxembourg	35	Montevideo	Uruguay	105
Dusseldorf	Germany	36	Kuwait City	Kuwait	106
Helsinki	Finland	37	Ankara	Turkey	107

(continued)

Table 3.3 (continued)

City	Country	SILVER RK	City	Country	SILVER RK
Eindhoven	Netherlands	38	Mexico City	Mexico	108
Wellington	New Zealand	39	St Petersburg	Russia	109
Linz	Austria	40	Guangzhou	China	110
Ottawa	Canada	41	Sao Paulo	Brazil	111
Marseille	France	42	Kiev	Ukraine	112
Los Angeles	United States	43	Shenzhen	China	113
Nice	France	44	Brasilia	Brazil	114
Málaga	Spain	45	Tunis	Tunisia	115
Liverpool	United Kingdom	46	Rio de Janeiro	Brazil	116
Vancouver	Canada	47	Panama City	Panama	117
Dublin	Ireland	48	Quito	Ecuador	118
Phoenix	United States	49	Hanoi	Vietnam	119
Auckland	New Zealand	50	Bangkok	Thailand	120
Antwerp	Belgium	51	Kuala Lumpur	Malaysia	121
Birmingham	United Kingdom	52	Casablanca	Morocco	122
Bilbao	Spain	53	Lima	Peru	123
Brussels	Belgium	54	Bogota	Colombia	124
Milan	Italy	55	Riyadh	Saudi Arabia	125
Osaka	Japan	56	La Paz	Bolivia	126
San Francisco	United States	57	Monterrey	Mexico	127
Yokohama	Japan	58	Mumbai	India	128
Rome	Italy	59	Guadalajara	Mexico	129
Boston	United States	60	Jakarta	Indonesia	130
Chicago	United States	61	Ho Chi Minh City	Vietnam	131
Seattle	United States	62	Cairo	Egypt	132
Dallas	United States	63	Bangalore	India	133
Miami	United States	64	Manila	Philippines	134
Houston	United States	65	Asuncion	Paraguay	135
Florence	Italy	66	New Delhi	India	136
Atlanta	United States	67	Medellín	Colombia	137
Washington, D.C.	United States	68	Durban	South Africa	138
Singapore	Singapore	69	Cape Town	South Africa	139
Porto	Portugal	70	Johannesburg	South Africa	140

Source: Authors' own from list of 140 top World Cities, we assign 50% weight to those best on Attractiveness (Ondiviela, 2020), and another 50% weight to the combination of most preferred components as obtained from the survey: Identity + Social Services/Health + Safety

3.5 Conclusions and Recommendations

People over 50 form a heterogeneous group, with very different nature, vital, social and assistance needs. They are much more valuable to society than what biased prejudices establish. They still can work and contribute. They are as well, a key pillar in family cohesion.

If 80% of the elderly will live in urban areas by 2050 (90% in Europe) and this age group will be the majority population worldwide, for the first time in history, then the challenges that society faces in terms of population ageing must be promptly addressed. An urgent action is more evident and clearer than ever! It seems that current governments are passing this hot potato to the next ones without paying attention or starting the needed investment. A strategic plan is needed, but very few cities have one for the next 10 years or more, the majority only works in the short term. A broad pact would be needed among all important political parties to design a strong and integrated strategy about the threatening tsunami of the aging problem.

The ratio of workers vs pensioners will decrease, posing a challenge for state-dependent retirement benefits. Demand for social services will increase and the national pension and social security system will collapse.

The COVID-19 pandemic has highlighted the weaknesses of the nursing home model and the need to rethink how we want them to live their last stage of life.

> No generation has a monopoly on justice, ethics, rights or responsibility. (Levine, 2020)

It must become a matter of State, not subject to electoral fluctuations, the development of long-term strategic plans from the local, regional and state levels.

Cities must be attractive enough for elderly so that they do not move to another city when their working period ends, since the silver economy generates great direct and indirect benefits.

The goal is to create inclusive, sustainable, safe and prosperous communities for all. *What's good for elderly is good for all.*

The ideal city for elderly (silver city) is a combination of the attractive city for talented citizen's drivers (Ondiviela, 2020) with specific additional adaptations as described by previously described Age-friendly world indicators. In other words, best silver city is a strong identity, top attractive city for talent, with special emphasis on social services/health, security and adapted urban mobility. In this way, societies will be inclusive environments for all ages.

The main ageing challenges our societies are facing: combat ageism in the professional sphere, gerontophobia, improve intergenerational relationships, change the pension model, achieve healthy ageing that reduces social assistance needs, adapt homes and cities (public spaces, transport, shops etc.).

And the main strategies are to understand the real elderly needs and the collaboration between the public and private sectors.

Acknowledgments Marta Fernández Bertos – RMIT University // Amaya Celaya Álvarez – Urban Resilience Hub // IMSERSO, Spain. Luis Barriga, Pablo Martínez and, specially, Mª Teresa Pozo // Benigno Lacort, Director Forum Economía Senior España // Prof. Tine Buffel. Manchester University.

Conflicts of Interest The authors declare no conflict of interest.

Funding This research received no external funding.

References

Baquero, M. T., & Higueras, E. (2019). Factores ambientales que influyen en el uso del espacio público para las personas mayores en Madrid. *Revista Urbano, 40*, 108–126. ISSN 0717 – 3997/0718-3607. https://doi.org/10.22320/07183607.2019.22.40.06

Barcelona. (2020). *Vincles-BCN City of Barcelona website*. Retrieved from https://ajuntament.barcelona.cat/vinclesbcn/en/vincles-bcn. Last accessed Oct 2020.

Bericat, E. (1998). *La integración de los métodos cuantitativo y cualitativo en la investigación social*. Ariel.

Collier, C. (2019). *Lonely people in big cities: How technology is both creating and solving the isolation crisis*. SmartCitiesConnect. Retrieved from http://smartcitiesconnect.org/lonely-people-in-big-cities-how-technology-is-both-creating-and-solving-the-isolation-crisis. Last accessed May 2019.

Donizzetti, A. R. (2019). Ageism in an aging society: The role of knowledge, anxiety about aging, and stereotypes in young people and adults. *International Journal of Environmental Research and Public Health, 2019*(16), 1329. https://doi.org/10.3390/ijerph16081329. www.mdpi.com/journal/ijerph

Ecosoc. (2008). *Resolution 2008/21* (p. 1). ECOSOC, The Economic and Social Council, UN. https://www.un.org/en/ecosoc/docs/2008/resolution%202008-21.pdf

Eurostat. (2011). *Archive: Population projections*. https://ec.europa.eu/eurostat/statistics-explained/index.php?title=Archive:Population_projections&oldid=59201. Retrieved by Jan 2019.

Eurostat. (2017). *People in the EU – Statistics on an ageing society*. Retrieved from https://ec.europa.eu/eurostat/statistics-explained/index.php/People_in_the_EU_-_statistics_on_an_ageing_society. Last accessed Apr 2018.

Felix, J. S. (2018). Economia da Longevidade, Gerontecnologia e o complexo econômico-industrial da saúde no Brasil: uma leitura novo-desenvolvimentista. *Revista Kairós – Gerontologia, 21*(1), 107–130. ISSNe 2176-901X. São Paulo (SP), Brasil: FACHS/NEPE/PEPGG/PUC-SP. https://doi.org/10.23925/2176-901X.2018v21i1p107-130

Glaeser, E. (2011). *Triumph of the city: How our greatest invention makes us richer, smarter, greener, healthier and happier* (p. 6). Penguin Press.

Grosvenor research. (2018). *Silver cities. Planning for an ageing population*. Retrieved from https://www.grosvenor.com/news-and-insight/all-articles/silver-cities-planning-for-an-ageing-population

Hamraie, A. (2018). A smart city is an accessible city. *The Atlantic*. Retrieved from https://www.theatlantic.com/technology/archive/2018/11/city-apps-help-and-hinder-disability/574963/. Last accessed Oct 2020.

Ibáñez, J. (2002). *Por una sociología de la vida cotidiana*. Siglo XXI.

Kano, M., Rosenberg, P. E., & Dalton, S. D. (2018). A global pilot study of age-friendly city indicators. *Social Indicators Research, 138*, 1205–1227. https://doi.org/10.1007/s11205-017-1680-7 link to Indicators Table: https://link.springer.com/article/10.1007/s11205-017-1680-7/tables/1

Kotelnikov, V. (2010). *Synergistic team*. Retrieved from http://1000ventures.com/business_guide/crosscuttings/team_synergistic.html. Last accessed Oct 2020.

Levine, J. (2020). As Covid slams us, elders want solidarity, not sentiment – Or blame. *TheNation*. Retrieved from https://www.thenation.com/article/politics/coronavirus-austerity-elderly/. Last accessed Nov 2020.

López, J., & Díaz, M. P. (2013). Modernización social de la vejez en España. *Revista Internacional de Sociología (RIS), 71*(1) Enero-Abril, 65–89. https://doi.org/10.3989/ris.2011.04.26

Moreno, C. (2019). *The 15 minutes-city: For a new chrono-urbanism!* Retrieved from http://www.moreno-web.net/the-15-minutes-city-for-a-new-chrono-urbanism-pr-carlos-moreno/. Last accessed Oct 2020.

OECD. (2015). *Governing cities*. Retrieved from http://www.oecd.org/gov/cities.htm. Last accessed Jan 2018.

Ondiviela, J. A. (2020). Beyond SmartCities: How to create an attractive city for talented citizens. *Kult-Ur, 7*(13), 205–232. https://doi.org/10.6035/Kult-ur.2020.7.13.8

Statista. (2019). *Number of doctor visits per capita in selected countries as of 2018*. STATISTA Reports. Retrieved from https://www.statista.com/statistics/236589/number-of-doctor-visits-per-capita-by-country/. Last accessed Dec 2019.

Survey. (2018, November). *Attractive cities survey*. SmartCityExpo. Retrieved from https://forms.office.com/Pages/DesignPage.aspx#FormId=v4j5cvGGr0GRqy180BHbR4L8YT1UMAdFgiXBzCdEJSBURTNWRjhDS05BMzNHMkdPUEw0VFFEUVpNTi4u. Last accessed Dec 2018.

SurveyMonkey. (2018). *SurveyMonkey calculator*. Retrieved from https://www.surveymonkey.com/mp/margin-of-error-calculator/. Last accessed Oct 2018.

UN. (2016). *Good practices of accessible urban development*. Department of Economic and Social Affairs. Social Inclusion. Retrieved from https://www.un.org/development/desa/dspd/2016/10/good-practices-of-accessible-urban-development/. Last accessed Dec 2020.

UN. (2017). *UN world population ageing report* (p. 1). Department of Economic and Social Affairs. Retrieved from https://www.un.org/en/development/desa/population/publications/pdf/ageing/WPA2017_Highlights.pdf. Last accessed Jan 2019.

UNdata. (2014). *Life expectancy at age 60 (years)*. Retrieved from http://data.un.org/Data.aspx?q=life+expectancy&d=WHO&f=MEASURE_CODE%3AWHOSIS_000015. Last accessed Jan 2019.

Varnai, et al. (2018). *The silver economy*. A study prepared for the European Commission, Directorate-General Communications Networks, Content & Technology by Technopolis Group and Oxford Economics.

Walsh, N. P. (2019). 10 technologies for accessible, affordable cities. *ArchDaily*. Retrieved from https://www.archdaily.com/923441/10-technologies-for-accessible-affordable-cities. Last accessed Oct 2020.

WHO. (2007). Global Age-friendly Cities: A Guide, 1.

WHO. (2020). "Age-friendly World". Retrieved from https://extranet.who.int/agefriendlyworld/. Last accessed Oct 2020.

WHO. (2015). *Medición del grado de adaptación de las ciudades a las personas mayores: guía para el uso de los indicadores básicos*. Ediciones de la Organización Mundial de la Salud. ISBN 978 92 4 350969 3 (Clasificación NLM: WA 380).

WHO. (2018). *La Red Mundial de Ciudades y Comunidades Adaptadas a las Personas Mayores: Revisar el último decenio y mirar con optimismo hacia el siguiente*. Organización Mundial de la Salud. Retrieved from https://www.who.int/ageing/publications/gnafcc-report-2018/en/

Zsarnoczky, M. (2016). Innovation challenges of the silver economy. *Vadyba Journal of Management* (Szent Istvan University, Enyedi György Doctoral School of Regional Sciences), *28*(1), 105–109.

Montserrat Calvo Social Researcher. Her main field of work focus on the interrelation between people and urban spaces. Social and Cultural Anthropologist by UNED. Expert in human-centred innovation methodologies and User Experience Designer. Author of papers and books on different topics. Former Director at Fundación Diseño e Innovación and Founder-Partner at the consulting firm Prosum Imagen y Comunicación. Master in Protocol and Institutional Relationships.

Jose Antonio Ondiviela is the Microsoft Western Europe Public Sector Government Industry Executive for Cities & Regions. Frequent speaker at International events like SmartCityExpo, Mobile World Forum, TEDTalk, EU Events… he is also UNESCO SmartCities SME Consultant. Telecommunications Engineer MS at UPM Madrid, Human Sciences MS at UFV Madrid, Enterprise and Institutional Communication Management MS at UAB Barcelona, and Financial Business MS online INSEAD. PhD at UFV, Madrid with Thesis "Beyond SmartCities: How to create an Attractive City for Talented Citizens". Associate Fellow Researcher and Director of WW Observatory for Attractive Cities at UFV (Universidad Francisco Vitoria, Madrid).

Chapter 4
Quality of Life, Living Arrangements, and Care Support for the Older Population

Carmen Rodríguez-Blázquez and Maria João Forjaz

Abstract Population ageing in most Western countries has attracted great attention in the knowledge on quality of life (QoL) in old age. QoL is a multidimensional construct, dynamic and centred on the person, which includes important dimensions such as health, psychological aspects, social relationships and activities, autonomy, and living circumstances. The chapter describes the main concepts of QoL, active and healthy ageing, age-friendly environments, and presents the most used questionnaires and rating scales to measure QoL in later life. The effect of the residential environment on QoL in older adults is particularly relevant since this group spends more time at home than younger adults and a supportive environment can reduce the impact of health problems. Although ageing in place, that is, remaining in their own home as long as possible, is the preferred living option for older adults, institutionalization in long term care facilities (LTCF) is useful when care at their own homes is no longer an option, although it also brings new challenges. In this chapter, we will review studies about the influence of the different living arrangements on older adults' QoL and identify pertinent areas for further research.

Keywords Quality of life · Older adults · Environment · Ageing in place · Long-term care

C. Rodríguez-Blázquez (✉)
National Centre of Epidemiology and CIBERNED, Carlos III Institute of Health, Madrid, Spain
e-mail: crodb@isciii.es

M. J. Forjaz
National Centre of Epidemiology and REDISSEC, Carlos III Institute of Health, Madrid, Spain
e-mail: jforjaz@isciii.es

E. Pozo Menéndez, E. Higueras García (eds.), *Urban Design and Planning for Age-Friendly Environments Across Europe: North and South*, Future City 19,
https://doi.org/10.1007/978-3-030-93875-8_4

4.1 Introduction

Quality of life (QoL) is defined as "an individual's perception of his or her position in life in the context of the culture and value system where they live, and in relation to their goals, expectations, standards and concerns" (The World Health Organization Quality of Life Assessment (WHOQOL), 1995). It is a multidimensional, dynamic, and person-centred concept that includes health, functional, cognitive, psychological, social, and environmental aspects of life.

A good QoL is key for ageing well and is paramount to the active and healthy ageing frameworks. *Active ageing* is understood as "the process of optimizing opportunities for health, participation and security in order to enhance quality of life as people age" (Fabre, 2015), while *healthy ageing* entails "the process of developing and maintaining the functional ability that enables wellbeing in older age" (WHO, 2015b). This concept of functional ability reflects the interaction between a person's physical and mental capacities and the environments where he or she inhabits. Thus, *age-friendly environments* are indispensable for achieving the goals of active and healthy ageing (WHO Regional Office for Europe, 2018). Moreover, impact on well-being and QoL are in the core set indicators proposed for age-friendly cities (WHO, 2015a).

4.1.1 Assessment of Quality of Life in Older People

QoL is usually assessed using standardized rating scales and questionnaires, and a broad range of instruments for older adults are available. Some of them have been designed for the general population but have been widely used in older ages. These are known as generic instruments and is the case of the *Schedule for the Evaluation of Individual Quality of Life* (SEIQoL) (O'Boyle et al., 1993), for individualized assessment of QoL, and the *EQ-5D* (EuroQoL Group, 1990) and the *SF-36* (Ware & Sherbourne, 1992), both for assessment of the aspects of QoL referred the health status (health-related QoL, HRQL). The *Personal Wellbeing Index* (PWI), the *Community Wellbeing Index* (CWI), and the *National Wellbeing Index* (NWI) form a set of instruments aimed at assessing different aspects of global QoL (International Wellbeing Group, 2006). The CWI is particularly relevant in this context, as it is focused on the individual's level of satisfaction with the local place of residence and it has been validated in older people (Forjaz et al., 2011).

However, QoL in older people may have characteristics and needs that are not covered by generic rating scales. The health component of QoL may be influenced by chronic diseases or disability, and the psychological and social aspects can be affected by retirement, changes in family and societal roles, and loss of family and friends. Lastly, the format and type of administration of rating scales could be hampered by sensory or cognitive impairments. Thus, there is also a set of instruments specifically designed for the older population. In this group, we can find the version

Table 4.1 Main instruments to assess the QoL of older adults

Name	Acronym	References	Number of items
Generic scales			
Schedule for the Evaluation of Individual Quality of Life	SEIQoL	O'Boyle et al. (1993)	5
EQ-5D	EQ-5D	EuroQol Group (1990)	5 + visual analogue scale
SF-36	SF-36	Ware and Sherbourne (1992)	36
Personal Wellbeing Index	PWI	International Wellbeing Group (2006)	8
Community Wellbeing Index	CWI	Forjaz et al. (2011)	10
National Wellbeing Index	NWI	International Wellbeing Group (2006)	6
Specific scales for the older population			
World Health Organization QoL questionnaire	WHOQOL-OLD	Power et al. (2005)	24
Older People's QoL questionnaire	OPQOL	Bowling (2009)	35
Control, Autonomy, Self-realisation and Pleasure Scale	CASP-19	Hyde et al. (2003)	19
Scales to assess QoL in dementia			
Quality of Life in Alzheimer's Disease scale	QOL-AD	Logsdon et al. (2002)	13
Quality of Life in Late-Stage Dementia	QUALID	Weiner et al. (2000)	11
DEMQOL DEMQOL-Proxy	DEMQOL DEMQOL-Proxy	Smith et al. (2005, 2007)	28 31

Source: Authors' own construction

for older people of the *World Health Organization QoL questionnaire* (WHOQOL-OLD) (Power et al., 2005), the *Older People's QoL questionnaire* (OPQOL) (Bowling, 2009), and the *Control, Autonomy, Self-realisation and Pleasure Scale* (CASP-19) (Hyde et al., 2003). The use of specific questionnaires for the assessment of QoL in the older population is strongly recommended. Table 4.1 lists the main instruments used to assess the QoL of older adults.

4.1.2 Influence of Environment in Quality of Life of Older Adults

Environments include the place where people live, their local community, their home, and where they work and play, as well as the access to health and social resources. When an environment offers protection from threats to health and enables people to expand their capabilities, it is a *supportive environment for health*

(WHO Regional Office for Europe, 2018). Supportive environments can help to alleviate or reduce the impact of conditions such as disability or frailty, contributing to improve or maintain QoL in older adults (Beard et al., 2009; Huang et al., 2019). In contrast, studies show that older adults residing in deprived environments experience more physical and mental health problems, follow less healthy lifestyles, and report lower levels of well-being (Barnett et al., 2018; Oiamo et al., 2015).

Researchers have analyzed what environment features contribute to older adults' QoL. The context and living conditions range from micro-level (proximal) features of older people's residence to macro-level (distal) characteristics of the neighborhood, city, or country (Grestorf et al., 2010). Proximal features include the geographical context, rural or urban environment, health, and social care services or transportation. At a micro level, accessibility to the building and available facilities (lift, bathroom, heating) of the place where older people live are also determinants of older people's QoL. Moreover, the environment can have a physical, social, and symbolic meaning, as people may have different experiences and expectations of a particular place (Wiles, 2005). These domains cannot be treated in isolation from personal factors such as age, gender, psychological or functional status, thus an interdisciplinary and ecological approach must be used when analyzing age-friendly environment and its contribution to QoL (Menec et al., 2011).

4.2 Living Arrangements

4.2.1 QoL and Community-Dwelling Older Adults

Most people live in their own family home as they age. *Ageing in place*, that is, remaining in their own home as one ages is the preferred housing option for most people, despite the presence of health problems and functional impairments or disability, and barriers (e.g., stairs, low maintenance, lack of accessibility, etc.) in the environment (Horner & Boldy, 2008). The term has been broadened to remaining in the current community and living relatively independently in the residence of one's choice, through the provision of appropriate services and assistance (Vanleerberghe et al., 2017). Ageing in place is a key element of QoL in older adults, as it favours a sense of control over their own life and facilitates social relationships with family, neighbours and friends (Prieto-Flores et al., 2011). Some elements of the *built environment*, defined as the objective and perceived characteristics of the physical context in which people spend their time (e.g., home, neighbourhood), are crucial for ageing in place. The built environment includes aspects of the design (e.g., sidewalks, seating places), traffic density and speed, locations for physical activity (e.g., streets and parks), and pollution, crime, and safety (Grimmer et al., 2015; Van Cauwenberg et al., 2011).

Several studies have analyzed the influence of these environmental factors in improving or maintaining QoL in community-dwelling older adults. In a qualitative study, participants stated that physical and mental health, family environment, and financial situation were the main aspects for healthy ageing in place with good QoL (Bosch-Farré et al., 2020). Participants were also able to identify several barriers and enablers to ageing in place. Financial problems and lack of access to health and social services were among the difficulties, whereas sense of belonging to the community, environmental and housing adaptation, and having family and social support were the facilitators of ageing in place.

Among the physical attributes of the environment, neighbourhood safety and quality, access to services, shops, and public transport and number of parks were positively associated with health-related QoL (Byles et al., 2014; Friedman et al., 2012; Zhang et al., 2019). Gobbens et al. pointed out what environmental factors facilitate ageing in place and thus, improve QoL of older adults (Gobbens & van Assen, 2018). In their study, they found that all QoL domains (physical, psychological, social, environmental) were associated with at least one of the assessed environmental factors: housing, facilities, nuisance, residents, neighbourhood, stench/noise, and traffic. The housing scale, which included indicators such as comfort, number of rooms, and facilities, was associated to all domains of QoL; while residents scale was associated to psychological, social relations, and environmental aspects of QoL; nuisance scale (safety concerns, pollution) was associated with physical health and environmental QoL; and neighbourhood quality (landscaping, paving, traffic lights), with environmental QoL.

Subjective perception of local community environment, assessed with the CWI, and its relationship with QoL has been the subject of a study by Giraldez-Garcia et al., which found that satisfaction with community services significantly contributed to better self-rated health and functional independence, and less depression and chronic medical conditions, all of them known determinants of QoL (Giraldez-Garcia et al., 2012).

Another component of the environment, the *social capital*, is defined as the features of social organization, such as trust, norms, and networks, that can improve the efficacy of society by facilitating coordinated actions (Murayama et al., 2012). Social capital has also an impact on health outcomes of older adults, particularly the components of social participation and engagement in the neighbourhood (Gobbens & van Assen, 2018).

Social capital has been found to influence QoL in two ways. One form is indirectly, mediated by health risk behaviours. Yang et al. observed that lower social capital score was observed in older people that smoked, were physically inactive, had unhealthy dietary behaviour, unhealthy weight, and sleep disorder than individuals without these behaviours, and that these individuals also showed lower QoL (Yang et al., 2020). Social capital can also act as a buffer against the effects of loneliness and poverty in older adults and as a result, can improve QoL (Cramm et al., 2013). On the other side, the structural (social network, civic engagement) and cognitive (social cohesion) components of social capital can directly influence QoL, as reported in a study in Singapore (Lane et al., 2020). Out of these two components, the structural aspect has been found more tightly associated to ageing in place than the cognitive component (Lu et al., 2020).

Social network and social support have been frequently included as domains of social capital, and they have been analyzed as determinants of QoL (Beridze et al., 2020; Chappell, 2014; Chemaitelly et al., 2013). *Social support* describes the network of social resources that an individual perceives, and it includes different forms of support, including informational, instrumental, and emotional support (Zhou, 2014). Older people with lower social support and higher degree of isolation exhibit poorer QoL (Beridze et al., 2020). Meaningful and satisfactory social relations buffer the negative impact of health decline and help older people to cope with illness and disability, playing an intermediate role between functional status and QoL (Puga et al., 2021). The number of friends living in the neighbourhood is associated with better well-being, particularly in older women (Liu et al., 2018). The effect of *social isolation* and the lack of support on health and QoL in older adults has been dramatically highlighted during the recent COVID-19 pandemic. Different strategies have been performed to allow the social connectedness of older people despite the social distancing restrictions (van Orden et al., 2020).

One of the functions of the social network is to deliver *care*, reflected in the provision of help for maintaining health, welfare, and protection of somebody. Care may be instrumental, including assistance with shopping, meal preparation, and housekeeping, for example, or personal, including bathing, dressing, and feeding, or emotional (Hellström et al., 2004). Receiving help from formal or informal caregivers is essential for ageing in place and is associated with higher QoL, particularly satisfaction with help and receiving enough help when needed (Vaarama, 2009).

Home and community-based care and assisted living services are critical for frail or vulnerable older adults living at home, many of them homebound due to chronic diseases and disability (Negrón-Blanco et al., 2016). These services can improve older persons' functional and health status, help them to live independently, and can delay institutionalization in a cost-effective way (De Coninck et al., 2017). In recent years, technologies have been developed for supporting older people in their living environment. *Assistive technologies* are equipment and services that are used to maintain or improve the physical and cognitive functions and enhance QoL of older adults (Ganesan et al., 2019), but there is a need to critically review their effectiveness in achieving these goals.

Older people can act as active agents of change in their environment to fit their needs and get higher levels of QoL (Stephens & Allen, 2021). In this sense, some researchers have introduced the idea of liveability of the environment, to highlight the notion of person-environment fit and to focus on the need of assessing older people's experiences of the quality of their home and neighbourhood environments and their satisfaction with their relationships (Stephens et al., 2019). Some interventions for creating age-friendly environments are more effective than others: in a review, Sánchez-González et al. found that changes in transportation and housing and increased participation were the interventions that better promote active ageing and thus improve older people's QoL (Sánchez-González et al., 2020).

Creating age-friendly environments requires the involvement and participation of older people in all decisions and processes, in a procedure that is frequently called "co-production" (WHO Regional Office for Europe, 2016). This idea has been implemented in the model of *cohousing*, as a form of community living that contains a mix of private and communal spaces with substantial self-managed common facilities and activities (Vestbro, 2010). A positive association between this model of housing and QoL has been found, but the quality of evidence is low, so more research is needed on the impact of cohousing on QoL of older adults (Carrere et al., 2020).

4.2.2 QoL of Older Adults Residing in Long-Term Care Facilities

Due to physical decline, frailty, or lack of support, ageing in place is not always possible and institutionalization in a *long-term care facility* (LTCF) may provide medical and social support, as well as meeting basic needs. A study in Switzerland identified the following factors as predictors for later institutionalization after post-acute care: frailty, impaired mobility, and being underweight (Tröster et al., 2020). Cognitive and/or functional impairment significantly predicted institutionalization in Finland (Salminen et al., 2020). The presence of a social network, visiting family and friends, and engaging in social activities acted as protective factors against moving into institutional settings in California, USA (Oh et al., 2019). Finally, a meta-analytical review identified the following risk factors for placement of people with dementia in LTCF: impairment in activities of daily living; reduced mobility; cognitive, behavioural, and psychological symptoms; multimorbidity; and caregiver burden (Toot et al., 2017).

LTCF may also be seen as a temporary living arrangement, where people recover from acute care or a period to increase functional independence, so that the transition back to the community may be possible. According to a scoping review of 36 articles, older adults with the following characteristics were more likely to transition from LTCF to the communities: being younger, a married woman who received intense therapy and expressed her desire to transition (Freeman et al., 2017). However, research methodology and study samples were very heterogeneous, and limited geographically mostly to the USA. More research is needed, especially from countries with high longevity.

A study about transitions in care of persons with dementia in the United States showed that 3 out of 4 patients who are admitted to a LTCF are actually transferred from hospitals (Callahan et al., 2012). When people with dementia return home from a LTCF, they will most likely lack home health care support. Only 19% of deaths of people with dementia occurred in LTCF, and 46% died at home and 35% in the hospital.

Although results vary by country, in general people living in LTCF have different characteristics than community-dwelling older adults. According to a study in Spain (León-Salas et al., 2015), the profile of residents with cognitive impairment is characterized by being a woman (83.4%), aged 85.6 (standard deviation, SD: 6.7) years, widowed (62.6%), with education level lower than primary (65%), lining in the institution for 3.6 (SD: 3.2) years, having in average 7.7 (SD: 2.8) chronic health conditions, and having severe disability or complete dependency (59%). This profile differs from people who no cognitive impairment who reside in LTCF, who are also mainly women (65.4%), widowed (54.7%), with low education level (42.3%), but present a lower age (mean, M: 81.0; SD: 7.1 years), lower number of chronic health conditions (M: 6.5; SD: 2.8), and moderate disability.

Measuring QoL of older adults living in LTCF is hindered by cognitive impairment of many residents, causing difficulty to understand the questions, or lack of enough insight to evaluate subjective aspects of themselves. In these cases, a proxy is used, meaning that the QoL questionnaire is answered by a caregiver who knows well the resident. The *EQ-5D*, a short, generic, frequently used QoL instrument, has been validated in older adults with dementia (Diaz-Redondo et al., 2014). Besides having a self-assessment version, there is also one completed by proxy. Results show that it is a valid and reliable instrument, with a greater inter-rater agreement for the most observable dimensions, mobility, self-care, and usual activities.

There are also questionnaires specifically designed to assess QoL of people with dementia (Table 4.1). The *Quality of Life in Alzheimer's Disease scale (QOL-AD)* is formed by 13 items, rated in a Likert-type scale from 0 to 4 (Logsdon et al., 2002). Higher scores indicate a better QoL. It has two versions, one for the patient and the other for the caregiver. The *Quality of Life in Late-Stage Dementia (QUALID)* aims at assessing QoL of people with advanced dementia, and it is completed by the caregiver (Weiner et al., 2000). The content of its 11 items focuses on observable behaviours and emotions. The response scale ranges from 1 to 5, with higher score reflecting a better QoL. The *DEMQOL* can be used in all dementia stages, and it is formed by 28 items answered in a 4-point response scale and administered by an interviewer to the person with dementia (Smith et al., 2005, 2007). There is also a 31-item version, answered by a caregiver (DEMQOL-Proxy). In both versions, items are scored in a 1–4 response scale. Some items are scored reversely, so that a higher score indicate better QoL.

Environmental or contextual factors greatly impact the QoL of older adults living in LTCF. A study of 14 facilities in Spain showed that 16.4% of the QoL variance may be explained by *residence factors*, such as the availability of a geriatrician and being public or private (Marventano et al., 2015). Concerning the individual factors, a better QoL was associated to functional independence, better health status, lower length of stay, gathering with family and friends, lower dementia severity and lower depression level. In addition, functional independence, previous QoL, and number of chronic health conditions predicted QoL assessed later on (18-month follow-up) (Castro-Monteiro et al., 2014). These results draw attention to the importance of multidisciplinary intervention to improve or maintaining the QoL of older adults in LTCF. In addition, *quality of care* and subsequent QoL of LTCF residents is highly impacted by nursing home staffing (Maas et al., 2008), calling for solutions of understaffing, as well as lack of gerontological training for both nurses and assisting staff. Interventions that include work-directed approaches have a greater and longer-lasting effect on staff burnout than person-directed approaches (Westermann et al.,

2014). In addition, a review supported the importance of advanced nursing practices, with positive health outcomes for the older adults, as well as family satisfaction (Donald et al., 2013).

There is a considerable amount of research about *psychosocial interventions* to increase the QoL of older adults living in LCTF. For instance, the therapeutic effects of horticultural or gardening therapy have been shown when comparing pre-post effects, but with small differences between intervention and control groups (Nicholas et al., 2019). Besides QoL, benefits were observed in physical, social, cognitive, and emotional areas. Interventions target maltreatment of older adults in LTCF are especially effective when aiming at reducing physical restraint (Ayalon et al., 2016). More research is needed about interventions that target neglect, the victims of maltreatment, family caregivers or paid carers who abuse, prevention of maltreatment, and public awareness. Other examples of interventions in LTCF, with good results in the health and QoL of residents, are: fall prevention (Senderovich & Tsai, 2020), advance nursing practices (Donald et al., 2013), hospice care in LTCF (Miller & Mor, 2002), and family inclusion programs (Puurveen et al., 2018).

4.3 Conclusions

As we have seen in this chapter, the characteristics of the place where older people live impact their health status and, consequently, their QoL. Building age-friendly environments is essential for healthy and active ageing, and the specific needs of older people should be considered when designing living facilities. Ageing in place is the preferred option for most persons as they age, due to the meaning places have for people. Remaining in their own homes gives older people a sense of control over their own life and influences their QoL. However, certain material and social conditions must be assured to allow people to age in place. The challenge for the society is to guarantee that the built environment possesses the characteristics that make it liveable and that favour the support and care older that people need. Recent developments such as assisted living technologies and new housing models like cohousing, help older people to remain at a comfortable and functional accommodation that covers their needs.

When ageing in place is no longer possible, LTCF can provide a supportive environment, compensating the loss of functional capacity, independence, and social network some older people may suffer, and thus, playing an essential role in maintaining their QoL. Some residential characteristics, having highly skilled staff that guarantee the quality of care, and implementing psychosocial interventions are the requirements for creating an environment that promotes healthy and active ageing and improves older persons' QoL.

In conclusion, designing age-friendly environments is a requirement for assuring older persons' QoL, and promoting the active participation of this group of population in all aspects of the living arrangements is a challenge for the society. It is mandatory to commit older adults in the design of age-friendly environments, as well as to promote research that evaluates the effect of interventions on older adults' QoL. Multidisciplinary actions and interventions aimed at maintaining and increasing the older population QoL will benefit not only this group, but also have a broader, beneficial effect to the whole society.

References

Ayalon, L., Lev, S., Green, O., & Nevo, U. (2016). A systematic review and meta-analysis of interventions designed to prevent or stop elder maltreatment. *Age and Ageing, 45*(2), 216–227. https://doi.org/10.1093/ageing/afv193

Barnett, A., Zhang, C. J. P., Johnston, J. M., & Cerin, E. (2018). Relationships between the neighborhood environment and depression in older adults: A systematic review and meta-analysis. *International Psychogeriatrics, 30*(8), 1153–1176. https://doi.org/10.1017/S104161021700271X

Beard, J. R., Blaney, S., Cerda, M., Frye, V., Lovasi, G. S., Ompad, D., Rundle, A., & Vlahov, D. (2009). Neighborhood characteristics and disability in older adults. *The Journals of Gerontology Series B: Psychological Sciences and Social Sciences, 64B*(2), 252–257. https://doi.org/10.1093/geronb/gbn018

Beridze, G., Ayala, A., Ribeiro, O., Fernández-Mayoralas, G., Rodríguez-Blázquez, C., Rodríguez-Rodríguez, V., Rojo-Pérez, F., Forjaz, M. J., & Calderón-Larrañaga, A. (2020). Are loneliness and social isolation associated with quality of life in older adults? Insights from Northern and Southern Europe. *International Journal of Environmental Research and Public Health, 17*(22). https://doi.org/10.3390/ijerph17228637

Bosch-Farré, C., Malagón-Aguilera, M. C., Ballester-Ferrando, D., Bertran-Noguer, C., Bonmatí-Tomàs, A., Gelabert-Vilella, S., & Juvinyà-Canal, D. (2020). Healthy ageing in place: Enablers and barriers from the perspective of the elderly. A qualitative study. *International Journal of Environmental Research and Public Health, 17*(18), 6451. https://doi.org/10.3390/ijerph17186451

Bowling, A. (2009). The psychometric properties of the older people's quality of life questionnaire, compared with the CASP-19 and the WHOQOL-OLD. *Current Gerontology and Geriatrics Research*, 298950. https://doi.org/10.1155/2009/298950

Byles, J. E., Mackenzie, L., Redman, S., Parkinson, L., Leigh, L., & Curryer, C. (2014). Supporting housing and neighbourhoods for healthy ageing: Findings from the Housing and Independent Living Study (HAIL). *Australasian Journal on Ageing, 33*(1), 29–35. https://doi.org/10.1111/j.1741-6612.2012.00646.x

Callahan, C. M., Arling, G., Tu, W., Rosenman, M. B., Counsell, S. R., Stump, T. E., & Hendrie, H. C. (2012). Transitions in care for older adults with and without dementia. *Journal of the American Geriatrics Society, 60*(5), 813–820. https://doi.org/10.1111/j.1532-5415.2012.03905.x

Carrere, J., Reyes, A., Oliveras, L., Fernández, A., Peralta, A., Novoa, A. M., Pérez, K., & Borrell, C. (2020). The effects of cohousing model on people's health and wellbeing: A scoping review. *Public Health Reviews, 41*, 22. https://doi.org/10.1186/s40985-020-00138-1

Castro-Monteiro, E., Forjaz, M. J., Ayala, A., Rodriguez-Blazquez, C., Fernandez-Mayoralas, G., Diaz-Redondo, A., & Martinez-Martin, P. (2014). Change and predictors of quality of life in institutionalized older adults with dementia. *Quality of Life Research, 23*(9), 2595–2601. https://doi.org/10.1007/s11136-014-0706-8

Chappell, N. (2014). Social capital and health inequalities. In A. C. Michalos (Ed.), *Encyclopedia of quality of life and well-being research* (pp. 6021–6024). Springer. https://doi.org/10.1007/978-94-007-0753-5_3955

Chemaitelly, H., Kanaan, C., Beydoun, H., Chaaya, M., Kanaan, M., & Sibai, A. M. (2013). The role of gender in the association of social capital, social support, and economic security with self-rated health among older adults in deprived communities in Beirut. *Quality of Life Research, 22*(6), 1371–1379. https://doi.org/10.1007/s11136-012-0273-9

Cramm, J. M., van Dijk, H. M., & Nieboer, A. P. (2013). The importance of neighborhood social cohesion and social capital for the well being of older adults in the community. *The Gerontologist, 53*(1), 142–152. https://doi.org/10.1093/geront/gns052

De Coninck, L., Bekkering, G. E., Bouckaert, L., Declercq, A., Graff, M. J. L., & Aertgeerts, B. (2017). Home- and community-based occupational therapy improves functioning in frail older people: A systematic review. *Journal of the American Geriatrics Society, 65*(8), 1863–1869. https://doi.org/10.1111/jgs.14889

Diaz-Redondo, A., Rodriguez-Blazquez, C., Ayala, A., Martinez-Martin, P., Forjaz, M. J., & Spanish Research Group on Quality of Life and Aging. (2014). EQ-5D rated by proxy in institutionalized older adults with dementia: Psychometric pros and cons. *Geriatrics & Gerontology International, 14*(2), 346–353. https://doi.org/10.1111/ggi.12108

Donald, F., Martin-Misener, R., Carter, N., Donald, E. E., Kaasalainen, S., Wickson-Griffiths, A., Lloyd, M., Akhtar-Danesh, N., & DiCenso, A. (2013). A systematic review of the effectiveness of advanced practice nurses in long-term care. *Journal of Advanced Nursing, 69*(10), 2148–2161. https://doi.org/10.1111/jan.12140

EuroQol Group. (1990). EuroQol – A new facility for the measurement of health-related quality of life. *The EuroQol Group. Health Policy, 16*(3), 199–208.

Faber, P. (Ed.). (2015). *Active ageing: A policy framework in response to the longevity revolution.* International Longevity Centre Brazil.

Forjaz, M. J., Prieto-Flores, M.-E., Ayala, A., Rodriguez-Blazquez, C., Fernandez-Mayoralas, G., Rojo-Perez, F., & Martinez-Martin, P. (2011). Measurement properties of the Community Wellbeing Index in older adults. *Quality of Life Research, 20*(5), 733–743. https://doi.org/10.1007/s11136-010-9794-2

Freeman, S., Bishop, K., Spirgiene, L., Koopmans, E., Botelho, F. C., Fyfe, T., Xiong, B., Patchett, S., & MacLeod, M. (2017). Factors affecting residents transition from long term care facilities to the community: A scoping review. *BMC Health Services Research, 17*(1), 689. https://doi.org/10.1186/s12913-017-2571-y

Friedman, D., Parikh, N. S., Giunta, N., Fahs, M. C., & Gallo, W. T. (2012). The influence of neighborhood factors on the quality of life of older adults attending New York City senior centers: Results from the Health Indicators Project. *Quality of Life Research, 21*(1), 123–131.

Ganesan, B., Gowda, T., Al-Jumaily, A., Fong, K. N. K., Meena, S. K., & Tong, R. K. Y. (2019). Ambient assisted living technologies for older adults with cognitive and physical impairments: A review. *European Review for Medical and Pharmacological Sciences, 23*(23), 10470–10481. https://doi.org/10.26355/eurrev_201912_19686

Gerstorf, D., Ram, N., Goebel, J., Schupp, J., Lindenberger, U., & Wagner, G. G. (2010). Where people live and die makes a difference: Individual and geographic disparities in well-being progression at the end of life. *Psychology and Aging, 25*(3), 661–676. https://doi.org/10.1037/a0019574

Giraldez-Garcia, C., Forjaz, M. J., Prieto-Flores, M.-E., Rojo-Perez, F., Fernandez-Mayoralas, G., & Martinez-Martin, P. (2012). Individual's perspective of local community environment and health indicators in older adults. *Geriatrics & Gerontology International.* https://doi.org/10.1111/j.1447-0594.2012.00872.x

Gobbens, R. J. J., & van Assen, M. A. L. M. (2018). Associations of environmental factors with quality of life in older adults. *The Gerontologist, 58*(1), 101–110. https://doi.org/10.1093/geront/gnx051

Grimmer, K., Kay, D., Foot, J., & Pastakia, K. (2015). Consumer views about aging-in-place. *Clinical Interventions in Aging, 10*, 1803–1811. https://doi.org/10.2147/CIA.S90672

Hellström, Y., Andersson, M., & Hallberg, I. R. (2004). Quality of life among older people in Sweden receiving help from informal and/or formal helpers at home or in special accommodation. *Health & Social Care in the Community, 12*(6), 504–516. https://doi.org/10.1111/j.1365-2524.2004.00519.x

Horner, B., & Boldy, D. P. (2008). The benefit and burden of "ageing-in-place" in an aged care community. *Australian Health Review, 32*(2), 356–365.

Huang, N. C., Chu, C., Kung, S. F., & Hu, S. C. (2019). Association of the built environments and health-related quality of life in community-dwelling older adults: A cross-sectional study. *Quality of Life Research, 28*(9), 2393–2407. https://doi.org/10.1007/s11136-019-02199-5

Hyde, M., Wiggins, R. D., Higgs, P., & Blane, D. B. (2003). A measure of quality of life in early old age: The theory, development and properties of a needs satisfaction model (CASP-19). *Aging & Mental Health, 7*(3), 186–194. https://doi.org/10.1080/1360786031000101157

International Wellbeing Group. (2006). *Personal Wellbeing Index* (4th ed.). Australian Centre on Quality of Life, Deakin University.

Lane, A. P., Wong, C. H., Močnik, Š., Song, S., & Yuen, B. (2020). Association of neighborhood social capital with quality of life among older people in Singapore. *Journal of Aging and Health, 32*(7–8), 841–850. https://doi.org/10.1177/0898264319857990

León-Salas, B., Ayala, A., Blaya-Nováková, V., Avila-Villanueva, M., Rodríguez-Blázquez, C., Rojo-Pérez, F., Fernández-Mayoralas, G., Martínez-Martín, P., Forjaz, M. J., & Spanish Research Group on Quality of Life and Aging. (2015). Quality of life across three groups of older adults differing in cognitive status and place of residence. *Geriatrics & Gerontology International, 15*(5), 627–635. https://doi.org/10.1111/ggi.12325

Liu, J., Chen, H., Chen, Y., & Li, Z. (2018). Emotional well-being and social support in social housing neighbourhoods in China. *The Journal of International Medical Research, 46*(8), 3209–3218. https://doi.org/10.1177/0300060518774718

Logsdon, R. G., Gibbons, L. E., McCurry, S. M., & Teri, L. (2002). Assessing quality of life in older adults with cognitive impairment. *Psychosomatic Medicine, 64*(3), 510–519.

Lu, N., Xu, S., & Zhou, Q. (2020). Social capital and preferences for aging in place among older adults living in rural Northeast China. *International Journal of Environmental Research and Public Health, 17*(14). https://doi.org/10.3390/ijerph17145085

Maas, M. L., Specht, J. P., Buckwalter, K. C., Gittler, J., & Bechen, K. (2008). Nursing home staffing and training recommendations for promoting older adults' quality of care and life: Part 1. Deficits in the quality of care due to understaffing and undertraining. *Research in Gerontological Nursing, 1*(2), 123–133. https://doi.org/10.3928/19404921-20080401-03

Marventano, S., Prieto-Flores, M.-E., Sanz-Barbero, B., Martín-García, S., Fernandez-Mayoralas, G., Rojo-Perez, F., Martinez-Martin, P., Forjaz, M. J., & Spanish Research Group on Quality of Life and Ageing. (2015). Quality of life in older people with dementia: A multilevel study of individual attributes and residential care center characteristics. *Geriatrics & Gerontology International, 15*(1), 104–110. https://doi.org/10.1111/ggi.12238

Menec, V. H., Means, R., Keating, N., Parkhurst, G., & Eales, J. (2011). Conceptualizing age-friendly communities. *Canadian Journal on Aging = La Revue Canadienne Du Vieillissement, 30*(3), 479–493. https://doi.org/10.1017/S0714980811000237

Miller, S. C., & Mor, V. N. T. (2002). The role of hospice care in the nursing home setting. *Journal of Palliative Medicine, 5*(2), 271–277. https://doi.org/10.1089/109662102753641269

Murayama, H., Fujiwara, Y., & Kawachi, I. (2012). Social capital and health: A review of prospective multilevel studies. *Journal of Epidemiology, 22*(3), 179–187. https://doi.org/10.2188/jea.je20110128

Negrón-Blanco, L., de Pedro-Cuesta, J., Almazán, J., Rodríguez-Blázquez, C., Franco, E., Damián, J., & DISCAP-ARAGON Research Group. (2016). Prevalence of and factors associated with homebound status among adults in urban and rural Spanish populations. *BMC Public Health, 16*, 574. https://doi.org/10.1186/s12889-016-3270-z

Nicholas, S. O., Giang, A. T., & Yap, P. L. K. (2019). The effectiveness of horticultural therapy on older adults: A systematic review. *Journal of the American Medical Directors Association, 20*(10), 1351.e1–1351.e11. https://doi.org/10.1016/j.jamda.2019.06.021

O'Boyle, C., McGee, H., Hickey, A., Joyce, C. R. B., Browne, J., O'Malley, K., & Hiltbrunner, B. (1993). *The Schedule for the Evaluation of Individual Quality of Life (SEIQoL). Administration manual* (Psychological reports). Royal College of Surgeons in Ireland. Retrieved from https://epubs.rcsi.ie/psycholrep/39

Oh, A., Patel, K., Boscardin, W. J., Max, W., Stephens, C., Ritchie, C. S., & Smith, A. K. (2019). Social support and patterns of institutionalization among older adults: A longitudinal study. *Journal of the American Geriatrics Society, 67*(12), 2622–2627. https://doi.org/10.1111/jgs.16184

Oiamo, T. H., Luginaah, I. N., & Baxter, J. (2015). Cumulative effects of noise and odour annoyances on environmental and health related quality of life. *Social Science & Medicine, 1982*(146), 191–203. https://doi.org/10.1016/j.socscimed.2015.10.043

Power, M., Quinn, K., Schmidt, S., & WHOQOL-OLD Group. (2005). Development of the WHOQOL-old module. *Quality of Life Research, 14*(10), 2197–2214. https://doi.org/10.1007/s11136-005-7380-9

Prieto-Flores, M.-E., Fernandez-Mayoralas, G., Forjaz, M. J., Rojo-Perez, F., & Martinez-Martin, P. (2011). Residential satisfaction, sense of belonging and loneliness among older adults living in the community and in care facilities. *Health & Place, 17*(6), 1183–1190. https://doi.org/10.1016/j.healthplace.2011.08.012

Puga, D., Fernández-Carro, C., & Fernández-Abascal, H. (2021). Multimorbidity, social networks and health-related wellbeing at the end of the life course. In F. Rojo-Pérez & G. Fernández-Mayoralas (Eds.), *Handbook of active ageing and quality of life: From concepts to applications* (pp. 609–628). Springer. https://doi.org/10.1007/978-3-030-58031-5_37

Puurveen, G., Baumbusch, J., & Gandhi, P. (2018). From family involvement to family inclusion in nursing home settings: A critical interpretive synthesis. *Journal of Family Nursing, 24*(1), 60–85. https://doi.org/10.1177/1074840718754314

Salminen, M., Laine, J., Vahlberg, T., Viikari, P., Wuorela, M., Viitanen, M., & Viikari, L. (2020). Factors associated with institutionalization among home-dwelling patients of Urgent Geriatric Outpatient Clinic: A 3-year follow-up study. *European Geriatric Medicine, 11*(5), 745–751. https://doi.org/10.1007/s41999-020-00338-7

Sánchez-González, D., Rojo-Pérez, F., Rodríguez-Rodríguez, V., & Fernández-Mayoralas, G. (2020). Environmental and psychosocial interventions in age-friendly communities and active ageing: A systematic review. *International Journal of Environmental Research and Public Health, 17*(22). https://doi.org/10.3390/ijerph17228305

Senderovich, H., & Tsai, P. M. (2020). Do exercises prevent falls among older adults: Where are we now? A systematic review. *Journal of the American Medical Directors Association, 21*(9), 1197–1206.e2. https://doi.org/10.1016/j.jamda.2020.05.010

Smith, S. C., Lamping, D. L., Banerjee, S., Harwood, R., Foley, B., Smith, P., Cook, J. C., Murray, J., Prince, M., Levin, E., Mann, A., & Knapp, M. (2005). Measurement of health-related quality of life for people with dementia: Development of a new instrument (DEMQOL) and an evaluation of current methodology. *Health Technology Assessment, 9*(10), 1–93, iii–iv.

Smith, S. C., Lamping, D. L., Banerjee, S., Harwood, R. H., Foley, B., Smith, P., Cook, J. C., Murray, J., Prince, M., Levin, E., Mann, A., & Knapp, M. (2007). Development of a new measure of health-related quality of life for people with dementia: DEMQOL. *Psychological Medicine, 37*(5), 737–746. https://doi.org/10.1017/S0033291706009469

Stephens, C., & Allen, J. (2021). Older people as active agents in their neighbourhood environments: Moving house can improve quality of life. *The Gerontologist*. https://doi.org/10.1093/geront/gnab065

Stephens, C., Szabó, Á., Allen, J., & Alpass, F. (2019). Livable environments and the quality of life of older people: An ecological perspective. *The Gerontologist, 59*(4), 675–685. https://doi.org/10.1093/geront/gny043

The World Health Organization Quality of Life assessment (WHOQOL): Position paper from the World Health Organization. (1995). *Social Science & Medicine, 41*(10), 1403–1409.

Toot, S., Swinson, T., Devine, M., Challis, D., & Orrell, M. (2017). Causes of nursing home placement for older people with dementia: A systematic review and meta-analysis. *International Psychogeriatrics, 29*(2), 195–208. https://doi.org/10.1017/S1041610216001654

Tröster, T., Thalmann, M., Fischer, K., Bieri-Brüning, G., Beeler, P. E., Bischoff-Ferrari, H. A., & Gagesch, M. (2020). Frailty, underweight and impaired mobility are associated with institutionalisation after post-acute care. *Swiss Medical Weekly, 150*, w20276. https://doi.org/10.4414/smw.2020.20276

Vaarama, M. (2009). Care-related quality of life in old age. *European Journal of Ageing, 6*(2), 113–125. https://doi.org/10.1007/s10433-009-0115-y

Van Cauwenberg, J., De Bourdeaudhuij, I., De Meester, F., Van Dyck, D., Salmon, J., Clarys, P., & Deforche, B. (2011). Relationship between the physical environment and physical activity in older adults: A systematic review. *Health & Place, 17*(2), 458–469. https://doi.org/10.1016/j.healthplace.2010.11.010

Van Orden, K. A., Bower, E., Lutz, J., Silva, C., Gallegos, A. M., Podgorski, C. A., Santos, E. J., & Conwell, Y. (2020). Strategies to promote social connections among older adults during "social distancing" restrictions. *The American Journal of Geriatric Psychiatry*. https://doi.org/10.1016/j.jagp.2020.05.004

Vanleerberghe, P., De Witte, N., Claes, C., Schalock, R. L., & Verté, D. (2017). The quality of life of older people aging in place: A literature review. *Quality of Life Research, 26*(11), 2899–2907. https://doi.org/10.1007/s11136-017-1651-0

Vestbro, D. U. (Ed.). (2010). *Living together – Cohousing ideas and realities around the world: Proceedings from the International Collaborative Housing Conference in Stockholm 5–9 May 2010*. Division of Urban and Regional Studies. Royal Inst. of Technology.

Ware, J. E., Jr., & Sherbourne, C. D. (1992). The MOS 36-item short-form health survey (SF-36). I. Conceptual framework and item selection. *Medical Care, 30*(6), 473–483.

Weiner, M. F., Martin-Cook, K., Svetlik, D. A., Saine, K., Foster, B., & Fontaine, C. S. (2000). The quality of life in late-stage dementia (QUALID) scale. *Journal of the American Medical Directors Association, 1*(3), 114–116.

Westermann, C., Kozak, A., Harling, M., & Nienhaus, A. (2014). Burnout intervention studies for inpatient elderly care nursing staff: Systematic literature review. *International Journal of Nursing Studies, 51*(1), 63–71. https://doi.org/10.1016/j.ijnurstu.2012.12.001

WHO Regional Office for Europe. (2016). *Creating age-friendly environments in Europe. A tool for local policy-makers and planners*. World Health Organization. Retrieved from https://apps.who.int/iris/bitstream/handle/10665/334252/9789289052122-eng.pdf

WHO Regional Office for Europe. (2018). *Age-friendly environments in Europe: Indicators, monitoring and assessments* (WHO/EURO:2020-1088-40834-55192). World Health Organization. Retrieved from https://apps.who.int/iris/handle/10665/334284

Wiles, J. (2005). Conceptualizing place in the care of older people: The contributions of geographical gerontology. *Journal of Clinical Nursing, 14*(8B), 100–108. https://doi.org/10.1111/j.1365-2702.2005.01281.x

World Health Organization. (2015a). *Measuring the age-friendliness of cities: A guide to using core indicators*. World Health Organization. Retrieved from http://www.who.int/kobe_centre/publications/AFC_guide/en/

World Health Organization. (2015b). *World report on ageing and health*. WHO Press. Retrieved from http://www.who.int/ageing/publications/world-report-2015/en/

Yang, Y., Wang, S., Chen, L., Luo, M., Xue, L., Cui, D., & Mao, Z. (2020). Socioeconomic status, social capital, health risk behaviors, and health-related quality of life among Chinese older adults. *Health and Quality of Life Outcomes, 18*(1), 291. https://doi.org/10.1186/s12955-020-01540-8

Zhang, C. J. P., Barnett, A., Johnston, J. M., Lai, P.-C., Lee, R. S. Y., Sit, C. H. P., & Cerin, E. (2019). Objectively-measured neighbourhood attributes as correlates and moderators of quality of life in older adults with different living arrangements: The ALECS Cross-Sectional Study. *International Journal of Environmental Research and Public Health, 16*(5). https://doi.org/10.3390/ijerph16050876

Zhou, E. S. (2014). Social support. In A. C. Michalos (Ed.), *Encyclopedia of quality of life and well-being research* (pp. 6161–6164). Springer. https://doi.org/10.1007/978-94-007-0753-5_2789

Carmen Rodríguez-Blázquez is a research technician at the National Centre of Epidemiology (Carlos III Institute of Health, Madrid, Spain). PhD in Public Health, Psychologist and MSc in Research Methodology. She has participated as a researcher in national and international projects and has more than 100 scientific publications on the fields of clinical and psychosocial aspects and quality of life of older people and in different health conditions (neurological diseases, multimorbidity, disability), and in the development and validation of rating scales in these fields. She is a research member in the Spanish Consortium on Neurodegenerative Diseases (CIBERNED) and in the European-funded Joint Action PHIRI and is an associate editor in BMC Neurology.

Maria João Forjaz is a scientific researcher at the National Centre of Epidemiology, Spanish National Institute of Health Carlos III. She graduated from the University of Lisbon and obtained her doctorate in Clinical Psychology in 2000 from the University of North Texas, USA, with a Fulbright scholarship. She participates actively in the training of junior researchers and medical residents in public health and preventive medicine. Her research interests include scale validation and health indicators, quality of life, ageing, multimorbidity, and impact of COVID-19. Dr. Forjaz is the author of more than 100 articles in peer-review journal and several books and book chapters. She is an associate editor of the journal *Quality of Life Research*, a member of the European-funded joint action PHIRI, and she is a group leader at the Spanish research network REDISSEC.

Chapter 5
Active Ageing and Age-Friendly Communities: Constructing an Image of Old Age and Ageing

Gloria Fernández-Mayoralas, Vicente Rodríguez-Rodríguez, Diego Sánchez-González, and Fermina Rojo-Pérez

Abstract This chapter analyses self-image among older adults, taking into account their different residential settings (family dwelling, cohousing, nursing homes), the relationship with pillars and determinants of active ageing, and the implications for the development of more inclusive settings for older adults. There is a qualitative analysis of discourses obtained during in-depth interviews and discussion groups held in 2017 with the older adult population in Spain. The findings showed that older adults construct different images of themselves according to their living environment, their social participation, and the availability of inclusive spaces. As a result, a public policy that is based on promoting age-friendly cities and communities (AFCCs), particularly seeking to reinforce the areas of *social participation*, *respect and social inclusion* and *civic engagement and employment*, could help to improve self-image among older people and facilitate their social and community inclusion.

Keywords Active ageing · Age-friendly cities and communities · Image of old age · Ageism · Social inclusion

---

G. Fernández-Mayoralas (✉)
Institute of Economics, Geography and Demography (IEGD); Spanish National Research Council (CSIC), Research Group on Ageing (GIE-CSIC), Madrid, España
e-mail: gloria.fernandezmayoralas@csic.es

V. Rodríguez-Rodríguez · F. Rojo-Pérez
Institute of Economics, Geography and Demography (IEGD); Spanish National Research Council (CSIC), Research Group on Ageing (GIE-CSIC), Ageing Network of the Latin American Population Association (ALAP), Madrid, Spain
e-mail: vicente.rodriguez@csic.es; fermina.rojo@csic.es

D. Sánchez-González
Department of Geography, Faculty of Geography and History, Universidad Nacional de Educación a Distancia (UNED), Madrid, España
e-mail: dsanchezg@geo.uned.es

E. Pozo Menéndez, E. Higueras García (eds.), *Urban Design and Planning for Age-Friendly Environments Across Europe: North and South*, Future City 19,
https://doi.org/10.1007/978-3-030-93875-8_5

5.1 Introduction

The policy framework of the World Health Organization (WHO) for active ageing (AA) (WHO, 2002) involved a series of paradigm shifts targeted at promoting a positive view of the phenomenon of ageing. The changes were intended first to treat people as active agents of their ageing, a responsibility to be shared with a society that has to supply appropriate political, legal, and social structures to optimise their health, participation, security, and learning conditions (Walker & Maltby, 2012), and second to interpret ageing as a process, in which older adults are a heterogeneous group (ILC-BR, 2015), and to incorporate the perspective of gender and equality of opportunities as a response to vulnerability and discrimination, which particularly affect women throughout their lives and are accentuated in old age (Fernández-Mayoralas et al., 2018).

Almost two decades after the holding of the Second World Assembly on Ageing in Madrid (UN, 2002), the phenomenon of ageing is still associated with a demographic problem (Pérez-Díaz & Abellán-García, 2018) and with the burden and economic unsustainability of pensions (Esteve et al., 2018), and old age is linked to a stereotyped image across each of the pillars of AA: deteriorating health, disability and dependency; reduced social and work participation; physical and financial insecurity; inability to learn; and general vulnerability and lack of autonomy.

AA policies are consistent with scientific developments, as shown by the WHO's assumption of the paradigm and its implementation in various regions in line with the Madrid International Plan of Action on Ageing, the United Nations Economic Commission for Europe (UNECE), and its construction of the Active Ageing Index (Del Barrio et al., 2018; Foster & Walker, 2013; Sidorenko & Zaidi, 2018) and other examples of regulatory development (Walker, 2018) such as those of the OECD and of the Council of Europe (Guillemard, 2013; Rodríguez-Rodríguez, 2018). However, this regulatory trend has not been exempt from criticism (Sánchez & Hatton-Yeoh, 2012).

The concept of AA has evolved with three fundamental constraints: (i) the more direct assessment of residential setting as a defining element for AA; (ii) the almost complete absence of older adults' opinions and experiences from applying AA to their lives as a necessary approach for improved quality of life (Ayala et al., 2021); and (iii) the consideration of ageism as a form of expression of the behaviour of the ageing person, the image they project and how it is seen by the people themselves and those comprising their community.

The place in which an older adult is living is fundamental to understanding how they interpret AA, because it is intertwined with their life course and gives meaning to their system of personal, social and community relationships. This is in line with the concept of ageing in place that has a long track record in environmental gerontology (Bigonnesse & Chaudhury, 2020) and significant for the application of public policies (Walker, 2021). The concept has also been recognised in the regulatory context via the Age-Friendly Cities and Communities (AFCC) project (WHO,

2007) and subsequent progress (Del Barrio et al., 2018; Sánchez-González et al., 2020).

One of the distinguishing features of the generation of knowledge regarding AA and the population's self-image is the limited development of analysis concerning the place in which people are living; their residential setting in both physical and social terms (Teater & Chonody, 2020). When analysis is focused on environmental gerontology, there are more abundant findings as regards definitions (Yeh, 2020), theoretical models and distinctive elements of the older adult's relationship with their residential setting, as in methodological approaches (Sánchez-González & Rodríguez-Rodríguez, 2016). Aspects inherent to the older adult (Bigonnesse & Chaudhury, 2021) such as adaptation and adjustment to limiting circumstances, developing specific activities, fostering personal autonomy, consolidating a better quality of life and constructing social meaning in the residential environment (Chum et al., 2020; Yeh, 2020), are essential to understanding the meaning of ageing in place as reference points in individual decision-making. However, this is not an easily managed concept owing to its complexity and the way in which it can be adapted to diverse situations (Bigonnesse & Chaudhury, 2021).

The construction of a self-image of old age cannot be detached from the place in which ageing happens, or from each person's view of ageing and their form of ageing. Stereotypes, prejudice and discrimination against people on the grounds of their age are the three dimensions that comprise ageism. Coined by Butler (1969), ageism is a complex phenomenon: a social construct based on the existence of preconceived and stereotyped images of old age and ageing, which present an obstacle to AA and contribute to the social exclusion of older adults (Walsh et al., 2017). This discrimination can be amplified by the isolation and self-exclusion of older adults as they internalise these stereotyped, pre-configured ideas about the ageing process and this stage of life (Swift et al., 2017). Self-perceptions regarding one's ageing can influence health, autonomy and longevity (Westerhof et al., 2014), performance of the activities of daily living, depression, and cognitive functioning, including dementia (Tully-Wilson et al., 2021 first online), as well as the well-being and quality of life that come from AA (Fernández-Mayoralas et al., 2021). In addition, a negative view of older adults is strongly associated with non-active forms of ageing (Fernández-Ballesteros et al., 2020). This evidence provides support for the consequences of ageing stereotypes, according to Levy's Stereotype Embodiment theory (Tully-Wilson et al., 2021 first online).

However, although older adults are living longer and in ever-improving physical and mental condition, the evidence reveals a certain paradox: rather than diminishing, age-based stereotypes have become increasingly negative (Levy, 2017).

The importance of these trends and the need to reverse them are reflected in challenge 7 of the policy framework for action in relation to AA (WHO, 2002), which is intended to forge a new and positive image of ageing and of older adults. Along the same lines and under the AFCC paradigm, respect and social inclusion are values that must be taken into account across all spaces and action frameworks, including buildings, transport, housing and social participation, and their interaction contributes to the empowerment of older adults (WHO, 2007). Both concepts – respect and social inclusion – adopt a positive approach and respectively take into account attitudes and behaviours with regard to older adults, and opportunities for social relationships and access to community resources (Ronzi et al., 2018). Strategies are also being promoted to combat age-based discrimination through awareness-raising campaigns, both in the media and directed at public and private decision-makers, professionals and society in general (WHO, 2015).

Collective residential settings can also contribute to encouraging the stereotypes that underpin their images as reflected in the media. In this respect, the representation of nursing homes as a marker of unsuccessful ageing depends on the degree of available choice and level of financial solvency (Rozanova et al., 2016).

The spatial experience of ageing in cities produces patterns of behaviour that affect attitudes, social relationships and identity in place, constituting a determining factor of longevity (Sánchez-González, 2009). Residential setting and personal circumstances can have a significant effect on the development and maintenance of a stereotyped image and on the formation of pre-configured ideas about old age and older adults. Degraded and marginal neighbourhoods could also contribute to environmental stigmatisation and the construction of stereotyped images of their older residents, aggravating their isolation and the deterioration in their quality of life (Hrybyk et al., 2012; Sánchez-González, 2014). In this regard, from a local and planning perspective, the promotion of the AFCC project, which supports AA via the fostering of "respect and social inclusion", refers to *"the attitudes, behaviour and messages of other people and of the community as a whole towards older people"* (WHO, 2007: 14).

There is currently limited knowledge regarding the relationship between residential settings, the image of old age and AA. This chapter analyses self-image among older adults taking into account different residential settings (family dwelling, cohousing, and nursing homes), as well as investigating the potential existence of relationships between said image and the pillars and determinants of AA. As a final aim, the study findings will be used to examine how one might encourage the development of more inclusive and age-friendly settings.

5.2 Source and Methodology

Material was used from in-depth interviews (IDIs) and discussion groups (DGs) held between April and June 2017 in the context of two research projects on AA, quality of life and gender (Fernández-Mayoralas et al., 2021). For the purposes of this chapter, the selection criteria were parity of gender and of residential setting in various Spanish locations (12 IDIs, average age 72.2 years; range 59–82 years), for cohousing (12 IDIs, average age 77.7 years; range 62–87 years), key informant members of the governing bodies for the aforementioned cohousing (7 IDIs), older adults in nursing homes (3 DGs with a total of 21 people, average age 81.2 years; range 68–98 years) and users of recreation centres for older adults (3 DGs, 24 people, average age 72.0 years; range 59–83 years). All interviewees living in family dwellings were or had been married and had children, while those living in cohousing included some women without children. A majority of people in both settings were retired, except for a woman who never worked outside the home, another woman who was unemployed and a third on a widow's pension, as well as a man who was still active, all of whom were living in family dwellings.

The methodological approach was based on a content analysis intended to show how participant opinions were shaped according to their residential setting, through the concepts (codes) that arose from their discourses, and to establish relationships among them according to the research aims. Several basic tasks were hence essential, involving the extraction of content linked to informant opinions, to the co-occurrence of certain aspects of this content and to the production of a thematic, concept map, used to interpret the content.

Semi-structured scripts were used for the IDIs and DGs with thematic blocks that followed the WHO's AA framework in terms of the health, lifelong learning, participation, and security pillars and the behavioural, personal, physical environment, social environment, economic, and health and social services determinants, in addition to two cross-cutting determinants: gender and culture, as well as a quality of life (QoL) block as an outcome variable. In addition to the foregoing, other codes were extracted that were specific to the image and self-image of old age and ageing, as well as a general code on global AFCCs. Finally, other codes were included that acted as conditioning factors in the construction of image. This process involved three strategies: searching for and exploring initial content and assessing it for later analytical phases; revising the codes in line with their significance in the research; and constructing groups of codes that were related to the residential settings subject to investigation.

For the qualitative network analysis, significance criteria were established for the co-occurrent relationship among codes, such as relationship strength measured via the percentage of citations with co-occurrent codes and its qualification (certain codes condition/influence others or are conditioned/influenced by others). Finally, the concept map was produced by implementing said relationships in a representative and interpretative space.

With this basic design, IDIs and DGs were analysed using version 8 of ATLAS. ti. This tool permitted the implementation of a well-established analytical sequence (Friese, 2019; Friese et al., 2018) with numerous references in the literature on qualitative methods (Tong et al., 2007; Wolff & Silver, 2017; Saldaña, 2021). Based on a preliminary analysis that facilitated a process of familiarisation with the data, the initial codes (which would require subsequent restructuring) were explored and generated. This phase can involve a strategy that is inductive or deductive as appropriate, based on the aim of the research. These tasks are mandatory when searching for topics that require a review prior to analysis. The entire process is ultimately targeted at the production of results.

The following tasks were carried out in the documentary analysis forming part of this study:

(a) Introduction of documents and creation of document groups based on residential setting. When used as filters, documents can be analysed specifically.
(b) Preliminary reading of documents with deductive application of codes generated in accordance with the AA paradigm (Fernández-Mayoralas et al., 2021). A provisional set of 25 adequate codes, similar to independent variables, was chosen for the explanation of image among older adults, from which four were discarded owing to lack of sufficient data (citations).
(c) Re-reading of documents for inductive extraction of content related to the self-image of older adults regarding their ageing process, assimilated to dependent variables in the analysis. The following expressions of concept of image were identified (codes), incorporating 1298 citations: (1/2) positive/negative: personal perception of a positive/negative image regarding themselves or regarding ageing and old age; (3) self-image: image and perception among older adults regarding themselves; (4) personal development: perception of image as self-realisation and personal development, as the desire to satisfy personal interests, tastes and ambitions; (5) competence: image related to the feeling of being competent or having expertise; (6) usefulness: image related to the feeling of being useful in general; (7) burden: image linked to feeling of being a burden, which older adults can believe themselves to be for family and those close to them; (8) image of others: Images that older adults perceive and receive from other people or from society as a whole; and (9) gender image: image related to gender patterns..
(d) Content emerging from the image codes, extracted from the documents analysed in the various settings, offered an expression of the frequency with which the different semantic content (words) appeared. Five word clouds were produced, with words coloured differently according to their scope of reference (gender, age, family circumstances, activities, residential setting, negative circumstances).
(e) The codes were linked in two different ways: first, the co-occurrent relationships among image codes were explored so as to establish an expression of the relationship among the defined components of the images; second, the image codes were linked to those referring to conditioning/influencing factors. This process was carried out with intermediate co-occurrence tables that were condensed to extract the strongest links based on the total percentage for each code and on the percentage for each co-occurrent cell.
(f) With this information, it was possible to draw the thematic maps (networks) to express the nature, strength and significance of these relationships on one plane and facilitate the interpretation of the thematic network of image codes and a comparison of the networks defined in five residential settings.
(g) The analysis was completed by incorporating into the interpretation of each network a selection of citations that help to better understand the meaning of the images that older adults have of themselves and that society projects onto them.

5.3 Results

5.3.1 *Semantic Content of Discourses on the Image of Old Age and Ageing*

Based on the image codes, a series of citations were extracted from the discourses of the participants who offered diverse semantic content. These are reflected in a simplified form as word clouds for the different settings (Fig. 5.1).

The semantic field related to activities clearly stood out in the discourses of older adults living in family dwellings (Fig. 5.1a); references were mainly general, but there were also many other specific citations: courses, university, memory, support, games (of cards), association, arts and crafts, computer, classes and gym. In contrast to these activities, which are more associated with free time, were others linked to obligations arising from work or school (such as helping children by caring for grandchildren). There was a notable reference to time, as a necessary resource for engaging in activities and a symbolic basis for the flow of daily life. Other important fields were related to the basic aspects that govern people's lives, their gender (men, women) and their age (older, years, age, old, ageing). The fundamental family and social network for participants are also covered: grandchildren/children, daughter, husband, but also mother, friends and people in general, as well as places for living: home, village, bar and older adult facility. Finally, other aspects were related to perceived problems, and particularly loneliness.

The word cloud based on citations from the discourses of users of older adult recreation centres (Fig. 5.1b) indicated a more notable role for gender (men and women) and age (years, older, age) and also places for living and relationships: the facility and home, where activities take place (learning, theatre, books), in contrast to work. Other semantic fields such as family and social network placed more importance on people in general than on grandchildren, husband, mother or daughter. Other general concepts with a symbolic content are also highlighted, such as freedom, society, rights or person. There was a residual appearance of problems such as loneliness, in general, or fear, in particular.

The word cloud for discourses arising among the groups of older adults in nursing homes (Fig. 5.1c) provided a contrast to the above. The semantic content in this context was primarily linked to the nursing home, assimilated to the home, and the dining room, as the daily meeting place. Activities were also important, specifically with reference to exercise, in addition to which gender (men and women) and age (years, older, old) occupied a significant position. On the contrary, family and social network appeared to have a smaller presence: Mother was more prominent, with family, daughter, sister (new), husband, friends and people behind it, and an absence of grandchildren/children. Problems appeared, including loneliness, illness, distress and depression. And last but not least, there was a relatively extensive semantic field regarding positive aspects such as motivation, time, enthusiasm and happiness.

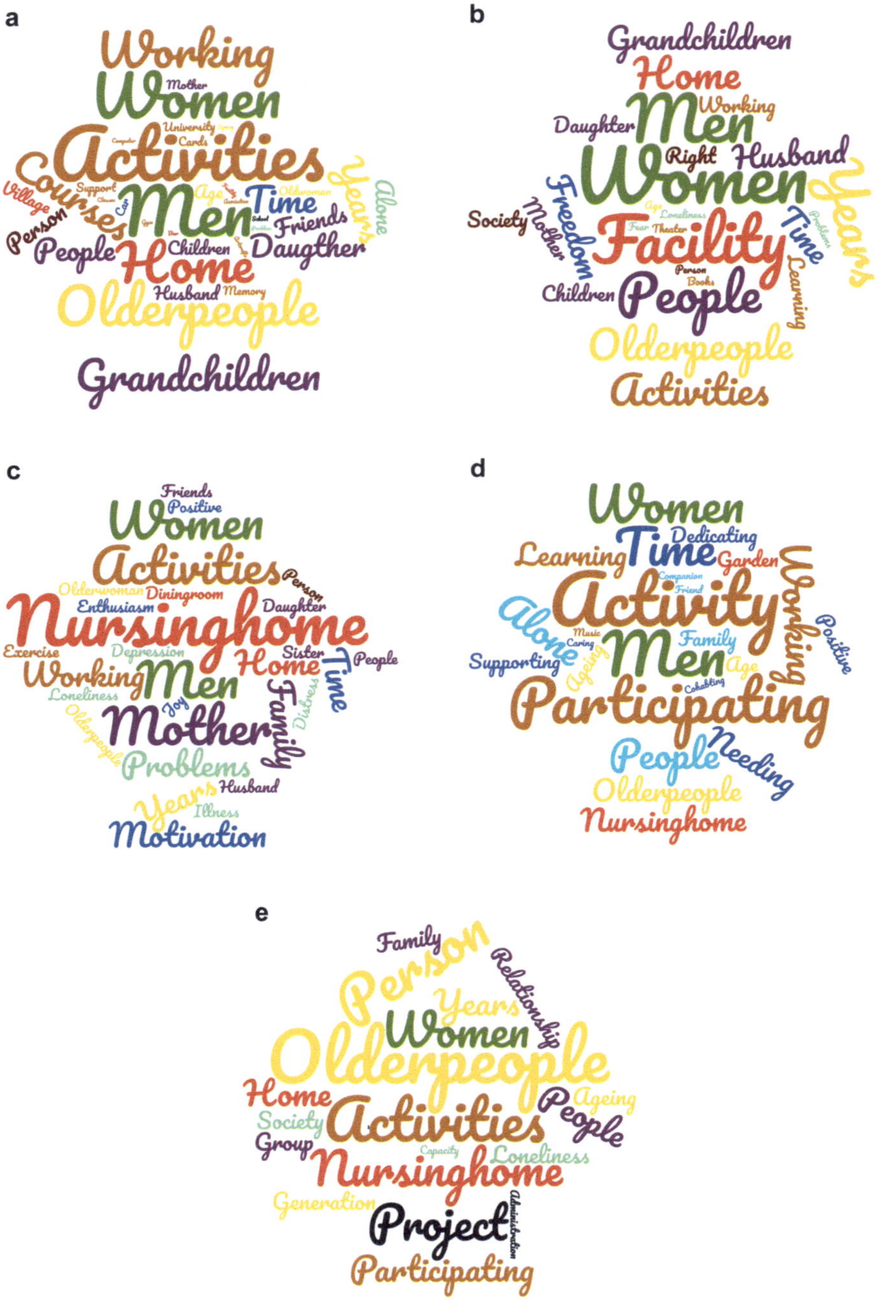

Fig. 5.1 Word clouds for discourses of older adults in different settings. (**a**) Family dwelling, (**b**) Older adult recreation centres, (**c**) Nursing homes, (**d**) Cohousing, (**e**) Cohousing key informants. Colour key Red = Setting Brown = Activities Green = Gender Purple = Support Blue = Social setting Yellow = Age Black = Task
(Source: Authors' own)

The word cloud extracted from the discourses of people living in cohousing (Fig. 5.1d) appeared to coincide with that observed among residents in family dwellings in terms of the importance of certain semantic fields. This was the case for activities, although this field included "more active activities" such as participating (and working), in addition to learning. Likewise, gender (men and women) and age (age, older, ageing) were important. Linked to activities was the semantic field of time, in this case, combined with positive aspects related to collaborative activities, such as dedicating, supporting, needing, caring and cohabiting. The relationship network appeared to extend to people, family, companion and friend, and places were limited to residential facility (cohousing) and garden. Loneliness (being alone) was a problem, and this is what the shared-responsibility way of life in cohousing is intended to avoid.

Finally, the central role of older adults (generation, years, ageing), and again, the semantic field of "more active" activities such as participation in general were notable in the discourses of housing collaborative key informants (Fig. 5.1e). Place, both nursing home and family dwelling, and collaborative residential project, including administration and management, were also of importance. In contrast, gender was only represented by women on this occasion. Networks were focused on relationship, with group, people and family, although loneliness was also present, as a problem that may have been a motive for choosing this kind of residential facility. Strikingly, there was a more notable presence of the semantic field related to person, society and capacity.

5.3.2 Construction of the Image of Old Age and Ageing

The relationship established among the nine codes, defined to describe and categorise the citations referring to aspects of image in the discourses of older adults, can be observed in Table 5.1, which only reflects co-occurrence of codes with a percentage in excess of 5% of citations.

One may observe that only six of the nine codes exceeded 5% of citations, with a significantly lower proportion for those referring to the image of old age and ageing related to gender patterns, the image of burden and the image that older adults perceive they are given by other people or society as a whole. However, these three contributed almost 12% of the citations, above all in co-occurrence with the codes for positive image, negative image and self-image.

The most citations were for the self-image code, in co-occurrence with personal development (7.7%) and positive image (7.6%), with a lower percentage for negative image (5.5%). The self-image that older adults describe for themselves is hence related to self-realisation and personal development (desire to satisfy personal interests, tastes and ambitions), a positive image, incorporating feelings of competence and capacity and of usefulness for the family, local or social setting.

Maintaining the same criterion of a percentage higher than 5% for the co-occurrence of codes in the citations, networks of relationships between these codes

Table 5.1 Co-occurrence of codes referring to image in old age (%)

	Positive image	Negative image	Self-image	Personal development	Competence	Usefulness	Other image codes (gender, other people, burden)
Positive image	0.0	0.5	7.6	5.9	1.8	2.1	2.2
Negative image		0.0	1.5	0.5	0.2	0.2	2.6
Self-image			0.0	7.7	3.6	2.7	2.8
Personal development				0.0	2.2	1.3	1.5
Competence					0.0	1.3	0.5
Usefulness						0.0	0.8
Total citations 1298 (%)	**20.0**	**5.5**	**25.9**	**19.1**	**9.5**	**8.4**	**11.6**

Source: Authors' own

were produced and analysed, constructing the self-image of older adults in each setting. The networks showing a self-image constructed based on more positive perceptions were those related to older adults in family dwellings, cohousing residents and users of older adult recreation centres (who were also living in family dwellings). In these three settings, together with positive images of old age and ageing, there was a prevailing feeling of personal development and self-realisation (personal interests, tastes and ambitions), added to a sense of competence and ability to learn among those living in cohousing and in family dwellings, in the latter case accompanied by a sense of usefulness in the family, local or social context. Certain citations from the three settings are illustrative of this[1]:

> You have to move your brain so you don't lose your rhythm (…) after the little culture I have (…) I have a really high capacity to understand things and … I also like learning, I like learning. (D86, CH, woman, 79, urban)

> I was dreaming of teaching those around me. That was my dream. (…). Almost at the end of my life, I'm doing what I've always dreamed of: teaching. (D2, DG-RC, mixed, man, 78, urban)

[1] The information included in the quotes is identified as follows to maintain interviewee anonymity: document number, residential context (FD: family dwelling; CH: cohousing; CH-Resp: cohousing key informant, not disclosing age; DG-NH: nursing home discussion group; DG-RC: older adult recreation centre discussion group), sex, age in years, type of area of residence. Words forming an implicit part of the discourse are included in square brackets. Ellipses in round brackets are used to omit a part of the discourse, whether a single word or a fragment.

The self-image network of those in nursing homes was mainly constructed around negative perceptions, based on their sense of being a burden to close family. This is the only setting in which gender image was related to a negative image of old age and ageing:

> I chose the home when my mother and my sister died (...). My daughter and my grandchildren live outside Spain. So I'm not going to bother anyone, I'm going to a care home, I'll be peaceful. (D4, DG-NH, mixed, woman, 81, urban)

The stereotyped images or ideas that other people or society might have did not appear to be significant for the construction of self-image in the discourses of older adults themselves. However, they were for the construction of image among cohousing key informants, who provided more complex discourses than those supplied by cohousing residents, incorporating the perception of the image that other people or society might have, and also placed importance on the relationship established between a sense of usefulness in one's surroundings and the personal development and self-realisation that it might imply in this collaborative setting.

> Anyone can make proposals (...), but first you have to take charge, and then obviously you have to conduct a study from a financial perspective, from a social perspective, from an environmental perspective, and so on (...). (D65, CH-Resp, male, urban)

The co-occurrence of codes between the image of old age and the pillars and determinants of AA can be analysed in order to understand how the AA framework is related to the construction of images regarding old age and ageing; this is shown in percentage form in Fig. 5.2. A majority of citations related construction of the image of old age and ageing to the concept of AA (13.5%), as defined by older adults in their discourses. The most significant factors and determinants were psychological (12.3%), gender (10.8%) and age-friendly residential setting (10.4%), as well as social and family networks (7.9% each). Only lifelong learning (7.7%) stood out among the AA pillars associated with the image of old age and ageing constructed by older adults.

5.3.3 Active Ageing: Concept, Pillars, and Determinants Providing the Basis for Self-Image in Old Age

Concept maps and thematic networks were produced to facilitate a detailed examination of the different relationships established between image and AA in the five settings subject to analysis. They were based on the co-occurrence tables, condensed to extract the strongest and most significant links and thereby achieve improved representation, comparability, and interpretation. The analysis was completed with the interpretation of each network and its elements, into which citations were incorporated to assist in understanding the meaning of the images that older adults have of themselves and those that society projects onto them.

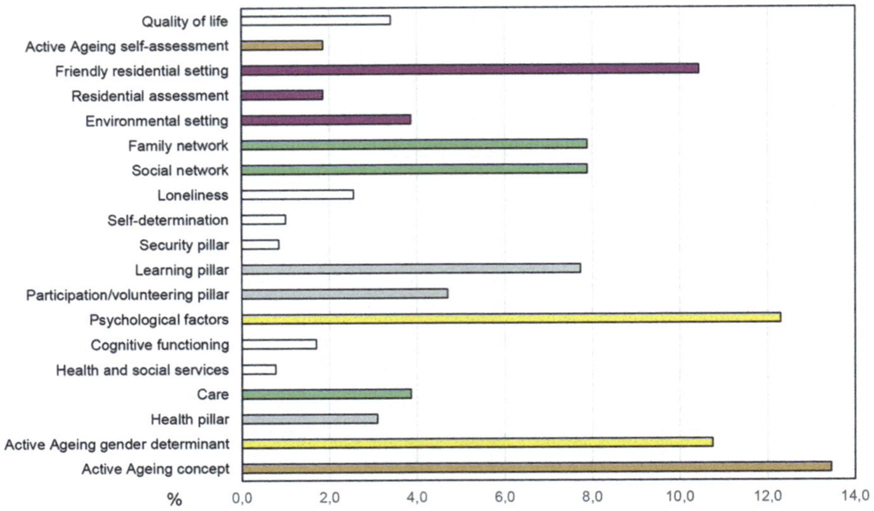

Fig. 5.2 Image of old age and ageing within the framework of active ageing. (The colours of the bars in this figure refer to those used later in the nodes relating to pillars and determinants of AA in the concept maps or thematic networks by setting (Fig. 5.3). The white bars are not coloured because they did not significantly appear in any of the settings. In Fig. 5.3, pale blue is added to the nodes referring to type of image; in these figures, the style and thickness of the lines indicates the strength of the association between nodes.) (Source: Authors' own)

The structures of the concept maps can be seen in Fig. 5.3a–e. They show clear differences according to the setting analysed, although all of the networks are organised around self-image as a central concept, with which other characteristic images are linked and the components of AA are related. The other images that are associated with self-image are generally positive (positive image, personal development, competence, and usefulness), except for the concept map for nursing homes, where the negative image and image of presumed burden for close family appear.

All of the concept maps incorporate the AA concept into the definition of self-image regarding old age, always in relation to a positive image of personal development, except for those living in nursing homes. Members of this group assessed their self-image and form of ageing, positively or negatively, as a matter of personal character, spirit and coping with the circumstances:

> Being enthusiastic, coming and going, moving and keeping going. Because if you sink, well (…). The time will come, but while you're eager to do what you want to, then go for it. (D1, GD-RES, woman, 90, urban)

In contrast, AA was a central aim for cohousing residents, and they used their form of accommodation to support it:

> I have plans to … travel (…), go to concerts, go to the theatre, take part in everything that's going on here … I don't know, really being a member of this big family, but actively. No, not isolated. We have our independence … which I really like … but taking part. (D105, CH, woman, 73, urban)

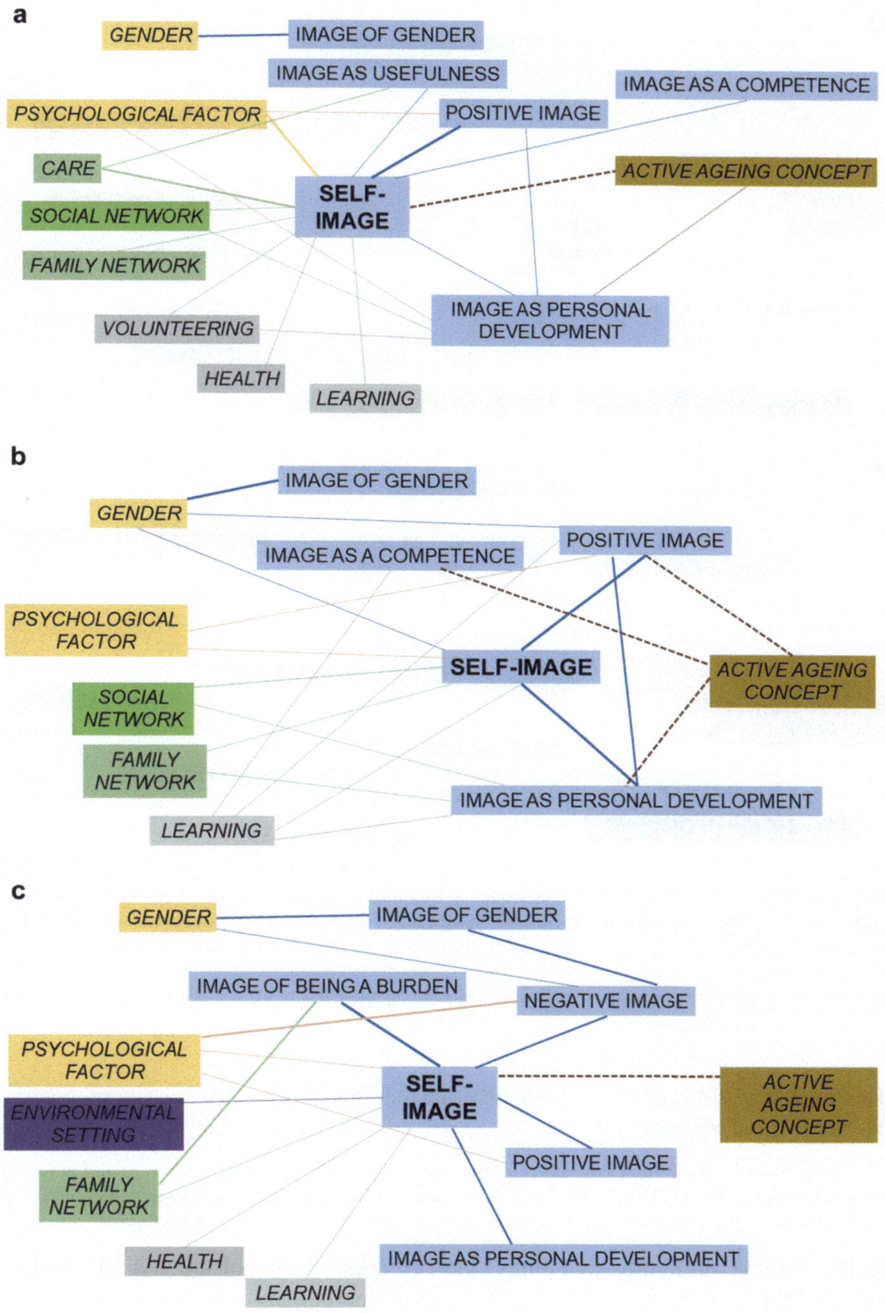

Fig. 5.3 Concept maps drawn up based on discourses extracted from each setting. (**a**) Family dwelling, (**b**) Older adult recreation centres, (**c**) Nursing home, (**d**) Cohousing, (**e**) Cohousing key informants

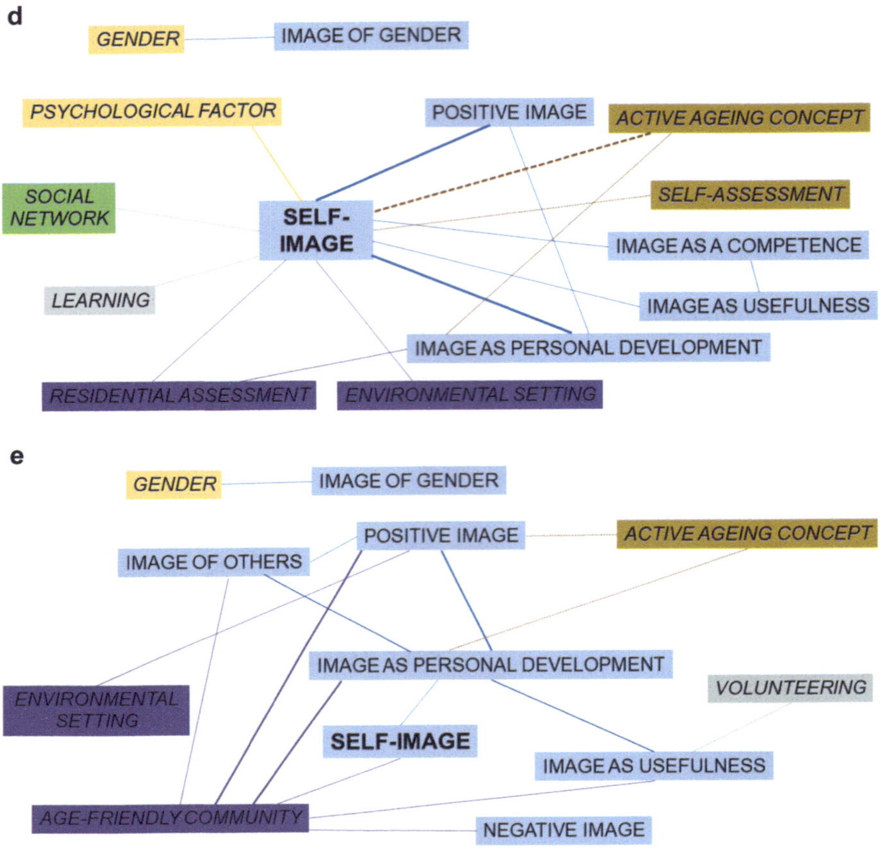

Fig. 5.3 (continued) The colours of the nodes in this figure refer to those used before for the bars in the Fig. 5.2. Also, pale blue is added to the nodes referring to type of image; in these figures, the style and thickness of the lines indicates the strength of the association between nodes. (Source: Authors' own)

The cohousing setting was also the only one that included a self-assessment regarding the form of ageing:

> I think that it's satisfying to the person; things you haven't been able to do before and now you can. You have to change, but you always have to be consistently doing new things (…), I think that people should always be active (…). (D87, CH, man, 75, intermediate)

In the context of family dwellings, the discourses that constructed self-image included three of the four pillars of AA: health, participation (voluntary) and lifelong learning. Health, or a lack of health, was a clear threat to an active self-image:

> My activity has always been going to the bar to play cards and that kind of thing, but three years ago I came down with this illness, so now I'm calmer and better at home, and I'm always worried about my wife, who I can't leave alone very much either. (D110, FD, man, 78, rural)

In contrast and whether voluntary or as an expression of lifelong learning, participation reinforced the self-image of personal development in the discourses of older adults living in family dwellings:

> Yes, the thing is that I always like to try to help (...) the first time I stopped working was when I looked, well, I offered to volunteer at the Association and you know, I was really happy there for some years. (D101, FD, woman, 61, urban)

This search for personal development was even clear among people living at home who joined older adult recreation centres as users, becoming involved in workshops, courses and activities that they had been saving for their retirement:

> And then I joined up with a group of friends who went to the facility to play cards. The time came when it wasn't enough for me and I said to myself that I'd rather study, do things, because (...) my priorities change. (D6, DG-RC, woman, 69, urban)

But in addition to personal development, this involvement in older adult recreation centres entailed a strengthening of the self-image of competence and motivation, which was closely related to performing activities for others:

> [In the] workshop, I find it gratifying to see people who are happy, who, even if they don't know each other at all, are able to exchange views, not just about the book we're borrowing and reading, no; about their personal lives, and they learn from one another. (D6, DG-RC, woman, 71, urban)

> (...) I don't give the classes, we share them out. We're not teachers here, none of us. There are people who don't know anything; fine, they can learn as they go along and otherwise, we'll all learn (...). (D2- DG-RC, mixed, man, 64, urban)

Learning was fundamental to defining the self-image of personal development and competence among users of older adult recreation centres, but it was also essential for older adults living in nursing homes, as settings that incorporate this kind of service as a therapeutic activity for residents:

> If you discover something now, it makes you feel more enthusiastic, more ... I don't know, more eager, more excited. (D4, GD-RES, mixed, woman, 88, urban)

Cohousers demanded learning as part of the aims of collaboration and AA that are pursued in these settings:

> You don't think about age here. You think about the person here. You're the person and you have ... a need to live. And so every day you're thinking from when you get up: I'm going to do this, I'm going to do that (...). (D97, CH, woman, 84, urban)

Participating as a volunteer was essential in the discourses supplied by cohousing key informants in relation to the image of usefulness of older adults:

> This volunteer group here, whenever there's anything going on they call, a bus comes for them and they go to the village, collect people who are worse off than us, take them out in

wheelchairs on trips all over the place, and we're taking part in that. (D67, CH-Resp, man, intermediate)

As a transversal determinant of AA, gender was only associated with the image of gender among older adults living in family dwellings or in cohousing or among cohousing key informants, and it was generally exclusively linked with the gender of the informant. However, gender played a key role in the construction of self-image among older adults who were users of older adult recreation centres and also among those living in nursing homes. For these latter two settings, the image of gender and the gender determinant were also related to a positive image in older adult recreation centres:

> I've had to take care of all the children and I haven't had any time for myself. Well now I'm going to chat with the people who go to the facility, I'll share my experiences for a while because I need to get them off my chest … And in short, to get to the point: [women] accept themselves as they are at that time, which is important for active ageing. (D6, DG-RC, woman, 71, urban)

> [Women] have more capacity for sacrifice and, I suppose we might say, a strength that we don't have. (D3, DG-RC, man, 75, urban)

They could be related to a negative image in nursing homes:

> [younger women] have more opportunities than we've had (…) working as a cleaner like I was, and looking after my children. And when he left work, he was off playing cards with his friends. That's more selfish. (D1, GD-RES, woman, 77, urban)

> Since [the women] are doing activities all day but we're walking around and … it's different. And they're always together, and there are problems between them. (D5, GD-RES, man, 72, urban)

Other determinants shaping self-image, such as psychological factors and personal/individual characteristics, were a key element in the four settings in which older adults were the subject-protagonists of their discourses, while they did not appear in the discourses of cohousing key informants. Therefore, whether respondents were living in family dwellings, used or did not use recreation centres, or lived in cohousing or nursing homes, the discourses in these four settings insisted on the need not to see old age as a limiting factor and on emphasising a positive image of ageing:

> Depending on the age of each member … they get involved in the activities, and the aches and pains don't matter, they need to take part in something to feel better, not to feel isolated, to have more self-confidence. (D6, DG-RC, woman, 71, urban)

> Yes, because I think I'm pretty positive and well, that's worth a lot to you … I think it's fantastic, cohabiting, for me it's been my salvation. (…) I still see myself as young. (D71, CH, woman, 77, intermediate)

> We have a group, we play and do our own things, and … the truth is, I couldn't be happier. (D1, GD-RES, woman, 90, urban)

Contextual determinants such as environmental setting (together with the residential assessment and age-friendliness of the setting being reported on) were important for older adults who had been forced to move to nursing homes or had decided to move to cohousing:

> I'm someone who gets depressed very often (...) the residential home was great for us, both my husband and I. And we're very happy. (D4, GD-RES, mixed, woman, 84, urban)

> I think the setting is great and the building ... everyone here, I honestly really love them now (...) I went away at Easter (...) and I missed this and the people here. (D71, CH, woman, 77, intermediate)

Cohousing key informants placed an emphasis on the age-friendliness of the community:

> The spirit in which we came here was to have independence and not be alone. Specifically, to coordinate the independence that our apartment gives us, where you have to ring the bell to come in, and the feeling of having the company of the others, because they're people with similar characteristics, culture and ideals to us (...) having company, but at the same time being independent. (D112, CH-Resp, woman, urban)

Some social determinants (family and social networks and the provision of care) were again present in the settings where older adults described their own experiences (family dwelling, cohousing, older adult recreation centre and nursing home), in contrast to the opinion on their experiences provided by the cohousing key informants. Nonetheless, the role of these determinants differed in each of the aforementioned settings.

The social network was key to defining a participative and collaborative self-image among cohousing residents:

> I've got my group (...) I get on very well with them, I get along perfectly (...) we have more similarities and more participation and I have a really good time. I'm also very ... very keen on accompanying those who can't, if I see that they're a little [down] ... I'm concerned. (D86, CH, woman, 79, urban)

Meanwhile, the family network was key for those living in nursing homes, an image linked to the burden that their families might face:

> In my case I'm 85, and I can see that my faculties are weakening. (...) So I've decided [to come] ... to the nursing home for that reason, because my kids can't, but I'm getting along wonderfully here. (D5, GD-RES, man, 84, urban)

Family and social networks were basic social determinants in defining the self-image of personal development and enrichment among users of older adult recreation centres:

> And this has given me the chance to meet people who are desperate, despite their age, to keep improving, to keep learning and to keep progressing as people. (D3, DG-RC, man, 73, urban)

And finally, three determinants (family network, social network and caregiving) defined self-image among older adults living in family dwellings, with a special place for caregiving as a marker of usefulness:

> I try to ignore my age and offer to do all I can for my children and to be involved, because if I don't get out and about (...) I have some friends here, lifelong friends, companions and

friends, and I try to get out with them (…) and to be as active as possible, so far as I can. (D90, FD, woman, 82, urban)

In short, the concept maps of self-image regarding old age and ageing appear to bear a degree of similarity in terms of a more active form of ageing for cohousing residents, users of older adult recreation centres and family dwelling residents, in contrast to a less active form and a more negative self-image among older adults in nursing homes. Meanwhile, cohousing key informants moderated the opinions of cohousing residents.

5.4 Discussion and Conclusions

In terms of concerns for the older population, respect and social inclusion were identified as one of the eight areas of the AFCC paradigm (WHO, 2007). According to this paradigm, "respect and social inclusion deals with the attitudes, behaviour and messages of other people and of the community as a whole towards older people" (WHO, 2007: 9), and it interacts with the other aspects than can potentially shape age-friendliness in the city and in the residential setting. In this context, this chapter has examined self-image, investigating its interrelationship with the pillars and determinants of AA, which is the seminal paradigm for the design and development of the concept of age-friendly cities to encourage the creation of inclusive environments. The chapter has confirmed that self-image among older adults is subject to certain differences based on the different residential settings in which they live.

First, people living in *nursing homes* perceive themselves as less competent. Many of them (mainly men) say that they have not learned anything new during this latter stage of their lives or that they cannot do certain things. This means that they do not show an interest or fully participate in activities or workshops, or seek to pursue their own initiatives. This latter aspect can be linked to an apparently lower interest in seeking personal development, but also to the type of motivation that leads this group to decide to participate, for example, in different workshops or activities. In this case, the main reasons are the mere fact of being distracted or having something to occupy their minds. In this respect, one of the most prevalent symptoms among this population is apathy or low motivation, characterised by a lack of interest, initiative, voluntary activity or capacity to respond to environmental stimuli and a state of blunted enthusiasm (Holtta et al., 2012). This apathy is associated with an increased rate of cognitive and functional impairment, lower quality of life and higher mortality (Vilalta-Franch et al., 2013).

People living in these residential facilities see themselves as more isolated and lonelier and worry that they are a burden to those around them. They are also aware of their health-related limitations and even of their dependency. These individuals have also been observed to incorporate more stereotyped attitudes or images in terms of older adults as a social group. The perceptions of older adults are also correlated with job satisfaction among caregivers (Glasspoole & Aman, 1990). In fact, stereotypes of residents among nursing home professionals (care providers, nurses) can have an adverse impact on effective care, well-being and health outcomes for older people (Reyna et al., 2007), in addition to influencing the type of care, professional behaviour and older adult functioning (Fernandez-Ballesteros et al., 2016).

People living in *family dwellings* have shown certain significant differences when their participation or collaboration in *older adult recreation centres* or other similar institutions or spaces is taken into account. In this case, self-image is generally observed to be more related to seeking self-realisation and personal development. Participation in various workshops or activities is guided precisely by a desire to satisfy personal wishes and interests. Other studies have shown that older adult participation is determined by satisfaction with social and family support, high self-esteem, an adequate socio-economic position and a lack of age-based discrimination in society (Aroogh & Shahboulaghi, 2020).

A greater sense of competence was also perceived among members of this group, who affirmed their ability and desire to learn new skills, and even to promote, take part in or organise various initiatives. Although these older adults also assume that this life stage is more solitary, they appear to try to alleviate this isolation and loneliness by participating in activities, which helps and encourages them to see themselves as relatively more active. In fact, older adults are known to be perceived as a competent group when they are involved in paid or volunteer work (Bowen & Skirbekk, 2013). They can also have a reduced sense of being a burden. Finally, it can be stated that these people appear to be aware of the stereotypes presented and reproduced in society as a whole regarding older generations and old age.

In the case of those who live in *family dwellings but do not attend older adult recreation centres* or participate in activities, there is a greater sense of self-identified loneliness and isolation, and a more pessimistic perception of their own ageing. The sense of competence is also clearly diminished in these cases. In this respect, greater barriers to social participation have been observed among lonely older adults, arising from the influence of social fears and identity (Goll et al., 2015).

Finally, *cohousing* residents report a self-image similar to that of residents in family dwellings who use older adult recreation centres or similar collaborative and participative spaces; cohousing settings offer and integrate into their own facilities many of these initiatives, and they are developed specifically in order to foster participative and solidarity-based AA (Rojo-Pérez et al., 2021). However, there is more enjoyment and express pleasure observed in this context as a result of living in a community, cohabiting, sharing and having daily communication. Studies specifically highlight the benefits of this residential alternative, which encourages AA by enhancing the living conditions of its older residents (autonomy, mutual support, communication, social participation, self-determination) (Torío et al., 2018). There are also more references and allusions to limitations in terms of health or accessibility, although a positive, can-do attitude always tends to be present.

The experiences of older adults are essential as a fundamental source of analysis via qualitative methods (Sánchez-González & Rodríguez-Rodríguez, 2016). It is undoubtedly impossible to understand AA or the construction of image among older adults without having access to their views and participation in research (Guell et al., 2018). The starting point is the observation that the older population is a heterogeneous group, that their view may depend on how information is obtained, and that this can be affected by multiple factors acting together (Teater & Chonody, 2020). It is also worth assessing the influence of life course on the development of a form of AA (Boudiny, 2013) and future expectations arising from integrating past experience into the present (Clarke & Warren, 2007). Analysing the opinion of older adults makes it possible to establish typologies of active behaviour through qualitative methodologies, which is more customarily done via quantitative analysis (Rodríguez-Rodríguez et al., 2018). And more importantly, while their opinions do not permit confirmation of relationships of cause (determinant) and effect/result (activities), they do make it possible to identify some of the challenges emerging from the discourses of those who are ageing (Stenner et al., 2011).

Self-image differs depending on the residential environment in which the person is living. These differences in self-perception might first be related to the *determinants* of AA; fundamentally, those of a personal, social and health-related nature, which might be playing an important role in the decision to live in one setting or another, or in whether the person takes part in activities, workshops and initiatives. On one hand is the effect of mechanisms related to *personality*, with personal aspirations and beliefs (self-enhancement and transcendence (Blawert & Wurm, 2021); responsibility in using experience when making fully autonomous decisions in economic settings (Moulaert et al., 2016); the symbol of autonomy and control over activities (Bigonnesse & Chaudhury, 2020), with perceived freedom and support in the setting; in contrast to factors external to the individual (Teater & Chonody, 2020). On the other hand, both *physical or health-related limitations* and the resulting dependency or required support also play a role. In this regard, nursing homes offer an appropriate context to assess resident needs and tasks to cope with their physical and functional abilities, often aimed at optimising social inclusion within the setting (Kim et al., 2021).

Other determinants, such as *economic* factors and the *physical environment*, could come into consideration when choosing the most appropriate residential option, which impacts self-image among older adults. For example, there may be issues regarding the possibility of covering the costs of living in one place or another, or the accessibility and adaptability of each place, or finding personal support and use of community services (Bigonnesse & Chaudhury, 2020). These are all aspects that underpin a residential strategy; that is, the decision to live in a family dwelling or in cohousing. These are essentially individual determinants, some functional and dependent on resources and abilities, routines and activities, and other symbolic and based on personality, values, and relationships (Boccacin & Bramant, 2015), without losing sight of the multilevel perspective in which the neighbourhood environment is important (Bigonnesse & Chaudhury, 2021; Moulaert et al., 2016).

The circumstances observed among those living in family dwellings and using older adult recreation centres, in addition to those living in cohousing, result in a more positive and competent self-image, linked to personal empowerment and a sense of purpose in life through activities, social interaction and mutual support (Lewis et al., 2020). Said self-image may contribute to a more satisfactory and even a less stereotyped life, in response to life circumstances constructed over the course of a lifetime (Clarke & Warren, 2007), and to personal motivations (Walker, 2018). It is not a universal solution, however (Bigonnesse & Chaudhury, 2020), as a result of difficulties with adaptation (Lewis et al., 2020). The same could probably be said for their perception of old age as a state or of their personal ageing process, which offers new opportunities for personal development (Blawert & Wurm, 2021), at the same time as constructing the "right place" to live (Sixsmith et al., 2017). Although it is not the most common situation, a certain positive image is possible in the context of nursing homes when personal autonomy is fostered, beyond physical limitations (Van Malderen et al., 2013), in order to develop rewarding leisure activities (Van Malderen et al., 2017) and AA (Fernández-Mayoralas et al., 2015).

In this regard, the WHO's plan for Global Age-Friendly Cities (WHO, 2007) could play a fundamental role as a paradigm involving both physical infrastructure and social aspects to facilitate ageing in place, in its broad conception of satisfaction of basic needs and community participation (Scharlach & Lehning, 2013). "Social participation" and "civic participation and employment" are significant areas of the plan owing to the mutual benefits for both older adults and the rest of the community. These areas refer to older adults being integrated into and participating in different kinds of activities and workshops and in volunteering or unpaid initiatives or opportunities (all closely connected to two pillars of AA: "participation" and "lifelong learning"). Specifically, these aspects have an important relationship with the development of a more positive and satisfied self-image among this group. The existence of and participation in these kinds of spaces, initiatives, and opportunities can foster the development of more positive attitudes and beliefs in society regarding older adults, insofar as they demonstrate and place value on the competence and ability of older adults to show initiative and commitment and to serve the community, as well as promoting their inclusion.

Second, there is a relationship with the area of "respect and social inclusion." This might be more precisely strengthened through the development of specific projects aimed at combatting discriminatory behaviour or ageism, which can directly impact self-image among older adults. Some studies have found this dimension to be strongly associated with the areas relating to a sense of community (Yu et al., 2019). This construct, in its broad meaning of a sense of identity, belonging and attachment to place, social and community participation, neighbourhood interaction, and satisfaction of needs (French et al., 2013; Talò et al., 2014), was a mediating factor between the material conditions of the residential setting and older adult well-being (Zhang et al., 2017). Well-being and quality of life have also been shown to be influenced by perceived built environment, together with sense of community (Guo et al., 2021) and urban design (Wood et al., 2010).

The development and promotion by public authorities of these three areas, in their interaction with the other areas of the AFCC model, would be highly useful and influential in terms of enhancing a *positive and non-stereotyped self-image* among older adults. As observed in this study, cohousing residents engage in these kinds of activities and initiatives and tend to be more open or to have a higher opinion of their freedom, decision-making ability and independence. This residential context, which remains rare in Spain, involves people whose profiles tend to favour creative activities, who foster relationships of solidarity and mutual support as part of "vital communities" (Pani-Harreman et al., 2021; Rodríguez-Rodríguez, 2021; Rojo-Pérez et al., 2021). To a lesser extent, older adults living in family dwellings, particularly in urban spaces, perceive and identify more opportunities as being available to them to foster their personal development or sense of competence.

Older adults living in family dwellings but in a rural setting refer to living in a friendlier environment as regards the areas of "housing" and "outdoor spaces and buildings": their residential environment is apparently better adapted and more accessible. Finally, it would be necessary to integrate nursing homes into their urban and/or rural residential settings, to promote the areas of "social participation", "respect and social inclusion" and "civic participation and employment".

This work contributes added value to the study of image among older adults in relation to how they express their AA, taking into account that the setting in which they live models their behaviour. Their personal abilities and motivation, the links they establish in their community and their activities are some of the factors that influence their image as ageing people. Their experiences are hence a point of reference for this construction of self-image, which requires an in-depth analysis to extract all the meanings that their experiences contain and the relationships that can be constructed around them. In contrast to other analytical approaches, a qualitative analysis offers this possibility.

Certain limitations of the study do not render the analysis less valuable, but should nonetheless be noted for future studies. One of the most common limitations of qualitative studies is sample selection, in this case intentional, but aimed at exploring the forms of AA among older adults depending on their place of residence. Although limited, the selection of informants complies with the established study criteria. In addition, all geographical areas are represented by size (rural, intermediate, urban). In any case, the sample offers a wealth of nuances that would not have emerged using another methodology. These findings contribute to an exploration of something rarely identified in the literature.

In conclusion, this study shows that its aim is appropriately expressed and of great use, in view of the findings obtained, for consideration in public policies targeted at fostering better, friendlier, and more inclusive settings for older adults. As such, there appears to be a need for more in-depth studies of this kind in the future. It would also be useful to monitor certain settings over time, to obtain longitudinal confirmation regarding how they evolve together with the older adults living in them.

Acknowledgements Research funded by: (i) R & D ENCAGEn-CM (CM, ESF-EC, ref. H2019/HUM-5698); (ii) R & D ENCAGE-CM (CM, ESF-EC, ref. S2015/HUM-3367); (iii) R & D & I ENVACES (MINECO-FEDER, ref. CSO2015-64115-R); (iv) R & D & I Long-Term Care Facilities and the COVID-19 Pandemic (Ref. 202010E158); (v) R & D & I QASP (AESI, PI18CIII/00046, ENPY 392/18).

References

Ayala, A., Rodríguez-Blázquez, C., Calderón-Larrañaga, A., Beridze, G., Teixeira, L., Araújo, L., … Forjaz, M. J. (2021). Influence of active and healthy ageing on quality of life changes: Insights from the comparison of three European countries. *International Journal of Environmental Research and Public Health, 18*(8), 4152. https://doi.org/10.3390/ijerph18084152

Bigonnesse, C., & Chaudhury, H. (2020). The landscape of "aging in place" in gerontology literature: Emergence, theoretical perspectives, and influencing factors. *Journal of Aging and Environment, 34*(3), 233–251. https://doi.org/10.1080/02763893.2019.1638875

Bigonnesse, C., & Chaudhury, H. (2021). Ageing in place processes in the neighbourhood environment: A proposed conceptual framework from a capability approach. *European Journal of Ageing*, 1–12. https://doi.org/10.1007/s10433-020-00599-y

Blawert, A., & Wurm, S. (2021). Shifting self-perceptions of ageing: Differential effects of value priorities on self-perceptions of ageing beyond age stereotypes. *European Journal of Ageing, 18*, 257–267. https://doi.org/10.1007/s10433-020-00578-3

Boccacin, L., & Bramanti, D. (2015). Active ageing: From individualistic withdrawal to social Generavity. *Familia, 50*, 111–128.

Boudiny, K. (2013). 'Active ageing': From empty rhetoric to effective policy tool. *Ageing and Society, 33*(6), 1077–1098. https://doi.org/10.1017/S0144686X1200030X

Bowen, C. E., & Skirbekk, V. (2013). National stereotypes of older people's competence are related to older adults' participation in paid and volunteer work. *The Journals of Gerontology, Series B: Psychological Sciences and Social Sciences, 68*(6), 974–983. https://doi.org/10.1093/geronb/gbt101

Butler, R. N. (1969). Age-ism: Another form of bigotry. *The Gerontologist, 9*(4), 243–246. https://doi.org/10.1093/geront/9.4_Part_1.243

Chum, K., Fitzhenry, G., Robinson, K., Murphy, M., Phan, D., Alvarez, J., … McGrath, C. (2020). Examining community-based housing models to support aging in place: A scoping review. *The Gerontologist*. https://doi.org/10.1093/geront/gnaa142

Clarke, A., & Warren, L. (2007). Hopes, fears and expectations about the future: What do older people's stories tell us about active ageing? *Ageing and Society, 27*(4), 465–488. https://doi.org/10.1017/S0144686X06005824

Dehi Aroogh, M., & Mohammadi Shahboulaghi, F. (2020). Social participation of older adults: A concept analysis. *International Journal of Community Based Nursing & Midwifery, 8*(1), 55–72. https://doi.org/10.30476/IJCBNM.2019.82222.1055

Del Barrio, E., Marsillas, S., Buffel, T., Smetcoren, A. S., & Sancho, M. (2018). From active aging to active citizenship: The role of (age) friendliness. *Social Sciences, 7*(8), 134. https://doi.org/10.3390/socsci7080134

Esteve, A., Devolder, D., Rentería, E., & Blanes, A. (2018). La sostenibilidad demográfica del sistema de pensiones en España. *Panorama Social, 28*, 161–172.

Fernandez-Ballesteros, R., Bustillos, A., Santacreu, M., Schettini, R., Diaz-Veiga, P., & Huici, C. (2016). Is older adult care mediated by caregivers' cultural stereotypes? The role of competence and warmth attribution. *Clinical Interventions in Aging, 11*, 545–552. https://doi.org/10.2147/CIA.S96235

Fernández-Ballesteros, R., Olmos, R., Pérez-Ortiz, L., & Sánchez-Izquierdo, M. (2020). Cultural aging stereotypes in European countries: Are they a risk to active aging? *PLoS One, 15*(5), e0232340. https://doi.org/10.1371/journal.pone.0232340

Fernández-Mayoralas, G., Rojo-Pérez, F., Martinez-Martin, P., Prieto-Flores, M. E., Rodríguez-Blázquez, C., Martín-García, S., … Forjaz, M.-J. (2015). Active ageing and quality of life: Factors associated with participation in leisure activities among institutionalized older adults, with and without dementia. *Ageing and Mental Health, 19*(11), 1031–1041. https://doi.org/10.1080/13607863.2014.996734

Fernández-Mayoralas, G., Schettini, R., Sánchez-Román, M., Rojo-Pérez, F., Agulló, M. S., & Forjaz, M. J. (2018). El papel del género en el buen envejecer. Una revisión sistemática desde la perspectiva científica. *Prisma Social, 21*, 149–176.

Fernández-Mayoralas, G., Rojo-Pérez, F., Rodríguez-Rodríguez, V., Sánchez-Román, M., Schettini, R., Rodríguez-Blázquez, C., … Abad Jiménez, F. (2021). Marco teórico y estudio de diseño e implementación de investigación cualitativa en Envejecimiento Activo, Calidad de Vida y Género. In G. Fernández-Mayoralas & F. Rojo-Pérez (Eds.), *Envejecimiento Activo, Calidad de Vida y Género. Las miradas académica, institucional y social* (pp. 47–125). Tirant lo Blanch.

Foster, L., & Walker, A. (2013). Gender and active ageing in Europe. *European Journal of Ageing, 10*(1), 3–10. https://doi.org/10.1007/s10433-013-0261-0

French, S., Wood, L., Foster, S. A., Giles-Corti, B., Frank, L., & Learnihan, V. (2013). Sense of community and its association with the Neighborhood built environment. *Environment and Behavior, 46*(6), 677–697. https://doi.org/10.1177/0013916512469098

Friese, S. (2019). *Qualitative data analysis with ATLAS.TI*. Sage.

Friese, S., Soratto, J., & Pires, D. (2018). Carrying out a computer-aided thematic content analysis with ATLAS.ti. *MMG Working Paper*, (18-02).

Glasspoole, L. A., & Aman, M. G. (1990). Knowledge, attitudes, and happiness of nurses working with gerontological patients. *Journal of Gerontological Nursing, 16*(2), 11–14. https://doi.org/10.3928/0098-9134-19900201-05

Goll, J. C., Charlesworth, G., Scior, K., & Stott, J. (2015). Barriers to social participation among lonely older adults: The influence of social fears and identity. *PLoS One, 10*(2), e0116664. https://doi.org/10.1371/journal.pone.0116664

Guell, C., Panter, J., Griffin, S., & Ogilvie, D. (2018). Towards co-designing active ageing strategies: A qualitative study to develop a meaningful physical activity typology for later life. *Health Expectations, 21*(5), 919–926. https://doi.org/10.1111/hex.12686

Guillemard, A. M. (2013). Le vieillissement actif: enjeux, obstacles, limites. *Retraite et Société, 2*, 17–38.

Guo, Y., Liu, Y., Lu, S., Chan, O. F., Chui, C. H. K., & Lum, T. Y. S. (2021). Objective and perceived built environment, sense of community, and mental wellbeing in older adults in Hong Kong: A multilevel structural equation study. *Landscape and Urban Planning, 209*. https://doi.org/10.1016/j.landurbplan.2021.104058

Hölttä, E. H., Laakkonen, M. L., Laurila, J. V., Strandberg, T. E., Tilvis, R. S., & Pitkälä, K. H. (2012). Apathy: Prevalence, associated factors, and prognostic value among frail, older inpatients. *Journal of the American Medical Directors Association, 13*(6), 541–545. https://doi.org/10.1016/j.jamda.2012.04.005

Hrybyk, G., Rubinstein, R., Eckert, K., Frankowski, A., Keimig, L., Nemec, M., … Doyle, P. (2012). The dark side: Stigma in purpose-built senior environments. *Journal of Housing for the Elderly, 26*(1–3), 275–289. https://doi.org/10.1080/02763893.2012.651384

ILC-BR – International Longevity Centre Brazil. (2015). *Active ageing: A policy framework in response to the longevity revolution*. International Longevity Centre Brazil. Retrieved from http://ilcbrazil.org/portugues/wp-content/uploads/sites/4/2015/12/Active-Ageing-A-Policy-Framework-ILC-Brazil_web.pdf

Kim, H. J., Shin, Y., Ha, J. H., & Kim-Knauss, Y. (2021). Active aging in long-term care facilities in Korea: Beyond the lexical meaning. *Journal of Gerontological Social Work*. https://doi.org/10.1080/01634372.2021.1948940

Levy, B. R. (2017). Age-stereotype paradox: Opportunity for social change. *The Gerontologist, 57*(S2), S118–S126. https://doi.org/10.1093/geront/gnx059

Lewis, N. A., Reesor, N., & Hill, P. L. (2020). Perceived barriers and contributors to sense of purpose in life in retirement community residents. *Ageing and Society*, 1–17. https://doi.org/10.1017/S0144686X20001749

Moulaert, T., Boudiny, K., & Paris, M. (2016). Active and healthy ageing: Blended models and common challenges in supporting age-friendly cities and communities. In Moulaert & Garon (Eds.), *Age-friendly cities and communities in international comparison* (pp. 277–304). Springer.

Pani-Harreman, K. E., van Duren, J. M., Kempen, G. I., & Bours, G. J. (2021). The conceptualisation of vital communities related to ageing in place: A scoping review. *European Journal of Ageing*, 1–14. https://doi.org/10.1007/s10433-021-00622-w

Pérez Díaz, J., & Abellán García, A. (2018). Envejecimiento demográfico y vejez en España. *Panorama Social, 28*, 11–47.

Reyna, C., Goodwin, E. J., & Ferrari, J. R. (2007). Older adult stereotypes among care providers in residential care facilities: Examining the relationship between contact, education, and ageism. *Journal of Gerontological Nursing, 33*(2), 50–55. https://doi.org/10.3928/00989134-20070201-08

Rodríguez-Rodríguez, V. (2018). El marco de las políticas de vejez en Europa. *Panorama Social, 28*, 147–160.

Rodríguez-Rodríguez, V. (2021). Participación en actividades para envejecer activamente. In G. Fernández-Mayoralas & F. Rojo-Pérez (Eds.), *Envejecimiento Activo, Calidad de Vida y Género. Las miradas académica, institucional y social* (pp. 231–266). Tirant lo Blanch. https://editorial.tirant.com/co/libro/envejecimiento-activo-calidad-de-vida-y-genero-9788418329142

Rodríguez-Rodríguez, V., Rojo-Pérez, F., Fernández-Mayoralas, G., & Prieto-Flores, M. E. (2018). ¿Cómo interpretan el envejecimiento activo las personas mayores en España?: Evidencias desde una perspectiva no profesional. *Aula Abierta, 47*(1), 67–78. https://doi.org/10.17811/rifie.47.1.2018.67-78

Rojo-Pérez, F., Fernández-Mayoralas, G., & Rodríguez-Rodríguez, V. (2021). Active ageing and quality of life: A systematized literature review. In F. Rojo-Pérez & G. Fernández-Mayoralas (Eds.), *Handbook of active ageing and quality of life: From concepts to applications* (pp. 63–98). Springer Nature. https://doi.org/10.1007/978-3-030-58031-5

Ronzi, S., Orton, L., Pope, D., Valtorta, N. K., & Bruce, N. G. (2018). What is the impact on health and wellbeing of interventions that foster respect and social inclusion in community-residing older adults? A systematic review of quantitative and qualitative studies. *Systematic Reviews, 7*(26), 1–22. https://doi.org/10.1186/s13643-018-0680-2

Rozanova, J., Miller, E. A., & Wetle, T. (2016). Depictions of nursing home residents in US newspapers: Successful ageing versus frailty. *Aging and Society, 36*(1), 17–41. https://doi.org/10.1017/S0144686X14000907

Saldaña, J. (2021). *The coding manual for qualitative researchers*. Sage.

Sánchez, M., & Hatton-Yeo, A. (2012). Active ageing and intergenerational solidarity in Europe: A conceptual reappraisal from a critical perspective. *Journal of Intergenerational Relationships, 10*(3), 276–293. https://doi.org/10.1080/15350770.2012.699819

Sánchez-González, D. (2009). Contexto ambiental y experiencia espacial de envejecer en el lugar: el caso de Granada. *Papeles de Población, 15*(60), 175–213.

Sánchez-González, D. (2014). Identidad del lugar, envejecimiento y presiones ambientales de la ciudad. Reflexiones desde la gerontología ambiental. In D. Sánchez-González & L. Á. Domínguez (Eds.), *Identidad y espacio público. Ampliando ámbitos y prácticas* (pp. 141–168). Gedisa.

Sánchez-González, D., & Rodríguez Rodríguez, V. (2016). *Environmental gerontology in Europe and Latin America. Policies and perspectives on environment and aging*. Springer.

Sánchez-González, D., Rojo-Pérez, F., Rodríguez-Rodríguez, V., & Fernández-Mayoralas, G. (2020). Environmental and psychosocial interventions in age-friendly communities and active ageing: A systematic review. *International Journal of Environmental Research and Public Health, 17*(22), 8305. https://doi.org/10.3390/ijerph17228305

Scharlach, A. E., & Lehning, A. J. (2013). Ageing-friendly communities and social inclusion in the United States of America. *Ageing and Society, 33*, 110–136. https://doi.org/10.1017/s0144686x12000578

Sidorenko, A., & Zaidi, A. (2018). International policy framework on ageing: Assessing progress in reference to the Madrid International Plan of Action on Ageing. *The Journal of Social Policy Studies, 16*(1), 141–154. https://doi.org/10.17323/727-0634-2018-16-1-141-154-

Sixsmith, J., Fang, M. L., Woolrych, R., Canham, S. L., Battersby, L., & Sixsmith, A. (2017). Ageing well in the right place: Partnership working with older people. *Working with Older People, 21*(1), 40–48. https://doi.org/10.1108/WWOP-01-2017-0001

Stenner, P., McFarquhar, T., & Bowling, A. (2011). Older people and 'active ageing': Subjective aspects of ageing actively. *Journal of Health Psychology, 16*(3), 467–477. https://doi.org/10.1177/1359105310384298

Swift, H. J., Abrams, D., Lamont, R. A., & Drury, L. (2017). The risks of ageism model: How ageism and negative attitudes toward age can be a barrier to active aging. *Social Issues and Policy Review, 11*(1), 195–231. https://doi.org/10.1111/sipr.12031

Talò, C., Mannarini, T., & Rochira, A. (2014). Sense of community and community participation: A meta-analytic review. *Social Indicators Research, 117*(1), 1–28. https://doi.org/10.1007/s11205-013-0347-2

Teater, B., & Chonody, J. M. (2020). How do older adults define successful aging? A scoping review. *The International Journal of Aging and Human Development, 91*(4), 599–625. https://doi.org/10.1177/0091415019871207

Tong, A., Sainsbury, P., & Craig, J. (2007). Consolidated criteria for reporting qualitative research (COREQ): A 32-item checklist for interviews and focus groups. *International Journal for Quality in Health Care, 19*(6), 349–357. https://doi.org/10.1093/intqhc/mzm042

Torío López, S., Viñuela Hernández, P., & García-Pérez, O. (2018). Experiencias de vejez vital. Senior cohousing: autonomía y participación/experiences of active aging. Senior cohousing: Autonomy and participation. *Aula Abierta, 47*(1), 79–86. https://doi.org/10.17811/rifie.47.1.2018.79-86

Tully-Wilson, C., Bojack, R., Millear, P. M., Stallman, H. M., Allen, A., & Mason, J. (2021 first online). Self-perceptions of ageing: A systematic review of longitudinal studies. *Psychology and Aging*, https://doi.org/10.1037/pag0000638.

UN – United Nations. (2002). *Political declaration and Madrid International Plan of Action on Ageing* (p. 58). United Nations.

Van Malderen, L., Mets, T., De Vriendt, P., & Gorus, E. (2013). The active ageing–concept translated to the residential long-term care. *Quality of Life Research, 22*(5), 929–937. https://doi.org/10.1007/s11136-012-0216-5

Van Malderen, L., De Vriendt, P., Mets, T., & Gorus, E. (2017). Active ageing in the nursing home: Could participatory action research provide the answer? *Action Research, 15*(3), 239–257. https://doi.org/10.1177/1476750316636668

Vilalta-Franch, J., Calvó-Perxas, L., Garre-Olmo, J., Turró-Garriga, O., & López-Pousa, S. (2013). Apathy syndrome in Alzheimer's disease epidemiology: Prevalence, incidence, persistence, and risk and mortality factors. *Journal of Alzheimer's disease: JAD, 33*(2), 535–543. https://doi.org/10.3233/JAD-2012-120913

Walker, A. (2018). Why the UK needs a social policy on ageing. *Journal of Social Policy, 47*(2), 253–273. https://doi.org/10.1017/S0047279417000320

Walker, A. (2021). Active ageing and quality of life: An agenda for scientific research. In F. Rojo-Pérez & G. Fernández-Mayoralas (Eds.), *Handbook of active ageing and quality of life: From concepts to applications* (pp. 252–270). Springer, International Handbooks of Quality-of-Life series). Springer Nature. https://doi.org/10.1007/978-3-030-58031-5_15

Walker, A., & Maltby, T. (2012). Active ageing: A strategic policy solution to demographic ageing in the European Union. *International Journal of Social Welfare, 21*(1), S117–S130.

Walsh, K., Scharf, T., & Keatinget, N. (2017). Social exclusion of older persons: A scoping review and conceptual framework. *European Journal of Ageing, 14*(1), 81–98. https://doi.org/10.1007/s10433-016-0398-8

Westerhof, G. J., Miche, M., Brothers, A. F., Barrett, A. E., Diehl, M., Montepare, J. M., … Wurm, S. (2014). The influence of subjective aging on health and longevity: A meta-analysis of longitudinal data. *Psychology and Aging, 29*(4), 793–802. https://doi.org/10.1037/a0038016

WHO – World Health Organization. (2002). Active ageing: A policy framework. World health. *Organization*. https://apps.who.int/iris/handle/10665/67215

WHO – World Health Organization. (2015). *World report on ageing and health 2015*. World Health Organization. https://www.who.int/ageing/events/world-report-2015-launch/en/

WHO – World Health Organization. (2007). *Global age-friendly cities: A guide*. World Health Organization. Retrieved from https://apps.who.int/iris/bitstream/handle/10665/43755/9789241547307_eng.pdf?sequence=1

Wood, L., Frank, L. D., & Giles-Corti, B. (2010). Sense of community and its relationship with walking and neighborhood design. *Social Science & Medicine, 70*(9), 1381–1390. https://doi.org/10.1016/j.socscimed.2010.01.021

Woolf, N. H., & Silver, C. (2017). Qualitative analysis using ATLAS. ti. In *The Five-Level QDATM Method*. Routledge.

Yeh, J. C. (2020). A critical analysis of age-friendly community initiatives (Doctoral dissertation, UCSF). https://escholarship.org/content/qt9z8944d4/qt9z8944d4.pdf

Yu, R., Wong, M., & Woo, J. (2019). Perceptions of neighborhood environment, sense of community, and self-rated health: An age-friendly city project in Hong Kong. *Journal of Urban Health, 96*(2), 276–288. https://doi.org/10.1007/s11524-018-00331-3

Zhang, Z., & Zhang, J. (2017). Perceived residential environment of neighborhood and subjective well-being among the elderly in China: A mediating role of sense of community. *Journal of Environmental Psychology, 51*, 82–94. https://doi.org/10.1016/j.jenvp.2017.03.004

Gloria Fernández-Mayoralas, PhD in Human Geography by the Complutense University of Madrid, is a Research Scientist in the Institute of Economics, Geography and Demography (IEGD), Centre for Human and Social Sciences (CCHS), at the Spanish National Research Council (CSIC). She works in the field of population studies, focusing on living conditions and well-being in old age, particularly on health conditions and physical and social environments for an active ageing with quality of life. Dr. Fernández-Mayoralas belongs to the Research Group on Ageing (GIE-CSIC) and she is also involved in the Spanish Research Group on Quality of Life and Ageing, a multidisciplinary network between CSIC and the Institute of Health Carlos III (ISCIII) aimed the study of the quality of life of older adults living in different residential contexts. From 2016 she coordinates the R & D Programme on Active Ageing, Quality of Life and Gender: promoting a positive image of ageing and old age (ENCAGE-CM and ENCAGEn-CM), a research consortium formed by public and private research institutions, as well as civil society organisations (http://encage-cm.es/); https://scholar.google.com/citations?hl=es & user=eWk2nZ4AAAAJ & view_op=list_works & sortby=pubdate

Vicente Rodriguez-Rodriguez holds a PhD in Geography from the Universidad Complutense, Madrid and currently he is Research Professor at the Centre for Humanities and Social Sciences, Spanish National Research Council. As a member of the Research Group on Ageing (GIE-CSIC), his research focuses on various aspects of ageing of the Spanish population and its social consequences (residential strategies, health and dependency, active ageing, quality of life, qualitative analysis). The latest publications (peer-reviewed papers, books, and book chapters) have focused mainly on active ageing and some other aspects of the social approach to ageing. Recently, Dr. Rodriguez is leading two research initiatives on older people care homes during the pandemic as a response to the CSIC calls on COVID-19. A general view of his CV can be accessed at: www.iegd.csic.es/es/personal/vicente.rodriguez; http://scholar.google.com/citations?user=VXSJA68AAAAJ

Diego Sánchez-González, PhD in Geography and MSc in Social Gerontology from the University of Granada, and associate professor in the Department of Geography at the National Distance Education University (UNED) (Spain). He has been a visiting professor at Spanish and Mexican universities, and a researcher at the Spanish Agency for International Development Cooperation (AECID). He has participated as responsible and evaluator of research projects in Spain, Holland, Switzerland, Israel, Mexico, Colombia, Chile and Uruguay. He has been an advisor to governments on ageing, urban planning, housing and climate change. Research lines: geography of ageing, environmental gerontology, environment, health and quality of life of the elderly; age-friendly cities and communities; health and social services planning; housing; processes of urbanisation and rural depopulation; social development and city; social vulnerability and adaptation to climate change.

Fermina Rojo-Pérez, PhD, is Research Scientist at the Institute of Economics, Geography and Demography (IEGD), in the Spanish National Research Council (CSIC), Madrid, Spain. She carries out her work in the Research Group on Ageing (GIE-CSIC) and in the Spanish Research Group on Quality of Life and Ageing. The domains related to age-friendly settings and living conditions, as facilitators or barriers for ageing well with quality of life, are among its priority research topics. She has coordinated several research projects and networks on active ageing and participates in projects on her subject of interest. Recently, she was the Principal Researcher of the scientific project on "The Construction of Active Ageing in Spain: Self-assessment, Determinants and Expectations of Quality of Later Life" (MINECO-FEDER, ref. CSO2015–64115-R). Currently, she is involved in the study of conditions and behaviour of old people in long-term care settings during the COVID-19 pandemic (CSIC, ref. 202010E158). She belongs to several scientific societies, as ISQOLS (member of the board of directors 2012–2016), AGE (member of the board of directors 2013–2017), ALAP, and SEGG. More information: http://iegd.csic.es/es/personal/fermina.rojo

Chapter 6
Comprehensive Perspective of Care Design for the Accompaniment of the Person During the Life Journey

Álvaro García-Soler, Heitor García-Lantarón, Sara Marsillas Rascado, Elena del Barrio Truchado, Cristina Buiza Bueno, and Pura Díaz-Veiga

Abstract The demographic change brought about by increased life expectancy and the decline in the number of births foresees a society with a higher proportion of older people in the coming decades. This situation poses not only challenges for the sustainability of a fragmented care system but also ethical and philosophical questions about what it means to care. Integrated care is posited as a broad framework to facilitate people receiving the care they need, maintaining continuity and transition of care as their situation changes, receiving holistic care and enabling them to develop their life and care plan. This chapter is approached from the positioning that all people need care at different times of life, as well as the civil and moral responsibility to care for others. Considering ageing not as a synonym of illness and dependence, but as a life process full of possibilities, several projects developed by the Matia Foundation and Matia Institute to accompany people in the ageing process are described. This life journey proposes a continuous line of care, from participation and prevention to care in situations of dependency, linking the objectives of the different projects with a focus on person-centred care. The design proposal aims to integrate citizen participation, friendliness, coordination of socio-health services, ageing in place and living like at home according to the interest of each person and alternative models of care for dependency.

Keywords Age-Friendly Cities and Communities · Ageing in place · Ageing like at home · Person-centred care · Long term care

Á. García-Soler (✉) · H. García-Lantarón · S. Marsillas Rascado · E. del Barrio Truchado · C. Buiza Bueno · P. Díaz-Veiga
Matia Fundazioa and Matia Research Institute, Donostia-San Sebastián, Spain
e-mail: alvaro.garcia@matiafundazioa.eus; heitor.lantaron@matiainstituto.net; sara.marsillas@matiafundazioa.eus; elena.barrio@matiafundazioa.eus; cristina.buiza@matiafundazioa.eus; pura.diaz-veiga@matiafundazioa.eus

E. Pozo Menéndez, E. Higueras García (eds.), *Urban Design and Planning for Age-Friendly Environments Across Europe: North and South*, Future City 19,
https://doi.org/10.1007/978-3-030-93875-8_6

6.1 Introduction

We are currently in the period of demographic transition (Pérez Díaz, 2000), this change is characterized by the increase in life expectancy at birth as opposed to the lower number of births. Thus, we find ourselves in a society with a growing number of people over 65 years old, an age segment that grows faster than the rest of the age structure groups (Instituto Nacional de Estadística, 2020). The percentage of the population aged 65 years and over, which currently stands at 19.6% of the total population in Spain, would reach a maximum of 31.4% around 2050. Life expectancy at birth would reach 85.8 years for men and 90 for women in 2069, with a gain of 4.9 and 3.8 years, respectively, compared to current values (Instituto Nacional de Estadística, 2020).

This population ageing will bring important social and economic opportunities to the country, but also significant challenges for the sustainability of the welfare state (Gobierno de España, 2021), especially with the retirement of the baby boomers on the one hand and the slowing labour workforce growth and rising costs of the pensions on the other (Bloom et al., 2011). The increased longevity of the population is associated with a growing need for long-term care, due to the relationship between age and dependency (Gobierno de España, 2021). While ageing is not synonymous with disease, age-related frailty and compression of morbidity in a society with longer life expectancy pose challenges to maintaining autonomy and participation. Furthermore, in Spain, the care situation falls mainly on the family unit and especially on women, historically overburdening a large part of the population on the basis of gender, on the one hand, and relying on a migrant woman's labour force in unrecognized and unprotected work, on the other hand (Durán, 2018).

The current situation is faced within different models of long-term care, most of them coming from the sanitary systems. The Chronic Care Model (CCM), and its subsequent developments, such as the Expanded Chronic Care Model (Barr et al., 2003), which evolved from the Primary Health Care (PHC) movement in 1978, are focused on the disease and contemplate the social situation and prevention as secondary factors. According to Goodwin et al. (2017), the PHC, CCM and other approaches have been significant steps towards integrated care, but many existing programs continue to use a vertical and disease-oriented approach to care.

In our context, the provision of care is articulated following the division of social, health and community services in a fragmentary manner, with scarce spaces for socio-health coordination and the non-inclusion of the community and the third sector in this action. Within this general framework, the administration has promoted developments to face the care needs of the society, usually from a fragmented biomedical or social point of view (Ministerio de Sanidad Política Social e Igualdad, 2011). The case of the Basque Country, framed in a Mediterranean model, shows the will to evolve towards a more community care model (Zuñiga & Arrieta, 2021), but showing practical difficulties to achieve it. These challenges are faced through coordination of service planning such as the Plan of Integrated Care of Euskadi (Osakidetza/Basque Health Service, 2013) or the Operative Socio-Sanitary Plan of

Bizkaia (2020). However, these plans continue to show a vertical coordination orientation by not structuring a form of coordination between social and health services and leaving out private initiative and the tertiary sector. Initiatives such as friendly cities, the change of the residential care model to a person-centred care model, services from the dependency care law, training programs throughout life, psychological interventions or voluntary action are dissimilarly endorsed in different municipalities.

The socio-sanitary emergency resulting from COVID-19 has deepened the challenges already identified as a consequence of the demographic transition. Thus, the crisis has accelerated a series of challenges for social policies, such as the coordination of systems, the promotion of a new model of person-centred care, the improvement of care quality and metrics to measure it, the reinforcement of home care, the generation of local care ecosystems and the challenge of redefining new professional profiles to offer innovative and efficient social services. In recent years, public policies in Gipuzkoa have taken up the concept of the care ecosystem as a comprehensive approach that embraces long-term care, attending to the premises of integrated care: socio-sanitary coordination, continuity of care and community.

This approach is built with the aim of providing a solution to the fragmentation of care and allowing people to participate in their own care, providing the person with the necessary and coordinated support at different levels of care. One key point in this approach is taking into consideration the perceived need to age at home receiving the necessary care, instead of going to a nursing home as assessed in the main surveys (Del Barrio et al., 2015). In addition, this approach is promising based on previous experiences in terms of the economic impact of a residential place as opposed to receiving care at home. In this way, people's needs coincide with the needs of the systems in terms of the sustainability of the healthcare system, which in its current approach shows signs of exhaustion and the prospects for economic and structural sustainability point to collapse with the ageing of the population and the increase in life expectancy.

In this chapter, we would like to position ourselves in an approach of potentiality instead of disability or dependency, understanding that although there are two sides of the same coin, the deficit approach can propose solutions for their compensation, but not considering the variables of potentiality makes their identification difficult and working with them something collateral. Potentiality should be considered in all stages of life including the most severe conditions. We will address possible life pathways throughout the ageing process and the kind of care, support or prevention needed at each point. In the description of these itineraries, although the increase in frailty associated with ageing and dependency situations is taken into account, emphasis is placed on healthy ageing and the potential that an ecosystem design of prevention and coordinated support has in order to develop a desirable life path from the point of view of health, well-being and participation. Likewise, we propose an itinerary of vital development not exclusively described by the passing of the years, but in which the increase in chronological age is the necessary background in which the meaning of life is framed and the life project is considered (Duppen et al., 2019; Hupkens et al., 2016).

In 2017, the World Health Organization (WHO) launched its Ageing and Health Strategy, which by 2020 aimed to support healthy ageing, align health systems to the needs of older populations, develop systems for providing long-term care, create age-friendly environments and improve measurement, monitoring and understanding of healthy ageing (World Health Organization, 2017).

The chapter raises the design of interventions from a point of view beyond the care understood as the reactive endowment in a situation of need. Adopting the term accompaniment as the actions provided externally from the physical, social and organizational context oriented to care, but also at the disposal of opportunities to lead a full life, with meaning and participation. The design of care as a contextual action that articulate the rights to care and the well-being must be both considered from the integral conception of the person at different levels of analysis: biological organism, behavioural entity interactive with the environment, participating member in terms of citizenship and subject of rights and duties from the ethical point of view. According to this consideration, care should aim to provide an integrated response to the specific objectives that may be raised at each level, since a unitary response is not feasible.

6.2 Purpose

As far as our knowledge reaches, there is no comprehensive design approach from the territorial scale to the dwelling that guides the support according to a life course perspective in which dependency situations are contemplated. The chapter presents the articulation of support in three blocks from the design of inclusive and friendly environments that foster the exercise of active citizenship of the person, the integrated care to provide the people the services they require according to their needs at home and the different residential models for autonomous living when home care is not possible. The life course perspective design is illustrated by the experience of Matia Foundation in developing projects for accompanying older people. Finally, an articulation of the different blocks is proposed with the aim of providing comprehensive support to the person in their life course, trying to link the different supports according to an ecosystemic and longitudinal logic.

6.3 Inclusive and Friendly Cities and Environments

Among the most overarching and preventive solutions in terms of caring for the people is the development of user-friendly environments. The World Health Organization (WHO) defined age-friendly cities as inclusive and accessible cities that optimize opportunities for health, participation and safety to ensure the quality of life and dignity of people as they age (WHO, 2007). This definition is based on the framework for active ageing (WHO, 2002), according to which material and social factors or determinants are involved in this ageing process, which in turn are reflected in urban environments and services (WHO, 2007).

In an age-friendly city, policies, services, settings and structures support and enable people to age well by recognizing the wide range of abilities and resources among older people, anticipating and responding with flexibility to ageing-related needs and preferences, respecting older people's choices and lifestyle options, protecting the most vulnerable and promoting the inclusion and contribution of older people in all areas of community life (WHO, 2007).

Other authors emphasize the participation of older people when defining friendly environments. For Fitzgerald and Caro (2014), an age-friendly community is defined as "a place where older people are actively involved, valued and supported with infrastructure and services that are effectively adapted to their needs". However, a truly age-friendly city does not focus on a single generation, but includes and embraces all generations (WHO, 2007). This approach is also reflected in the design principles of the universal design concept (van Hoof & Kazak, 2018). Ideally, an age-friendly city should be inclusive and provide opportunities for all people living in it, not just older people (van Hoof & Kazak, 2018). Therefore, we understand friendliness as that one characterized by enabling mutually reinforcing social and physical environments, by a participatory model of collaborative governance and, above all, by inclusiveness (Lui et al., 2009).

The Age-Friendly Cities and Communities (AFCC) initiative launched by the WHO in 2007 has the general aim of implementing at the local level the so-called active ageing paradigm (Del Barrio et al., 2018) based on two trends characteristic of the contemporary world population: demographic ageing and the urbanization process. Both converge as a result of an increasing number of older people residing in urban areas worldwide (UN, 2011, 2017). The friendliness movement was born as an attempt to respond to these two trends by making the most of the potential offered by older citizens (WHO, 2007). It is intended to promote a movement of citizen participation led by older people. A leading process from diagnosis of needs to implementation of actions, both in the physical and in the social environment, for generating well-being in the municipality at the same time that cities get adapted to the ageing process (Kalache, 2016; Del Barrio et al., 2016). This type of community intervention program provides a space where people can develop and contribute to society by playing a role in which they recognize themselves, where they continue to maintain their identity.

The development of AFCC adapted to older people's needs has become an important area of work in the fields of health, ageing and public policy. This evolution reflects several broader trends, including the complexity of the demographic change, with the emergence of a wide spectrum of needs among different cohorts in the 50+ year age group, the recognition of the role of the physical and social environment in maintaining older people's quality of life and healthy ageing (Buffel et al., 2018; Noordzij et al., 2019) and the emphasis on community care and neighbourhood support as major goals of health and social care (Buffel et al., 2018). As early as the 1980s, the physical environment was described as one of the domains determining people's well-being (Lawton & Nahemow, 1973), such as accessible proximity services (Wahl et al., 2012) or the provision of nearby bus stops (Zeitler et al., 2012). Current research evidence suggests that this relationship between physical environment and well-being is maintained in old age and that this is an important factor mediating experiences and opportunities in ageing (Lui et al., 2009). Other neighbourhood social characteristics, such as social capital, social cohesion, social support (van Dijk, 2015), belongingness and relationships with neighbours (Cramm & Nieboer, 2015; Norstrand et al., 2012), have also been studied in their relationship with well-being.

AFCC initiatives are driven from the increased participation of citizenship and democratic engagement of older people in the city (Moulaert & Garon, 2016), following Lefebvre's postulates on "right of the city" (Buffel et al., 2012; Phillipson, 2010). It is a fundamental right of human beings and citizenship, which implies the motivation of civil society to recreate the city as part of a common and collective "mission" (Costes, 2011). Therefore, based on a full conception of citizenship, in which all people are recognized (Subirats, 2011), older people have the right to enjoy full citizenship and to participate actively and integrally in society, claiming to be social subjects and being present in the social and political dynamics of each city and community. On this basis, it can be understood that the notion of active citizenship in old age represents a step forward from active ageing to a broader concept that integrates the principle of equality among people regardless of gender, age and need for support (Del Barrio et al., 2020).

The methodology for the promotion of citizen participation in the friendly cities project is based on participatory action research (PAR). In this way, the aim is to promote the participation of community members in the search for co-design solutions to their problems and to help them increase their degree of control over relevant aspects of their lives (increase of power or empowerment) (Balcázar, 2003). Such participation starts from the expression of concerns and the specification of the characteristics of the services and resources needed for a friendly city, thus defining a bottom-up built governance process (Lui et al., 2009).

Matia Institute's experience in this initiative began in 2009 with the diagnosis of the first city in Spain to join AFCC, Donostia-San Sebastian. In 2012, a collaborative work was initiated together with the Basque Government in the implementation of the Euskadi Lagunkoia Network (translated from Basque as "friendly Basque Country"). The work approach of Euskadi Lagunkoia aimed to promote this movement in all towns, environments and cities of the Basque Country, supporting smaller municipalities to join the initiative and promoting a network that integrates municipalities, entities and individuals from the private, public and social spheres in a cross-cutting manner.

An outstanding milestone in the development of the project was established at the end of 2014 when a work plan was defined in collaboration with the Territorial Federation of Provincial Associations of Pensioners and Retirees of Alava, Gipuzkoa and Bizkaia (EuskoFEDERPEN). Its aim was to promote and develop the project from the leadership of the older adults. From that moment on, the older people's associations at territorial level took the lead in generating the impulse in the local associations so that they could assume the leading role and claim the involvement of the city council in the adhesion to the program, following the bottom-up methodology. At present, 67 municipalities in the Basque Country joined this program and work to improve the quality of life of citizens. Each municipality develops an independent, autonomous and flexible process, following some common points for the good development of the project following the PAR methodology. Some of these keys refer to the fact that the origin is from the bottom-up, being the citizens themselves the ones who ask the city council to join this program; it arises from a shared perceived need. This organization is called a promoting group, led by older people, who work in a self-managed and autonomous way; it combines scientific and popular knowledge with technical support when necessary; it is a dialectic process from knowing to intervening; and it is aimed at solving specific problems with the result of empowering the people involved.

The work is carried out in collaboration with city councils and associations, giving a leading role to citizens in general and to older people in particular. In this way, it goes beyond the barriers of active ageing towards active citizenship, promoting the active role of the older people in social and political decisions in a specific territorial framework.

In addition to the work at the local level, in Euskadi Lagunkoia, different actions are developed to advance the friendliness of the territory, involving agents from the private sector, such as shops or pharmacies; from the social sphere, associations of relatives of patients with Alzheimer's disease, associations of older people and others and different public administrations, city councils, provincial councils and government. Some of these initiatives are the friendly establishments and pharmacies program, working through a workshop on the awareness and the keys to a friendly establishment and the community role in detecting risk situations in pharmacies; the construction of dementia-friendly communities, with a citizen awareness brochure on what dementia is, how to detect some symptoms or behaviours and recommendations; or the guide for friendly housing, with simple and low-cost tips to adapt homes to the ageing process and for people with dementia.

However, the AFCC project faces some challenges, including the criticism that the friendliness movement is overly focused on active ageing and that, for instance, it excludes interventions towards people with dementia or physical disabilities (Cieśla, 2019). These identified challenges in this type of projects can only be overcome by developing new forms of empowerment in working with older people; more effective partnerships with key actors at local, regional and national levels; and taking into account aspects such as social inequality, participation, co-production and co-design, multisector and multidisciplinary collaboration and integration of research into public policies (Buffel et al., 2018).

6.4 Community Resources for Integrated Care

The chapter's approach is that during the life itinerary, people should know their rights and be empowered to exercise full citizenship, which is not incompatible with receiving or providing care at some point. According to Sancho (2016), we live in a caring society, and this will become more accentuated over time, so she proposed that caring and being cared for should be included in the daily life personal philosophy. According to data, approximately 30% of older people with dependency live in Spain, having received a dependency degree. In a great part of the cases, the situation of care is dyadic in which someone needs care and a relative becomes a caregiver, usually a woman of middle age (Artiaga Leiras, 2021; Durán, 2018). At this point of the itinerary, people need support to care or be cared. In the case of a person who needs some support or intensive care, the interventions should aim to improve the well-being, make possible for the person to participate as a full citizen in case and prevent further impairment. In the case of the caregiver, the system should make it possible to care minimizing the effects of caregiving such as burden, overload and grief and prevent developing chronic conditions related to care, making possible the exercise of rights, access to the labour market, citizen participation, leisure and free time.

The Law for the Promotion of Personal Autonomy and Care for Dependent Persons (LAPAD 39/2006) (frequently addressed to as the "Dependency Law") (Boletín Oficial del Estado, 2006) although it provides a framework of action for care in terms of rights and sharing of responsibilities, giving a leading role to the state (Martinez Virto & Pérez Eransus, 2018), has limitations both intrinsically and in its application: the appeal to individual responsibility for the management of dependency; the limitation of resources in terms of coverage, rigidity and latency in their endowment; and the scarce promotion of personal autonomy (Artiaga Leiras, 2021; Martínez-Virto & Hermoso-Humbert, 2021) pose practical problems for the comprehensive care of the person.

The main challenge is the operational articulation of the collaborative model from the concept of governance that allows, through the coordination of the different agents, the satisfaction of the main preference of dependent persons to remain at home receiving the care they need. This challenge, which is the main objective identified at national and international level by organizations such as the European Social Network (ESN) or the International Foundation for Integrated Care (IFIC), is a complex challenge in developed societies, which does not have a single solution, nor proven approaches in mature systems with sufficient consistency to propose a gold standard for its implementation.

The preference of living at home in the community for as long as possible also avoids the costly option of institutional care and is therefore favoured by policy makers, health providers and many older people themselves (Wiles et al., 2012). This concept of living at home is usually referred to as ageing in place (AIP). AIP is understood as people's main preference to live in one's own home and community safely, independently and comfortably, regardless of age, income or ability level (Centers for Disease Control and Prevention, 2013). In fact, the preference to age at home stands out as a main issue in the demands of the older people in Europe, since 91% of Europeans believe it is very important to use public budgets for support services that allow the older people to remain at home for as long as possible; 84% of older people in the Basque Country would like to receive care at home in case of need (Del Barrio et al., 2015). This demand is combined with the use of public money to help financially to adapt homes to the new period of life (Abellán &

Esparza, 2009). People's demand on AIP has been addressed from different countries and organizations, becoming part of several initiatives and policies. Since the1980s, most European countries have focused on improving home-care services from a personcentred perspective while reducing and specializing the provision of institutional long-term care facilities (Houben, 2001; Sancho & Rodriguez, 2001).

Aiming to test the integrated care approach for AIP from the social services perspective, the Etxean Bizi project (Matia Instituto Gerontológico, 2020), funded by the Diputación Foral de Gipuzkoa, which translates as "Living at Home", was carried out during 2018–2020. The objective of the project was to identify a sustainable model of integrated social and healthcare focused on people living at home, from the coordination of those services and support necessary to obtain the best quality of life for these people and their family environment. The project was conceived as the natural evolution of Etxean Ondo ("Being Well at Home"), which was a pioneer project (funded by the Basque Government) in implementing the person-centred care approach in Spain (Díaz-Veiga et al., 2014).

The project aimed to integrate and incorporate all the participating parties, all the agents involved in care: families, social services, home care service, primary and specialized healthcare, the important and invisible sector of domestic and care workers, personal assistants, volunteers and community network, proximity services and other unidentified agents that could have relevance. The main coordination effort was carried out through the incorporation of the case manager role into the social services of the municipalities. According to the case management definition of the Case Management Society of America (2021): "The case manager helps identify appropriate providers and facilities throughout the continuum of services, while ensuring that available resources are being used in a timely and cost-effective manner in order to obtain optimum value for both the client and the reimbursement source". In Etxean Bizi, also according to the conception of France and the Netherlands (Schrijvers & Somme, 2017), the case manager only works with complex cases, which, in the life course itinerary, could be broadly defined as the people (including both care recipient and caregiver) who stop to being able to organize their care and life plan due to the complexity of the social situation, the chronic condition or both situations in interaction. The complex case was identified in healthcare terms as highly complex patients who are the least frequent of the population with chronic conditions but the ones who generate the highest cost and healthcare burden.

The relationship between the people exercising their rights and the social services was reformulated according to the new role of the case manager based on respect for the dignity and fulfilment of the person's rights, interests and preferences and not only in the administrative/bureaucratic management of the provision of resources. To this end, the person is "the protagonist". Autonomy is prioritized in its broadest concept. The transformation of the role of the case managers was achieved through intensive accompaniment for them in order to identify complex cases, make the service provision more flexible, search for socio-sanitary coordination spaces and co-create with the person the life and care plan within a PCC approach. Participants in the project improved their quality of life, perceived health and satisfaction with social services among other variables (Matia Instituto Gerontológico, 2020) It is also expected that this model will be cost-effective, generate employment and improve the quality of life of the people who need help, their families and the territories they live in, promoting a movement of friendliness between citizens and their dependent population. In short, the key to the project lies in the coordination and optimization of resources from a conceptual framework centred on people.

6.5 Nursing Homes and Living Like at Home

If a person faces dependency in their life course (29.6% over 65 years of age in Spain) (Ministerio de Sanidad, 2021), there are several chances that he or she needs professional daily support. Cognitive impairment or dementia, number of chronic conditions, three or more activities of daily living dependencies, urinary incontinence, behavioural disturbances or a poor social network are the main predictors of nursing home admission (Banaszak-Holl et al., 2004; Friedman et al., 2005; Pinzón-Pulido et al., 2016).

When ageing at home is not possible, or is not the desired option, nursing homes are one of the most extended living options. Admission to a nursing home requires adaptation to the new environment and lifestyle, which is probably very different from the life the person led at home. However, different authors state that there are good reasons to argue that living in a nursing home and "feeling at home" is not a contradiction (van Hoof et al., 2016). If the wishes of each person are taken into account and individualized plans are offered, people living in residential care can create a feeling of home (Cooney, 2011; Van Hoof et al., 2015), which is a very important aspect in the quality of life of any older person (Del Barrio et al., 2015).

For more than 10 years, we have witnessed a major transformation in the model of care for the older people living in institutions in Spain. Dissatisfaction with the hotel-hospital model has led to a shift to person-centred care approaches, which, without abandoning the provision of professional care, place aspects such as the promotion of autonomy and the preferences of the older people in a very relevant place (Díaz-Veiga et al., 2014; Díaz-Veiga & Sancho, 2013). It becomes necessary to have an alternative model to traditional institutional care, which provides the person living in it with an experience as similar as possible to "being at home" and which guarantees the basic principles of dignity, autonomy and well-being. In this context, when AIP is not possible or desirable, the concept of "ageing like at home" (ALH) is central to the lives of the older people, fundamentally because of the value of the meaning we all attach to home.

In this sense, at Matia Foundation, a transition of care model towards person-centred care has been underway since 2010. Matia Foundation has been exploring both AIP and ALH in different projects during the last 10 years. Different interventions have been designed, developed and implemented to help and accompany professionals, people living in the centres and their families to adapt to the new philosophy and way of working. Regarding ALH, the approach moved from environmental interventions in existing nursing homes (Etxean Ondo Project) to the development of a new model of long-term facility (Usurbil Care Homes).

6.5.1 Etxean Ondo (2011–2014)

Etxean Ondo (which translates as "living well at home") was probably the first project in Spain aiming to respond to the preferences of older people in a situation of dependency in relation to their wishes to live "as at home" in a residential care setting within a PCC approach. It had the main objective of increasing the quality of life of the people living in residential care settings through achieving home-like, comfortable, safe and accessible environments that facilitate the daily life of the residents, integrating their preferences, customs and significant activities. In order to achieve all this, the project was carried out in three steps: awareness-raising; collection of information from residents, caregivers and environments; and implementation of changes. According to the information gathered, an intervention design was developed in three areas: environment, organization and professional behaviour.

In relation to the physical environment, physical spaces were selected that could be adapted to the characteristics of domestic environments, favouring the incorporation of furniture and other significant decorative objects in both public and private spaces.

In relation to the organizational dimensions, volunteer direct care professionals (assistants) were identified to work in the cohabitation units, and the role of reference personnel was incorporated. The ratio of assistants was increased, reducing staff rotation between the different areas of the centres.

With respect to professional aspects, periodic meetings were held between the multidisciplinary technical team and the auxiliary teams, modifying the decision-making process in relation to care. The preferences and tastes of the residents were identified and the development of activities of daily living with meaning and significance for people based on their life history was promoted. Continuous training was provided on the characteristics of person-centred care, for example, individualized training sessions were developed using our own audiovisual material (with viewing of real video recordings of assistant-resident interactions) in order to facilitate the acquisition of competencies that favour the independence and autonomy of the residents.

The organization of the living units revolves around the daily life. To this end, the preferences and tastes of the residents were identified and activities of daily life, with meaning and significance for them, have been incorporated. Thus, to obtain information on values, interests, habits, customs and tastes, the life history of each resident was elaborated. In the event that the person did not have sufficient communicative resources, the collaboration of a close relative was requested. The information obtained, together with that related to the residents' abilities and needs, from the observations made by the assistants in relation to the choices and preferences of the residents in their daily lives, encouraged, initiated or maintained the participation of the residents (taking into account their functional and cognitive abilities). Activities of daily living related to the generation of feeling like at home, such as putting the washing machine on, drying clothes, setting and clearing the table and folding and ironing clothes, are routines that can be closely related to memories of past events or future expectations. Therefore, any type of activity or routine in the home has a high component of belonging as it contributes to creating a sense of familiarity. Through routines on the use of space, people develop a certain intimacy with the physical aspects of the home environment. The concept of "choreography of being-in-place" describes these daily routines, rituals, daily actions, relationships and meanings (Rowles, 1978, 2000). The importance of this concept lies in how it affects the older people due to the changes related to ageing (Werner, 1985). When routines disappear, there is a loss of identity, freedom, self-esteem and relationships with the environment and the community, with a consequent alteration in the perception of home (Dahlin-Ivanoff et al., 2007).

The research approach to this intervention made it possible to identify and develop interventions to adapt the care provided to dependent older people living in gerontological centres to their wishes, tastes and preferences. In addition, it has been possible to establish the effects that the interventions carried out have had on the quality of life of the participating residents, family members and professionals involved. Furthermore, the development of the project has favoured the elaboration of didactic materials, reports and audiovisual documents that constitute resources for the generalization of the new professional practices associated to the preferences of the elderly citizens in a situation of dependency.

6.5.2 Instruments and Training to Foster PCC in Residential Care Settings

In the experience of the Matia Foundation, the change of the residential model implies the development of practices at different levels and a flexible approach with a broad position of idea, trial and error. This approach has its most systematic exponent in the development of applied research in a natural environment, usually following an observational methodology (given the profile of residents in a more advanced deterioration situation). In this approach to generating knowledge on the change of model, challenges also arise such as the absence of instruments dedicated to the assessment of variables that were not contemplated in the traditional model (well-being, engagement, PCC environmental keys), the need to approach changes from a non-hierarchical model working in a bottom-up participatory manner and with funding that is normally scarce. Below are presented several practices, procedures and instruments development which aimed to improve the quality of life of people who lives in a residential care setting.

In conjunction with the development of environmental PCC, it is becoming increasingly important that the activity carried out in the centres is meaningful, individualized and based on the life history, likes and preferences of each person (Buiza & Diaz-Veiga, 2019). There is evidence that there are three important factors that make an activity meaningful: active participation, activity content related to participants' past interests and roles and activities that meet basic psychological needs of identity and belonging (Genoe & Depuis, 2012; Harmer & Orrell, 2008; Phinney et al., 2007).

The professional activity – communication abilities, empathy, respect for dignity, autonomy and promotion of the well-being of the people they care for – is one of the main areas of interest. The professional behaviour is the key element in generating well-being to residents. The work from Kitwood (1997) and further developments such as the Dementia Care Mapping (Bradford Dementia Group, 2005) pointed out the importance of the positive and negative behaviours called in this instrument: malignant social psychology (infantilization, objectification, mockery, etc.) and positive person work (collaboration, validation, recognition, etc.). The Matia Foundation work aiming to promote "good care" in healthcare professionals, in the framework of PCC, was to promote the reflection in professional caregivers working in nursing homes, on their practices in daily routines (hygiene, bedding, food administration, etc.) generating reflection through videos and role-playing exercises, generating a procedure of co-research which aims to share procedures and generate knowledge from different contexts (Matia Instituto Gerontológico, 2021). From this project and based in the knowledge generated in the research we carried out in dining rooms, we developed training materials with ludic component (García-Soler et al., 2020) with which the reference staff would impulse activities to change the environment, the organization and the relational behaviour of their colleagues during the lunch context.

The lunch context is considered of crucial relevance in our approach of PCC. Included as part of a broader institutional project in which actions have been incorporated at the organizational and architectural levels in order to design "at home" environments that enhance the autonomy and well-being of residents and professionals, this research aimed to identify and promote professional practices for personalization and promotion and increase the well-being and quality of life in the dining rooms of residential centres. The practice was developed in the dining rooms of dementia units of seven gerontological centres of the Matia Foundation in Gipuzkoa, in which supporting teams were created for the nurse assistants in order to generate from the bottom new behaviours in line with the PCC. The study was carried out through observational methodology and mapping tools for recording the professional's behaviour. Analysing the video recordings, the results showed an increase of PCC verbal interactions of the professionals with the people with dementia. Related to this, a second study, once achieved the behavioural change through the accompaniment, was aimed to increase the well-being, nutrition and hydration of the people living in the residences. One of the keys identified was developing a participatory action research approach that includes the creation of mixed professional care and research teams. Each centre carried out the implementation of the practices sequentially. A meal session was recorded every 2 weeks for 4 months with a total of eight accompanying recordings, thus allowing time for interventions to take effect. Nutritional status and hydration improved, and several practices were documented in terms of the aim, procedure and results; satisfaction increased in several centres, while not in others, showing different results which can be linked to the different implementation of the practices.

On the other hand and in this same context of PCC, there is a need for the development and validation of instruments for the assessment of well-being; engagement; satisfaction in people with dementia, from a perspective that aims to highlight the effects that the physical and social environment has on the person; and how they are reflected in their well-being. To this end, Matia Foundation designed the LIBE Scale (List of Well-Being Indicators) (Buiza et al., 2020) and RIE (Engagement Recording Instrument) (García-Soler et al., 2014). The instruments included variables relevant to the PCC model that were not contemplated in the more geriatric models, making it possible to assess the efficacy of the interventions that exclusively relied previously in the professional judgment (Table 6.1).

6.6 New Models of Long-Term Care Facilities: Usurbil Care Homes (2017–Present)

Usurbil Project's main focus is on developing a new model of long-term facility, an alternative to the existing institutional model, largely extended in Spain. This change in the focus is clearly evident when we talk about the accommodation itself: We move from shared rooms to individual dwellings. This paradigm shift has been introduced in European countries more than 40 years ago, responding to the real demand of people to protect their privacy and independence. Thus, through this project, Matia Foundation aims to incorporate the reality of how older people want to live into the Basque LTC model in collaboration with the Diputación Foral de Gipuzkoa.

This project is based on the environmental and organizational experiences acquired in Matia Foundation through the different projects previously described. In addition, it is complemented by other European experiences, deeply studied and analysed. The innovative aspect is how, during an interdisciplinary process, the model has been adapted to the Basque Country context and reality on LTC provision. Thus, references, experience and adaptation define the Usurbil Innovative Centre for Homes with Care.

The building is located in the small village of Usurbil (around 6000 inhabitants), which is part of a larger area, Comarca de Buruntzaldea, close to San Sebastian. Thus, it is planned to cover the need of care in the whole area. Usurbil Care Homes offers 110 accommodations for 127 people, divided in seven different living units. Each unit includes between 15 and 20 dwellings and is designed so it can be divided in the future, scaling down to half-size units. The project includes a day centre,

Table 6.1 Matia person-centred care research

Name/description	Area of influence	Key aspects	Results
"Engagement recording instrument"	Behaviour	Observation of the engagement behaviour (Table 6.1)	Instrument for mapping the engagement behaviour of the residents with the social and physical environment (García-Soler et al., 2014)
"Meaningful everyday activity"	Behaviour, environment, organization	Participation, activity content related to participants' past interests and roles and activities	Guide to support the creation of meaningful everyday activity in people with dementia (Buiza & Díaz-Veiga, 2020)
"Well-being assessment"	Residents' behaviour, environment	Observation of well-being: interaction with the environment, interaction with other people, choices and emotional expression	LIBE Scale (List of Well-Being Indicators) (Buiza et al., 2020)
"Put yourself in my position"	Professionals' behaviour	Promoting good care from a PCC point of view: respect for dignity, autonomy and promotion of the well-being of the people they care for	Training materials (Matia Instituto Gerontológico, 2021) Awareness videos (Prieto Sancho, 2017)
"Put yourself in my dining room"	Behaviour, environment, organization	Changing the context and practices in dining rooms to improve quality of life during mealtime	Training materials (García-Soler et al., 2020)
"Eat at ease, live with quality"	Behaviour, environment, organization	Promoting professional practices aimed at personalization and well-being. Verbal interactions	Documentation about practices and procedures that have influence in the behaviour of professionals (García-Soler et al., 2018a)
"Good practices menu"	Behaviour, environment, organization	Bottom-up generation of PCC practices for improvement of mealtimes	Documentation about practices and procedures that have influence on nutritional status and hydration (García-Soler et al., 2018b)

Source: Authors' own elaboration

common spaces for the local community and a café. In addition, as it is located at the end of the village, it is organized around an open space, a new plaza (Fig. 6.1).

The three main guidelines on which the architectural proposal is based are as follows: integration within the urban and social context, social interaction through the common areas and, finally, fostering independence, autonomy and personal identity through the design of individual dwellings. The implications that these three criteria have had on the development of the project are detailed below:

Regarding integration, several strategies have been developed. First, in order to integrate such a large facility in a small rural village, the building has been divided into four smaller buildings. All of them are connected by glass elements, which

Fig. 6.1 General view of the Usurbil Care Homes Innovative Centre. (Source: Authors' own)

allow it to work both as a whole facility and each building independently. In addition, the volumes, scale and architectural language follow the traditional style of farm buildings in the area, assuring a familiar aspect for the residents and the rest of the village community. In essence, including traditional and local materiality, colours and scale, it supports the sense of belonging to the place while it helps create a continuity on lifestyle for future users of the facility. Second, the previously mentioned new plaza creates an intergenerational space by providing good-quality open areas in a part of the village lacking these kinds of spaces. In doing so, the day centre, common spaces for the local community and the café are located in the ground floor, allowing the rest of the citizens of the village, especially those living in the surrounding areas, to benefit from the use of the facilities. The location and quality of these spaces, together with the plaza, allow the area to become attractive. The aim is to facilitate integration by supporting interaction with the local community.

Besides the social interaction supported by the plaza, all the common spaces inside the living units play an important role. The main objective is to improve the quality of life of the older people by reinforcing their autonomy and independence, respecting their privacy and previous life experiences, but at the same time encouraging common activities and social relationships. To this end, the design of the common areas is characterized by flexibility in their use. There are areas for common activities among all the users of the living unit, but also small spaces for individual enjoyment. The objective is to offer different ways of using the space depending on the capabilities of each user, or simply on what they feel like doing at any given moment. The activities to be carried out by the users are related to everyday life, thus maintaining normality and continuity in the way of life. All spaces, both private and common, enjoy one of the great qualities of this location: the views.

When old age no longer allows walking through the neighbourhood, through the village, the visual contact with the outside becomes a fundamental resource to continue enjoying. All common areas are connected to an outdoor terrace that can be used for different activities. In addition, natural lighting is guaranteed in all spaces, with the consequent benefit in the quality of life. The different views towards the plaza, the village or the valley benefits people who, due to their dependency, spend a lot of time indoors.

The new conceptual framework introduces the concept of ALH, replacing the concept of room by individual dwellings. This switch guarantees the existence of a private space in which communication and accompaniment with families, friends or volunteers acquire the necessary sense of "privacy" that the traditional institution cannot achieve. The dwellings follow design criteria largely tested on similar facilities in Europe. First, there is an easily accessible small area with space to store the wheelchair, walker or other aids, which also supports personalization with individual's objects and furniture. This especially helps orientation and identification to those people with dementia. In addition, a small kitchenette is included, which is not used for cooking, but to prepare food or drinks for family and friends. It is a central element in the construction of the meaning of home. The size of the dwelling is determined by the need of space for providing care and support continuity in the lifestyle of the user, as the relationship with the place is established through functional and symbolic transactions. Thus, the main space of the dwelling consists of the bedroom and living room, separated by a closet to promote privacy, but avoiding the use of partitions that reduce space and the feeling of spaciousness. All the space can be furnished and decorated by the user, since personal objects also play an important role on the meaning of home. Finally, the bathroom is located next to the bed and is adapted with a sufficient size to help caregivers develop the necessary tasks.

6.7 Discussion

The design of support for the care of people during their life course is highly complex given the different areas involved, the levels of analysis and the agents and beneficiaries. The scope of the task makes it difficult to be covered by a single initiative in terms of design, but also in terms of financing, political alignment, coordination of agents, citizen participation and outcomes (Tsiachristas & Rutten-van Mölken, 2017).

This challenge usually forces the design itself to be fragmented, setting objectives for specific areas and attending to variables at a single level of analysis. In this constrained context, design decisions can be made that favour integral design even if they do not start from it. Taking into account the needs of the beneficiaries and looking with them at the solutions from a co-creative design, in many cases, will show the requirements for coordination of care, thus incorporating them into the design objectives. Establishing relationships between different agents in the shaping of the projects and structuring procedures that make these relationships possible during development could help open ways for integration. The participatory action research approach, although costly in resources, develops the growing design according to the needs and interests of the people, including into it the idiosyncrasy of the people involved.

Other design approaches can make use of the PCC approach with the intention of starting from a framework of work with a growing corpus of knowledge that contemplates the needs and provides empirical evidence of the interventions developed. The PCC framework also poses an ideological and ethical positioning that must be considered in the development of any type of design, which, in integrative approaches, obliges the designers to value interactions with other agents (e.g. the labour rights of personal assistants and the requirements of people in a situation of dependency).

It is not strictly necessary in terms of intervention, but it could be useful to design interventions with a participatory research approach in order to document developments in a descriptive way but also to establish objectives, indicators and metrics that allow us to monitor the development of the project, its success and foster decision-making from the micro level to the political and administrative level. The theory of change methodology could be helpful to this end for including a structured but flexible way to co-design a joint venture taking into account the voice of the beneficiaries and the requirements and knowledge of the relevant stakeholders in the micro, meso and macro levels (Breuer et al., 2019).

The conceptual approach from a hypothetical life pathway shows us key points of coordination of transversal and longitudinal care, where the need for coordination becomes evident. Integrating the needs of the older person using co-creative instruments such as the customer journey (Tomitsch, 2018) could show the pain points, interaction with the system and agents in his or her life, where the person feels that something could be better. Involving the different people in creative design could provide ideas for tentative solutions that can be tested in the project. The life course itinerary serves us a customer journey approach in which we describe, based on existing gerontological knowledge, possible life situations of a person during the ageing process in which the person could need stimulation, support and care. It is understood that all people need care at some point, but ageing should not be travelled along the path of dependency. The ideal itinerary is the itinerary of healthy and successful ageing, an itinerary that can be enhanced by providing the appropriate services and support at all times so as not to have to require support that could have been foreseen.

In the life course itinerary design, we can go as far as we want, but with prevention in mind, we should not design for old age starting at age 65 years or older. Ageing, from the biological point of view, starts approximately at the end of the 20s, and the main point of pain here could be the difficulty at the societal level of generating healthy habits of physical exercise and cognitive exercise to build cognitive reserve, good nutrition, control over stress or care of social networks to minimize the chances of developing chronic diseases and disorders and avoiding frailty situations. Promoting an active life in which we exercise our rights and duties as citizens to create a better society through participation, co-production of services or co-responsibility and voluntary action to support in the pain points of other people life journeys. At this point, although the older people would not need care in the most functional sense, they are discriminated against or stereotyped and relegated to a secondary role in the labour market and social environment. Ageist behaviours are deeply rooted in society and pose a problem in people's daily lives based on arbitrary criteria associated with living a certain number of years. In such a way, ageism violates people's dignity and can exogenously establish situations of isolation that can cause major problems in the future. The design in these situations has to contemplate the empowerment for citizen participation, continuous training, early awareness through intergenerational projects, the possibilities of developing a healthy life and an urban design that facilitates them, such as super blocks (Moreno-Colom, 2021), or Euskadi Lagunkoia, and Usurbil city design.

According to the contextual models, if people have the opportunities and a stimulating environment to use them, the probability of developing healthy behaviours increases, and thus, the risk of developing frailty is minimized; the design should be aimed at this objective and at attending to frailty and mild degrees of dependence once they exist. In this circumstance, actions are proposed to modify the frailty situation in order to prevent future hospitalizations, dependency or death. Comprehensive individualized plans of a preventive nature, empowering the person in the management of their health, facilitating rehabilitation, identifying risks at key moments (such as falls or hospitalizations) and having a diagnostic category that allows protocols to be articulated are some of the key interventions found in the strategic plans in our environment (Bizkaia, 2020; Hendry et al., 2018a, b; Osakidetza/Basque Health Service, 2013). In addition to the necessary vertical health coordination, alternative gateways to primary medicine such as the case manager in the social services, or community agents, and the third sector are shown to be relevant for socio-sanitary and community coordination. The design should contemplate the creation of socio-sanitary spaces (whether physical, virtual or temporary), shared language and intervention algorithms for case management or efficient referral in a proactive model instead of a reactive model (Ministerio de Sanidad Política Social e Igualdad, 2011).

In case of dependency developed through the ageing process, design should take into account not only the state's leading role through the Dependency Law, but also the coordination of services for integrated care with the aim of AIP. Coordination is feasible if there is political will and initial participation in the design by at least the social and health services. With a view to generating care in the community, it is also necessary to include the private initiative, voluntary action and families. Likewise, case management from the social services is proposed as an effective methodology for the coordination of services and the change of perspective "returning to grassroots social work", attending to the individual under the PCC model (García-Soler et al., 2019; Prieto Sancho, 2020).

Designing interventions in residential settings aimed at changing the residential model to become a person-centred care environment is complex as there are different areas that affect each other (environment, organization and behaviour), and changes in one area could imply undesired changes in the others, if the design does not address integrity. It is also time- and effort-consuming, as it usually involves several people who have been working with other quality standards in a geriatric model for periods of time lasting up to 30 years. Whether making changes to a traditional center or creating a new center, one of the basic concepts around which the design revolves is the conception of the home, the identity and feeling of belonging. However, different authors state that there are good reasons to argue that living in a nursing home and "feeling at home" is not a contradiction (van Hoof et al. 2016; Rubinstein, 1989; Rubinstein & Parmelee, 1992; Despres, 1991; Cooney, 2011); moreover, the integrated design should take into account not only intramural actions but also working with the community. In addition to coordination processes, any integral design should address health aspects within the environment including the urban scale. Projects are usually designed and in turn limited by economic funding and the willingness to act of the different agents.

Ultimately, the design has to take into account aspects such as the need for time for the establishment and consolidation of practices, so it is advisable to design with a conservative time frame and develop sequential projects articulated under a broad framework that allows the development of long-term strategies.

References

Abellán, A., & Esparza, C. (2009). Las demandas de los mayores en Europa. In *Informes Portal Mayores* (p. 94). IMSERSO and CESIC.

Artiaga Leiras, A. (2021). Cuidados comunitarios y gobierno común de la dependencia: las viviendas colaborativas de personas mayores. *Revista Española de Sociología, 30*(2), a29. https://doi.org/10.22325/fes/res.2021.29

Balcázar, F. E. (2003). Investigación acción participativa (IAP): Aspectos conceptuales y dificultades de implementación. *Fundamentos en Humanidades, 4*, 59–77.

Banaszak-Holl, J., Fendrick, A. M., Foster, N. L., Herzog, A. R., Kabeto, M. U., Kent, D. M., & Langa, K. M. (2004). Predicting nursing home admission: Estimates from a 7-year follow-up of a nationally representative sample of older Americans. *Alzheimer Disease and Associated Disorders, 18*(2), 83–89. https://doi.org/10.1097/01.wad.0000126619.80941.91

Barr, V. J., Marin-Link, B., Underhill, L., Dotts, A., Ravensdale, D., & Salivaras, S. (2003). The expanded chronic care model. *Healthcare Quarterly, 7*(1), 73–82.

Bloom, D. E., Boersch-Supan, A., McGee, P., & Seike, A. (2011). Population aging: Facts, challenges, and responses. *Benefits and Compensation International, 41*(1)(71), p. 22.

Boletín Oficial del Estado. (2006). *BOE-A-2006-21990 Ley 39/2006, de 14 de diciembre, de Promoción de la Autonomía Personal y Atención a las personas en situación de dependencia.* Retrieved from https://www.boe.es/buscar/act.php?id=BOE-A-2006-21990

Bradford Dementia Group. (2005). *Dementia care mapping DCM* (8th ed.). University of Bradford.

Breuer, E., Comas-Herrera, A., Docrat, S., Freeman, E., Schneider, M., & STRiDE. (2019). *Theory of change workshops: Guidance and resources*. WP1: Theory of Change, (1).

Buffel, T., Phillipson, C., & Scharf, T. (2012). Ageing in urban environments: Developing 'age-friendly' cities. *Critical Social Policy, 32*, 597–617.

Buffel, T., Handler, S., & Phillipson, C. (2018). *Age-friendly cities and communities: A global perspective*. Policy Press.

Buiza, C., & Díaz-Veiga, P. (2019). La actividad cotidiana significativa para personas con demencia. *Neurama, 16*(1), 59–65.

Buiza, C., & Díaz-Veiga, P. (2020). *Guía para apoyar la creación de Actividad Cotidiana Significativa en personas con demencia.* Retrieved from https://www.matiainstituto.net/es/publicaciones/guia-para-facilitar-la-realizacion-de-actividades-cotidianas-significativas-con

Buiza, C., García-Soler, A., Díaz-Veiga, P., Arriola, E., & Fernández, E. (2020). Desarrollo y validación de un instrumento observacional para la evaluación del bienestar en personas con demencia desde la perspectiva de la atención centrada en la persona: Listado de Indicadores de Bienestar. *Revista Española de Geriatría y Gerontología, 55*(3), 147–155. https://doi.org/10.1016/j.regg.2019.06.006

Case Management Society of America. (2021). *What is a case manager.* Case Management Society of America. Retrieved from https://cmsa.org/who-we-are/what-is-a-case-manager/. Last access 18 June 2021.

Centers for Disease Control and Prevention. (2013). *Healthy places terminology: Aging in place.* Retrieved from http://www.cdc.gov/healthyplaces/terminology.htm. https://doi.org/10.2307/3090233

Cieśla, A. (2019). Book review: Age-friendly cities and communities in international comparison: Political lessons, scientific avenues, and democratic issues. *Frontiers in Sociology, 4.* https://doi.org/10.3389/FSOC.2019.00002

Cooney, A. (2011). 'Finding home': A grounded theory on how older people 'find home' in long-term care settings. *International Journal of Older People Nursing, 7,* 188–199. https://doi.org/10.1111/j.1748-3743.2011.00278.x

Costes, L. (2011). Del 'derecho a la ciudad' de Henri Lefebvre a la universalidad de la urbanización moderna. *Urban NS02,* 89–100.

Cramm, J. M., & Nieboer, A. P. (2015). Social cohesion and belonging predict the Well-being of community-dwelling older people. *BMC Geriatrics, 21,* 15–30. https://doi.org/10.1186/s12877-015-0027-y

Dahlin-Ivanoff, S., Haak, M., Fange, A., & Iwarsson, S. (2007). The multiple meaning of home as experienced by very old Swedish people. *Scandinavian Journal of Occupational Therapy, 14,* 25–32.

del Barrio, E. (2016). *Ciudades y comunidades amigables con las personas mayores. La Visión Territorial y Sostenible Del Desarrollo Local: Una Perspectiva Multidisciplinar,* Vol. 2, 2016 (Sociedad, Gobierno, Métodos, Herramientas), ISBN 978–84–370-9933-0, Págs. 141–163, 2, 141–163. https://dialnet.unirioja.es/servlet/articulo?codigo=5633166

Del Barrio, E., Pinzón, S., Sancho, M., & Peña, F. G. (2020). Ciudadanía activa y personas mayores: viejos conceptos, nuevos abordajes. Una revisión sistemática y metasíntesis cualitativa. *Revista Española de Geriatría y Gerontología, 55*(5), 289–299.

Del Barrio, E., Mayoral, O., & Sancho, M. (2015). *Estudio sobre las condiciones de vida de las personas 55 y más años en Euskadi.* Retrieved from http://www.bibliotekak.euskadi.eus/WebOpac

Del Barrio, E., Marsillas, E., Buffel, T., Smetcoren, A. S., & Sancho, M. (2018). From active aging to active citizenship: The role of (age) friendliness. Active aging and wellbeing: Advancement of interdisciplinary research. *Social Science, 7*(8), 134. https://doi.org/10.3390/socsci7080134

Despres, C. (1991). The meaning of home: Literature review and directions for future research and theoretical development. *The Journal of Architectural and Planning Research, 8*(2), 96–115.

Díaz-Veiga, P., & Sancho, M. (2013). Residencias, crisis y preferencias de las personas mayores. *Revista Española de Geriatría y Gerontología, 48*(3), 101–102. https://doi.org/10.1016/j.regg.2013.01.003

Díaz-Veiga, P., Sancho, M., García-Soler, Á., Rivas, E., Abad, E., Suárez, N., & Yanguas, J. (2014). Efectos del Modelo de Atención Centrado en la Persona en la calidad de vida de personas con deterioro cognitivo de centros gerontológicos. *Revista Española de Geriatría y Gerontología, 49*(6), 266–271. https://doi.org/10.1016/j.regg.2014.06.003

Diputación Foral de Bizkaia. (2020). *Plan Operativo Sociosanitario de Bizkaia.*

Duppen, D., Machielse, A., Verté, D., Dury, S., De Donder, L., & Consortium, D. S. (2019). Meaning in life for socially frail older adults. *Journal of Community Health Nursing, 36*(2), 65–77. https://doi.org/10.1080/07370016.2019.1582160

Durán, M. Á. (2018). *La riqueza invisible del cuidado* (Vol. 7, p. 520). Universitat de València.

Fitzgerald, K., & Caro, F. (2014). An overview of age-friendly cities and communities around the world. *Journal of Aging & Social Policy, 26*, 1–18.

Friedman, S. M., Steinwachs, D. M., Rathouz, P. J., Burton, L. C., & Mukamel, D. B. (2005). Characteristics predicting nursing home admission in the program of all-inclusive care for elderly people. *Gerontologist, 45*(2), 157–166. https://doi.org/10.1093/geront/45.2.157

García-Soler, Á., Etxaniz, N., & Díaz-Veiga, P. (2020). *Ponte en mi comedor*. Instituto Matia. Retrieved from https://www.matiainstituto.net/es/investigacion/investigacion-colaborativa/ponte-en-mi-comedor. Last access 20 June 2021.

García-Soler, A., Buiza, C., Vaca, R., & Ansorena, X. (2014). Development of a tool for recording involvement/engagement in activities in the systematic observation of elderly persons with cognitive impairment. *Revista Española de Geriatría y Gerontología, 49*(2). https://doi.org/10.1016/j.regg.2013.02.003

García-Soler, Á., Etxaniz, N., Díaz Veiga, P., & Sancho, M. (2018a). *Menú de buenas prácticas*. Instituto Matia. Retrieved from https://www.matiainstituto.net/es/proyectos-de-investigacion/proyectos/menu-de-buenas-practicas. Last access 20 June 2021.

García-Soler, Á., Etxaniz, N., Díaz-Veiga, P., & Sancho, M. (2018b). *Comer bien, vivir con calidad*. Instituto Matia. Retrieved from https://www.matiainstituto.net/es/proyectos-de-investigacion/proyectos/comer-bien-vivir-con-calidad. Last access 20 June 2021.

García-Soler, Á., Etxaniz, N., Azurmendi, M., Zapiain, A., & Sancho, M. (2019). Etxean Bizi. Estar en casa bien: integración de servicios para envejecer en casa. *International Journal of Integrated Care, 19*(4), 480. https://doi.org/10.5334/ijic.s3480

Genoe, M. R., & Dupuis, S. L. (2012). The role of leisure within the dementia context. *Dementia, 13*, 33–58.

Gobierno de España. (2021). *España 2050. Fundamentos y propuestas para una estrategia nacional a largo plazo.*

Goodwin, N., Stein, V., & Amelung, V. (2017). What is integrated care? In *Handbook integrated care* (pp. 3–23). Springer. https://doi.org/10.1007/978-3-319-56103-5_1

Harmer, B. J., & Orrell, M. (2008). What is meaningful activity for people with dementia living in care homes? A comparison of the views of older people with dementia, staff and family carers. *Aging and Mental Health, 12*(5), 548–558.

Hendry, A., Carriazo, A. M., Samaniego, L. L., Espinosa, J. M., O'caoimh, R., Liew, A., & Giammarchi, C. (2018a). *Models of care for frailty-report on good practices.*

Hendry, A., Carriazo, A. M., Vanhecke, E., & Rodríguez-Laso, Á. (2018b). Integrated care: A collaborative advantage for frailty. *International Journal of Integrated Care, 18*(2), 1–1. https://doi.org/10.5334/ijic.4156

Houben, P. (2001). Changing housing for elderly people and coordination issues in Europe. *Housing Studies, 16*(5), 651–673.

Hupkens, S., Machielse, A., Goumans, M., & Derkx, P. (2016). Meaning in life of older persons. *Nursing Ethics, 25*, 1–19.

Instituto Nacional de Estadística. (2020). *Proyecciones de Población 2020–2070, 2070* (pp. 1–19). Retrieved from https://www.ine.es/prensa/pp_2020_2070.pdf

Kalache, A. (2016). Active ageing and age-friendly cities—A personal account. In T. Moulaert & S. Garon (Eds.), *Age-friendly cities and communities in international comparison* (pp. 65–77). Springer.

Kitwood, T. M. (1997). *Dementia reconsidered: The person comes first*. Open University Press.

Lawton, P., & Nahemow, L. (1973). Ecology and the aging process. The psychology of adult development and aging. In L. Eisdorfer (Ed.), *The psychology of adult development and aging* (pp. 619–674). American Psychological Association.

Lui, C., Everingham, J.-A., Warburton, J., Cuthill, M., & Bartlett, H. (2009). What makes a community age-friendly: A review of international literature. *Australasian Journal on Ageing, 28*, 116–121.
Martinez Virto, L., & Pérez Eransus, B. (2018). El modelo de atención primaria de Servicios Sociales a debate: Dilemas y reflexiones profesionales a partir del caso de Navarra. *Cuadernos de Trabajo Social, 31*(2), 333–354. https://doi.org/10.5209/cuts.55168
Martínez-Virto, L., & Hermoso-Humbert, A. (2021). Hacia un modelo público de cuidados en la comunidad. *Revista Española de Sociología, 30*(2), a26. https://doi.org/10.22325/fes/res.2021.26
Matia Instituto Gerontológico. (2020). *ETXEAN BIZI*.
Matia Instituto Gerontológico. (2021). *Ponte en mi lugar*. Retrieved from https://www.matiainstituto.net/es/investigacion/investigacion-colaborativa/ponte-en-mi-lugar. Last access 20 June 2021.
Ministerio de Sanidad. (2021). Informe Anual del Sistema Nacional de Salud 2019. *INFORMES, ESTUDIOS E INVESTIGACIÓN 2020*.
Ministerio de Sanidad Política Social e Igualdad. (2011). *Libro blanco de la coordinación sociosanitaria en España*. Ministerio de Sanidad, Política Social e Igualdad.
Moreno-Colom, S. (2021). Construyendo comunidad desde lo público: el caso de las Superilles Socials. *Revista Española de Sociología, 30*(2), a27. https://doi.org/10.22325/fes/res.2021.27
Moulaert, T., & Garon, S. (2016). Introduction: Toward a better understanding of AFCC. In *Age-friendly cities and communities in international comparison* (Vol. 14). Springer.
Noordzij, J. M., Beenackers, M. A., Roux, A. V. D., & Van Lenthe, F. J. (2019). Age-friendly cities: Challenges for future research. *Bulletin of the World Health Organization, 97*, 436–437.
Norstrand, J. A., Glicksman, A., Lubben, J., & Kleban, M. (2012). The role of the social environment on physical and mental health of older adults. *Journal of Housing for the Elderly, 26*, 290–307.
Osakidetza/Basque Health Service. (2013). *Plan de Atención Integrada en Euskadi*. Retrieved from https://www.osakidetza.euskadi.eus/contenidos/informacion/buen_gob_planes/eu_def/adjuntos/Plan_de_atencion_integrada_en_Euskadi.pdf
Pérez Díaz, J. (2000). *Cuidados de larga duración* (p. 7). Retrieved from https://www.msssi.gob.es/estadEstudios/estadisticas/sisInfSanSNS/pdf/Cuidados_de_Larga_Duracion.pdf
Phillipson, C. (2010). Ageing and Urban Society: Growing old in the "century of the City". In C. Phillipson & D. Dannefer (Eds.), *The SAGE handbook of social gerontology* (pp. 597–606). Sage Publications Ltd.
Phinney, A., Chaudhury, H., & O'Connor, D. L. (2007). Doing as much as I can do: The meaning of activity for people with dementia. *Aging and Mental Health, 11*(4), 384–393.
Pinzón-Pulido, S., Garrido Peña, F., Reyes Alcázar, V., Lima-Rodríguez, J. S., Raposo Triano, M. F., Martínez Domene, M., & Alonso Trujillo, F. (2016). Factores predictores de la institucionalización de personas mayores en situación de dependencia en Andalucía. *Enfermería Clínica, 26*(1), 23–30. https://doi.org/10.1016/j.enfcli.2015.08.003
Prieto Sancho, D. (2017). *(9) "Ponte en mi lugar" – Trailer – YouTube*. Retrieved from https://www.youtube.com/watch?v=ldq2Z6Gf1H4. Last access 20 June 2021.
Prieto Sancho, D. (2020). *Acercamiento cualitativo al período experimental de implementación del programa Etxean Bizi*. Informe de Resultados.
Rowles, G. (1978). *Prisoners of space? Exploring the geographical experience of older people*. Westview Press.
Rowles, G. (2000). Habituation and being in place. *The Occupational Therapy Journal of Research, 20*, 52–67.
Rubinstein, R. (1989). The home environments of older people: A description of the psychosocial processes linking person to place. *Journal of Gerontology: Social Sciences, 44*(2), 545–553.
Rubinstein, R., & Parmelee, P. (1992). Attachement to place and the representation of the life course by the elderly. In *Place Attachement* (pp. 139–163). Plenum Press.

Sancho, M. (2016). *(16) Construyendo la Sociedad de los Cuidados*. Mayte Sancho. TEDxUDeusto Madrid – YouTube. Retrieved from: https://www.youtube.com/watch?v=DkHoKIwQfXA. Last access 21 June 2021.

Sancho, M., & Rodríguez, P. (2001). Envejecimiento y protección social de la dependencia en España. Veinte años de Historia y una Mirada hacia el futuro. *Intervención Psicosocial, 10*(3), 259–275.

Schrijvers, G., & Somme, D. (2017). Case-managers and integrated care. In *Handbook integrated care* (pp. 55–71). Springer. https://doi.org/10.1007/978-3-319-56103-5_4

Subirats, J. (2011). Ciudadanía y personas mayores. In I. P. Causapié, A. Balbontin, M. Porras, & A. Mateo (Eds.), *Libro blanco Envejecimiento Activo*. Instituto de Mayores y Servicios Sociales (IMSERSO).

Tomitsch, M. (2018). *Design. Think. Make. Break. Repeat: A handbook of methods*. Retrieved from http://designthinkmakebreakrepeat.com/. Last access 21 June 2021.

Tsiachristas, A., & Rutten-van Mölken, M. P. M. H. (2017). Evaluating complex interventions. In *Handbook integrated care* (pp. 297–313). Springer. https://doi.org/10.1007/978-3-319-56103-5_18

United Nations (UN). (2011). *World population ageing: Profiles of ageing 2011. CD-ROM*. United Nations.

United Nations (UN). (2017). *World population prospects: The 2017 revision*. Department of Economic and Social Affairs, Population Division.

van Dijk, H. (2015). Neighbourhoods for Ageing in Place. PhD Thesis, Erasmus University Rotterdam, Rotterdam, The Netherlands.

Van Hoof, J., & Kazak, J. K. (2018). Urban ageing. *Indoor and Built Environment, 27*, 583–586.

Van Hoof, J., Verhagen, M. M., Wouters, E. J. M., Marston, H. R., Rijnaard, M. D., & Janssen, B. M. (2015). *Picture your nursing home: Exploring the sense of home of older residents through photography* (p. 312831). Journal of Aging Research. https://doi.org/10.1155/2015/31293139

Van Hoof, J., Verbeek, H., Janssen, B. M., Eijkelenboom, A., Molony, S. L., Felix, E., & Wouters, E. J. M. (2016). A three perspective study of the sense of home of nursing home residents: The views of residents, care professionals and relatives. *BMC Geriatrics, 16*, 169. https://doi.org/10.1186/s12877-016-0344-9

Wahl, H.-W., Iwarsson, S., & Oswald, F. (2012). Aging well and the environment: Toward an integrative model and research agenda for the future. *The Gerontologist, 52*(3), 306–316.

Werner, C., Altman, I., & Oxley, D. (1985). Temporal aspects of homes: A transactional perspective. In Altman & Werner (Eds.), *Home environments* (Vol. 8, pp. 1–32). Plenum Press.

Wiles, J. L., Leibing, A., Guberman, N., Reeve, J., & Allen, R. E. S. (2012). The meaning of "aging in place" to older people. *The Gerontologist, 52*(3), 357–366. https://doi.org/10.1093/geront/gnr098

World Health Organisation (WHO). (2002). *Active ageing: A policy framework*. World Health Organization.

World Health Organization (WHO). (2007). *Vancouver Protocol*. World Health Organization.

World Health Organization. (2017). Global strategy and action plan on ageing and health. Retrieved from http://apps.who.int/bookorders.

Zeitler, E., Buys, L., Aird, R., & Miller, E. (2012). Mobility and active ageing in suburban environments: Findings from in-depth interviews and person-based GPS tracking. *Current Gerontology and Geriatrics, 2012*. https://doi.org/10.1155/2012/257186

Zuñiga, M., & Arrieta, F. A. (2021). *Analizando la función de la comunidad en el sistema de organización social de los cuidados en Euskadi*. Gizarte Zerbitzuetarako Aldizkaria. Revista de Servicios Sociales.

Álvaro García-Soler PhD in Psychology from the Universidad Autónoma de Madrid (UAM) (2013) and an MSc in Behavioural Therapy from the Therapeutic Institute of Madrid (2006). His thesis was titled: "Development of an observational instrument for researching the behavioural disorders in older institutionalized people". This dissertation was conducted in the UAM Behavioural Sciences program in collaboration with Matia Institute. He has worked as a researcher at the UAM and as a collaborator at the Universidad Rey Juan Carlos. Álvaro is author and co-author of several journals in clinical psychology on the learning processes that take place in the clinical environment, behaviour and cognition in the elderly people and elderly people adapted devices' development. He is currently working at Fundación Matia Instituto as head of methodology researching on issues related to the gerontological field: person-centred care model, dementia, development of well-being programs, creation and validation of assessment tools and user experience, as well as in various national and international projects of integrated long-term care.

Heitor García-Lantarón Specialist in Ageing and PhD Architect from the Technical School of Architecture of Madrid (ETSAM) of the Polytechnic University (UPM), with a thesis on Housing Models for the Elderly. He is currently head of the architectural design at the Matia Foundation, researcher at Matia Institute (San Sebastian, SP) and faculty at the Danish Institute for Study Abroad, DIS (Copenhagen, DK). His actual academic and professional interests are focused on design challenges related to the ageing process, by addressing them as a great opportunity for enhancing environment quality for any age and (dis)ability. He has developed his teaching work as an assistant at the ETSAM and the Francisco de Vitoria University (UFV). He has participated in several conferences and lectures at IMSERSO (Spanish Institute for Social Services), CEAPAT (centre depending on the Ministry of Health, Social Services, and Equality), ETSAM, UPV (University of Basque Country) in Spain and the North Urban and Housing Research Network in Denmark. Heitor has developed his professional career in different prestigious studios such as Herzog & de Meuron (Switzerland) or Nieto y Sobejano Arquitectos (Spain) in addition to more than 6 years working independently.

Sara Marsillas Rascado PhD in Psychology from Santiago de Compostela University in 2016 and a master's degree in Organizational Psychology, Forensic Law and Social Intervention in 2012. In 2015, she was granted by the Barrié Foundation for a stay in the Belgian Ageing Studies research group at the Vrije Universiteit Brussel. Between 2012 and 2015, she worked as a research professional in training and, in 2016, as a postdoctoral researcher (USC-Psicom). Her thesis project consisted in the development of a validated measurement tool for active ageing from an individual point of view, referring to the important dimensions for people. During this period, she has developed and participated in projects related to active ageing, with its measurement, the use of new technologies by older people, the monitoring of the behaviour of older people living alone or the evaluation of pilot tele-assistance programmes, among others. During the same period, she has acquired knowledge of co-creation methodologies such as Design Thinking. She has also worked as a teacher in the Designthinking.gal project and VermisLAB as a teacher of creativity, lateral thinking and intra- and interpersonal emotional intelligence.

Elena del Barrio Truchado Degree in Sociology (UCM, 1998–2002), master in Social Gerontology (UAM, 2003–2004), postgraduate fellow at UAM (2004–2005), and predoctoral fellow at the CSIC (2005–2008). In 2008, she worked as a technician in the "Elderly Portal" (IMSERSO-CSIC). From June 2008, she has been working at Matia Institute developing projects related to active ageing, living conditions, use of time of the elderly and working on the Age-Friendly Cities and Communities projects. She was the principal investigator of the Donostia-San Sebastián Age-Friendly City's diagnosis (2009) and the living conditions of the elderly in Euskadi's study (2010 and 2014). She coordinates the Euskadi Lagunkoia project since 2012. From 2013 to June 2016, she has been a member of the Social and Behavioural Sciences at the Spanish Society of Geriatrics and Gerontology (SEGG) and the coordinator of the Image and Ageing Working Group from 2012 to 2020.

Cristina Buiza Bueno PhD in Health Psychology from the Autonomous University of Madrid, with the doctoral thesis "Evaluation and conceptualization of cognitive functioning in advanced Alzheimer's disease", master in Gerontology from the University of the Basque Country, master in Clinical Neuropsychology from the Autonomous University of Barcelona and master in Contextual and Third-Generation Therapies from the ACT Institute of Madrid. Cristina has a professional career of 20 years in the field of the elderly and dementias, with special interest in the field of advanced dementias. She combines research with intervention and clinical practice in both the institutional and hospital fields. She has numerous publications on this subject in national and international scientific journals and conferences. She currently combines her work as a researcher at Matia Institute with her work as a psychologist at the Julián Rezola Gerontological Centre in Matia Foundation.

Pura Díaz-Veiga Degree in Psychology and diploma in Advanced Studies from the Autonomous University of Madrid. Professor in masters and courses related to gerontology and disability in Spanish and Latin American institutions. She has been part of the Board of Directors of the Spanish Society of Geriatrics and Gerontology. She has worked in Gerontology Research teams at the Autonomous University of Madrid and as a Clinical Psychologist in a Spanish Institution for blind persons (ONCE). She has participated in the development and coordination of various national and European research projects related to gerontology and disability and is the author of more than 40 publications. She has worked at Matia Institute since 2009 as a researcher and coordinated the Etxean Ondo Residences Project, the first in the Spanish state to identify the effects of person-centred care in residential centres for older people. She is the director of Matia Institute and currently supports dissemination and transfer of knowledge.

Chapter 7
Resilient Cities and Built Environment: Urban Design, Citizens and Health. Learning from COVID-19 Experiences

Miguel Padeiro, Ângela Freitas, Cláudia Costa, Adriana Loureiro, and Paula Santana

Abstract The COVID-19 pandemic has severely hit urban areas around the world, where approximately 55% of the world's population lives. It has tested the ability of their authorities and residents to respond to multiple challenges and not fail in terms of public health, social, economic and political aspects. This chapter questions the role of urban environments in the context of the COVID-19 crisis. In particular, we examine (i) how urban environments and their characteristics have accelerated or slowed the progression of the virus and are associated with a greater or lower incidence of the virus; (ii) how their characteristics have contributed to the strengthening or to the reduction of pre-existing socio-spatial inequalities of their inhabitants, with particular attention to social and psychological resilience; and (iii) what interventions have been carried out locally to enhance the resilience of cities when responding to future pandemics.

Keywords Resilience · Built environment · Urban health · COVID-19 · Urban Environment

7.1 Introduction

In a globalized world, where 39 million flights take off each year connecting more than 41,000 airports and where 55% of the population live in urban settings, urban areas very quickly found themselves at the centre of the pandemic: born at the core of an agglomeration of more than 11 million inhabitants, the virus quickly affected

M. Padeiro (✉) · Â. Freitas · C. Costa · A. Loureiro · P. Santana
Centre of Studies in Geography and Spatial Planning (CEGOT), Department of Geography and Tourism, University of Coimbra, Coimbra, Portugal
e-mail: jmnp@uc.pt; angela.freitas@uc.pt; claudiampcosta@uc.pt; adrianalour@gmail.com; paulasantana@uc.pt

E. Pozo Menéndez, E. Higueras García (eds.), *Urban Design and Planning for Age-Friendly Environments Across Europe: North and South*, Future City 19,
https://doi.org/10.1007/978-3-030-93875-8_7

the main Asian megalopolises, as confirmed on December 31, 2019. Reaching Europe, it was primarily concentrated in the urbanized and industrialized region of northern Italy and in the industrial regions of other countries, as the examples of north-east France, north Rhine-Westphalia in Germany and north-west Portugal demonstrate. This predominance of urban areas in the COVID-19 geography during the first wave does not mean that rural areas have been spared. As the virus progressed, small- and medium-sized towns were affected by the virus, and rural areas also experienced hotbeds of infection. As the virus spread across the nations, urban areas were faced with their strengths and vulnerabilities as well as with their resilience, which is defined as the 'capacity of a city or community to prepare for, respond to and adapt from dangerous and disruptive events' (Figueiredo et al., 2018).

The role of the built environment in public health is now widely recognized. It has been the subject of numerous epidemiological studies (Lindsay et al., 2017; Prussin et al., 2020; Sarkar et al., 2014). In fact, this role has been recognized for so long, despite the persistence of many knowledge gaps, that health issues have often had a significant impact on major breaks in urban planning paradigms. Residential locations, the places we frequent every day or with great regularity, and the possibilities to access services, care and leisure, as well as economic, social and training opportunities condition everyone's chances of early morbidity and mortality to a large extent. It is therefore hardly surprising that the characteristics of urban areas and the built environment may have contributed to the spatial pattern of the spread of the virus.

This chapter questions the role of urban environments in the context of the COVID-19 crisis. In particular, we examine (i) how urban environments and their various characteristics have accelerated or slowed the progression of the virus and are associated with a greater or lower incidence of the virus; (ii) how their characteristics have contributed to the strengthening or to the reduction of pre-existing socio-spatial inequalities of their inhabitants, with particular attention to social and psychological resilience; and (iii) what interventions have been carried out locally to enhance the resilience of cities in response to future pandemics. Finally, we consider the role of urban environments in the anticipation of possible, and even probable, pandemic scenarios as well as the role of such crises in the acceleration of pre-existing projects and paradigms – and the associated risks. At the same time, we elucidate the lessons that contemporary urban planning will have to draw from this public health crisis, particularly in an ageing world.

7.2 How Urban Space Slows or Accelerates the Spread of the Virus

A virus like SARS CoV-2 is spread through people-to-people contact. Researchers are a long way from identifying and interpreting all the factors that favour the spread of the virus, from the global to the individual level. Historical and institutional differences, varying political responses, the activation capacities of the health system, cultural differences governing relationships between people in society and wearing a mask (Baehr, 2008; Horii, 2014), economic and labour differences, the relative weight of services in terms of employment, the feasibility of teleworking, climatic differences, and air quality all come into play. Questioning the role of the urban in the spread of a virus fundamentally comes down to disentangling the territorial and spatial factors to understand why certain territories (e.g., certain cities within a country or a region and certain neighbourhoods within an agglomeration) are more affected than others.

As the virus spread across territories, the most important measures concerned reducing mobility. Borders were closed, airports were controlled and flights were restricted. The new global geography of human and capital flows largely governs the global trajectories of the virus, and airports functioned from the outset as a gateway for the virus to enter national territories (Keil & Mahon, 2010), as was the case for Hong Kong, Singapore and Toronto in 2003. The cities also had to restrict local and regional public transport services, avoid the formation of crowds in parks and gardens and limit contact between traders, services and citizens by closing all activities.

During the pandemic, changes have taken place. Public transport has been abandoned by the population who could afford to do so, and the frequency of service has often been reduced, constituting an even greater barrier to its use. The modal share of the private car increased during the period (Eisenmann et al., 2021) as well as that of taxis and paratransit. Bicycles, pedestrian travel and motorized two-wheelers have asserted themselves, not only because of the sharp reduction in the frequency of use of public transport but above all because of the generalized perception of the role of public transport in the transmission of the virus (Troko et al., 2011; TUMI, 2020).

7.2.1 Densities and Urban Forms

One of the points that has generated the most ink is that of the role of urban density as a facilitator of transmission of the virus. The existence of a link between population density and the risk of contagion is rather intuitive and is based on the notion that social interactions are a function of the co-presence of many people in a small portion of the territory. The presence of numerous public transport services, the concentration of jobs in central areas and the poorer ventilation often associated with older housing in a tighter urban fabric also contribute to forging this idea.

However, these premises hardly stand up to observation. As indicated by Khavarian-Garmsir et al. (2021), the idea of density as an impact factor is mainly based on comparisons between urban and rural areas or between towns belonging to different regions. Correlations have been observed between urban density and cases of COVID-19 (Almagro & Orane-Hutchinson, 2020; Coşkun et al., 2021; Wheaton & Kinsella Thompson, 2020), but the authors remain cautious about the causal relationship since unobservable and confounding factors were not taken into account.

Furthermore, urban densities may not be considerably representative, partly because of the definition and the various realities covered by the concept of density. The notion of density can be misleading. The same density, measured in inhabitants per km^2 or per hectare, can reflect significantly varied situations ranging from high-rise buildings to residential areas (Urban Task Force, 1999). Likewise, there is not much in common between a population density of a district (or a city) inhabited by rich social classes and that of an area where a rather poor population lives: the patterns of mobility are different, social interactions in public space are diverse, working and housing conditions vary, and the degree of social and family support and people's prerequisite health conditions differs. Density as a vector of contagion more probably consists of a simplification of diverse realities and an aggregation of diverse factors. In addition, population density is probably less important than the more fluctuating density of working days, which includes commuters and would make it necessary to consider certain rural or suburban territories that are classified as sparsely dense in the region and made of low residential densities in another way. As Almagro and Orane-Hutchinson (2020) assert, crowded spaces seem to play a greater role than residential density.

A solution to this problem is to ensure adequate control of the respective variables in existing studies. However, several studies have implemented this precaution and have arrived at results that diverge from the initial intuition (Carozzi, 2020; Hamidi & Hamidi, 2021; Hamidi et al., 2020; Teller, 2021). The most densely populated places are the first receptacles of the virus because of their connection to other central places, but density is less important than age structure (Carozzi, 2020) and the predominance of low-income groups and minorities (Hamidi & Hamidi, 2021). Moreover, the densest places are just as associated with a decrease in distances travelled and greater compliance with physical distancing recommendations (Hamidi et al., 2020; Khavarian-Garmsir et al., 2021). Cities like Singapore or Hong Kong have been hit with less severity than a number of rural areas (Salama, 2020). Current evidence tends to contradict the notion that densities played a role in the pandemic.

7.2.2 Socio-Spatial Equity, Territorial Justice

Beyond the major patterns of mobility and urban density, the geography of the transmission and incidence of the virus is largely related to that of deprivation. Current research confirms that socio-economically deprived people are at greater risk of exposure to the virus, which had already been identified with previous illnesses such as tuberculosis (Duarte et al., 2018), the 2003 SARS, the H1N1 virus (Kumar et al., 2012), and HIV. Thus, low-paid people, minorities, low-skilled workers and marginalized people are the most exposed to the virus, for various reasons. For example, DiMaggio et al. (2020) identified an almost fivefold increase in the risk of COVID-19 in Black/African American populations.

The first reason is unequal access to information. The most vulnerable groups are those to whom information arrives more slowly and with poorer quality, partially because of the digital divide, which is particularly evident in the countries of the South. These groups are more likely to lack knowledge regarding public health and precautions regarding the need for physical distancing. This is why discrepancies in compliance with the measures recommended by the health authorities of different countries have been noted between social groups (Wright et al., 2020). Compliance was also higher where political trust was higher (Bargain & Aminjonov, 2020; Nivette et al., 2021).

The second major category of reasons relates to living conditions. The risk of contracting the virus decreases with better ventilation of homes, a smaller number of co-residents and greater access to basic infrastructure such as electricity, water and sanitation. The absence of these conditions particularly complicates the adoption of hygiene measures, and the most vulnerable groups are those whose houses have fewer of such conditions. The case of informal settlements, in the Global South but not only, is particularly evocative as well as that of the homeless because of their greater difficulty in finding refuge. Finally, the scenario of older adults living in often overcrowded retirement homes with staff in precarious situations, often from low-wage neighbourhoods with a high proportion of minorities, has tended to overexpose the older adults living in such facilities to the virus. Older people living in nursing homes were often overlooked in the initial plans, as they were thought to be relatively protected by the lack of day-to-day contact in the community.

The third reason – and perhaps the most important according to several studies (Almagro & Orane-Hutchinson, 2020) – is linked to economic insecurity and working conditions. Instead of teleworking, numerous low-skilled workers working in industrial areas (e.g., northern Italy and Portugal) and in the meatpacking industries (Andersen et al., 2021) had to comply with the obligation to work at their usual place of employment. These sites were often in a poor condition, which reduced the convenience of the use of a mask and ability to maintain a physical distance from colleagues. Others working in the agricultural sectors (e.g., migrant seasonal workers) also had to move. In addition to the conditions in the workplace, these workers are often captive of public transport and do not have easy access to alternative forms of mobility, which has made them even more exposed to the virus. In general, economic precariousness has often forced greater mobility challenges during the pandemic, as individuals who perform manual labour are often forced to travel greater distances than others, since their commute is affected by the reduction in the frequency of public transport.

7.3 The Mediating Role of Urban Design in the Face of the Crisis

It is not only in terms of impact that the built environment exerts its influence on COVID-19 geography. The severity of the pandemic, in health and social terms, is not only the result of the intrinsic vulnerabilities of people and their belonging to a risk group defined in medical terms. Rather, the severity also depends to a large extent on the environment of the place where people live, work or play. The multiple crises (e.g. social, economic, familial, psychological) that people have faced so far throughout the pandemic and those they will face in the near future are at least as much the result of socio-spatial inequalities as that of the virus itself. Therefore, in addition to the infection rate, it is important to consider to what extent residential environments and their spatially patterned distribution contribute to a triple fragility of increased vulnerability to the virus itself, through comorbidities, which then results in a higher mortality rate; greater vulnerability to pre-existing health conditions, being themselves impacted by the pandemic crisis; and greater social, economic and psychological vulnerabilities.

7.3.1 The Public Health Danger of Urban Sprawl

During the first wave of COVID-19, part of the population of large cities made the choice to move to second homes located in rural areas that were supposed to be outside the main zones of contamination. An increase in real estate demand in rural areas has been noted in several countries. At the same time, many people spent more time tending to their gardens during the enforced quarantine (Corley et al., 2021; Mullins et al., 2021), and the idea of living in a single-family home located far from the city centre in a suburban or peri-urban area with a garden gained ground. According to some authors, this scenario entails a growing preference for a residential location outside of urban centres that is linked both to fears of future pandemics and other risks and to the desire to rediscover the attractions of the open air and the new developments and possibilities of teleworking. This development is admittedly contested by other authors, who consider it as a cyclical situation and believe that the necessities of co-presence and other location advantages, that the agglomeration represents for households, will prevail. Furthermore, this shift carries the seeds of other problems and raises the question of the link between the present urban configurations and the impacts of the pandemic.

The potential distance from households is thus largely associated with a wide range of adverse effects on society: greater dependence on cars; higher levels of pollution and greenhouse gas emissions (e.g. NOx, VOCs, CO, SO_2, $PM_{2.5}$); greater energy expenditure; higher collective costs in terms of infrastructure and public services; greater artificialization of soils, correlated with an increase in surface temperatures; lower access to resources and opportunities, such as healthcare, education and employment; and a lower level of physical activity.

However, the health effects of urban sprawl in general (Ewing et al., 2003; Frumkin, 2016), and of each of the elements mentioned, are becoming more understood. People with health conditions such as cardiovascular, cerebrovascular or respiratory disease are particularly at risk of death after contamination. However, these conditions are largely influenced by automobile-related pollution. Likewise, low levels of physical activity and limited access to a balanced diet are also associated with several comorbidities, which present a great risk in the face of COVID-19.

Finally, avoidable comorbidities are socially patterned and are reflected in the racial and social discrepancies of the COVID-19 death rates. For example, in Chicago, higher COVID-19 mortality was detected in neighbourhoods with a higher percentage of Black and Hispanic/Latino residents (Scannell Bryan et al., 2021). Gross et al. (2020) also came to the same conclusion in a statewide study of the United States, and the same observation was made in São Paulo, Brazil (Ribeiro et al., 2021).

7.3.2 *The Role of Green and Open Spaces*

Contemporary urban planning had partially broken with previous approaches more favourable to automobility – an unfinished fight in view of numerous urban planning operations and the difficulties of territorial planning to mitigate urban sprawl. Many interventions have multiplied in recent years with the aim of improving the conditions of walkability in public spaces. The role of the different social, physical and service dimensions of the residential environment in physical and mental health and well-being has already been stated many times (Padeiro et al., 2021), and observations made during the pandemic have above all confirmed their importance.

Thus, the small size of dwellings, their overcrowding and their poor quality (e.g. sound system, humidity, air conditioning or heating) as well as the absence of an outdoor space is associated with more negative results in terms of mental health (Amerio et al., 2020; Chen et al., 2021). The existence of an outdoor space, whether it be a balcony, terrace, patio or garden (or vegetable patch) has generally allowed residents to practice an activity with at least moderate physicality and to have privileged access to the open air and nature, which has positive consequences on the reduction of stress and anxiety. Even a view of a green space from the apartment's windows has been associated with a decrease in feelings of loneliness, anxiety and depression (Guzman et al., 2021; Soga et al., 2021), and a similar effect was seen in people with green plants inside the home (Pérez-Urrestarazu et al., 2021). People in confinement generally spent more time than usual maintaining their indoor or outdoor plants (Pérez-Urrestarazu et al., 2021).

For residents with no outdoor space, the quality of public space played an important role in mitigating or improving the multiple family, social and economic tensions to which individuals were subjected. The relationship between access to and the quality of green spaces and health has been well established for many years. The presence of green spaces offers residents opportunities to practice physical activity (e.g. walking, running, using physical exercise equipment), contemplate nature, enjoy the calm or observe passers-by, and these activities offer benefits in terms of mental health. Green spaces also provide opportunities for social interactions, which are also beneficial for mental health and for the development of social support that can be mobilized when needed.

Outdoor recreational activity increased by 291% during the lockdown in Oslo, Norway (Venter et al., 2020). Several studies have highlighted the reduction in psychological distress linked to COVID-19 when people live in a quiet neighbourhood with sufficient sunshine (Chen et al., 2021). Alternatively, in many contexts, the ban on sitting on public benches and the removal of certain elements of street furniture – in particular gaming tables and playgrounds – has probably led to a drop in the outdoor mobility of the elderly and the young. This removal raises questions about the fairness of the measures, since during the confinement the parks and gardens become more reserved for a young and active population and closed to older people.

7.4 Advancing Urban Health Resilience Through Innovative and Adaptive Urban Planning

Cities have been on the front line of coping with pandemics and their impacts for centuries. Past infectious disease outbreaks triggered many innovations in city planning: in the fifteenth century, the bubonic plague prompted Leonardo Da Vinci's concept of an 'anti-epidemic city' that inspired Renaissance cities to open larger squares and public spaces; in the eighteenth and nineteenth centuries, the response to cholera, smallpox and yellow fever brought large boulevards to Paris and improved water systems to London; in the twentieth century, tuberculosis, typhoid, polio and Spanish flu breakouts prompted modern urbanism itself, with radical reforms in housing and urban form, including slum clearance, waste management, light and airy spaces and single-use zoning (e.g. separating residential and industrial areas), among others (Hall, 2014; Liu & Wang, 2021; Sarkar et al., 2014).

Following Meerow et al. (2016, p. 45) definition:

> urban resilience refers to the ability of an urban system and all its constituent socio-ecological and socio technical networks across temporal and spatial scales to maintain or rapidly return to desired functions in the face of a disturbance, to adapt to change, and to quickly transform systems that limit current or future adaptive capacity.

Looking to urban resilience in the face of a pandemic, special attention is given to urban planning, namely, how changes in the urban form can increase the city's ability to respond, recover and adapt.

Recent research on the impacts of COVID-19 in cities have offered opportunities to understand and identify which urban interventions have the potential to enhance the resilience of cities in future pandemics. Existing knowledge highlights similar domains informing urban interventions and post-COVID adaptive urban planning: physical and built environmental conditions, socioeconomic factors, transportation, urban design and city governance. Depending on the pandemic-related phase (response, mitigation and preparedness), each domain can be addressed in different ways.

Lak et al. (2020) propose a framework to enhance urban resilience considering three scales – housing, neighbourhood/public spaces and city – that focus on the physical (e.g. urban design, access, infrastructure, land use and natural environment) and nonphysical aspects (e.g. sociocultural, economic and political factors) of resilient urban form. Following this framework, examples of urban interventions in each phase are presented in Table 7.1.

A review made by Sharifi and Khavarian-Garmsir (2020) on the impacts of COVID-19 highlights the main long-term lessons for the design of safe, resilient and healthy urban places in the following four thematic domains: environmental

Table 7.1 Current urban strategies and interventions addressing the COVID-19 pandemic by phase

Phase	Urban interventions	Intervention level
Response	Provision of sufficient shared spaces indoors for quarantine and lockdown for large urban families	Housing
	Provision of spaces in residential areas for providing care to ill people	
	Subsidies and economic support for suitable ventilation to maintain optimal humidity (range of 40–60%) and provision of filtering indoor air for affected families	
	Enhancing infrastructures for more effective use of cell phones, smart apps, and the internet in the monitoring and surveillance of disease trends by community centres	Neighbourhood
	Promoting biking and walking for local trips	
	Controlling the use of parks, community gardens and playgrounds according to social distancing protocols of local authorities	
	Decontaminating and sanitizing public places and urban furniture in neighbourhoods frequently	
	Placing homeless people in vacant buildings such as schools, sport facilities and churches	
	Restricting large public gatherings, cancelling programmes and events and closing indoor recreational facilities	
	Closing theatres and concert venues, museums, restaurants, amusement parks and other places of leisure	
	Provision of facilities for teleworking in neighbourhoods	
	Establishing epidemic prevention areas based on public health spatial units	
	Providing a well-developed, reliable and transparent information system to increase public awareness	
	Providing local community aid groups to support vulnerable groups (e.g. elderly)	
	Supporting healthcare services and access to medical facilities	City
	Mobilizing the response coordination system of NGOs	
	Establishing comprehensive and accurate databases of transportation, land and infrastructure planning information systems, such as websites or phone hotlines as well as monitoring technologies	

(continued)

Table 7.1 (continued)

Phase	Urban interventions	Intervention level
Mitigation	Improving the quality of housing, facilitating the filtration, ventilation and sanitation system of buildings and homes	Housing
	Designing semi-open spaces in housing, like balconies for planting and pleasure	
	Providing semi-public and semi-private or shared open spaces in residential buildings for planting, playing and working out in pandemic situations	Neighbourhood
	Increasing the number of open and green spaces in neighbourhoods to enable people to socialize without being packed	
	Designing flexible bike-friendly and walkable neighbourhoods and cities	Neighbourhood City
	Providing wide sidewalks and bike paths, entryways and public transit stations	
	Designing transit neighbourhood development that supports uncrowded public transportation with social distancing, to help make public transportation and mobility safer	
	Implementing decreasing heavy traffic measures	
	Planning for well-designed roads and streets and changing passengers' behaviours via signs and pavement markings to show safe distances between riders and passengers	
	Developing infrastructures and data platforms to democratize smart city technology and improve city governance	City
	Enhancing the quality and quantity of the healthcare services as a multilayer system (all parts of the city/municipality should be under service of health services)	
	Providing supportive and equitable amenities for health: reducing inequity and improving conditions in underserved neighbourhoods, planning and designing decentralized amenities within neighbourhoods and city centres, well distributed public open spaces, distributing retails within neighbourhood instead of concentrating them in the city centre, and ensuring quality and accessibility to hospitals, internal care units and hospital beds	
	Providing adequate quality and quantity of public open and green spaces as contexts of social interaction to allow citizens to engage in outdoor physical activities in lockdown situations and to practice social distancing	

(continued)

Table 7.1 (continued)

Phase	Urban interventions	Intervention level
Preparedness	Reinforcing the decision-making and the planning process by developing smart city technology, taking advantage of Decision Support Systems (DSSs), and Geographic Information Systems (GIS), along with setting up urban dashboards, to help local government to decide more precisely	City
	Developing a surveillance system using digital and web-based information along with disease spread models	
	Develop a local Strategic Preparedness and Response Plan to future pandemics	

Source: Lak et al. (2020)

The response phase is defined as the provision of emergency services and public assistance during or immediately after a pandemic to reduce health impacts by controlling disease transmission as well as ensuring public safety and meeting the basic subsistence needs of the affected people. The mitigation (prevention) phase is known as the structural/non-structural measures and the activities undertaken to limit the adverse impact of epidemics. The preparedness phase aims to increase the preparedness and response capabilities of cities against outbreaks

quality, transportation and urban design, socioeconomic impacts and management and governance. The recent work of Rojas-Rueda and Morales-Zamora (2021) provides a set of built environment and transport interventions that can help support strategies in the mitigation phase. Figure 7.1 shows this framework as well as the pathways through which urban interventions influence health determinants and health outcomes, for example, by improving air and noise quality, traffic safety, access to services and physical activity opportunities.

7.4.1 Tactical Urbanism: A Short-Term Tool for Long-Term Healthy Urban Planning

While many temporary urban design changes have been made to support COVID-19 containment and mitigation strategies, the pandemic has resulted in a broad acceleration of many urban planning strategies that contribute to more sustainable and healthier cities.

From the perspective of an emerging new and adaptive urban planning paradigm, it is worthy to mention the 'soft' urban planning approaches linked to tactical urbanism. Soft approaches are urban space transformations implemented through small-scale, short-term, low-cost, temporary and community-based projects. They are intended to improve built environments (Lydon & Garcia, 2015). These projects are known as bottom-up interventions since they are community-led processes where local residents have an active role in transforming public space. Across the globe, streets and public spaces have been reclaimed by people. Streets have helped channel the movement of emergency vehicles and other essential services, and

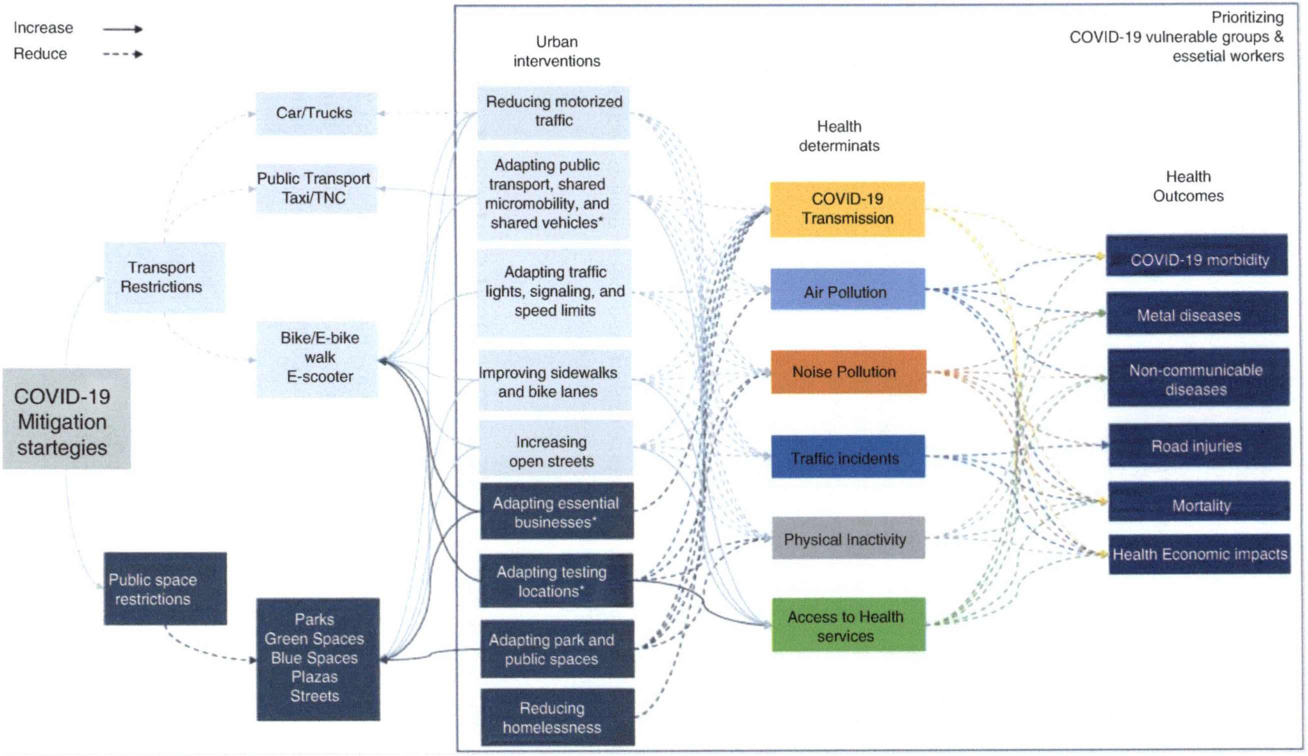

Fig. 7.1 Built environment and transport interventions in the context of COVID-19 mitigation strategies with long-term health equity vision. Urban interventions marked with an asterisk (*) refer to recommendations that apply during the pandemic; all others apply to a post-pandemic context. (Source: Rojas-Rueda & Morales-Zamora, 2021, Fig. 4)

neighbourhood open spaces have served as hubs of pandemic operations, such as food distribution, testing centres, temporary markets and other needs.

The COVID-19 pandemic has brought a new life to tactical urbanism, making urban leaders and residents reconsider many aspects of urban daily life and their benefits in a more permanent manner, beyond current times. Although tactical urbanism interventions were implemented to be temporary, they could be designed with a long-term vision to support a more permanent healthy urban design and long-term health equity vision, by improving conditions in disadvantaged and unserved neighbourhoods. Tactical urbanism interventions also improve air quality, reduce noise levels, encourage physical activity, reduce the risk of traffic accidents, promote mental health and prevent non-communicable diseases (Rojas-Rueda & Morales-Zamora, 2021).

Examples of tactical urbanism interventions from around the world include (i) the collaborative creation of public squares or plazas with local artists and residents (Fig. 7.2a), (ii) the transformation of streets once dominated by cars into bike lanes (as well as temporary runners' lanes that are spray-painted) and the creation of safe routes for children walking and cycling to school (Fig. 7.2c), (iii) the widening of sidewalks that provide more space and amenities for people and are usually installed in parking lanes and use one or more parking spaces (Fig. 7.2d), (iv) the closing of streets to cars and its conversion into public space (Fig. 7.2b, e) and (v) the creation of pop-up parks, which are usually small green spaces usually nestled in empty urban areas or parking lots (Fig. 7.2f).

Fig. 7.2 Examples of tactical urbanism interventions. (**a**) *Planteia*, Ílhavo, Portugal. (**b**) Calgary, Canada. (**c**) Safe route to school, New Jersey, USA. (**d**) Public Parklet, Stockholm, Sweden. (**e**) Poblenou 'Superblock', Barcelona, Spain. (**f**) Pop-up plaza, Santiago Chile, Chile. (Source: (**a**) 23 Milhas (http://www.23milhas.pt/); (**b**) Federation of Calgary Communities (https://activateyyc.calgarycommunities.com/); (**c**) Street Plans Collaborative (http://blog.beopenfuture.com/2018/10/28/tactical-urbanism/); (**d**) Tengbom (https://archello.com/pt/project/public-parklet); (**e**) Centre de Cultura Contemporània de Barcelona (https://www.publicspace.org/works/-/project/k081-poblenou-s-superblock); (**f**) OpenCity Projects (http://opencityprojects.com/okuplaza-san-diego/))

In summary, and based on the work of Rojas-Rueda and Morales-Zamora (2021), ten general tactical urbanism interventions are recommended to boost healthy urban planning and urban resilience in the face of future public health crises: (1) implementing tactical urbanism interventions early; (2) expanding new sidewalks and bike infrastructure to promote soft and active mobility (biking and walking); (3) implementing everyday open streets and expanding the existing ones; (4) adapting parks and public spaces and expanding small open and green spaces at the neighbourhood level; (5) adapting traffic lights and signalling that favour pedestrians and cyclists and lowering streets speed limits; (6) adapting public transport (e.g. creating user schedules that favour vulnerable populations, such as the elderly) and integrating shared micromobility to public transit and bike lanes; (7) concentrating motorized transport within a few streets and removing traffic lanes and parking spaces to support car-free streets, active transportation and traffic safety; (8) adapting essential businesses, for example, by designating bike shops as essential services and expanding sidewalks for on-sidewalk queuing, restaurants eating and outdoor markets; (9) providing transport options to access testing sites and other health services beyond cars; and (10) providing and adapting homeless shelters and services suited for physical distancing and COVID-19 precautionary measures.

7.5 Conclusion

The lessons that urban planning will have to draw from this public health crisis are not only numerous but also ambiguous, complex and even variable in space. Moreover, planning measures must incorporate global, intergenerational and socially inclusive dimensions. The improvement of the local public space and the reduction of the place of the automobile must be carried out in such a way as to promote the use of spaces not exclusively used by certain ages or social categories. This challenge is often overshadowed by the often-positive changes in the local landscape.

The current ageing of the population prompts a review of all urban planning paradigms. This process of demographic change will indeed make populations more sensitive in the future to the disruptive dangers of new public health crises. This process implies an approach that is both multi-sectoral and collaborative, including improving housing conditions, enhancing the accessibility to public and private buildings, promoting formal and informal social support, developing infrastructure and essential services and improving neighbourhood design. This approach should constitute an integrative process co-constructed with older people themselves.

One obstacle remains, however. While it is important to rethink the shape of cities, the proximity of services and inclusive design, we must not forget that urban production is now mainly done through models favourable to the power of the private sector. The new modalities of urban management and development are characterized by several major trends: (i) the domination of entrepreneurial managerialism; (ii) the almost flawless acceptance of touristification, at least in the most attractive cities; and

(iii) the financialization of the land and real estate market. These three interconnected trends imply that improvements in public and private spaces go hand in hand with an increase in land rent and with the emergence of significant social risks. These risks are often accompanied by processes of real or symbolic evictions, with consequences for the resilience of the most vulnerable groups. Affected people are at risk of moving to less accessible neighbourhoods where routines are more difficult to implement, where the conditions are worse and where access to services and amenities are more reduced. Ultimately, a process of improving urban design can improve the conditions of space, not the conditions of people. One of the great challenges therefore lies in reconciling urban planning imperatives with social needs, which implies in particular a revaluation of public power and its collective service mission.

Acknowledgements This text received support from the Centre of Studies in Geography and Spatial Planning (CEGOT), funded by national funds through the Foundation for Science and Technology (FCT) under the reference UIDB/04084/2020. Angela Freitas is a recipient of an Individual Doctoral Fellowship funded by national funds through the FCT, under the reference SFRH/BD/123091/2016.

References

Almagro, M., & Orane-Hutchinson, A. (2020). JUE insight: The determinants of the differential exposure to COVID-19 in New York City and their evolution over time. *Journal of Urban Economics, 127*(3), 103293.

Amerio, A., Brambilla, A., Morganti, A., Aguglia, A., Bianchi, D., Santi, F., … Capolongo, S. (2020). COVID-19 lockdown: Housing built environment's effects on mental health. *International Journal of Environmental Research and Public Health, 17*(16), 5973. Retrieved from: https://www.mdpi.com/1660-4601/17/16/5973

Andersen, L. M., Harden, S. R., Sugg, M. M., Runkle, J. D., & Lundquist, T. E. (2021). Analyzing the spatial determinants of local Covid-19 transmission in the United States. *Science of the Total Environment, 754*, 142396.

Baehr, P. (2008). City under siege: Authoritarian toleration, mask culture, and the SARS crisis in Hong Kong. In *Networked disease: Emerging infections in the Global City* (pp. 138–151).

Bargain, O., & Aminjonov, U. (2020). Trust and compliance to public health policies in times of COVID-19. *Journal of Public Economics, 192*, 104316. https://doi.org/10.1016/j.jpubeco.2020.104316

Carozzi, F. (2020). *Urban density and COVID-19*. Institute of Labor Economics.

Chen, Y., Jones, C., & Dunse, N. (2021). Coronavirus disease 2019 (COVID-19) and psychological distress in China: Does neighbourhood matter? *Science of the Total Environment, 759*, 144203.

Corley, J., Okely, J. A., Taylor, A. M., Page, D., Welstead, M., Skarabela, B., … Russ, T. C. (2021). Home garden use during COVID-19: Associations with physical and mental wellbeing in older adults. *Journal of Environmental Psychology, 73*, 101545. https://doi.org/10.1016/j.jenvp.2020.101545

Coşkun, H., Yıldırım, N., & Gündüz, S. (2021). The spread of COVID-19 virus through population density and wind in Turkey cities. *Science of the Total Environment, 751*, 141663.

DiMaggio, C., Klein, M., Berry, C., & Frangos, S. (2020). Black/African American Communities are at highest risk of COVID-19: Spatial modeling of New York City ZIP Code–level testing results. *Annals of Epidemiology, 51*, 7–13.

Duarte, R., Lönnroth, K., Carvalho, C., Lima, F., Carvalho, A., Muñoz-Torrico, M., & Centis, R. (2018). Tuberculosis, social determinants and co-morbidities (including HIV). *Pulmonology, 24*(2), 115–119.

Eisenmann, C., Nobis, C., Kolarova, V., Lenz, B., & Winkler, C. (2021). Transport mode use during the COVID-19 lockdown period in Germany: The car became more important, public transport lost ground. *Transport Policy, 103*, 60–67.

Ewing, R., Schmid, T., Killingsworth, R., Zlot, A., & Raudenbush, S. (2003). Relationship between urban sprawl and physical activity, obesity, and morbidity. *American Journal of Health Promotion, 18*(1), 47–57.

Figueiredo, L., Honiden, T., & Schumann, A. (2018). *Indicators for resilient cities*. OECD Publishing. https://doi.org/10.1787/6f1f6065-en

Frumkin, H. (2016). Urban sprawl and public health. *Public Health Reports, 117*(3), 201–217.

Gross, C. P., Essien, U. R., Pasha, S., Gross, J. R., Wang, S. Y., & Nunez-Smith, M. (2020). Racial and ethnic disparities in population-level Covid-19 mortality. *Journal of General Internal Medicine, 35*(10), 3097–3099.

Guzman, V., Garrido-Cumbrera, M., Braçe, O., Hewlett, D., & Foley, R. (2021). Associations of the natural and built environment with mental health and wellbeing during COVID-19: Irish perspectives from the GreenCOVID study. *The Lancet Global Health, 9*, S20.

Hall, P. (2014). *Cities of tomorrow: An intellectual history of urban planning and design since 1880*. Wiley.

Hamidi, S., & Hamidi, I. (2021). Subway ridership, crowding, or population density: Determinants of COVID-19 infection rates in New York City. *American Journal of Preventive Medicine, 60*(5). https://doi.org/10.1016/j.amepre.2020.11.016

Hamidi, S., Sabouri, S., & Ewing, R. (2020). Does density aggravate the COVID-19 pandemic? *Journal of the American Planning Association*, 1–15. https://doi.org/10.1080/01944363.2020.1777891

Horii, M. (2014). Why do the Japanese wear masks? *Electronic Journal of Contemporary Japanese Studies, 14*(2).

Keil, R., & Mahon, R. (2010). *Leviathan Undone? Towards a political economy of scale*. UBC Press.

Khavarian-Garmsir, A. R., Sharifi, A., & Moradpour, N. (2021). Are high-density districts more vulnerable to the COVID-19 pandemic? *Sustainable Cities and Society, 70*, 102911.

Kumar, S., Quinn, S. C., Kim, K. H., Daniel, L. H., & Freimuth, V. S. (2012). The impact of workplace policies and other social factors on self-reported influenza-like illness incidence during the 2009 H1N1 pandemic. *American Journal of Public Health, 102*(1), 134–140.

Lak, A., Asl, S. S., & Maher, A. (2020). Resilient urban form to pandemics: Lessons from COVID-19. *Medical Journal of the Islamic Republic of Iran, 34*, 71.

Lindsay, S. W., Wilson, A., Golding, N., Scott, T. W., & Takken, W. (2017). Improving the built environment in urban areas to control Aedes aegypti-borne diseases. *Bulletin of the World Health Organization, 95*(8), 607.

Liu, H., & Wang, P. H. (2021). Research on the evolution of urban design from the perspective of public health under the background of the COVID-19. *The International Journal of Electrical Engineering & Education*, 0020720921996598.

Lydon, M., & Garcia, A. (2015). *Tactical Urbanism: Short-term Action for Long-term Change*. Island Press.

Meerow, S., Newell, J. P., & Stults, M. (2016). Defining urban resilience: A review. *Landscape and Urban Planning, 147*, 38–49.

Mullins, L., Charlebois, S., Finch, E., & Music, J. (2021). Home food gardening in Canada in response to the COVID-19 pandemic. *Sustainability, 13*(6), 3056. Retrieved from: https://www.mdpi.com/2071-1050/13/6/3056

Nivette, A., Ribeaud, D., Murray, A., Steinhoff, A., Bechtiger, L., Hepp, U., … Eisner, M. (2021). Non-compliance with COVID-19-related public health measures among young adults in

Switzerland: Insights from a longitudinal cohort study. *Social Science & Medicine, 268*, 113370. https://doi.org/10.1016/j.socscimed.2020.113370

Padeiro, M., de São José, J., Amado, C., Sousa, L., Roma Oliveira, C., Esteves, A., & McGarrigle, J. (2021). Neighborhood attributes and well-being among older adults in urban areas: A mixed-methods systematic review. *Research on Aging*, 0164027521999980. https://doi.org/10.1177/0164027521999980

Pérez-Urrestarazu, L., Kaltsidi, M. P., Nektarios, P. A., Markakis, G., Loges, V., Perini, K., & Fernández-Cañero, R. (2021). Particularities of having plants at home during the confinement due to the COVID-19 pandemic. *Urban Forestry & Urban Greening, 59*, 126919. https://doi.org/10.1016/j.ufug.2020.126919

Prussin, A. J., Belser, J. A., Bischoff, W., Kelley, S. T., Lin, K., Lindsley, W. G., ... Bibby, K. (2020). Viruses in the Built Environment (VIBE) meeting report. In *BioMed central*.

Ribeiro, K. B., Ribeiro, A. F., Veras, M. A. D. S. M., & de Castro, M. C. (2021). Social inequalities and COVID-19 mortality in the city of São Paulo, Brazil. *International Journal of Epidemiology, 50*, 732–742.

Rojas-Rueda, D., & Morales-Zamora, E. (2021). Built environment, transport, and COVID-19: A review. *Current Environmental Health Reports*. https://doi.org/10.1007/s40572-021-00307-7

Salama, A. M. (2020). Coronavirus questions that will not go away: Interrogating urban and socio-spatial implications of COVID-19 measures. *Emerald Open Research, 2*, 14.

Sarkar, C., Webster, C., & Gallacher, J. (2014). *Healthy cities: Public health through urban planning*. Edward Elgar Publishing.

Scannell Bryan, M., Sun, J., Jagai, J., Horton, D. E., Montgomery, A., Sargis, R., & Argos, M. (2021). Coronavirus disease 2019 (COVID-19) mortality and neighborhood characteristics in Chicago. *Annals of Epidemiology, 56*, 47–54.e45. https://doi.org/10.1016/j.annepidem.2020.10.011

Sharifi, A., & Khavarian-Garmsir, A. R. (2020). The COVID-19 pandemic: Impacts on cities and major lessons for urban planning, design, and management. *The Science of the Total Environment, 749*, 142391–142391. https://doi.org/10.1016/j.scitotenv.2020.142391

Soga, M., Evans, M. J., Tsuchiya, K., & Fukano, Y. (2021). A room with a green view: The importance of nearby nature for mental health during the COVID-19 pandemic. *Ecological Applications, 31*(2), e2248.

Teller, J. (2021). Urban density and Covid-19: Towards an adaptive approach. *Buildings and Cities, 2*(1), 150–165.

Troko, J., Myles, P., Gibson, J., Hashim, A., Enstone, J., Kingdon, S., ... Van-Tam, J. N. (2011). Is public transport a risk factor for acute respiratory infection? *BMC Infectious Diseases, 11*(1), 1–6.

TUMI. (2020). *The COVID-19 outbreak and implications to sustainable urban mobility – Some observations*. Retrieved from https://www.transformative-mobility.org/news/the-covid-19-outbreak-and-implications-to-public-transport-someobservations

Urban Task Force. (1999). *Towards an urban renaissance (Final report)*. Routledge.

Venter, Z. S., Barton, D. N., Gundersen, V., Figari, H., & Nowell, M. (2020). Urban nature in a time of crisis: Recreational use of green space increases during the COVID-19 outbreak in Oslo, Norway. *Environmental Research Letters, 15*(10). https://doi.org/10.1088/1748-9326/abb396

Wheaton, W. C., & Kinsella Thompson, A. (2020). *The geography of COVID-19 growth in the US: Counties and metropolitan areas*. Available at SSRN 3570540.

Wright, A. L., Sonin, K., Driscoll, J., & Wilson, J. (2020). Poverty and economic dislocation reduce compliance with covid-19 shelter-in-place protocols. *Journal of Economic Behavior & Organization, 180*, 544–554.

Miguel Padeiro is an Assistant Professor in the Department of Geography and Tourism, at the University of Coimbra, and member of the Centre of Studies in Geography and Spatial Planning (CEGOT). His research focuses on urban planning, mobility and the relationship between built environment and older adults' health and well-being.

Ângela Freitas is a Geographer, PhD Research Fellow in Human Geography at the University of Coimbra and member of the Centre of Studies in Geography and Spatial Planning (CEGOT). Her dissertation explores the impact of urban policies on health equity and the role that local governments can play in promoting population health and reducing inequities in health determinants. Her research interests lie in the relationship between local environment and health, urban planning, policy health impact assessment and participatory research.

Cláudia Costa is a Geographer, Master in Science, PhD student in Human Geography at the University of Coimbra and member of the Centre of Studies in Geography and Spatial Planning (CEGOT). Her dissertation explores the impact of the territory on mortality. Her research interests lie on spatial epidemiology, public health and the utilization of quantitative methodologies to identify the relationship between contextual determinants and health outcomes.

Adriana Loureiro is a Geographer, PhD student in Human Geography at the University of Coimbra and member of the Centre of Studies in Geography and Spatial Planning (CEGOT). Her dissertation explores the impact of the local environment on mental health. Her research interests lie on strategic assessment of neighbourhood environmental impacts. She has participated in several research projects focusing on geography of health, healthy urban planning and health impact assessment.

Paula Santana is a Senior Researcher and Full Professor of Health Geography in the Department of Geography and Tourism, at the University of Coimbra, and member of the Centre of Studies in Geography and Spatial Planning (CEGOT). She has more than 30 years of experience in studying the connection between place and health, focusing on the geographical disparities in health, urban health, healthy urban planning and health-promoting policies, at European, national and local levels.

Part II
An Inclusive Environment: Alternatives and Challenges for Ageing in Place in North and South European Cities

Chapter 8
Framework for Inclusive Residential Projects for All: Further Research Lines

Rose Gilroy

Abstract House and home. *Home* is a place for intimacy, where we can take off the mask we wear in public life. *House* brings to mind planning and financial mechanisms and the belief, at least in the UK, that years of financial struggle will be rewarded by profit from rising property values. Increasingly, the discourse in the UK is of the exchange value of our housing not the use value of our homes.

The UK housing market is dominated by a small group of developers who design and target housing at narrowly understood life-course groups. City centre apartment blocks are homes for young professionals; suburban houses with gardens are targeted at families with children, while age segregated apartment blocks are for older people.

The response to the COVID-19 pandemic has provided an opportunity to reflect on the critical importance of housing quality and of neighbourhoods invested with human exchange and meaningful relationships.

This chapter explores whether we can break away from separated and segregated housing to explore the possibility of alternatives that promote reciprocity and intergenerational living. It draws on two decades of research to ask what happens if we stop thinking about houses as commodities but as bundles of relationships.

Keywords Intergenerational living · Housing development · Co-housing · Relationships · Use value

8.1 The UK Housing Landscape

The British housing landscape is dominated by a normative expectation of home ownership such that 65% of English households are owner occupiers, of which more than half own their house outright reflecting an ageing population who have

R. Gilroy (✉)
Newcastle University, Newcastle upon Tyne, UK
e-mail: r.c.gilroy@newcastle.ac.uk

E. Pozo Menéndez, E. Higueras García (eds.), *Urban Design and Planning for Age-Friendly Environments Across Europe: North and South*, Future City 19,
https://doi.org/10.1007/978-3-030-93875-8_8

paid off their mortgages. The housing policy debate has been dominated for some years by mechanisms to assist young people "get on the housing ladder," many of which have fuelled house price increases (Stephens, 2021).

The ascendancy of the market system means that in the UK, very few people ever have the opportunity to make their own housing but are reliant on developers who see commodities, not homes (Dove, 2020). The market, of course, needs people to move frequently and, to that end, has developed a set of dwelling types congealed around stereotypical ideas of the life course. City centre apartment blocks are seen as homes for young professionals; suburban houses with gardens are targeted at families (nuclear not extended) with children, while older people are judged as needing supported accommodation in age-segregated apartment blocks. Increasingly, this leads to our cities becoming spatially segregated with the inner city dominated by young people's housing including purpose-built student accommodation, while families and older people are housed in suburbia. The increasing diversity of UK households points to a growing gap between the housing on offer and how we might choose to live. Is it possible to challenge these housing cultures and break away from this linear approach?

8.1.1 The New Spotlight on Relationships

The lockdown response to COVID-19 has demonstrated the significance of place and community spirit at the neighbourhood level. Drawing on ONS weekly research into the social impacts of the pandemic, Stansfield et al. (2020) reported that:

- Nearly two-thirds of adults say other local community members would support them if they needed help during the pandemic.
- Three in four adults say they thought people are doing more to help others since the pandemic.
- Nearly two in three adults had checked in on neighbours who might need help at least once in the last 7 days.
- Over one in three adults had gone shopping or done other tasks for neighbours.

Findings from the Relationship Project (Robinson, 2021) suggest nine million people had been caring for others and 70% want to continue to do that including 39% who did little or no caring before that. A National Lottery Community Fund found that 46% of respondents felt that the pandemic will have had lasting impact on the amount people care about others. Can this community spirit outlast the pandemic? As Matthew Taylor (2020) asks, "can a crisis lead to a long-term positive change?" Does this reawakening of community spirit in neighbourhoods suggest an appetite for greater social connection and could we use this to develop spaces that promote interaction between age groups' (Biggs & Haapala, 2015, p. 2)? Is it also time to re-evaluate the considerable investment of time and love by elders in volunteering and family care work to finally shift our societal understanding of older people from passive beneficiaries to valued contributors to this more community-oriented new society? A truly age-friendly society not only benefits but also relies on older people as one older man eloquently states:

The idea of an elder friendly city is that people have to do things for you, you know put seats in the shops, it does not say anything about what older people can do about quality of life in the city by coming together and finding things they can take charge of. In a way if we look at the park group, we could see that as a way of enabling older people in this area to have a sense of togetherness. Community cohesion tracks out of this sort of thing. (Gilroy, 2009)

If cohesion and relationships become the objective of design, then what might our neighbourhoods look like?

A central focus on personal well-being, as the Scottish Government has pledged, can lead to exploring the challenge set by Architecture and Design Scotland's (2019) *A Caring Place* "to re-imagine our town centres as places to live meaningfully and to feel connected. Through rethinking what it means to care for our people, we can also take the opportunity to care for our places". The focus here is on social inclusion and reduction of loneliness that prior to the pandemic had emerged as new public health concerns. Our new increased awareness of mental health issues may provide a new impetus to building happier futures in pro well-being neighbourhoods.

In many ways, the time/distance templates adopted by many cities are a model here. With some cities labelling these as 15- or 20-min neighbourhoods (Seattle and Paris to name but two) or complete neighbourhoods (Portland, Oregon) or lifetime neighbourhoods (London), the intention is to create places that everyone can call home with easy-to-reach amenities.

Where access to public transport, basic amenities, local shops, cultural facilities, places to meet and relax, and green and open spaces are within easy reach of homes, and where facilities such as public toilets and seating are consciously planned into proposals at the outset, help to build cohesive, successful, and sustainable communities, and achieve social sustainability. (Mayor of London, 2016, p. 284)

Easy access to physical amenities is not enough, however, to create an inclusive, supportive neighbourhood, and the relational resources of good neighbours and friends in a place where you are recognised or at least not excluded is vital and is a project that involves everyone.

Reflections on the possible future place in a post-COVID-19 era are suggesting a renewed democratic offer in which town leaders "bring their communities together for an open, honest conversation about the future" in order to solve divides and build inspiring visions (Glover et al., 2020). On the small scale, there are many examples of creative collaborations with local people that enrich neighbourhoods and at the same time build confidence in the community's ability to be a voice for change.

If we flow these principles into housing, what might be the result, co-housing?

8.1.2 Co-housing: Providing Support and Purpose

Co-housing has a long history. Sir Thomas More, in his vision *Utopia* (1517), discusses the physical and social benefits of communal eating at the end of a day's labour for a society that works to promote mutual prosperity and welfare. The early American pioneers documented by Hayden (1976) developed co-housing and collaborative living arrangements that reflected their ideas of communitarianism and challenges to the established gender assignment of household tasks. In short, using architecture and design to reflect new ideologies. Since then, successive groups of people looking for better ways of living have taken up the concept of co-housing. Variants on co-housing, such as the kibbutzim in the newly founded state of modern Israel, were pioneering attempts to develop collaborative models of living that challenged patriarchy and the dominance of the traditional family structure. The Scandinavian feminists in the 1970s saw co-housing as a way of supporting family life by sharing domestic tasks. Recent reviews (see Tummers, 2016) suggest that co-housing can be a means of establishing local identities in the face of globalization. She also identifies how it supports a range of values from what is termed the "weconomy" of sharing, cooperatives and slow food and equal citizenship particularly around gender mainstreaming and ageing to climate change around energy and water management, recycling and conscious dampened consumption of goods. In the United States, the work of architects Durrett and McCamant (1989) has helped familiarize people with the concept and support practical decision-making through their extensive publications. More recently, the debates about the brittleness of the welfare state and "the fragile boundaries between positive and negative sides of self-reliance" (Tummers, 2016, p. 2029) have led to the emergence of both senior co-housing developments (Choi, 2004) and those which are consciously intergenerational.

In the UK, in spite of the central government setting aside grant aid to support what it terms "the community led housing sector" (Homes England, 2018), the development of co-housing is slight. Colquhoun (2020) found 20 active co-housing groups delivering 266 units in 2015 compared to the much more established housing co-operative sector where 836 groups are delivering about 170,000 units of housing. However, the ability of co-housing to promote a plethora of beliefs, lifestyle choices and ideologies suggests that in the face of an ageing demographic, concern over climate emergency and a desire to build on the community spirit shown in the pandemic, co-housing may become a more significant part of British housing.

In the UK, one of the best-known senior co-housing developments is New Ground in North London developed by Older Women's CoHousing (OWCH) that after a very long gestation period finally partnered with a housing association[1] to create 25 apartments with private and shared gardens, shared laundry and common room. What is most significant is that design decisions and day-to-day management decisions are made by the women themselves. The mission is to look out for each other rather than look after each other so individuals are expected to meet and arrange their own care needs. Every woman is expected to contribute to the running of New Ground to the best of her abilities from bookkeeping to garden work to assisting with the communal meal-making. Everyone can have a role to play and is needed. This is a fundamental step change from the expectations laid upon older people in the more familiar housing models. It is worth noting that the residents at New Ground reported that they fared well during the pandemic, in part thanks to the

[1] A registered social landlord is a not-for-profit organization developing and managing rented and low-cost home ownership dwellings.

Fig. 8.1 New Ground with shared and private outdoor space designed to support community interaction. (Source: Brenton, 2021 OWCH. More information: https://www.owch.org.uk/)

well-established relationships and mechanisms for mutual support. This is in marked contrast to older people's housing managed by others where the experience has been one of isolation with communal facilities often locked (Fig. 8.1).

More recently, an intergenerational model, Marmalade Lane in Cambridge, has attracted awards for its design and commitment to sustainability. The project offers 42 apartments and houses together with a common house that includes a gym, workshops and flexible spaces for meetings, eating together, classes, hobbies and children's play. A play street running through the development provides safe interaction space, while a shared garden provides areas for relaxation, play and food growing. Like all co-housing developments, the brief was created with the future co-housing members that develops not only housing and shared space that people want to live in but also builds relationships and a deeper sense of ownership of the community as a whole. This presents highly attractive housing that is built on the premise that people and community matter.

8.2 Housing That Works for All of Us

The path of private space tracks from crib to coffin with young people leaving home to find their own independent home and then moving to family-making and the potential to buy more internal and external space in higher value locations. Those ageing without children may have a different set of experiences and expectations of what they would like and can afford (Ageing without Children Consultancy, 2021). In later life, the message is "downsizing" to fewer rooms with more support and then the en-suite room of a nursing home, a normative journey up and down a ladder, but could we stay put and flex our housing around changing needs meeting the challenge set by Bimiendra Jha?

> *If we really want to challenge the status quo there needs to be a paradigm shift in the way housing is considered – not as a series of rungs which the consumer must climb from starter homes upwards, and then back down again in later life, but to a building type capable of adaptation, of morphing to support a growing family, and then accommodating an ageing one. This can only be achieved by looking at the needs of the occupant and how they change over time.* (Agile Ageing Alliance, 2019, p. 20)

One such approach was offered by the Joseph Rowntree Foundation that developed a 16-point standard that became known as Lifetime Homes (Sopp & Wood, 2001). These could, with minimal adaptation, accommodate a through floor lift for a wheelchair user and incorporate a bedsitting room with shower room on the ground floor from a large toilet and a dining room for an adult offspring. The intention was to allow family changes – the shift from young to adult children and the inclusion of an older parent or the acquiring of a disability by any family member – to be met. Since then, changes in UK building regulations have promoted these standards, but a more recent shift from mandatory to voluntary standards threatens the mainstreaming of lifetime homes that would make a major positive in the face of an ageing society.

A more recent and sophisticated version of lifetime homes that offers greater internal flexibility and attention to sustainability issues is the development goal of the Future Homes Alliance. The result is flexible housing templates with moveable walls that allow householders to alter their housing as lives change. These coupled with Passivhaus[2] standards and a commitment to low or zero carbon will ensure dwellings that promote quality of life, longevity of occupation and reduced running costs. What is innovative is not just the fusion of grey and green concerns but the determination to develop relationships and to commit to iterative learning loops drawing on knowledge transfer.

Relationships are at the heart of the organization, which was set up as a quadruple helix model in which the business sector, the academia, the municipality and the community sector might shed their single organization perspective and take a problem-centric and creative approach to the two societal challenges of our time: demographic shift to an ageing population and sustainability. Further relationships were developed from a rich programme of co-design and collaboration that drew in and drew out many different forms of expertise from older people and parents of young children to housing middle managers and occupational health therapists (Gilroy et al., 2021; Fig. 8.2).

---

[2] Passivhaus is a voluntary standard for energy efficiency in a building, which reduces the building's ecological footprint. It results in ultra-low-energy buildings that require little energy for space heating or cooling.

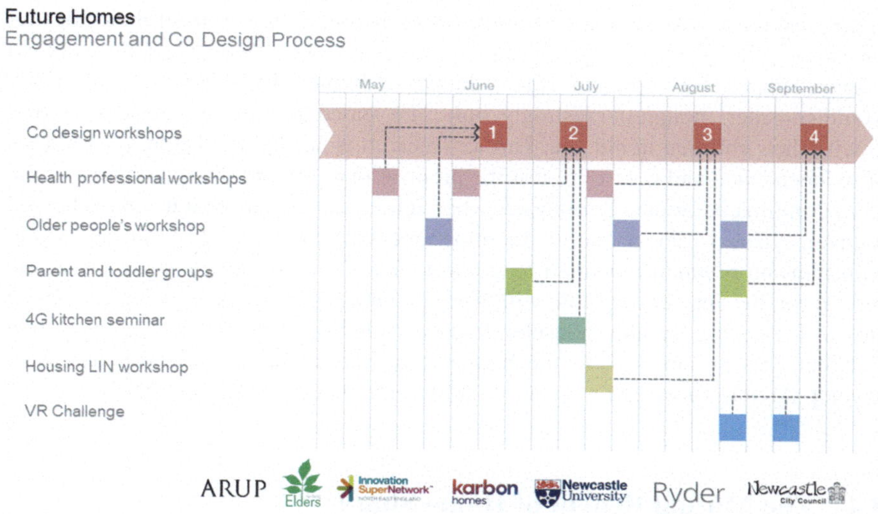

Fig. 8.2 The breadth of design involvement in developing Future Homes. (Source: Ryder Architecture Ltd., 2019)

Many of these voices are largely unheard in housing design, challenging the existing silo approach and creating greater opportunities for reaching an understanding of what people want from their housing. The commitment to relationships goes beyond the initial design phase to recast householders as mutual learners who will input their own experiences of living in these dwellings. Together with streamed environmental data, the organization grows an evidence base to test its approach and challenge itself to continuous improvement. This innovative method of designing houses creates templates that offer a new inclusive approach with a balance of product and process. Can this be mainstreamed?

8.3 A New Appreciation of Family

One of the fundamental building blocks of any community is family. This can no longer be adequately encompassed by the idea of a couple with two children. One of the trends is the percentage of young adults (under 35) who are living with their parents. In 2017, this was 26% of that age group and a predicted rise to 33% by 2040. Another trend is the slight increase in multigenerational living understood as parents and adult children; parents, adult children and their children; or adults with their older parents. There is considerable evidence that this extended family might improve older people's sense of worth and the mutually beneficial exchange of support and practical assistance. Statistics on the rise of multigenerational living are muddled and, at best, demonstrate a slow drift upward (Cambridge Centre for Housing and Planning Research, 2017). Glover's (2019) research that included a polling exercise gives new insights into what the public wants from their housing and what principles might guide and future-proof our dwellings. This included nascent support for caring of older relatives at home which 31% of participants would consider, though interestingly, only 21% stated they would consider moving in with their children in old age. Further research is needed to explore the cultures of independence and control to test the premise that "we are more receptive to the idea of multigenerational households when it does not require that *we* move home" (Glover, 2019, p. 28). If one of the impediments is the belief that current, newly built housing is much too small to support such living arrangements, then, in the short term, the answer might lie with local planning authorities reducing the obstacles to increasing liveable household space including conversion of garages into so-called granny annexes. In the longer term, the challenge to developers is to develop flexible space that responds to household diversity (NHBS, 2017).

8.4 The Mutual Benefit of Home Share

The rising number of older people, particularly those ageing without children, and the lack of affordable options for younger people can both be addressed by what is called Home Share. This is a highly sustainable model since it does not rely on new build or creating new communities such as co-housing, but which enables a different use of existing housing and space. Like the OWCH development at New Ground, the ethos is that both the younger and older person may look out for each other and, in the case of the younger person, may advocate for their older host. The older person provides a room and shared facilities free of charge to the younger person who in return provides companionship and about 10 h of help with agreed tasks that might include shopping, gardening, support with technology or housework. Older people can stay in neighbourhoods they are familiar with and where they have relationships/known acquaintances, which is evidenced as being beneficial to health and well-being (Thomese et al., 2018). The younger person has a home to return to and someone to share their thoughts and feelings. Evidence, particularly through the lockdown response to COVID-19, has amply demonstrated that loneliness levels were higher among young people (Mental Health Foundation, 2021), and one organization reports a 23% increase in householders registering to share their home (Cooke, 2021). This model that relies entirely on building relationships again points to the need for homes where space can be used flexibly and challenges established concepts of what "supported living" might be like and indeed who it is for.

8.5 Discussion

In an ageing society, there is a challenge of responding positively to the increasing heterogeneity of older people and their family structures. This chapter has drawn on examples to consider whether they offer new ways of thinking about housing and homes by decentring the debate to focus on relationships. At the heart of these must be the need to draw on relationships that in turn may drive better ways of thinking about the place and the housing within it. Such transformation can deliver benefit not simply to older people but people of all ages. For this to happen, we need to reinvigorate public debate about the use of space and how we may develop or repurpose our homes to live the best lives we can.

References

Agile Ageing Alliance. (2019). Why is a steel company interested in housing. In Agile Ageing Alliance (Ed.), *Neighbourhoods of the future* (p. 20). Retrieved from: https://www.agileageing.org/page/neighbourhoods-future/

Ageing without Children Consultancy. (2021). *The invisibility of people without children is a life course issue which has a drastic impact on policy around age and ageing*. Retrieved from: https://ageingwithoutchildrenconsultancy.com/2021/02/25/the-invisibility-of-people-without-children-is-a-life-course-issue-which-has-a-drastic-impact-on-policy-around-age-and-ageing/

Biggs, S., & Carr, A. (2015). Age-and child-friendly cities and the promise of intergenerational space. *Journal of Social Work Practice, 29*(1), 99–112.

Cambridge Centre for Housing and Planning Research. (2017). *Multigenerational Living: An opportunity for UK House Builders?* Final Report to NHBC Foundation. Retrieved from: https://www.housinglin.org.uk/_assets/Resources/Housing/OtherOrganisation/Report-Multigenerational-Living.pdf

Choi, J. S. (2004). Evaluation of community planning and life of senior cohousing projects in northern European countries. *European Planning Studies, 12*(8), 1189–1216.

Colquhoun, G. (2020). *Housing by the community, for the community: An assessment of the value for money of community led housing in England*. Capital Economics. Retrieved from: https://www.communityledhomes.org.uk/sites/default/files/resources/files/2021-02/final-report-capital-economics-housing-community-community-sept-2020.pdf

Cooke, C. (2021). *In good company: The need for Homeshare and intergenerational living has never been greater*. Blog, Housing Learning, and Improvement Network. Retrieved from: https://www.housinglin.org.uk/blogs/In-good-company-the-need-for-Homeshare-and-intergenerational-living-has-never-been-greater/

Dove, C. (2020). *Radical housing*. RIBA.

Durrett, C., & McCamant, K. (1989). *Cohousing: A contemporary approach to housing ourselves*. New Society Publishers.

Gilroy, R. (2009). *Elder Count*. Copy available from author r.c.gilroy@ncl.ac.uk

Gilroy, R., Aitken, D., & Miller, P. (2021). Future homes: Developing new responses through new organisations. In S. Davoudi, M. Steer, M. Shucksmith, & L. Todd (Eds.), *Hope under neoliberal austerity* (pp. 187–202). Policy Press.

Glover, N., Carr, H., Smith, J., & Phillipson Brown, S. (2020). *The future of Towns*.

Glover, B. (2019). *Future homes*. Demos.

Hayden. (1976). *Seven American Utopias: The architecture of communitarian socialism, 1790–1975*. MIT Press.

Homes England. (2018). *Community housing fund*. Ministry of Housing, Communities, and Local Government. Retrieved from: https://assets.publishing.service.gov.uk/government/uploads/system/uploads/attachment_data/file/772919/CHF_prospectus_-_FINAL_updated_16.1.19.pdf

Mayor of London. (2016). *The London Plan: The spatial development strategy for London consolidated with alterations since 2011*. Retrieved from: https://www.london.gov.uk/sites/default/files/london_plan_march_2016_malp_-_frontispiece_and_ch1_context_and_strategy.pdf

Mental Health Foundation. (2021). *Loneliness during coronavirus*. Retrieved from: https://www.mentalhealth.org.uk/coronavirus/loneliness-during-coronavirus

NHBC Foundation. (2017). *Multigenerational Living: An opportunity for UK House Builders*. Retrieved from: https://www.nhbcfoundation.org/wp-content/uploads/2017/06/NF77-WEB.pdf

Robinson, D. (2021) *Turning to the Light*. Relationships Project. Retrieved from: https://relationshipsproject.org/reports/turning-to-the-light/

Sopp, L., & Wood, L. (2001). *Consumer and industry views of lifetime homes*. Joseph Rowntree Foundation. Retrieved from: https://www.jrf.org.uk/report/consumer-and-industry-views-lifetime-homes

Stansfield, J., Mapplethorpe, T., & South, J. (2020). *The community response to coronavirus (COVID-19)*. Public Health Matters, Public Health. Retrieved from: England. https://publichealthmatters.blog.gov.uk/2020/06/01/the-community-response-to-coronavirus-covid-19/

Stephens, M. (2021). Building back better' after Covid: Lessons for housing policy from the 'lost decade. In M. Stephens, J. Perry, P. Williams, G. Young, & S. Fitzpatrick (Eds.), *UK housing review 2021*. Chartered Institute of Housing.

Taylor, M. (2020). *The path from crisis*. Royal Society of Arts. Retrieved from: https://medium.com/bridges-to-the-future/the-path-from-crisis-6d3f83c96d0b

Thomese, F., Buffel, T., & Phillipson, C. (2018). Neighbourhood change, social inequalities, and age friendly communities. In T. Buffel, S. Handler, & C. Philipson (Eds.), *Age friendly cities and communities* (pp. 33–49). Policy Press.

Tummers, L. (2016). The re-emergence of self-managed co-housing in Europe: A critical review of co-housing research. *Urban Studies, 53*(10), 2023–2040.

Rose Gilroy is Professor of Ageing, Planning and Policy in the School of Architecture, Planning and Landscape, Newcastle University, UK. Her work focuses on the transactional relationship between older people and their place that embraces how people individually and collectively may influence their environments. Her latest book "Planning for an Ageing Society" was published by Lund Humphries in February 2021. Rose is committed to change and much of her work is action-oriented. She is the founder of the Future Homes Alliance.

Chapter 9
Housing for an Active Ageing: Learning and Outcomes from the Danish Paradigm and the Application to the Spanish Context Lugaritz Lifetime Homes in Donostia – San Sebastián

Heitor García-Lantarón, Mayte Sancho Castiello, and Maider Azurmendi Ormazabal

Abstract The aim of this chapter is to describe the process that has been carried out for the adaptation of the housing for the elderly model to the Spanish context. Under the perspective of architectural design, new solutions have been implemented for defining an alternative to institutional models. The interest of Lugaritz lifetime homes lies in its innovative approach, which works from two different perspectives, home care and design, both with a common purpose: ageing in place. It is also important to highlight its aim to create knowledge that makes this project a replicable and scalable example from a perspective based on participation, inclusion and design for change.

Keywords Senior housing · Lifetime homes for the elderly · Active ageing · Home care · Design for change · Inclusive design

---

H. García-Lantarón (✉)
Architectural Design Department, Matia Fundazioa, Donostia-San Sebastián, Spain
e-mail: heitor.lantaron@matiainstituto.net

M. Sancho Castiello
Matia Research Institute, Donostia-San Sebastián, Spain

M. Azurmendi Ormazabal
Community Co-operation, Matia Fundazioa, Donostia-San Sebastián, Spain

E. Pozo Menéndez, E. Higueras García (eds.), *Urban Design and Planning for Age-Friendly Environments Across Europe: North and South*, Future City 19,
https://doi.org/10.1007/978-3-030-93875-8_9

9.1 Introduction

9.1.1 Housing and Active Ageing: Two Converging Trends

The relationship between architecture and ageing is very recent within the historical process by which humanity has shown interest in old age. Within this relationship, the influence of the social and health sciences has had a great impact on the physical environment. Through their contributions to understanding the ageing process, architectural design has also been influenced. However, this impact has been more focused on the institutional-type care home models that were already largely developed.

From a more general perspective, the importance of the contributions made by the social sciences does not lie exclusively in the understanding of the ageing process but also in how these currents have implemented active policies that support the demands of the elderly, as well as their full participation in society. The current trend is defined by active ageing (WHO, 2002), which includes and expands on the concepts used by previous theories developed by gerontology.

Active ageing (AA) stresses the importance of older people's participation in decision-making and therefore on maintaining control over their lives. It is the result of an evolution from a first model based on the fragility of the person, according to the ecological theory of ageing the environment became a fundamental element in the ageing process (Lawton & Nahemow, 1973), to a more balanced vision, such as successful ageing (Rowe & Kahn, 1987) and emphasising personal competence as another essential element for successful ageing, the selection-optimisation-compensation model (Baltes & Baltes, 1990). In other words, conceptually, it has evolved from an environment (E) model to a person-environment (P-E) model and finally to a person-environment-activity (P-E-A) model.

For the World Health Organisation, AA illustrates the optimal idea of ageing and in its definition, subjective and objective aspects were included. It relates ageing to the individual, i.e. to his or her personal characteristics, whether physical or psychological, as well as to his or her social capacities. It includes the relationship established with the physical or social environment and the influence or support that this may entail. It also incorporates a fundamental factor that has emerged strongly during the COVID-19 crisis: the importance of long-term care (LTC) and the weakness of social and healthcare services, particularly home care provision, in most countries. Finally, AA is related to the "new paradigm" in ageing research that emerged in the last decades of the twentieth century and which is centred on a much more positive view of ageing (Fernández-Ballesteros et al., 2012), incorporating people's initiatives and participation as a collective value with great social significance.

Within the previously mentioned relationship between architecture and ageing, housing is the most recent alternative. It is also part of the current debate on how housing has been largely overtaken by the complexity of the contemporary user. The city and the house are linked to lifestyles. They are both the result and the driving force, the result of a general process that began almost a century ago, during which housing has been specialised and optimised, clearly improving functionality and comfort conditions. But at the same time, it has been the driving force behind a process of homogenisation of the domestic and public space by ignoring the continuous processes of change in lifestyles. Both current housing and cities offer homogeneous solutions based on a simplified and abstract user type, practically non-existent, a young and healthy type of user. Furthermore, the housing model is also based on a nuclear family model that is increasingly limited, as the ways of living together are changing. In parallel and as an alternative, the current housing model offers specific solutions to less "conventional" users. This has an impact on social differentiation and stigmatisation of specific population groups represented by quotas imposed through mandatory regulations.

In the specific case of housing for the elderly, there has been a significant increase in the number of solutions in recent years. The increase in life expectancy and the so-called "age revolution" has generated new user profiles that demand solutions in line with their needs and desires, as an alternative to care homes. Consequently, the market has noticed the potential of this group, representative of the baby boom and therefore, attractive for a model based on sales volume.

This process began in the 1960s, when institutional care models were strongly questioned, especially in Northern Europe. The main reasons for this were due to a combination of two factors: on the one hand, a greater sensitivity to the importance of autonomy, privacy and the older adults' right to choose (Houben, 2001). On the other hand, during the 1970s, it became clear that the model of care homes was not economically sustainable in the long term. The main reason for this was that with increasing life expectancy, there was a shift from short to much longer care, with the added factor of dementia emerging associated with most dependency situations. New policies in the Nordic countries resulted in a reduction in the number of elderly people living in institutional models. Since the end of the 1980s, the housing model has been strongly promoted, supported by a main focus on person-centred home care services. At the same time and following an opposite trend, Southern European countries started to grow economically, so the institutional models were gradually introduced. In the 1990s, in all European countries, the trend was towards greater implementation of home care and other solutions for ageing in one's own familiar surroundings or in the place of choice (Houben, 2001).

It is important to highlight that, albeit exceptionally, the housing for the elderly model has been existing for almost a century. This type of housing has been scarce, generally included within the quotas imposed by social housing regulations. The first examples can be found in Denmark from the 1930s onwards when housing emerged as a social service. At that time, the main goal was to respond to the need for improving housing living conditions. In the case of the elderly, the first examples were promoted by the trade unions to provide decent accommodation for their pensioner members. Thus, the right for living in those homes was linked to the user's working life, so that in order to have access to them, one had to be retired. Known as "pensioner housing", this conventional housing model inaugurated one of the current trends: housing not designed for the elderly but aimed at them and with limited access.

Housing for the elderly, as a housing model, and active ageing, as a social and political objective, represent two recent, current and converging trends.

Contributing to this trend, the aim of this chapter is to describe the process that has been carried out for the adaptation of this housing model to the Spanish context. Under the perspective of architectural design, new solutions have been implemented for defining an alternative to institutional models. The interest of Lugaritz's lifetime homes lies in its innovative approach, which works from two different perspectives, home care and design, but both with a common purpose: ageing in place. It is also important to highlight its aim to create knowledge that makes this project a replicable and scalable example from a perspective based on participation, design for change and inclusion.

9.1.2 The Housing for the Elderly Model: Definition and Types

From a general point of view, what characterises the elderly housing model is that the care and services received are separated from the domestic space, turning them into external services that are adapted to the needs of the users. This requires a commitment to home care services that is not sufficiently extended in most developed countries. Without this indispensable and determining condition, it is impossible to transform the model towards a housing model and to respond to the demand for ageing in place.

In general, the housing for the elderly model share a number of characteristics: promoting independence, reducing social isolation, providing an alternative to residential or institutional models of care, providing residents with a home for life and improving the residents' quality of life (Croucher et al., 2006). Thus, the model can be defined as:

> The homes designed with the ability to adapt to the individual changes, as well as to the social and healthcare support or needs derived from the ageing process; allowing greater autonomy and independence for longer time (García & Soler, 2015).

These are attractive housing models that promote an independent lifestyle and social participation and encourage active ageing. In this trend, defined as the third paradigm (Houben, 1997), older people seek to live in appropriate, self-contained accommodation with common facilities and freedom for users to share daily activities and help each other. In other words, they combine private living, common facilities and shared activities (Lindstrøm, 1997).

As mentioned before, in the last two decades, a number of these housing initiatives have emerged. Although they go by different names, they can be classified according to their main focus:

- Focus on social relationships within the community for sharing experiences and fight loneliness. The example that best illustrates this trend is self-promoted models such as senior cohousing. In this type of accommodation, the users' group participates very actively in the process of defining the model. Moreover, it is a group of seniors who want to form and live in a community with strong social ties.

- Focus on the services they offer, targeting a young-old user with a huge potential and attractiveness for the real estate market. The example that best illustrates this trend is the 50+ initiatives. It is related to the interest that this model is awakening among private investors and developers, who are trying to attract this age group. They focus on a series of services in line with the needs of the elderly, which can be changed and increased according to the user own demands.
- Focus on designing a suitable environment for ageing at home. The example that best illustrates this trend is the lifetime homes model, which is based on the assumption that all the difficulties currently encountered in ordinary housing can be eliminated with simple design modifications. The aim is to ensure that no one is excluded due to badly designed solutions leading to architectural discrimination. Its approach to the problem is through architectural design, for example, the Joseph Rowntree Foundation created a set of standards in 1989. It is not that we should always stay in that home but that if we want to, we can. It is therefore important to note that lifetime housing is not special housing but conventional housing that is suitable for almost anyone. Their approach is completely inclusive. Along the same lines and with a similar philosophy, the Humanitas Foundation coined the term *levensloopbestendige*, which has spread and is in common use in the Netherlands.

In general, in every type, whatever the country, the main objective is to be able to age in an environment that responds to your needs. They can exist under different formulas, either rented or owned, promoted by the public sector, non-profit associations or the private sector. They are normally demanded by people who have not yet retired but see the need to find a model that meets their demands and that they cannot find in the conventional market.

Significantly, these three different approaches rarely coexist and, although they respond to the trends described above, they remain being minority initiatives. The majority option is still to remain at one's own home, although it is important to highlight that there are indicators that relate this option to the lack of alternatives (Fernández-Carro & Evandrou, 2014). Therefore, the housing model for the elderly emerges as an alternative to conventional housing and institutional care models. It is aimed at responding to the demands of the elderly who refuse to allow dependency, or the need for care, to determine their way of life and where to live.

9.1.3 Reality in Spain

The end of the Second World War marked the moment in which the paths of accommodation for the elderly in Spain began to diverge from those followed in most of Europe. Among the many issues that make the Spanish model different, it is worth highlighting those related to cultural and family structure aspects but also to historical, economic and political aspects. While in Europe the development of social services evolved rapidly after the war, in Spain the Franco regime delayed the process. Although it introduced numerous specific social action measures, it did so with a paternalistic mentality and with great organisational dispersion. As a result, during this period of our history, Spain did not yet have a level of social services coverage similar to those available in other European countries (Alemán, 1993). Since the 1978 Democratic swift, the system focused on the consolidation of social services with a universal character, although in practice and due to the political and economic events that have occurred since then, this process has also been conditioned.

On the subject of accommodation for the elderly, while in the 1980s and 1990s in Europe the model was being transformed towards a progressive deinstitutionalisation and a commitment to person-centred services at home, in Spain we lived through the golden age of the construction of high-capacity residential care homes, despite the devastating evaluations of psychiatric and elderly care institutions that our neighbours managed (Sancho & Rodriguez, 2001). This trend has continued in Spain to the present day. While in the rest of the developed countries the tendencies are in the opposite direction, in our country institutional model is still the most widespread. At this point, it is important to highlight how much the COVID-19 has impacted on the Spanish system and society, opening to a general discussion on questioning the residential care model.

The number of residential places has experienced a clear growth, especially since the implementation of the Law for the Promotion of Personal Autonomy and Care for Dependent Persons (LAPAD 39/2006). It currently stands at 384,251 (Abellán et al., 2021), which means a coverage rate of around 4.2%, close to the European average. The current situation of the residential care model in Spain is characterised by the great variety of centres in terms of size, ownership, architectural design as well as in the care and services they offer.

Moreover, in Spain, the residential care models promoted by public administrations and managed mostly by private initiatives, whether for-profit or not-for-profit, are almost entirely intended for people in a situation of severe or very severe dependency, around 75% of whom suffer from cognitive impairment. Admission to a residential centre is still somehow perceived as a "surrender" of the family responsibility for care, traditionally assigned to the women in the family. Ultimately, care was settled in the domestic, invisible, feminine sphere.

Both the unstoppable process of women's incorporation into the world of work and the evolution of situations of dependency, with their consequent complexity and intensity, are promoting a much greater acceptance of the need for professional services. All of it within the framework of the recognition that care today is a matter for society in general.

There are few alternatives to the majority option of the institutional model. There are timid initiatives such as sheltered housing for the elderly, which still represent an insignificant percentage compared to residential homes. These homes, whose name denotes a protectionist conception of old age (*viviendas tuteladas*), are aimed at autonomous elderly people who have socio-economic or housing problems and who must leave them when dependency appears. Currently, around 10,000 people use these homes.

Regarding home care services, they developed very little, with a generalist vocation and dedicated almost exclusively to cleaning tasks. Since the approval of the LAPAD, a clear development has been observed. However, home care services still appear too often as an "isolated" service that is not complemented transversally with other interventions in terms of housing, accessibility within the home and in the immediate environment, support products, voluntary accompanying action and, in general, the promotion of a collaborative and protective community environment that makes the daily life friendly to dependent persons (Sancho & Martínez, 2021). On the other hand, its hourly intensity continues to be scarce and with significant territorial differences, which means that it still does not meet all needs. Although it is a service that has improved greatly in its professionalisation, it has not yet faced – at least in a generalised manner – the necessary conceptual, organisational and management changes. It is characterised by atomisation and lack of coordination between the different administrations responsible for care. Furthermore, the budgetary imbalance between home care services (300.36 €/month) and residential care (1,723.81 €/month) is excessive.

On the other hand, and following the trend in other developed countries, in Spain, we also find self-promoted housing alternatives. The number of examples is still far from that of any other European country, and their dispersion over time, as well as territorially, makes it difficult to evaluate the impact of this phenomenon in Spain. What characterises the Spanish model is the lack of analysis of the Scandinavian experience, where the sociocultural characteristics, as well as the solid development of a universal and extensive home care system, are decisive keys to the progress of this model, in accordance with the wishes and preferences of its citizens. When this analysis does not exist, and therefore there is no questioning and understanding of the other countries' experiences, they become intermediate models between one's own home and the nursing home. In other words, senior cohousing in Spain does not usually prevent from moving into an institutional model. In any case, they are still in the minority and the projects that have been successfully built are few and far between.

9.2 Process

The definition of the Lugaritz lifetime homes has been nurtured from various aspects along the itinerary followed during the definition process. This journey began in 2017 and is still ongoing. From the design point of view, it is based on the European experiences, especially from the Danish ones. But it also draws on the experience in terms of the care system in those European countries where the commitment to the housing model is clear and is also supported by a home care system. On the other hand, it has been necessary to adapt it to the context in which it is located, which has been contrasted and constructed together with different local agents and experts through a participative process.

9.2.1 State of the Art

Denmark is considered a reference due to its large experience in implementing the model of housing for the elderly, supported by a clear commitment to home care services. What sets Denmark apart from other countries is that the Danish government established the first ever Commission on Ageing (*Ældrekommisionen)* in 1979. The aim was to respond to the rapid ageing of the population while maintaining the quality of services and paying attention to the views and wishes of the elderly. The *Ældrekommisionen* worked on developing a unified and coherent policy for the elderly, reviewing existing policies and setting out future lines of action. All aspects of living conditions were addressed, and housing was therefore a key issue for this commission. Therefore, Denmark brings experience in both the transformation of the care model and the design of housing alternatives for the elderly.

With regard to the transformation of the care model, the main objective set by the *Ældrekommisionen* was to enable the elderly to maintain as much as possible a quality life according to their own choices, understanding old age as a continuation of the person's own life and thus ensuring that they could choose where they wanted to live and for as long as possible (Danish Ministry of Housing, Urban and Rural Affairs, 2010). For doing so, the provision of public services became a key aspect:

> The key word is flexibility, both in housing and services. This means that institutions should be replaced by appropriate housing for the elderly and provided by external and flexible care services that meet the needs of the elderly (Gottschalk, 1995).

In three reports presented by the commission, specific recommendations were detailed to improve the living conditions of the elderly, including housing, health and social services, as well as the training of qualified personnel to provide such services. These three reports together provide the necessary guidelines for reconstructing policies for the elderly, focusing on continuity, improving self-determination and increasing independence through a person's own resources (Lindstrøm, 1997).

Since then, efforts have focused on ensuring that people who preferred to could stay in their own homes as long as possible (Gottschalk, 1995). This principle was reinforced by the Accommodation for the Elderly Act, *Ældreboligloven*, of July 1987, which aimed at the gradual replacement of existing old people's homes by new, modern social housing for the elderly. Furthermore, in January 1988, the right to build new homes was permanently revoked (Gottschalk, 1993). This was in line with the wishes of future generations of elderly people and increased the interest in alternative housing models. Due to this policy of separating housing and care functions, the system changed from an institutional long-term care model to a wide range of home care services with special focus on the local environment. Since then, housing and care services are considered independent entities and therefore are kept separate.

In this process of transformation of the care model, another key factor was the increase of people with dementia. People with dementia do not cope well in big institutions and nursing homes. They benefit from living in an easily recognisable environment, in close contact with a limited number of residents and caregivers (Lindstrøm, 1997). Most patients over 85 years of age continue to live at home, but from that age onwards, the dependency ratio increases exponentially, so there is a great demand for accommodation specially prepared to treat those who can no longer live independently. Consequently, and as a result of this demand, the Danish care housing reform (*plejeboligreformen*) was introduced in 1996. Its aim was to increase the number of specialised housing units for people with needs for continuous care and a high degree of dependency (*plejebolig*) (Lindstrøm, 1997).

In summary, the process developed by Denmark for changing its care model reflects the two trends followed by the rest of most of the European countries over the last 30 years, on the one hand, a commitment to personalised home care services to promote ageing in place and on the other, the specialisation of long-term care institutional models.

Throughout this process and as a result of the work developed by the *Ældrekommisionen*, it became clear that there was an absolute need for housing in accordance with the demands and wishes of the population. As a result of this process and the characteristics of the Danish system, today coexist a number of different housing alternatives for the elderly sharing common design features. It is difficult to make a comparison between the Danish and the Spanish system, but it is important to note that beyond the characteristics of each welfare model, the interest lies in the approach:

The Danish welfare state model is based on a coordinated and locally provided range of social services that are free, universal and public. Therefore, home care, whether temporary or permanent, does not imply any kind of social stigma or loss of status. On the contrary, it is a right acquired through years of helping to make it possible. This has a direct impact on the characteristics of housing for the elderly. On the one hand, it contributes to making them attractive models, with a high degree of demand. In fact, in some cases, the waiting list for access can be several years long. On the other hand, neither are they residual models within the architectural design and housing market. This encourages a multitude of architects to participate in the ongoing design process, contributing to the high quality of their architectural definition. Both attractiveness and quality mean that this type of accommodation is perfectly integrated into the housing options and thus does not become a ghetto for the elderly.

Due to the Danish welfare state and their support to ageing in place, as the main goal of the care model, only a very small percentage of elderly people decide to change to a new home. In most cases, the decision has to do with the need to combat loneliness and to find people with the same interests. This has a major impact on the design, as elderly people's housing models are characterised by responding to the need of common spaces for performing shared activities. Regardless of the size of the building and the number of users, these types of spaces share two characteristics, flexibility in the use of these spaces and diversity in size and location. Both the shared activities and the common spaces favour interaction between users, as well as with the rest of the community.

9.2.2 Participation and Stakeholders

As a complement to the study of examples and initiatives in other countries, a participatory process is being developed. This is aimed at detecting local needs and also at adapting and specifying a housing project that provides a solution to these needs. Different stages have been covered, from the design criteria to be implemented to those relating to the management and type of services offered. The process is still open and continues incorporating proposals from the community, with the understanding that the solutions adopted must respond not only to actual needs but also be flexible enough to adapt to future conditions. The process involves different agents from the public sector, associations of potential users and their families, local community (neighbours) and experts in this field. Specific workshops and in-depth interviews have been carried out.

With regard to design, the needs and interests identified affect various scales of the project:

Urban context

- Solve mobility and accessibility of the neighbourhood
- Quality of public space
- Include commerce and services that complements actual neighbourhood offer
- Include an intergenerational approach
- Integrate housing within the rest of the community

Building

- Encourage social networking and shared activities
- Mixed-use common spaces as a meeting point with the community
- Flexible and adaptable common spaces according to common interests

Housing

- Total accessibility
- Spacious and flexible spaces
- Quality of materials
- Importance of light and orientation
- Need for garage and storage room

With regard to the management and services offered, the main needs identified are:

- Include existing community services
- Include volunteers

- Affordable prices
- Inclusive access criteria
- Include a case manager
- Possibility of adding à la carte services
- The need to make common areas more dynamic
- Need for support and respite for carers

A team made up of professionals from Matia has been appointed to make this project a reality. It brings a multidisciplinary approach that changes and evolves with the process. This team aims to ensure an inclusive and replicable solution. Thus, it works on aspects such as the design of the environment, economic viability and future maintenance, as well as future home care services management and research.

9.3 Result: Lugaritz Lifetime Homes for the Elderly

The Lugaritz lifetime homes for the elderly is the result of previously described process for adapting the model to the context of the city of Donostia-San Sebastián. The aim of this project is to demonstrate that there are alternatives to institutional care models that respond to the demand for ageing in place and the need for long-term care. It aims to be a possible solution within the diversity of offers that must exist. It is a project that identifies opportunities in accordance with the wishes and preferences of the elderly in the physical and sociocultural context in which it is located. With the premise of separating accommodation from care, the aim is to research and generate knowledge about the characteristics that makes possible ageing in place in a sustainable way, even when a high degree of support and care is needed. It is also proposed as an urban solution, as an initiative for advancing in the construction of a friendly city to the elderly but also to the rest of citizens.

Lugaritz lifetime homes is promoted by Matia, a non-profit Foundation, whose mission is to accompany people in their ageing process with support and person-centred services, as well as to generate knowledge through applied research for contributing to elderly people's well-being. It is an innovative project in the field of housing for the elderly in Donostia, San Sebastián, co-financed by the IMSERSO (Spanish Institute of Social Services) and in collaboration with the City Council and the Regional Council.

The whole project incorporates different uses in 14,400 sqm: housing, day-care centre, offices, commerce and parking. It includes 51 homes, 43 two-room apartments and 8 three-room apartments, with parking places and storage rooms, as well as common facilities and outdoor areas. It is currently under construction, with an expected completion date of summer 2022. Therefore, this is an opportunity to assess the effectiveness of this type of initiatives, which are currently non-existent in Spain and the Basque Country.

These are affordable rented homes for couples, families and cohabitation units, in which one of the persons is in a situation of dependency, with the wish that this will be their home until the end of their lives. The aim is to support continuity of the life project, facilitating the process through design, care and a protective environment:

- Design: design and how it can facilitate and improve the conditions of each person during the ageing process. A design with a person-centred and inclusive approach, understanding that ageing is a process and therefore design must be oriented towards change. The scales at which action is taken range from the home itself to shared communal spaces and the urban environment.
- Care: person-centred care and support will be provided from outside and will be co-funded according to the legal benefits and services established:
  - The benefits and services that people living in Lugaritz are entitled to receive under the LAPAD.
  - Local services provided by the private sector that can complement previously mentioned services. Matia Foundation will be able to offer services of these characteristics while services provided by other agents in the neighbourhood and the city will be also available. Users will choose the best option for their well-being in Lugaritz.
  - The involvement of community agents and voluntary associations will have a very important place and role in the life of Lugaritz. The social fabric of the neighbourhood and the city will be part of the life and support of this care provision. In this way, the integration of Lugaritz into the neighbourhood and the city will be achieved.
- Protective environment: the third aspect that characterises these homes is the fact that they are located in a protective environment. The experience accumulated by Matia Foundation throughout its existence and the knowledge generated in projects carried out for the different public administrations document the introduction of professional profiles such as the Case Manager. This professional plays a coordinating role between the people and the administration, ensuring that the motto "one person, one plan", i.e. person-centred care, becomes a reality.

The homes respond to the challenge of long-term care in episodes where dependency and the necessary support is of high intensity. The condition for access to them is that one of the persons is in a situation of dependency. Dependency may be physical or generated by processes of cognitive deterioration such as dementia, the consequence of a stroke, illnesses such as Parkinson's disease, etc. Furthermore, during the analysis process, a new need has been detected: couples who, due to their age, are very challenged to continue taking care of dependent children.

The aim is to offer solutions for a normalised daily life, integrated into the environment, which improves people's quality of life. The aim is to provide a protective environment that makes it easier for them to continue with their life project. Both, the person in a situation of dependency and the caregiver.

The reason for limiting access to other types of users is based on the need to explore the solutions on LTC to be adopted, with the understanding that by offering solutions to those with the greatest needs, the rest benefit. This is a pilot project aimed at evaluating the impact of alternatives to residential care homes within the current care system. It will help to build an alternative that is not yet recognised in the portfolio of services.

In short, it is a housing project for people who need care, caregivers who need an environment that helps them to continue caring and, finally, at those people who need a protective environment to continue with their life project. It is an alternative to residential care homes, focused on people with the same support needs as those who currently live in them. They have a person-centred, inclusive design that allows for continuity even when the intensity of care is increasing and can last up to 24 h.

9.3.1 Architectural Design Criteria

The principal design criteria is that these homes can adapt to the changes experienced by the person during life, as well as to the needs they may have. The aim is to guarantee home care and for a longer period of time. In addition, they provide greater autonomy and independence in the personal sphere, together with the possibility of strengthening social network and interaction with the community. The characteristics that make this possible at the different scales of action are described below, starting with the most singular and innovative element, which are the homes and their common areas, up to the urban scale of the surroundings.

9.3.1.1 Homes

The aspect that had the greatest influence on the housing design was change, understood as the standard life condition. During lifetime our physical and mental capabilities change and therefore can be challenged by unappropriated design solutions. The physical design, layout and construction of buildings and places can confront people with hazards and barriers that make the built environment inconvenient, uncomfortable or unsafe for everyone to use and may even prevent some people from using it at all (Hanson, 2001). Design for change puts architectural design focus on human lifespan progress. The aim is how design for change support us to live in our own home regardless of disability, age, need of care or any other personal condition. This means expanding users' capability by concentrating on usability and inclusiveness.

On supporting this approach and answering a specific demand located during the definition process, the size of the homes also plays an important role. The different homes have between two and three rooms and were not sized down due to two main arguments:

- We all have some kind of emotional attachment to our personal objects, including furniture. They are containers of our memories and experiences, and therefore we should not be forced to give them up because of lack of space. At the same time, space is necessary for continuing being hosts and playing a central role in the family. Space should be sufficient and not reduced just because we are getting older.

- In addition, in Spain, when permanent care is needed, it is common to hire a live-in carer, who must have their own bedroom and personal space. On other occasions, due to the care that one of the couples must receive, the double bed will have to be replaced by a hospital bed, reducing the space in the bedroom and therefore making it necessary for the other person to occupy a second bedroom.
- Finally, in addition to the diversity of possible users, we must add the need detected for an elderly couple with a disabled child in their care. In this case, the family unit may need double support and care, which is why a certain number of dwellings with three bedrooms are included.

The second aspect that has had the greatest influence on the definition of housing has been the need to reconcile the idea of being able to receive care with the need to design a domestic environment that is as normalised as possible. In this balance between private living and care, the home layout plays a central role. The dwellings follow a fairly conventional and widespread layout (see Fig. 9.1), consisting of a first band of more public or common areas (entrance, kitchen and living/dining room) and a second parallel band of private areas (bedrooms and bathrooms). What makes this layout special is the fact that the dimensions not only meet the minimum requirements of the corresponding regulations but also go further and incorporate experience and knowledge from Matia's nursing homes on the provision of care. In this sense, it is worth highlighting the location of the bathroom in the main room, aligned with the bed, precisely to facilitate the transfer from the bed to the bathroom while preserving privacy. This route can become a risk factor if the bathroom is far from the bedroom or if the route is narrow.

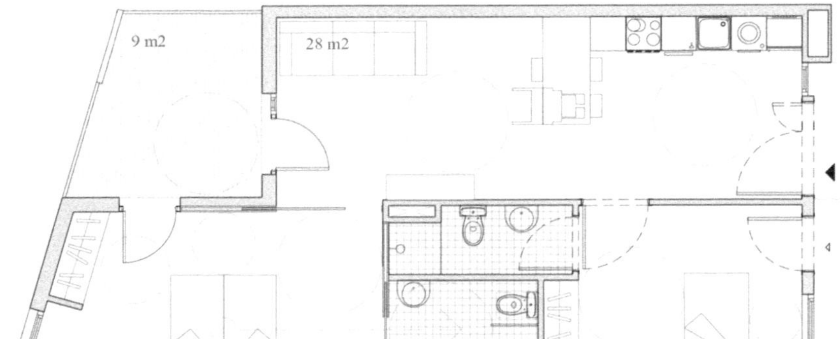

Fig. 9.1 Standard home layout. Sizes vary from 68–97 sqm in two-room apartments to 86–114 sqm in three-room apartments. Still, the space organisation is similar to all of them, as well as the two accesses (main and secondary). (Source: Heitor G. Lantarón)

With regard to the provision of care, it is important to mention a rather innovative aspect related to the second room: it has its own access from the outside (common areas). The aim is to give independence to the caregiver (either a family member who assumes the care or a hired person) who needs his or her space of privacy, so that also his or her quality of life is benefited (see Fig. 9.1). The importance of having one's own space, being able to invite someone over and maintain social relations without interfering with the life of the rest of the household are essential aspects to improve conditions that are already quite demanding from a psychological and emotional point of view. On the other hand, this alternative access helps to sectorise the dwelling, so that in situations such as the COVID-19 pandemic, the provision of care through this second bedroom is greatly facilitated without compromising the safety of the rest of the home.

Another design criterion that has determined the layout of the house has been the importance of favouring relations, both between the different rooms and with the exterior. In order not to limit living options and thus improve the quality of life, there is a very direct relationship between the master bedroom and the rest of the house, especially the living-dining room. The aim is to include a person who spends a lot of time in bed to continue being part of everyday life. In addition, there is direct access to a terrace to facilitate access to natural light, fresh air and contact with the outdoors. To facilitate such contact with the outdoors, all windows have also been designed lower than normal, allowing views for someone in bed or sitting in a wheelchair.

Among the design strategies to foster relationships with the home and the outside world, the most innovative is the one that has to do with creating a boundary between the private (housing) and the public (common areas) that is much more permeable than in conventional housing. The aim is to help create links between neighbours by giving them the opportunity to communicate with each other, to look at each other, in short, to relate to each other. To this end, the thresholds of the dwellings have been deeply designed and windows have also been included towards the communal areas.

With regard to the thresholds, it should be noted that the design criteria also respond to the needs detected in users with cognitive impairment, such as dementia, so they are aimed at facilitating "colonisation" with personal objects that help to recognise the space as one's own. In addition, the main door is clearly marked and made of a different material to the walls to ensure that people with reduced visibility can distinguish it, while at the same time improving and aiding understanding of where the main access to the home is. Concerning the second access, the strategy is the opposite; materiality and design hide the door for those users that might not use it.

The windows towards the common areas are characterised by their size and low height, so that they allow looking out from the inside of the room, but prevent looking in, respecting privacy. This is a well-documented strategy of how by putting eyes in the common areas we create a safer and more controlled environment for the neighbours. In this way, they become spaces of exchange and relationship. The kitchens and part of the rooms are turned towards the common spaces, facilitating spontaneous conversations, greetings or simply glances between neighbours.

9.3.1.2 Common Areas and Common Rooms

In addition to the possibility of receiving specialised home care, one of the main objectives is to fight loneliness. To this end, the design criteria that characterise the communal spaces and common rooms are directly aimed at maintaining and reinforcing social ties by facilitating shared activities. In this sense, and as described in the previous point, the limits of the dwellings have been deeply addressed but also the conditions that make the common spaces a friendly, comfortable and inviting environment for socialising. In short, it is a question of going beyond the simple condition of connecting spaces between the lift and the door of the home. In addition, the location, character and dimensions of the common rooms have been worked on, establishing a range of shared uses, from the most private to the most public.

With regard to the common spaces, the design has focused on encouraging users to appropriate them through their use and enjoyment, colonising them through their objects and plants. Living and changing spaces that reflect the personality of the tenants while at the same time allowing social interaction between them. The fundamental objective is to facilitate and reinforce contact with neighbours in order to avoid situations of loneliness and abandonment.

To this end, three zones have been differentiated on each floor according to their dimensions and design, so that the bridge (A), the square (B) and the street (C) have been created (see Fig. 9.2), according to their character as living or passageways but also including an orientation component. In this sense, the courtyards are essential elements to organise and characterise the space, so that they become points of

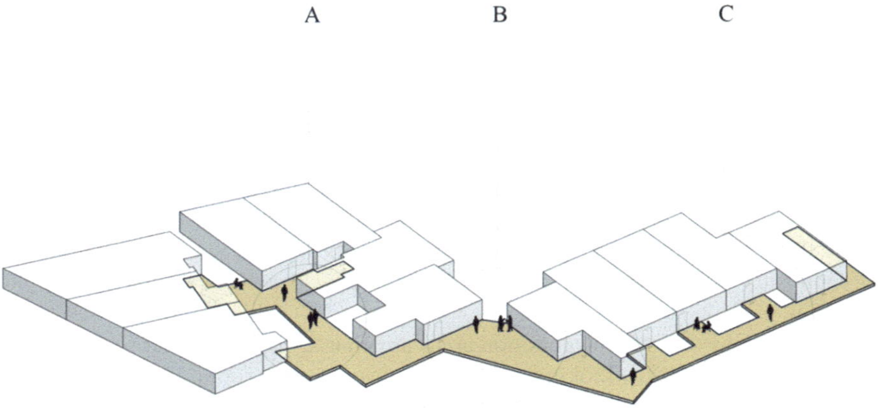

Fig. 9.2 Common areas and their characteristics. (A) The bridge, (B) the plaza, (C) the street. The bridge is located in six levels, the plaza is located in three levels and the street in three levels. (Source: Heitor G. Lantarón)

reference. The aim is that they help in spatial orientation but also in temporal orientation, introducing natural light, the passage of time and the seasons.

Both El Puente and La Calle share a character of passage and of connection between the accommodations, but they are also defined by the openings in the floor slabs that allow a visual relationship between the different floors. These are spaces for seeing and being seen, for talking to neighbours.

In the plaza we find a completely different space, a living space, characterised by its location between the two groups of homes. Precisely this location and its dimensions allow for meeting activities between neighbours, so it hosts some of the common rooms. Its large window to the outside offers beautiful views while illuminating the space with natural light.

On the other hand, a series of rooms for common use have been located on the different floors in response to a need identified during the participatory process to facilitate meetings and activities. Their design is based on flexibility in order to respond to future uses to be defined by the users and that will change over time. In addition, it consists of spaces of different sizes and different locations within the building. In this way, a gradient of privacy is established that defines the rules of use and the type of activities to be carried out.

There are three semi-private rooms for internal use of the building. In particular, they can be used as a place for shared activities that provide respite for carers. In any case, it is envisaged that they can be used occasionally for events or activities that include people from outside the building.

The semi-public rooms are two spaces located on different floors of the building, but with a common feature: they are intended as spaces for connection and relationship with the community. Both have their own independent access from the outside and a kitchen to facilitate activities. One is located in the urban plaza at the entrance to the housing building and the other in a more discreet position with access from a side street.

With regard to the public spaces, these are a day centre and a cafeteria, also located in the urban plaza. These are uses that respond to the needs of the neighbourhood detected during the analysis process. Their management is out of the Lugaritz proposal, emphasising that they are elements that serve the community and respond to the needs of the context, not only of the building.

9.3.1.3 Urban Environment

One of the fundamental aspects of the Lugaritz lifetime homes is the desire to create an innovative benchmark not only in the life that takes place inside the building but also in the relationship it establishes with the exterior. This building is located at a strategic point within the urban fabric, with the potential to solve a major mobility problem in the city. This fact allows for the exploration and innovation of solutions within the urban environment, an area that is normally beyond the scope of any housing project. It also becomes an opportunity to better integrate the building into the urban environment in which it is located.

This is an inclusive urban project that explores new strategies for integration with the physical, social and community environment. These strategies can be summarised in two main approaches: those related to the design of the environment, defined by the type of user they are aimed at. And those related to the integration of the building itself.

The COVID-19 has, once again, highlighted the need to act on public space to improve the well-being of citizens. Within this reflection, the aim of the public space in Lugaritz is to ensure an inclusive environment that leaves no one behind and allows use and enjoyment regardless of personal conditions. The design strategy employed is defined by a type of user who is the most limited when it comes to making use of the public environment: people with dementia. From the approach of this proposal, dementia is understood as a complement to other urban design approaches, taking a further step towards a more accessible and friendly city. Therefore, this project proposes to create a dementia-friendly urban environment while generating knowledge and evidence within the urban scale.

The design of the environment includes a new urban route to resolve mobility in this part of the city (Fig. 9.3). This is a fundamental need as there is a difference up to 36 m height between the two parts of the neighbourhood. It is an urban route designed for people who may suffer from spatial and temporal disorientation, as well as sensory difficulties in perceiving space.

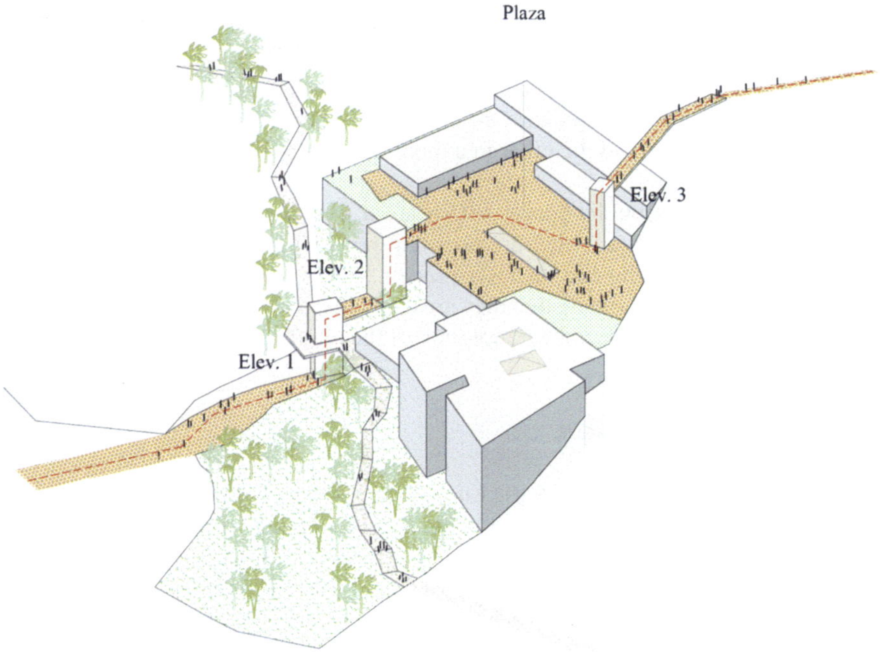

Fig. 9.3 Urban route connecting the two neighbourhood levels while creating a hub of social activity under an inclusive approach. (Source: Heitor G. Lantarón)

The route is entirely accessible thanks to three strategically placed urban lifts, which not only resolve the difference in height but also provide access to two intermediate points: a park and a plaza. In other words, the route becomes an element of social and community interaction. The new plaza becomes a fundamental element in this route. It is an urban element that does not exist right now in this part of the city, so it can facilitate relations between the residents of the neighbourhood. It provides access to the homes, the day centre and the rest of the building's programme, so it is intended to be a hub of social activity. The aim is to give life and activity to the square, so that it becomes not only a transition space but also a place to stay and a place for intergenerational encounters.

The entire design of this new urban route responds to criteria of inclusion, allowing anyone to stay and use it, regardless of their age or physical condition. The aim is to go beyond mere accessibility and facilitate the use of the space by designing an environment that is easier to understand and navigate. In this sense, the urban lifts act as landmarks, distinctive references that facilitate to move along the route. In addition, the materiality of the floors helps to easily follow the path, avoiding disorientation. As a complement, special attention has been paid to comfort and safety as fundamental criteria for a satisfactory use of the environment. Finally, it has street furniture, materials and lighting that vary according to the purpose, use or characteristics of the environment but are always appropriate for the type of user with dementia.

With regard to the building, urban integration has been worked on through building dimensions, volume and appearance. The main design criteria have been, on the one hand, to respect the dimensions and volumes of housing, especially those of the area, so that the building, despite its large size, is physically integrated into the context. In addition, the use of different materials in different areas helps to break up the scale of the building. In this way, the whole becomes a series of smaller elements related to each other. On the other hand, local materials have been used, i.e. a familiar architectural language that ensures the aesthetic inclusiveness of the proposal.

9.4 Future Challenges for Housing the Elderly

The demographic revolution, coupled with the complex social reality, is transforming generational boundaries making current markers of ageing obsolete. Therefore, the proposed accommodation models must necessarily continue to adapt to a constantly changing and evolving process. Solutions must be as varied as demand. In order to respond to the challenge of ageing, it is necessary to diversify, providing housing alternatives to institutional care models that respond to the demand for a successful ageing. In the case of Spain, the need is even greater due to the lack of experiences and options. But on the other hand, it allows us to act from the knowledge and analysis of those models that have been implemented in other countries.

The importance of the elderly does not lie exclusively in their complexity, implicit in their condition as a group that reflects the rest of society, but in the demographic potential they represent and the great opportunity that arises for us as professionals: to participate in the definition of a housing model that, although originally aimed at the elderly, by extension might become mainstream. Furthermore, the ageing of the ageing process makes necessary to find solutions that include external care and inclusive urban design as essential conditions for ageing in place.

Although the trends in housing for the elderly are clear, they are still in the minority. In general, in Europe and specifically in Spain, the solutions that are starting to emerge suffer from the same limitations as the rest of housing types: a shortage of alternatives and, those that do exist, are too specific to a particular type of user and therefore exclusive. Design is a determining factor in the improvement of living conditions and the integration of long-term care, but without a change in the care model in favour of home care, housing models alone will not solve the challenges arising from dependency.

It is necessary to generate housing alternatives for the elderly that are viable, sustainable and adapted to the context in which they are located. It is a matter of incorporating an inclusive approach that does away with specific solutions for each group and that focuses on the use of the physical environment and on the capacities of the users. It is therefore important to include the diversity of people, as well as ageing as a constantly changing process.

References

Abellán, A., Aceituno, M., Ramiro, D., & Castillo, A. (2021). Residence statistics: Distribution of centres and residential places by province. Data from September 2020. *Envejecimiento en red reports, 27*.

Alemán, M. (1993). A perspective of social services in Spain. *Alternativas. Social Work Notebooks, 2*, 195–205.

Baltes, P., & an Baltes,. M. (1990). Psychological perspectives on successful aging: The model of selective optimization with compensation. In *Successful aging. perspectives from the behavioral sciences* (pp. 1–34). Cambridge University Press.

Croucher, K., Hicks, L., & Jackson, K. (2006). *Housing with care for later life: A literature review*. York.

Danish Ministry of Housing, Urban and Rural Affairs. (2010). *Fact Sheet on Danish Housing for the elderly*.

Fernández-Ballesteros, R., Robine, J. M., Walker, A., & Kalache, A. (2012). Active aging: A global goal. *Current Gerontology and Geriatrics Research, 2013*, 1–4.

Fernández-Carro, C., & Evandrou, M. (2014). Growing old at home: Housing preference or lack of alternatives? An analysis of the European context. In C. Heriberto & F. Lucila (Eds.), *Crisis and change. Proposals from sociology* (Vol. II, pp. 694–704). Actas del XI Congreso Español de Sociología.

García, H., & Soler, A. (2015). Universal accessibility: Housing for life. *Aenor, revista de la normalización y la certificación, 297*, 28–33.

Gottschalk, G. (1993). Danish case study: Modernization and adaptation of pensioners flats in Copenhagen. In *SBI town and planning 68. Better housing and living conditions for older people. Case studies from six European cities*. Danish Building Research Institute.

Gottschalk, G. (1995). Housing and supportive services for frail elders in Denmark. In J. Pynoos & P. Liebig (Eds.), *Housing frail elders: International policies, perspectives and prospects*. The Johns Hopkins University Press.
Hanson, J. (2001). From sheltered housing to lifetime homes: An inclusive approach to housing. In *Lifetime housing in Europe: Proceedings of the European seminar and colloquium living tomorrow starts today*, Leuven, Belgium, 4–5. December 2000.
Houben, P. (1997). Challenges in the modernisation of Dutch housing and care for the elderly. *Housing Studies, 12*(3), 355–366.
Houben, P. (2001). Changing housing for elderly people and coordination issues in Europe. *Housing Studies, 16*(5), 651–673.
Lawton, P., & Nahemow, L. (1973). Ecology and the aging process. The psychology of adult development and aging. In L. Eisdorfer (Ed.), *The psychology of adult development and aging* (pp. 619–674). American Psychological Association.
Lindstrøm, B. (1997). Housing and service for the elderly in Denmark. *Ageing International, 23*(3–4), 115–132.
Rowe, J., & Kahn, R. (1987). Human aging: Usual and successful. *Science, 237*, 143–149.
Sancho, M., & Rodríguez, P. (2001). Envejecimiento y protección social de la dependencia en España. Twenty years of history and a look into the future. *Intervención Psicosocial, 10*(3), 259–275.
Sancho, M., & Martínez, T. (2021). The future of long-term care in the face of the Covid-19 crisis. In *Informe España 2021* (in press).
World Health Organization. (2002). *Active ageing, a policy framework. Ageing and the life course*. WHO Press.

Heitor García-Lantarón Specialist in Ageing, Ph.D. Architect from the Technical School of Architecture of Madrid (ETSAM) of the Polytechnic University (UPM), with a Thesis on Housing Models for the Elderly (2016). He is currently Head of Architectural Design at the Matia Foundation, Researcher at Matia Institute (San Sebastian, SP) and Faculty at the Danish Institute for Study Abroad, DIS (Copenhagen, DK). His actual academic and professional interests are focused on design challenges related to the ageing process, by addressing them as a great opportunity for enhancing environment quality for any age and (dis)ability.

Mayte Sancho Castiello Psychologist and gerontologist with expertise in gerontological planning. Dedicated to research in various institutions, with multiple publications in different dimensions of gerontology. She has been Vice-President of the Spanish Society of Gerontology and Geriatrics and Scientific Director of the Matia Institute until 2017. She is currently working independently on several projects in the field of gerontology for international organisations and public administrations in Spain, focusing mainly on long-term care and its future design.

Maider Azurmendi Ormazabal Diploma in Social Work from the University of the Basque Country in 1997 and Degree in Social Work from the UNED in 2017. Specialisation in gerontology and psychogeriatrics (Matia Fundazioa 2001). University specialist in direction and management of gerontological and socio-sanitary programmes and centres (School of Business. University of the Basque Country 2006). She currently works at Matia Fundazioa as Head of Community Cooperation and is the coordinator of the home care area. She has worked as Social Worker in different Care Homes. She has been director of the Lamourous and Argixao centres of Matia Fundazioa and participated in the Etxean Bizi project.

Chapter 10
Development and Management of Cohousing Initiatives for a Friendly Ageing in Spain

Fermina Rojo-Pérez, Diego Sánchez-González, Vicente Rodríguez-Rodríguez, and Gloria Fernández-Mayoralas

Abstract There is growing interest in promoting active and healthy ageing environments through age-friendly communities within the scientific and public administration sectors. Older people's health and quality of life expectations are structured around living for as long as possible in their usual environment, home and neighbourhood, so they can maintain autonomy and independence. Different residential alternatives have emerged in recent years, including cohousing, which allows older people to live at home in an environment with shared-use services. Outstanding features of this model include its encouragement of autonomy for ageing in place, help in preventing loneliness, development of social bonds and assistance in optimising financial resources and use of socio-health services. However, there is limited theoretical and methodological knowledge about cohousing for older people in Mediterranean Europe. This chapter examines the cohousing development and management process in Spain, as well as the implications of this strategy in promoting active and healthy ageing in the context of age-friendly communities. The data are sourced from in-depth interviews conducted with key informants involved in senior cohousing in Spain. Thematic analyses are applied using the Atlas.ti program.

F. Rojo-Pérez (✉) · V. Rodríguez-Rodríguez
Institute of Economics, Geography and Demography (IEGD); Spanish National Research Council (CSIC), Research Group on Ageing (GIE-CSIC), Ageing Network of the Latin American Population Association (ALAP), Madrid, Spain
e-mail: fermina.rojo@csic.es; vicente.rodriguez@csic.es

D. Sánchez-González
Department of Geography, Faculty of Geography and History, Universidad Nacional de Educación a Distancia (UNED), Madrid, Spain
e-mail: dsanchezg@geo.uned.es

G. Fernández-Mayoralas
Institute of Economics, Geography and Demography (IEGD); Spanish National Research Council (CSIC), Research Group on Ageing (GIE-CSIC), Madrid, Spain
e-mail: gloria.fernandezmayoralas@csic.es

E. Pozo Menéndez, E. Higueras García (eds.), *Urban Design and Planning for Age-Friendly Environments Across Europe: North and South*, Future City 19,
https://doi.org/10.1007/978-3-030-93875-8_10

The results can provide a basis to design currently non-existent specific regulations that would facilitate the development of this residential category.

Keywords Residential environments · Age-friendly communities · Senior cohousing · Qualitative methods

10.1 Introduction: Backgrounds and Objectives

Environmental context (both physical and social) affects health and quality of life, in addition to the experience of ageing at home (Schalard & Diaz-Moore, 2016; Wahl & Gerstorf, 2020; Fernández-Mayoralas et al., forthcoming; Rojo-Pérez et al., 2007; Rojo-Pérez & Fernández-Mayoralas, 2021a, b; Rojo-Pérez et al., 2020). The location, design and conditions of neighbourhood and home impact on significant aspects of security, comfort and accessibility for older people, including social relationships and access to services, which determine autonomy and independence in old age (Rojo-Pérez & Fernández-Mayoralas, 2011; Sánchez-González, 2015; García-Valdez et al., 2019). In response to the new challenges for ageing in place associated with various political and social contexts, particularly those involving vulnerability (Buffel et al., 2018; Greenfield et al., 2019), community-based residential alternatives are being created whose social capital is a key resource in promoting active and healthy ageing (Kaspar et al., 2015).

The scientific literature contains limited knowledge regarding the preferences and behaviours of older people when faced with residential decisions (Roy et al., 2018; Golant, 2020). Studies indicate that residential decision-making processes in old age are influenced by environmental experiences relating to residential comfort and control (Sánchez-González, 2009), as well as by adaptive coping strategies aimed at achieving residential normality (Golant, 2011). These strategies arise from processes of agency and person-environment links over time (Wahl & Gerstorf, 2018; Chaudhury & Oswald, 2019). The perception of housing is related to beliefs regarding control and independence in daily activities during old age (Lawton & Simon, 1968; Oswald et al., 2007; Oswald & Kaspar, 2012).

Older people prefer to age in their home, neighbourhood and community; these are sources of positive emotions arising from residential ties and social bonds (Wiles et al., 2012). These places offer a support network based on close and trusted interpersonal relationships. The daily environment also influences the capacities to live independently and enjoy old age (Coleman & Wiles, 2020).

People experience a series of age-related changes in old age (retirement, loss of partner, ill health and physical limitations) (Diehl & Wahl, 2010), which can limit their adaptation to place. However, these people resist moving as a change in residence is often perceived as a physically and emotionally difficult challenge and sometimes a painful and stressful experience (Ekerdt, 2020). Older people therefore attempt to adapt to housing, even where it has deficiencies (Wahl & Oswald, 2016) that may be incompatible with maintaining an independent lifestyle in old age

(Greenfield, 2012). In fact, many older people who want to stay in their homes do not know how to adjust them to suit their different needs and demands (Golant, 2015). People who are less autonomous, disabled or dependent are also more vulnerable to environmental pressures and present more limitations in terms of their ability to proactively adjust to inadequate domestic environments (Wahl et al., 2012; Chávez-Alvarado & Sánchez-González, 2016).

In recent decades, there has been discussion of the need for public policies to promote ageing in place (Golant, 2015). Emphasis has been placed on the importance of improving understanding of the specific challenges that older people face in terms of satisfying their desire to age in place (Fausset et al., 2011). In this regard, it has been noted that it would be useful to redirect design efforts and interventions by promoting residential strategies and new housing categories aimed at fostering safety, autonomy and independence among older people in line with their experiences, preferences and needs (Vasara, 2015).

There is growing interest in supporting age-friendly housing initiatives among the scientific community and public authorities (Beck, 2020). In this respect, affordable, well-designed and safe housing options, offering accessibility to social and health services and good connectivity to community, can enable older people to continue living in their current places of residence and avoid relocation (Luciano et al., 2020). With this purpose in mind and as part of the age-friendly cities and communities programme, the World Health Organization (WHO, 2007, 2018) has encouraged the development of environmental and psychosocial interventions based around housing and has promoted various residential strategies intended to favour active and healthy ageing (WHO, 2015; Sánchez-González et al., 2020; Fernández-Mayoralas & Rojo-Pérez, 2021).

Cohousing is an accommodation model developed in the 1970s in Denmark, which has particularly expanded in Northern Europe, the United States, Australia, New Zealand and Japan (Vestbro, 2000; Ache & Fedrowitz, 2012). This residential model is generally located in suburban and rural areas and to a lesser extent in urban areas. It comprises different types of apartments, houses, residential developments and villas, based on new constructions or even on the refurbishment of existing properties, and can accommodate small or large groups of self-organised residents (Jones et al., 2010). In addition to the widespread use of the term cohousing, collaborative housing is also used (Czischke et al., 2020), although the latter is considered an umbrella term referring to a broad range of forms of self-organised and community-oriented housing (Bresson & Labit, 2020; Lang et al., 2020). Together with "intentional community", "senior cohousing" and "collective housing", these terms are also used in Spain (López Gómez & Estrada Canal, 2016; López de la Cruz & Sánchez Medina, 2017; del Monte Diego, 2017) to describe various residential projects based on collaboration, solidarity and mutual support among residents. Collaborative housing projects can be intergenerational or for specific groups (older people, ethnic minorities) (Fromm, 2012). These collaborative models offer a form of community life based on cohabiting in private spaces and sharing others of common use, with self-managed facilities and activities organised by the residents themselves (Vestbro & Horelli, 2012).

The various published literature reviews reflect a surge in studies regarding the collaborative housing model (Fromm, 1991; Lang et al., 2020). Cohousing interprets housing as a social good rather than a market value. One aim of this residential model would be to reduce problems in access to housing for disadvantaged groups and their adverse impacts on health (Hagbert et al., 2020). In many countries such as Austria, a growing number of older people have low incomes and cannot afford to buy or rent housing that fits their needs (Baldwin et al., 2019). However, the form of tenure may affect the success of the model insofar as it can exclude potential low-income residents (Larsen, 2019). It has also been stated that the implementation of this model tends to involve gentrification processes that marginalise less affluent families (Tummers, 2016).

This self-managed residential alternative may imply a reduction in public spending (health, care) and improvements in health and quality of life for the older population (Labit, 2015). Most studies point to potential benefits for perceived physical and psychological health among cohabitees in communities of older people (Glass, 2009). However, some research has reported a deterioration in physical and/or mental health after people have moved into collective housing (Wasylishyn & Johnson, 1998).

The cohousing model can accommodate collective and solidarity-based approaches to care in old age. A sense of community and mutual support is perceived as an important reason for choosing this residential alternative, since residents provide mutual and solidarity-based support by offering household help and care on a personal level (Glass, 2009; Durrett, 2009; del Monte Diego, 2017).

It is possible that not all residents will feel these inherent senses of community, control and social support, which can cause conflicts within the cohousing arrangement (Wasylishyn & Johnson, 1998; Markle et al., 2015). The model should encourage social inclusion and respect for the cultural diversity of its residents through the establishment of agreed group guidelines and rules for conduct and communication in the resolution of disputes (Linares, 2018).

The design and management of buildings and shared spaces will determine the community life and collective perception of their residents (Marcus, 2000; Ruiu, 2015; Tummers, 2017). In this regard, design can encourage or discourage links and interactions between residents and nature (Sanguinetti, 2014; Lies et al., 2017), potential environmental sustainability practices (Chatterton, 2013) and the possible implications for socially inclusive urban development (Droste, 2015). It has specifically been noted that well-designed common areas foster social relationships and the organisation of activities among older residents (Tyvimaa, 2011). Notwithstanding this, there may be disputes owing to a lack of consensus regarding architectural design (Motevasel, 2006). A participative and consensual design process is hence necessary, with professionals (architects) playing more of an integrating role that takes into account the experience and creativity of the potential older residents (Awan et al., 2011; Hammond, 2018).

These residential initiatives promote community governance and group decision-making, community design, sustainable construction methods as well as financial and legal structures (Williams, 2005). In France, for example, the social housing sector has started to support collaborative housing as a way of improving neighbourhood solidarity and encouraging new forms of housing management revolving around the residents themselves (Bresson & Labit, 2020). In Sweden, meanwhile, these intergenerational collaborative housing projects (older people, the young and refugees) have facilitated the promotion of solidarity-based social ties, the development of resilient coping strategies among residents during the COVID-19 pandemic and social integration (Arroyo et al., 2021).

The public policy impact of the paradigm shift in relation to collaborative housing for older people remains modest. In Spain, for example, residential initiatives for older people have been developed outside the context of housing policies and social services, under an institutionalised model (López de la Cruz & Sánchez Medina, 2017). Uncoordinated housing, social services and healthcare policies have also restricted the creation and development of satisfactory housing solutions for older people.

In most high-income countries, including German, Spain and the United Kingdom, there tends to be little public funding for these residential initiatives, which are generally private, not-for-profit and promoted by cooperatives (Fernández Cubero, 2015). This lack of public funding risks increasing social and health inequalities among the older population (Motevasel, 2006). However, a higher amount of public funding in line with that seen in Sweden and Denmark could mitigate these inequalities and make it possible for disadvantaged groups to benefit from the positive health and social effects of the cohousing model (Carrere et al., 2020).

The limited conceptual and empirical knowledge in this area is associated with the difficulty of understanding the great range of designs and forms of cohousing organisation over time (Beck, 2020). In fact, a dominance of transversal approaches and case studies that do not easily admit extrapolation has meant that there is little empirical evidence to facilitate the definition and reproduction of successful cohousing models on an international level (Czischke et al., 2020).

Against this background, this chapter emphasises the views of the residents themselves, or cohousers (Sanguinetti, 2014), regarding (i) the senior cohousing development process in Spain and (ii) the management and maintenance of cohousing and of life after entering a senior cohousing environment. The study focuses on collaborative housing implemented on a self-managed basis since the start of the 2000s (Keller & Ezquerra, 2021), with the purpose of responding to the concerns and desires of many people regarding where and how to spend their old age. This study forms part of a broader research aimed at identifying the understanding of active ageing and the relationship between its pillars and determinants from a lay perspective (Fernández-Mayoralas & Rojo-Pérez, 2021).

10.2 Source and Methodology

There are no documentary databases in Spain to address the aims of this chapter. As a result, material was used that had been taken from a broader study on active ageing and quality of life in different residential contexts (Fernández-Mayoralas &

Rojo-Pérez, 2021). The study design was qualitative, with 58 in-depth interviews conducted with the older population living in different types of residence (family homes, cohousing, elderly care homes). Both residents and key informants, 33 in total, participated in the case of cohousing. The interviews conducted with key informants, or those with extensive knowledge or experience, were used for the purposes of this chapter, as the informants were generally represented by the president of the collaborative housing governing body or additionally by the director-manager.

The information was gathered on the basis of a semi-structured script common to the areas under study (Fernández-Mayoralas et al., 2021) but with specific additional questions for those selected as key informants. The questions used in this chapter were related to the development and management of senior cohousing for older people: "we would like to know your opinion on how this residential facility came into being and how it is managed in relation to all of the aspects of residents' daily lives".

An intentional sample was designed so that seven interviews were conducted with senior cohousing key informants, which are representative of the first-wave collaborative housing in operation in Spain at the time of data collection (Project MOVICOMA, 2015; López Gómez & Estrada Canal, 2016). The senior cohousing involved in the study was located in the Spanish provinces of Cuenca, Jaen, Madrid, Valladolid (all inland) and Malaga (on the south Mediterranean coast), with the highest number of senior cohousing projects located in the latter province (three) (Table 10.1).

There was also diversity in terms of size of residential setting. One house was located in a rural area (fewer than 2,000 inhabitants), two were in areas of intermediate size (2,001–10,000 inhabitants) and four were in urban areas (more than 10,000 inhabitants). In the latter case, the houses were located on the outskirts of the provincial capital. The key informants were made up of three women and four men. It is worth noting that at the time the information was collected, there was no cohousing involving older people living together with people from other generations.

The interviews were conducted at each of the residential establishments by the researchers participating in the scientific project, recorded on audio, transcribed and entered into the Atlas.ti programme for codification and analysis. A single list of primary codes was designed for use by all researchers in the first codification phase (Fernández-Mayoralas et al., 2021). This is not a multiplatform programme, meaning that the codification was first independently implemented by the researchers, who then sent their copy to the project administrator in order for the parts to be combined into the master project. This master project was reviewed by all of the team members and any discrepancies were discussed until an agreement was reached.

The primary analytical codes used were development and management. The first refers to the collaborative housing development process, while the second covers references to aspects concerning its management, administration and maintenance. These level one codes were used as a basis to review the transcripts and generated a

Table 10.1 Case studies by geographical area

Cases	Location: municipality (province)	Size of residential area[a]	Website	Characteristics
Cooperative TRABENSOL, Social Centre for Coexistence, Assistance and Services for older people	Torremocha del Jarama (Madrid)	Rural	https://trabensol.org/	A bioclimatic and full-accessible two-storey building. It consists of 54 private apartments, as well as other areas and services for common use (kitchen, dining room, social room, multi uses room, TV room, laundry, cleaning services, gym, garden, therapeutic bathroom, care room health and special care, room for various activities, community management office, car park)
Cooperative CONVIVIR, residential centre for older people	Cuenca	Intermediate	https://apartamentosconvivir.com/	Two-storey building with three types of apartments (1 or 2 bedrooms, studio type), which uses biomass as an energy source and has various types of services (dining room, cleaning, laundry, medical and nursing care, physiotherapy, social work, occupational therapy, hairdressing, podiatry, speech therapy, as well as activities of leisure and social participation)
Residential "Valle del Azahar"	Pizarra, Zalea (Málaga)	Intermediate	https://www.residencialvalledelazahar.es/	It consists of 204 apartments under the Cooperative regime. Among the common services, it will provide with an indoor pool, spa, relaxation area, expansion of gardens with recovery of native plants and other community services

(continued)

Table 10.1 (continued)

Cases	Location: municipality (province)	Size of residential area[a]	Website	Characteristics
"Fuente de la Peña", residential centre for older people	Jaén (Jaén)	Urban	https://www.mayoresdejaen.com/	It is a residential environment designed and managed by older persons under a Cooperative regime. It is organised in a three-storey building with private accommodations and a wide range of therapeutic, recreational and other services (dining room, cleaning, laundry, convalescence, functional recovery and rehabilitation, post-operative care, palliative care, podiatry, social work, hairdressing, laundry)
"Los Milagros", Andalusian Cooperative Society ("Residential Santa Clara")	Málaga (Málaga)	Urban	https://residencialsantaclara.es/	The building is located on a hillside in the Montes de Málaga, close to the city centre and the beach and enjoys a Mediterranean climate. It consists of 76 private apartments with sea views and social and health services (gardens, swimming pool, library, chapel, hairdresser, occupational therapy rooms, gym, Wi-Fi, rooms for the sick, medical assistance)
Residential "Puerto de la Luz"	Málaga (Málaga)	Urban	https://www.residencialpuertodelaluz.com/	It is a self-managed senior cohousing in a cooperative regime, made up of 60 independent private houses and other common services (medical consultation, gym and physiotherapy, swimming pool and gardens, library, dining room, hairdresser and aesthetics, car park, orchard). "To live the individuality in community" is its residential motto
PROFUTURO, residential area	Valladolid (Valladolid)	Urban	https://topmayores.es/precios-opiniones/cohousing-profuturo-valladolid/ https://cohousingcoop.es/complejo-residencial-profuturo/	Under a cooperative regime, this centre was inaugurated in 2011 and its motto is "to live in your own house", adapted to the residents' needs, with a coverage of assistance services (common areas, gardens, gym, social club, restaurant). It has 127 apartments of two types, for two or one person (with 50 or 37 square metres each)

Source: Authors' own

[a]Rural: ≤2,000 inhabitants; intermediate: 2,001–10,000 hab.; urban: ≥10,001 inhabitants

range of development subcodes (time, promoting individuals or group, setting, sociodemographic profile, ownership, initial cost, housing type, aspects relating to transfer of ownership in the event of death, previous commitments or community participation, bureaucracy, regulation, problems) and management subcodes (autonomy in decision-making, shared services, costs, governing board, self-management, healthcare services, management problems, solidarity).

This study pursued a thematic analysis, known as an "information processing method in qualitative research that permits the identification, organisation, detailed analysis and reporting of patterns or topics based on a careful reading and re-reading of the information gathered, in order to infer results that facilitate a proper understanding/interpretation of the phenomenon under study" (Mieles Barrera et al., 2012, p. 217). This involved the completion of several phases: transcription and reading of the information, codification, topic search and identification and analysis of results.

The results are accompanied by a selection of quotes from interviewees. In order to secure their anonymity, their document number (Dn), sex and type of residential habitat (rural, intermediate, urban) were recorded. Likewise, names of participants, residences, cities and businesses were removed. Implicit parts of discourse were annotated in brackets. Bracketed ellipses (…) were used to omit parts of discourse, whether a word or a fragment.

10.3 Findings

10.3.1 Why Age in a Cohousing Environment?

In addition to describing a residential setting, cohousing represents a shared living model aimed at optimising active ageing and solidarity among residents. This was implicit in informant discourses relating to the implementation of this type of residence for older people in Spain:

> We are looking to create the conditions, the necessary setting so that these relationships can develop and so that active and healthy ageing becomes a reality. (D63, male, intermediate)

The main reasons cited for living in a setting inhabited by older people and with collective or shared services concerned the desire to age somewhere beyond the traditional long-term residential care facilities:

> We felt that there was a significant social shortfall in terms of treatment of older people, that there are basically care facilities, the classic elderly long-term care homes and other projects: home care, etc. But we thought that what we have created, which is an alternative to the classic elderly care homes, didn't exist (…) this isn't related to the old asylums or to the old poorhouses and all that. No, it is a way of (…) living perfectly well, and with the human plans that each person has and in the best conditions for older people. (D63, male, intermediate)

Informant discourses referred to both modern long-term care facilities and old asylums. The latter operated in Spain until the beginning of the 1970s, serving as places to accommodate individuals without means or family or who were unable to care for themselves (Guillén Sádaba, 1994) and were fundamentally staffed by members of religious orders (Barenys Pérez, 1992). There was a clear desire to age differently to previous generations and also to avoid institutionalisation.

The reasons for the development of senior cohousing also commonly included autonomy in terms of the choice of where to age, as a form of self-determination and to avoid a potential future situation in which others, such as children or other family members, are the ones to make the decision:

> I (…) am choosing my own home while I still have capacity to decide and I'm choosing a home where I want to be, with my friends, with people who function well in a centre like this, and that's good for me, because I will find conditions that are suitable for me, and it is good for my children, so they do not have to worry about. (D63, male, intermediate)

Added to a feeling of independence in decision-making was the desire not to be or feel alone, or to be in the company of one's peers:

> The spirit in which we came here was one of independence and not being alone. Coordinating that independence that our own apartment gives us, where you have to ring a bell to come in and feeling that we are in the company of all the others, because they are people with characteristics similar to ours, culture, ideals (…) being accompanied but independent at the same time. (D112, female, urban)

Another important reason was the previous home's lack of suitability for the current family situation, as a result of ageing and children departing. Along the same lines, there were references to inadequate post-retirement financial circumstances, which meant that a decision to move to senior cohousing made it possible to optimise pension income and reduce the costs of housing that had become unsuited to the circumstances of old age:

> The fact that today you're using things you do not need, like big houses, (…) the costs of running that household, and [then] you are using something that is really what you need, which is this, and which costs less than a house. (D65, male, urban)

A further reason impacting on the decision was the cost of living in a family household, or even in a long-term care facility, compared to the cost of senior cohousing. Long-term care homes would be more expensive, whether in the case of an individual or for a couple:

> It is always going to be cheaper for us [than] an elderly care home, because (…) you only pay for services here, [but] in an elderly care home you're paying for the service, plus the investment, plus the commercial profits. (D65, male, urban)

The director-manager of a cohousing facility in central Spain provided a well-expressed summary of reasons for choosing this type of residential setting:

> It is the future, really. The public or public-private residential facilities are always there, but people… for you to choose the place where you want to grow old, with your group of friends, without bothering your children, surrounded by a healthcare team and where if anything happens to you then you're looked after and everything, and it is all at an affordable price. (D104, male, urban)

10.3.2 A Lengthy Cohousing Development Period

Cohousing development takes place over a long period of time. It is necessary to take into account the need to create a group of people who would initially inhabit the cohousing and give it legal form, design the idea, search for, select and acquire the appropriate land plot to develop the residential complex, go through the relevant legal channels with the respective regional public authorities and complete the architectural, construction and habitability design. According to our findings, these tasks took a total period of no less than 10 years. And often added to this lengthy administrative and management development period were stances involving a lack of belief in the idea, such as external criticism, which also had the potential to hinder the process. Various participants explained this:

> We started in 2000 (…) with what was an idea, (…) until it went further in 2009, convincing the authorities, (…) and sometimes dealing with people who called us dreamers because here, in Spain, [they told us] it would never work. (D63, male, intermediate)

Other informants described several stages leading up to the occupation of the residential facility, at which time a new journey began:

> The three important stages have already been covered, we might say: one was the sale of ideas, then the execution, and then the operation.... (D65, male, urban)

It is noteworthy that some informants explained that they had had the idea when they were still young:

> I explained the idea I had as a girl (…) a few friends picked up on it and we started the project (…). Now, the realisation (…) I'd say, to be precise it was in (…) 1991 (…), in 2000 it was inaugurated. (D113, female, urban)

This lengthy development period is combined with a belief from some informants that cohousing is an ongoing project, which will grow with its own use and residential life:

> Because we tell people who are coming, too, that this is a project (…) that is constantly growing and there are always improvements to make, they can always take part in moving this forward. (D113, female, urban)

10.3.3 The Members of the Promoter Group

Our findings showed that the relationship binding together the members of the seminal group that started the cohousing development tended to be one of friendship from youth. It could also involve friends or work acquaintances, particularly from some large Spanish companies, or even neighbours or people encountered at a previous residential facility:

> At a meeting with friends, people who have known each other for a while and where maybe the common denominator is a degree of social sensitivity, we proposed doing something along the lines of what [this residential facility] has become (…) the promoter group was

made up of 10 or 12 people, which later increased with our friends, family, so people that we know, and as we showed them (…) our project they signed up to it. (D63, male, intermediate)

People who were already interested started to develop it. The promoter group, the core group that promoted (name of residential facility) is a group of people (…) who have lots of initiatives, who are very entrepreneurial. (D70, female, rural)

In terms of the sociodemographic and functional profiles of those who were first involved and subsequently became residents, a series of criteria were established (age, functional capacity and financial means). First, a minimum age was set, and sometimes also a maximum (e.g. between 50 and 80 years), which varied a little depending on the specific senior cohousing project. Some informants highlighted that residents were not rejected on the grounds of sex or race and were merely required to be of a certain age. There is a trend for more single women to live in these residential settings than men, as well as a trend for more married or cohabiting couples to be involved, at least at the beginning of their residential cohabitation. In general, participants were professionals, civil servants and teachers during their working lives. Socioeconomic status was high and medium-high, which enabled residents to cover development and management costs:

When someone lives here, we have to ask them to have financial means, some means, at least (…) Not because we are elitist, just because we cannot keep ten people here on a pension of six hundred euros. (D112, female, urban)

Functional capacity was another selection criterion. At least at the outset of the development of the residential facility or at the start of cohabitation, it is ensured that people are functionally able or autonomous, although the ageing process may give rise to disability and functional dependence:

We do not have a limit, provided that the individual is able to carry out all the activities of daily living. (D113, female, urban)

Many of them had capacity when they arrived and have become largely dependent. (D104, male, urban)

10.3.4 Physical and Environmental Setting

The geographical location of the residential facility was settled based on the opportunity to find a financially accessible space, with large amounts of time and effort invested in searching for the location. Others focused on the desire to age in a specific place:

Almost all of us who started the project [were thinking of] a place where we could be close to where our lives had unfolded, our childhood. (D113, female, urban)

In terms of the characteristics of the physical setting, the residential facilities tended to be located on large plots, commonly on the outskirts of areas and generally in highly valuable and beautiful landscapes (Fig. 10.1). It is also noteworthy that in some cases, it was necessary to go through the process of securing planning reclassifications from rural status to enable the land to be developed:

> There's loads of space (...) we are in a setting that's really a paradise, because we are surrounded by nature. (D67, male, intermediate)

> Magnificent location, with a great site because we are in the countryside and we are linked to the city by public bus at the same time, because we are within the centre of [name of city]. (D112, female, urban)

The residential facilities were varied in terms of building style. Some cases involved dwellings consisting of a detached building with several floors, and others had two floors. The number of dwellings or residential units was also variable. All had individual terraces and gardens (Fig. 10.1):

> We made three blocks of forty apartments. [In addition] there are the shared parts, with the administrative area, infirmary, ... games room, chapel, utility room, (...) gym, ... laundry room, (...) forty per cent of it is shared. (D65, male, urban)

> There are sixty dwellings (...) each fifty metres squared, plus twelve metres of terrace space. (D112, female, urban)

> Various sizes, there's one for a single person that we call a studio, and then another for (...) two people, and then there's one that has two bedrooms and a living room. (...) It is a bright building, (...) some people say this is paradise, others call it a twenty-first century castle. (D63, male, intermediate)

As these are recent constructions, it was explained that they were adapted to the circumstances of the residents and to the current situation (Figs. 10.2 and 10.3). Different models have been constructed at some residential facilities based on functional capacity:

Fig. 10.1 Panoramic view of the Puerto de la Luz cohousing, located in the outskirts of the city of Malaga, Spain. Opened in 2,010, it has 50 private apartments, in addition to community facilities for the use and enjoyment of the cohousers. (Source: https://www.residencialpuertodelaluz.com/blog/page/3/)

Fig. 10.2 The Trabensol cohousing is located in the rural municipality of Torremocha del Jarama (Region of Madrid), about 70 kms from the city capital. It is a pioneering initiative in Spain (opened in 2013), together with Santa Clara (Malaga). Corridors to the apartments are identified by colours for a better perception and orientation of the residents. (Source: https://trabensol.org/galerias/)

Fig. 10.3 A two people apartment in the Profuturo cohousing (Valladolid, Spain), with adapted and bright spaces is shown, consisting of a bedroom, a bathroom and the entrance-living room, with a small American kitchen. (Source: https://cohousingcoop.es/complejo-residencial-profuturo/)

> The residential architecture, location and concept are completely different to any other residential facility (…) it is modern architecture and although the building is 10 years old, they did it so you can see a large new building, with lots of potential (…) we have got two residential facilities inside, one for those who are more dependent and another for those who are less. (D104, male, urban)

10.3.5 Legal Framework and Form of Ownership and Use

The senior cohousing in this study was structured in the form of cooperatives:

> It is (…) shared ownership, we haven't implemented commonhold division, the flats aren't owned by each of the residents, instead the cooperative is the owner of [name of residential facility]. And the rest (...) we have the right to use the apartment and the other facilities. We did it like that because we want to avoid speculation. (D63, male, intermediate)

In some of the cohousing projects studied, there was even assignment of the right to use through person-to-person agreements, subject to compliance with any relevant requirements (age, functional capacity, others) but not on a profitmaking basis for the cooperative:

> I can assign to another person (…) all we require at the cooperative is a type of person, of a certain age and so on, who has to undergo the (…) medical assessment, has to be authorised, (…) and it is approved by the [Governing] Body. (D65, male, urban)

A package of shares is acquired under the cooperative senior cohousing system. This entitles the holder to personal use of the dwelling and to the enjoyment of the other shared services at the residential facility. The share package can be bequeathed in the event of death, but if the heirs reject the package then the cooperative must acquire and manage it once again, allocating it among those who have requested the package and satisfy the requirements for access and use:

> Their heirs inherit those shares, and if (…) they do not want them, the cooperative always has to buy them. They cannot negotiate with a third party. (D104, male, urban)

The cooperatives handled the process of arranging and purchasing the land on which the cohousing was to be constructed, which required much time and effort owing to the highly specific nature of a transaction with these characteristics. In some exceptional cases, even the municipal council was involved:

> The project was managed in the area by purchasing the land and having a good understanding with the local mayor. (D70, female, rural)

There is no specific framework for structuring the cohousing model included in either State cooperative law or the regional laws based on it, which regulate the incorporation and operation of cooperatives (Simón Moreno, 2020). This is something that senior cohousing residents in Spain have long demanded from the authorities:

> We still do not have our own section, [we are classified] as a consumer cooperative (…), some cooperatives of this kind might have registered as housing, but we are not housing either, so we really do not have our own section. (D112, female, urban)

> That's the problem, because this kind of residential facility (…) in Spain they want to treat us like nursing homes, but we are not a nursing home. (D113, female, urban)

In this context, all senior cohousing facilities are fighting for official recognition of their status. For example, the Andalusian regional authorities were asked to include this residential category in their law on care and protection for older people (President of the Regional Government, 1999), through an addendum incorporating the category of "older adults residential cooperative" into the "other options" subcategory (see category III, article 14) (c.f. D113, female, urban).

In the light of this absence of a legal framework for senior cohousing, some informants explained that the physical construction of residential facilities has taken place based on the regulation applicable to elderly care homes:

> We comply with all the criteria here, and what is more, we are classified as an elderly care home, although we are free, but we are classified as an elderly care home, so if someone from here ends up in a situation of dependency it is not necessary… [for them to be moved]. (D63, male, intermediate)

In any case, and as a conclusion in this respect, a key informant described their understanding of a senior cohousing cooperative as an all-embracing idea, emphasising in particular the way of life that it involved:

> We are not just owners of a share [in the cooperative] that covers an apartment; it is the sum of all of our experiences [that] makes this bigger and better. (D65, male, urban)

And it was emphasised that under this system:

> They [the residents] have the power and make the decisions on everything that is done here. (D104, male, urban)

However, development and maintenance problems were also reported, such as oversizing for the space available, problems or shortfalls in terms of construction quality, suspension of works and subsequent recommencement with other professional teams, related problems concerning financial resources and other associated setbacks, as seen from the following discourses:

> We bought a large amount of land (…) An initial plan was made (…) That plan had to be abandoned (…) [The] current project (…) has been reformed (…) To fit within the budget (…) lots of things were removed that are not here today, and they're missed (…) Money was also saved on construction materials (…) The works have been suspended for almost three years, (…) all of the shortfalls we have are a result of [all] that. (D67, male, intermediate)

> Some negative aspects have an influence because [the cohousers] have never been used to paying a monthly instalment of money that they're paying here, and with it being less than we should be paying the instalment has increased. (D67, male, intermediate)

> We have had some problems with VAT (…) they charged us VAT on the (…) cash contributions for construction, the cooperative capital contributions (…) We made some appeals, then the director general ruled in our favour (…). But what is more, we have now got the problem that they do not want to refund us the VAT on the construction (…) We have appealed (…) it is been almost two years with the regional economic and administrative court, (…) and we have money frozen there. (D63, male, intermediate)

10.3.6 Residential Facility Management and Participation

The residents placed a very high value on residential independence in their daily lives, compared to the regulated lifestyle experienced in long-term care facilities or elderly care homes:

> People are independent because each individual lives in their home, nobody tells them when they have to get up like in an elderly care home, or when they have to go to bed; they have their own kitchen, (...) So it is independent but you're not alone (...) we are often closer to the other people than (...) they are to their own family. (D65, male, urban)

In addition to the independence that comes from having individual use of a dwelling, cohousers tend to manage all their residential matters by means of a specific structure:

> We have a structure: the general assembly, the governing board, the management committee. Because it is not single-person management here, it is a management committee where there's a technical director, there's a residents' representative and there's a representative of the governing board. And then there are different committees, there are working committees. We want to increase participation by doing that; as I have told you, it is a participative enterprise. We chose a cooperative precisely because we think it is the most participative model. We are "co-managers" here, we do things ourselves. (...) That is what we understand [as] active ageing, being active in governance, administration and activities. And there are people who volunteer in different areas, because the directors do not charge a penny here and neither does anyone else, it is all altruistic, it is all genuinely for free. (D63, male, intermediate)

And management takes place subject to internal governance rules:

> Approved by everyone; the rules tell you what you can and cannot do. (D113, female, urban)

Interestingly, given the gender-based factor that affects active ageing, both women and men were involved in cohousing management tasks:

> There are women and men on the governing board; we have had a female president and we currently have a male president, and the other members have been men and women, I haven't noticed any difference. (D70, female, rural)

There is sometimes a two-pronged approach based on self-management:

> Everything is managed by us. We have had a lawyer who has helped us and a financial administrator, nothing else. (...) All our work and management is done on a completely voluntary and altruistic basis. (D112, female, urban)

But in this case, self-management is combined with the hiring of external personnel for specific activities, including financial management and administration. As in all areas of life, some tasks have to be performed by professionals and experts:

> It is hard for them to understand that professionals have to be specialists (...) sometimes we have really significant problems between management and the governing board regarding how human resources are managed in this facility. (D70, female, rural)

And these problems often revolve around the aim of independence and autonomy in all aspects that provided the basis for senior cohousing coming into conflict with the fact that, as stated, many tasks have to be performed by specialists:

> I have found it hard to understand [name of facility] because I came from a more standardised model, with highly planned and organised procedures and so on, so now maybe I'm still coming from the perspective of greater planning [of autonomy in] management. (D70, female, rural)

Given that senior cohousing initiatives in Spain have so far been private, the cooperative members bear the costs of processing, development, maintenance and services. External support has been received on a few occasions:

> We had 106,000 euros of support for the installation of the canteen, the kitchen and possibly the laundry room from the LEADER programme, through (…) a development company from [name of region] (…). (D63, male, intermediate)

Maintenance costs can be paid on an individual or shared basis:

> We have (…) shared costs, (…) but the individual costs (…) [are paid] by each individual (…) one person does not pay for what another wants to consume. (D65, male, urban)

And in the context of solidarity, another of the reasons for cohousing development, financial resources were also contributed to create a support fund for people facing unanticipated needs:

> We (…) [have] a healthcare contribution that each partner has made every month since the start of the cooperative (…) so if anyone gets to a point where they cannot cover the cost that this involves, we have a fund that would help that person so they never have to leave [name of cohousing facility]. (D113, female, urban)

> [Name of cohousing facility] does not have any external financial support (…) the most we have achieved is for (…) banking entities (…) to give us a charitable sum for the solidarity fund. (D70, female, rural)

Despite all their efforts, cohousers complain of a lack of recognition from public authorities, citing a lack of specific regulation in addition to scarce financial support bearing in mind that this residential category is offering social services that can mitigate the burdens and problems that authorities are facing:

> We are offering a very important social service, which they should appreciate much more, because we are taking a problem off the authorities' hands. The authorities do not have facilities of this kind; I'd say they do not even have them for (…) lower-income people. We are doing this first of all because we do not have public facilities like these, and [private] facilities like these charge three or four times what we are paying here. (D63, male, intermediate)

10.4 Discussion and Conclusions

In recent decades, the phenomenon of ageing and related social change has caused us to re-examine our models of the residential facilities where we age (Schalard & Diaz-Moore, 2016; Golant, 2020). A literature review reveals limited knowledge regarding the designs and forms of organisation of cohousing facilities, as well as their implications for the health and quality of life of older people (Beck, 2020; Czischke et al., 2020).

The senior cohousing findings analysed in the Spanish context indicate a prevalence of private initiatives, which are not restricted to the cooperative housing model that is underpinned by a form of ownership and the availability of shared spaces and services (Torío-López, 2018; Keller Gargantè & Ezquerra Samper, 2021). Senior cohousing thus reflects a broader conception of the cooperative in the topic of residential project management, based on shared living, the use of common resources and services and even solidarity among residents (Etxezarreta et al., 2018). In this respect, cooperative members are granted use of a dwelling without the transfer of ownership. The cooperative retains ownership, while residents have the right to private use of their residential units and of shared spaces (Louredo Casado, 2021). Specifically, senior cohousing in Spain is a form of shared habitation (Fuentes-Lojo Rius, 2020), falling within the category of a cooperative on assignment of use (Simón Moreno 2020). Under this model, partners or residents maintain their right of use by making a financial contribution, but ownership remains with the cooperative (Etxezarreta et al., 2018).

Informants stated that cohousing development was a lengthy process owing to the requirement to create a group of people interested in a common project, the legal and bureaucratic hurdles to overcome and difficulties with construction projects. In this regard, collaborative accommodation proposals for older people face various political, economic, legal and cultural obstacles that can slow or deter their materialisation (Allison, 2020). As described above, there is still no legal framework for this shared residential model in Spain (Fuentes-Lojo Rius, 2020), and only one in every ten cohousing initiatives comes to fruition owing to problems with financing, lack of management experience and disagreements among participating older people during the process (Crabtree, 2016). In addition, the participative nature of the process increases the time it requires and there are higher risks of increased project costs for residents and developers (Scanlon & Fernández, 2015).

The informants' narrative, regarding their experiences of the development and management of residential projects, facilitates an understanding of who they are and how they have come to this point with control over their own lives and fates. Emphasis is placed on the desire to decide on a place to age actively and healthily, based on personal experiences and motivations such as anticipating foreseeable problems with unsuitable housing, dependency and relocation to a long-term care facility, which may imply a burden for their relatives (Jaggard, 2018). Also important is the desire to achieve a residential solution with services (social, health) suited to old age, adapted to older inhabitants with social and cultural kinship. These models include various alternatives for elderly care, based on mutual support among residents and professional services providing support for dependency (Gelardin & Muscat, 2009).

Positive perceptions of quality of life and well-being among cohousing residents are associated with psychosocial factors such as personal autonomy, community solidarity (Labit, 2015), social support (Rodriguez-Alonso & Comas-d'Argemir,

2017), a feeling of safety and a sense of community (Ruiu, 2015). A sense of community encompasses belonging, trust and togetherness of residents. It is the outcome of a cohabitation process involving like-minded people exchanging experiences and support, performing community tasks and promoting reciprocal learning (Jolanki & Vilkko, 2015). As a result, belonging to a group based on positive social relationships can contribute to physical and psychological well-being (Cram and Nieboer, 2015). These social relationships do not always arise from the start of a project; rather, they can be gradually cultivated by establishing closer connections with other residents. Belonging also tends to imply a reduction in the importance of individuality and the promotion of solidarity-based actions, which entail a lower risk of psychological problems (depression, anxiety) and loneliness (Nostrand and Glicksman, 2015). However, some authors observe limited potential for respite from loneliness among the residents of these communities (Rusinovic et al., 2019).

The dominant selection criteria for potential residents are age and functional capacity in terms of being able to carry out activities of daily living, as well as having the financial status to cover residential facility development and management costs. This model is based on a fragile balance between active people (potential carer-residents) and dependants, which restricts the entry of people of advanced ages (Durrett, 2009). Mutual support among residents often partly frees children from care-related tasks, but it does not replace formal and informal care (Rusinovic et al., 2019). Moreover, economic and cultural barriers such as income levels and the feminisation of care may undermine the democratising potential of these residential models (Keller & Ezquerra, 2021). In addition, the informants' narratives identified the financial factor as decisive in the choice of cohousing as a residential alternative, insofar as it limits access to a facility or affects the location and construction of the project. In this respect, various studies refer to a loss of purchasing power in old age and higher financial vulnerability when it comes to facing the personal challenges of health, care and accommodation (Grundy, 2006; Sánchez-González & Egea, 2011; Kwan & Walsh, 2018).

The location and construction features of residential facilities are determined by financial, planning (availability of land), cultural and geographical (climate, landscape) factors. In particular, experiences with nature (observing the natural environment) offer a transcendental and inspiring connection to the residential setting, enriching the daily lives of elderly residents (Sánchez-González et al., 2018; Freeman et al., 2019).

Senior cohousing management outcomes show the importance of governance as a decision-making process based on negotiating priorities, finding consensus on the administration and maintenance of shared spaces and services and, above all, conflict resolution, based on the encouragement of social participation, collaboration and solidarity. Collaborating on the cleaning of shared areas and helping other residents are also essential components that give meaning to cohabitation and shape personal expectations for the future. In addition to relationships between residents and cooperation with the community, senior cohousing can also offer privacy and autonomy, and all of this helps to satisfy the needs and rights of individuals in balance with the group (Glass, 2012). In this regard, keys to daily residential life are the promotion of social capital (Ruiu, 2015) and self-management and decision-making processes (Renz, 2006; Sanguinetti, 2012; Tummers, 2016). Nonetheless, problems can be experienced with cohabitation arrangements and with a lack of solidarity among residents, such as conflicts associated with the use and maintenance of shared areas (buildings with a single entrance or courtyard) and budgetary management, which can discourage and deter senior cohousers (Labit and Dubost, 2016).

This research was subject to various limitations. First, the data used came from a study with broader aims, such as understanding active ageing in different residential contexts (Fernández-Mayoralas & Rojo-Pérez, 2021). This means that the analysis and findings in this chapter do not come from an ad hoc study. Second, in the context of the social sciences perspective that has informed the examination of this matter, we have not addressed other fields such as law in detail. This perspective has been addressed through several studies in Spain (Etxezarreta et al., 2018; Fuentes-Lojo Rius, 2020; Keller Garganté & Ezquerra Samper, 2021; López De la Cruz & Sánchez Medina, 2017; Louredo Casado, 2021; Simón Moreno, 2020). Third, this study is representative of the first wave of senior cohousing residential facilities, constructed between 2000 and 2015. But residential facilities from the second wave of construction (from 2016 onwards) have not been included; these facilities are characterised by a diversity of occupant groups and particularly by the entry of businesses dedicated to the development of cooperatives (López Gómez & Estrada Canal, 2016). The reason for this lack of inclusion is that these residential facilities are not yet in operation.

In terms of future lines of research, a priority is to develop a more detailed understanding of the range of factors that influence the development and management of senior cohousing, based on interdisciplinary and longitudinal studies with a multi-dimensional perspective. As is natural for a qualitative analysis, there is a need to compare the opinions and discourses of a larger number of informants so as to achieve information saturation. Quantitative approaches would facilitate other interpretations of the topic through the use of representative samples, environmental assessments and statistical analyses in order to disentangle the factors that explain the development of or are decisive in cohousing that is favourable to active and healthy ageing (del Monte Diego, 2017). There is also a need for further understanding of the cohousing model's implications for health and quality of life. In addition, more rigorous analysis is required to obtain detailed information regarding the characteristics of the physical and social environments of residential facilities (Fernández-Carro, 2016). It is also important to develop a better understanding of senior cohousing from a gender perspective (Fernández Cubero, 2015; Labus, 2016) and to analyse the potential adverse effects of the implementation of the model on social and health inequalities for aged populations at the entities involved (Carrere et al., 2020). There is a demand for higher levels of understanding and collaboration among researchers, policymakers, service providers and older people as part of the collective approach to this residential model in order to promote active and healthy ageing in place (Puplampu, 2020).

In the context of the difficult reality in all aspects relating to age-friendly environments, there is a pressing need to collect and share relevant data and information to provide evidence, strengthen the knowledge base for decision-making and facilitate optimal access to cohousing for older people, insofar as it can clearly be interpreted as an age-friendly community. This category can foreseeably expand as a residential alternative for a growing aged population in the coming decades.

Acknowledgements Research funded by (i) R & D & I ENVACES (MINECO-FEDER, ref. CSO2015-64115-R); (ii) R & D ENCAGE-CM (CM, ESF, ref. S2015/HUM-3367); (iii) R & D ENCAGEn-CM (CM, ESF, ref. H2019/HUM-5698); (iv) R & D & I Long-Term Care Facilities and the COVID-19 Pandemic (Ref. 202010E158); and (v) R & D & I QASP (AESI, PI18CIII/00046, ENPY 392/18).

References

Ache, P., & Fedrowitz, M. (2012). The development of co-housing initiatives in Germany. *Built Environment, 38*(3), 395–412. https://doi.org/10.2148/benv.38.3.395

Allison, R. (2020). *Cohousing for life. A practical and personal story of earthsong eco-neighborhood.* Mary Egan Publishing.

Arroyo, I., Montesino, N., Johansson, E., & Yahia, M. W. (2021). Social integration through social connection in everyday life. Residents' experiences during the COVID-19 pandemic in SällBo collaborative housing, Sweden. *International Journal of Architectural Research Archnet-IJAR, 15*(1), 79–97. https://doi.org/10.1108/ARCH-10-2020-0236

Awan, N., Schneider, T., & Till, J. (2011). *Spatial agency: Other ways of doing architecture.* Routledge.

Baldwin, C., Dendle, K., & McKinlay, A. (2019). Initiating senior co-housing: People, place, and long-term security. *Journal of Housing for the Elderly, 33*(4), 358–381. https://doi.org/10.1080/02763893.2019.1583152

Barenys Pérez, M. P. (1992). Las residencias de ancianos y su significado sociológico [Nursing homes and their sociological meaning]. *Papers: Revista de Sociología, 40*, 121–135.

Beck, A. F. (2020). What is co-housing? Developing a conceptual framework from the studies of Danish intergenerational co-housing. *Housing, Theory and Society, 37*(1), 40–64. https://doi.org/10.1080/14036096.2019.1633398

Bresson, S., & Labit, A. (2020). How does collaborative housing address the issue of social inclusion? A French perspective. *Housing, Theory and Society, 37*(1), 118–138. https://doi.org/10.1080/14036096.2019.1671488

Buffel, T., Handler, S., & Phillipson, C. (Eds.). (2018). *Age-friendly cities and communities: A global perspective.* Policy Press.

Carrere, J., Reyes, A., Oliveras, L., Fernández, A., Peralta, A., Novoa, A. M., Pérez, K., & Borrell, C. (2020). The effects of cohousing model on people's health and wellbeing: A scoping review. *Public Health Review, 41*, 22. https://doi.org/10.1186/s40985-020-00138-1

Chatterton, P. (2013). Towards an agenda for post-carbon cities: Lessons from lilac, the UK's first ecological, affordable cohousing community. *International Journal of Urban and Reginal Research, 37*(5), 1654–1674. https://doi.org/10.1111/1468-2427.12009

Chaudhury, H., & Oswald, F. (2019). Advancing understanding of person-environment interaction in later life: One step further. *Journal of Aging Studies, 51*, 100821. https://doi.org/10.1016/j.jaging.2019.100821

Chávez-Alvarado, R., & Sánchez-González, D. (2016). Vulnerable aging in flooded households and adaptation to climate change in cities in Latin America: The case of Monterrey. *Papeles de Población, 22*(90), 9–42. https://doi.org/10.22185/24487147.2016.90.033

Czischke, D., Carriou, C., & Lang, R. (2020). Collaborative housing in Europe: Conceptualizing the field. *Housing, Theory and Society, 37*(1), 1–9. https://doi.org/10.1080/14036096.2020.1703611

Coleman, T., & Wiles, J. (2020). Being with objects of meaning: Cherished possessions and opportunities to maintain aging in place. *The Gerontologist, 60*(1), 41–49. https://doi.org/10.1093/geront/gny142

Crabtree, L. (2016). Build it like you mean it: Replicating ethical innovation in physical and institutional design. In R. Lane & A. Gorman-Murray (Eds.), *Material geographies of household sustainability* (pp. 157–174). Routledge.

Cramm, J. M., & Nieboer, A. P. (2015). Building social capital may protect against loss of well-being among older people. In F. Nyqvist & A. K. Forsman (Eds.), *Social capital as a health resource in later life: The relevance of context* (pp. 145–154). Springer.

del Monte Diego, J. (2017). *Cohousing. Modelo residencial colaborativo y capacitante para un envejecimiento feliz [Cohousing. Collaborative and enabling residential model for a happy aging*]. Estudios de la Fundación Pilares. Retrieved from: http://www.fundacionpilares.org/docs/publicaciones/fpilares-estudio04-delmonte-cohousing.pdf

Diehl, M. K., & Wahl, H. W. (2010). Awareness of age-related change: Examination of a (mostly) unexplored concept. *The Journals of Gerontology. Series B, Psychological Sciences and Social Sciences, 65*(3), 340–350. https://doi.org/10.1093/geronb/gbp110

Droste, C. (2015). German co-housing: An opportunity for municipalities to foster socially inclusive urban development? *Urban Research & Practice, 8*(1), 79–92. https://doi.org/10.1080/17535069.2015.1011428

Durrett, C. (2009). *The senior cohousing handbook: A community approach to independent living*. New Society Publishers.

Ekerdt, D. (2020). *Downsizing: Confronting our possessions in later life*. Columbia University Press.

Etxezarreta, A., Cano, G., & Merino, S. (2018). Las cooperativas de viviendas de cesión de uso: Experiencias emergentes en España [Housing cooperatives on assignment of use: Emerging experiences in Spain]. *CIRIEC-España, Revista de Economía Pública, Social y Cooperativa, 92*, 61–86. https://doi.org/10.7203/ciriec-e.92.9266

Fernández Cubero, A. (2015). *Cuerpo, género y vejez en las viviendas colaborativas para personas mayores* [Body, gender and old age in senior cohousing]. Universidad del País Vasco.

Fernández-Carro, C. (2016). Ageing at home, co-residence or institutionalisation? Preferred care and residential arrangements of older adults in Spain. *Ageing and Society, 36*(3), 586–612. https://doi.org/10.1017/S0144686X1400138X

Fernández-Mayoralas, G., & Rojo-Pérez, F. (Eds.). (2021). *Envejecimiento Activo, Calidad de Vida y Género. Las miradas académica, institucional y social* [Active Ageing, Quality of Life and Gender. The academic, institutional and social views]. Tirant lo Blanch.

Fernández-Mayoralas, G., Rojo-Pérez, F., Rodríguez-Rodríguez, V., Sánchez-Román, M., Schettini, R., Rodríguez-Blázquez, C., … Abad Jiménez, F. (2021). Marco teórico y estudio de diseño e implementación de investigación cualitativa en Envejecimiento Activo, Calidad de Vida y Género [Theoretical framework, design study and implementation of qualitative research on active ageing, quality of life and gender]. In G. Fernández-Mayoralas & F. Rojo-Pérez (Eds.), *Envejecimiento Activo, Calidad de Vida y Género. Las miradas académica, institucional y social* (pp. 48–126). Tirant lo Blanch.

Fernandez-Mayoralas, G., Rojo-Perez, F., & Rodriguez-Rodriguez, V. (forthcoming). Housing and ageing to enhance quality of life. In F. Maggino (Ed.), *Encyclopedia of quality of life and well-being research*. Springer.

Fausset, C. B., Kelly, A. J., Rogers, W. A., & Fisk, A. D. (2011). Challenges to aging in place: Understanding home maintenance difficulties. *Journal of Housing for the Elderly, 25*(2), 125–141. https://doi.org/10.1080/02763893.2011.571105

Freeman, C., Waters, D. L., Buttery, Y., & van Heezik, Y. (2019). The impacts of ageing on connection to nature: The varied responses of older adults. *Health & Place, 56*, 24–33. https://doi.org/10.1016/j.healthplace.2019.01.010

Fromm, D. (1991). *Collaborative communities: Cohousing, central living, and other new forms of housing with shared facilities*. Van Nostrand Reinhold.

Fromm, D. (2012). Seeding community: Collaborative housing as a strategy for social and neighbourhood repair. *Built Environment, 38*(3), 364–394. https://doi.org/10.2148/benv.38.3.364

Fuentes-Lojo Rius, A. (2020, January 8). Reformas legales necesarias para adaptar nuestra legislación al 'cohousing' [Legal reforms necessary to adapt our legislation to 'cohousing']. *Expansión*, 1–6.

García-Valdez, M. T., Sánchez-González, D., & Román-Pérez, R. (2019). Aging and adaptation strategies to urban environments from environmental gerontology. *Estudios Demográficos y Urbanos, 34*(1), 101–128. https://doi.org/10.24201/edu.v34i1.1810

Gelardin, S., & Muscat, E. (2009). The emerging workforce caregiving housing dilemma. *Business Renaissance Quarterly, 4*(2), 59–84.

Glass, A. P. (2009). Aging in a community of mutual support: The emergence of an elder intentional cohousing community in the United States. *Journal of Housing for the Elderly, 23*(4), 283–303. https://doi.org/10.1080/02763890903326970

Glass, A. P. (2012). Elder co-housing in the United States: Three case studies. *Built Environment, 38*(3), 345–363. https://doi.org/10.2148/benv.38.3.345

Golant, S. M. (2011). The quest for residential normalcy by older adults: Relocation but one pathway. *Journal of Aging Studies, 25*(3), 193–205. https://doi.org/10.1016/j.jaging.2011.03.003

Golant, S. M. (2015). *Aging in the right place*. Health Professional Press.

Golant, S. M. (2020). The distance to death perceptions of older adults explain why they age in place: A theoretical examination. *Journal of Aging Studies, 54*, 100863. https://doi.org/10.1016/j.jaging.2020.100863

Greenfield, E. A. (2012). Using ecological frameworks to advance a field of research, practice, and policy on aging-in-place initiatives. *The Gerontologist, 52*(1), 1–12. https://doi.org/10.1093/geront/gnr108

Greenfield, E. A., Black, K., Buffel, T., & Yeh, J. (2019). Community gerontology: A framework for research, policy, and practice on communities and aging. *The Gerontologist, 59*(5), 1–8. https://doi.org/10.1093/geront/gny089

Grundy, E. (2006). Ageing and vulnerable elderly people: European perspectives. *Ageing and Society, 26*(1), 105–134. https://doi.org/10.1017/S0144686X05004484

Guillén Sádaba, E. (1994). Vejez, pobreza y algunas alternativas a las residencias de ancianos ante la nueva situación social [Old age, poverty and some alternatives to nursing homes in the face of the new social situation]. *Cuadernos de Trabajo Social, 7*, 73–85.

Hagbert, P., Larsen, H. G., Thörn, H., & Wasshede, C. (Eds.). (2020). *Contemporary co-housing in Europe. Towards sustainable cities?* Routledge.

Hammond, M. (2018). Spatial agency: Creating new opportunities for sharing and collaboration in older people's cohousing. *Urban Science, 2*(3), 64. https://doi.org/10.3390/urbansci2030064

Jaggard, P. (2018). Are older adults a burden? *Medical Ethics, 19*(2), 18. https://doi.org/10.1016/j.carage.2018.01.014

Jolanki, O., & Vilkko, A. (2015). The meaning of a "sense of community" in a finnish senior cohousing community. *Journal of Housing for the Elderly, 29*(1–2), 111–125. https://doi.org/10.1080/02763893.2015.989767

Jones, A., Howe, A., Tilse, C., Bartlett, H., & Stimson, B. (2010). *Service integrated housing for Australians in later life*. Australian Housing and Urban Research Institute.

Kaspar, R., Oswald, F., & Hebsaker, J. (2015). Perceived social capital in self-defined urban neighborhoods as a resource for healthy aging. In F. Nyqvist & A. K. Forsman (Eds.), *Social capital and health resource in later life: The relevance of context* (pp. 109–125). Springer.

Keller Gargant é, C., & Ezquerra Samper, S. (2021). Senior cohousing: Democratize elderly care. *REVESCO. Revista de Estudios Cooperativos, 1*(137), 1–22. https://doi.org/10.5209/REVE.71867

Kwan, C., & Walsh, C. A. (2018). Old age poverty: A scoping review of the literature. *Cogent Social Sciences, 4*, 1. https://doi.org/10.1080/23311886.2018.1478479

Labit, A. (2015). Self-managed co-housing in the context of an ageing population in Europe. *Urban Research & Practice, 8*(1), 32–45. https://doi.org/10.1080/17535069.2015.1011425

Labit, A., & Dubost, N. (2016). Housing and ageing in France and Germany: The intergenerational solution. *Housing, Care and Support, 19*(2), 45–54. https://doi.org/10.1108/HCS-08-2016-0007

Labus, A. (2016). Senior cohousing in cross-cutting research. *Architecture Civil, Engineering, Environment, 9*(4), 21–31.

Lang, R., Carriou, C., & Czischke, D. (2020). Collaborative housing research (1990–2017): A systematic review and thematic analysis of the field. *Housing, Theory and Society, 37*(1), 10–39. https://doi.org/10.1080/14036096.2018.1536077

Larsen, H. G. (2019). Three phases of Danish cohousing: Tenure and the development of an alternative housing form. *Housing Studies, 34*(8), 1349–1371. https://doi.org/10.1080/02673037.2019.1569599

Lawton, M. P., & Simon, B. B. (1968). The ecology of social relationships in housing for the elderly. *The Gerontologist, 8*(2), 108–115. https://doi.org/10.1093/geront/8.2.108

Lies, M. M., Kang, M., & Sample, R. K. (2017). Place attachment and design features in a rural senior cohousing community. *Housing and Society, 44*(1–2), 41–63. https://doi.org/10.1080/08882746.2017.1394130

Linares, R. (2018). Growing inclusivity in cohousing: Stories and strategies. *Communities, 178*, 20–23.

López de la Cruz, L., & Sánchez Medina, J. A. (Eds.). (2017). *Soluciones habitacionales para el envejecimiento activo: Viviendas colaborativas o cohousing* [Housing solutions for an active ageing: Collaborative housing or cohousing]. Tirant lo Blanch.

López Gómez, D., & Estrada Canal, M. (2016). ¿Cómo avanzan las dinámicas de senior cohousing en España? [How are senior cohousing dynamics progressing in Spain?]. In S. Ezquerra, M. Pérez Salanova, M. Pla, & J. Subirats (Eds.), *Edades en transición. Envejecer en el siglo XXI* (pp. 227–237). Ariel. Retrieved from: https://tinyurl.com/yh59kp6a

Louredo Casado, S. (2021). Las cooperativas de viviendas en régimen de cesión de uso como cauce jurídico Para los nuevos modelos habitacionales [The housing cooperatives with a regime of assignment of use as a legal tool for the new housing patterns]. *CIRIEC-España, Revista Jurídica de Economía Social y Cooperativa, 37*, 167–206. https://doi.org/10.7203/ciriec-jur.37.17396

Luciano, A., Pascale, F., Polverino, F., & Pooley, A. (2020). Measuring age-friendly housing: A framework. *Sustainability, 12*(3), 848. https://doi.org/10.3390/su12030848

Marcus, C. C. (2000). Site planning, building design and a sense of community: An analysis of six cohousing schemes in Denmark, Sweden, and the Netherlands. *Journal of Architectural and Planning Research, 17*(2), 146–163.

Markle, E. A., Rodgers, R., Sanchez, W., & Ballou, M. (2015). Social support in the cohousing model of community: A mixed-methods analysis. *Community Development, 46*(5), 616–631. https://doi.org/10.1080/15575330.2015.1086400

Mieles Barrera, M. D., Tonon, G., & Alvarado Salgado, S. V. (2012). Investigación cualitativa: el análisis temático para el tratamiento de la información desde el enfoque de la fenomenología social [Qualitative research: Thematic analysis for the treatment of information from the approach of social phenomenology]. *Universitas Humanística, 74*, 195–225.

Motevasel, I. N. (2006). Senior housing in Sweden—A question of class differences and collective aging. *Journal of Housing for the Elderly, 20*(3), 77–93. https://doi.org/10.1300/J081v20n03_06

Norstrand, J. A., & Glicksman, A. (2015). Influence of living arrangements of community dwelling older adults on the association between social capital and health. In F. Nyqvist & A. K. Forsman (Eds.), *Social capital as a health resource in later life: The relevance of context* (pp. 89–107). Springer.

Oswald, F., & Kaspar, R. (2012). On the quantitative assessment of perceived housing in later life. *Journal of Housing for the Elderly, 26*(1–3), 72–93. https://doi.org/10.1080/02763893.2012.673391

Oswald, F., Wahl, H. W., Schilling, O., & Iwarsson, S. (2007). Housing-related control beliefs and independence in activities of daily living in very old age. *Scandinavian Journal of Occupational Therapy, 14*(1), 33–43. https://doi.org/10.1080/11038120601151615

Puplampu, V. (2020). Forming and living in a seniors' cohousing: The impact on older adults' healthy aging in place. *Journal of Aging and Environment, 34*(3), 252–269. https://doi.org/10.1080/02763893.2019.1656134

Presidencia de la Junta. (1999, July 29). LEY 6/1999, de 7 de julio, de atención y protección a las personas mayores [LAW 6/1999, of July 7, on the care and protection of the elderly]. *Boletín Oficial de la Junta de Andalucía*, (87), 6. https://www.juntadeandalucia.es/boja/1999/87/2

Proyecto MOVICOMA. (2015). *Mapa de viviendas colaborativas de mayores* [Map of collaborative housing for the elderly]. Retrieved from: http://movicoma.blogs.uoc.edu/mapa/

Renz, M. A. (2006). Paving consensus: Enacting, challenging, and revising the consensus process in a cohousing community. *Journal of Applied Communication Research, 34*(2), 163–190. https://doi.org/10.1080/00909880600574088

Rodriguez-Alonso, N., & Comas-d'Argemir, D. (2017). La construcción social del cuidado comunitario en La Muralleta, una cooperativa autogestionada para gente mayor [The social construction of community-based care at La Muralleta, a self-managed cooperative for the elderly]. *Quaderns-e de l'Institut Català d'Antropologia, 22*(2), 183–198.

Rojo-Pérez, F., & Fernández-Mayoralas, G. (Eds.). (2011). *Calidad de vida y envejecimiento. La visión de los mayores sobre sus condiciones de vida* [Quality of life and ageing. The old people views about their living conditions]. Fundación BBVA.

Rojo-Pérez, F., & Fernández-Mayoralas, G. (2021a). Comprensión y autovaloración de la Calidad de Vida como medida de resultado del Envejecimiento Activo [Understanding and Self-assessment of Quality of Life as an Outcome Measure of Active Ageing]. In G. Fernández-Mayoralas & F. Rojo-Pérez (Eds.), *Envejecimiento Activo, Calidad de Vida y Género. Las Miradas Académica, Institucional y Social* (pp. 743–801). Tirant lo Blanch. https://editorial.tirant.com/co/libro/envejecimiento-activo-calidad-de-vida-y-genero-9788418329142

Rojo-Pérez, F., & Fernández-Mayoralas, G. (Eds.). (2021b). *Handbook of active ageing and quality of life. From concepts to applications*. Springer. https://doi.org/10.1007/978-3-030-58031-5

Rojo-Pérez, F., Fernández-Mayoralas, G., Rodríguez-Rodríguez, V., & Rojo-Abuín, J. M. (2007). The environments of ageing in the context of the global quality of life among older people living in family housing. In H. Mollenkopf & A. Walker (Eds.), *Quality of life in old age. International and multi-disciplinary perspectives* (Social indicators research series) (Vol. 31, pp. 123–150). Springer. https://doi.org/10.1007/978-1-4020-5682-6_8

Rojo-Pérez, F., Fernández-Mayoralas, G., Rodríguez-Rodríguez, V., Lardiés-Bosque, R., Prieto-Flores, M. E., Gallar-do-Peralta, L. P., … Schettini, R. (2020). Contextos Residenciales, Envejecimiento Activo y Calidad de vida. Un Análisis a Microescala en España [Residential contexts, active ageing and quality of life. A micro-scale analysis in Spain]. In J. D. Sempere Souvannavong, C. Cortés Samper, E. Cutillas Orgilés, & J. R. Valero Escandell (Eds.), *Población y territorio. España tras la crisis de 2008* [Population and territory. Spain after the 2008 crisis] (pp. 191–208). COMARES. Retrieved from: https://www.comares.com/libro/poblacion-y-territorio_117828/

Roy, N., Dubé, R., Després, C., Freitas, A., & Légaré, F. (2018). Choosing between staying at home or moving: A systematic review of factors influencing housing decisions among frail older adults. *PLoS One, 13*(1), e0189266. https://doi.org/10.1371/journal.pone.0189266

Ruiu, M. L. (2015). The effects of cohousing on the social housing system: The case of the threshold centre. *Journal of Housing and the Built Environment, 30*, 631–644. https://doi.org/10.1007/s10901-015-9436-7

Rusinovic, K., Bochove, M. V., & Sande, J. V. (2019). Senior co-housing in the Netherlands: Benefits and drawbacks for its residents. *International Journal of Environmental Research and Public Health, 16*(19), 3776. https://doi.org/10.3390/ijerph16193776

Scharlach, A. E., & Diaz-Moore, K. (2016). Aging in place. In V. L. Bengtson & R. A. Settersten (Eds.), *Handbook of theories of aging* (pp. 407–426). Springer.

Sánchez-González, D. (2009). Environmental context and the spatial experience of ageing in the place of origin: The case of granada. *Papeles de Población, 15*, 162–198.
Sánchez-González, D. (2015). Physical-social environment and ageing of the population from environmental gerontology and geography. Socio-spatial implications in Latin America. *Revista de Geografía Norte Grande, 60*, 97–114. https://doi.org/10.4067/S0718-34022015000100006
Sánchez-González, D., Adame, L. M., & Rodríguez-Rodríguez, V. (2018). Natural landscape and healthy ageing in place: The case of the Cumbres de Monterrery National Park in Mexico. *Boletín de la Asociación de Geógrafos Españoles, 76*, 20–51. https://doi.org/10.21138/bage.2514
Sánchez-González, D., & Egea, C. (2011). Social Vulnerability approach to investigate the social and environmental disadvantages. Its application in the study of elderly people. *Papeles de Población, 17*(69), 151–185.
Sánchez-González, D., Rojo-Pérez, F., Rodríguez-Rodríguez, V., & Fernández-Mayoralas, G. (2020). Environmental and psychosocial interventions in age-friendly communities and active ageing: A systematic review. *International Journal of Environmental Research and Public Health, 17*(22), 8305. https://doi.org/10.3390/ijerph17228305
Sanguinetti, A. (2012). The design of intentional communities: A recycled perspective on sustainable neighborhoods. *Behavior and Social Issues, 21*(1), 5–25. https://doi.org/10.5210/bsi.v21i0.3873
Sanguinetti, A. (2014). Transformational practices in cohousing: Enhancing residents' connection to community and nature. *Journal of Environmental Psychology, 40*, 86–96. https://doi.org/10.1016/j.jenvp.2014.05.003
Scanlon, K., & Fernández, M. (2015). Development of new cohousing: Lessons from a London scheme for the over-50s. *Urban Research & Practice, 8*(1), 106–121. https://doi.org/10.1080/17535069.2015.1011430
Simón Moreno, H. (2020). Las cooperativas de viviendas en régimen de cesión de uso: ¿una alternativa real a la vivienda en propiedad y en alquiler en España? [Housing cooperatives under the assignment of regime: A real alternative to home ownership and rental in Spain]. *REVESCO. Revista de Estudios Cooperativos, 134*, e69165. https://doi.org/10.5209/reve.69165
Torío López, S., Viñuela Hernández, P., & García-Pérez, O. (2018). Experiencias de vejez vital. Senior Cohousing: autonomía y participación [Experiences of active aging. Senior cohousing: Autonomy and participation]. *Aula Abierta, 47*(1), 79–86. https://doi.org/10.17811/rifie.47.1.2018.79-86
Tummers, L. (2016). The re-emergence of self-managed co-housing in Europe: A critical review of co-housing research. *Urban Studies, 53*(10), 2023–2040. https://doi.org/10.2307/26151181
Tummers, L. (2017). *Learning from co-housing initiatives: Between passivhaus engineers and active inhabitants*. Delft University of Technology. https://doi.org/10.7480/abe.2017.14
Tyvimaa, T. (2011). Social and physical environments in senior communities: The Finnish experience. *International Journal of Housing Markets and Analysis, 4*(3), 197–209. https://doi.org/10.1108/17538271111152997
Vasara, P. (2015). Not ageing in place: Negotiating meanings of residency in age-related housing. *Journal of Aging Studies, 35*, 55–64. https://doi.org/10.1016/j.jaging.2015.07.004
Vestbro, D. U. (2000). From collective housing to cohousing – A summary of research. *Journal of Architectural and Planning Research, 17*(2), 164–178.
Vestbro, D. U., & Horelli, L. (2012). Design for gender equality: The history of co-housing ideas and realities. *Built Environment, 38*(3), 315–335. https://doi.org/10.2148/benv.38.3.315
Wasylishyn, C., & Johnson, J. L. (1998). Living in a housing co-operative for low income women: Issues of identity, environment and control. *Social Science & Medicine, 47*(7), 973–981. https://doi.org/10.1016/s0277-9536(98)00170-1
WHO- World Health Organization. (2007). *Global age-friendly cities: A guide*. World Health Organization. Retrieved from: https://tinyurl.com/y482zruf
WHO-World Health Organization. (2015). *World report on ageing and health*. World Health Organization. Retrieved from: https://tinyurl.com/yff7jhle

WHO-World Health Organization. (2018). *The global network for age-friendly cities and communities looking Back over the last decade, looking forward to the next*. World Health Organization. https://apps.who.int/iris/bitstream/handle/10665/278979/WHO-FWC-ALC-18.4-eng.pdf?sequence=1&isAllowed=y

Wahl, H. W., & Gerstorf, D. (2018). A conceptual framework for studying Context Dynamics in Aging (CODA). *Developmental Review, 50*(Part B), 155–176. https://doi.org/10.1016/j.dr.2018.09.003

Wahl, H. W., & Gerstorf, D. (2020). Person-environment resources for aging well: Environmental docility and life space as conceptual pillars for future contextual gerontology. *The Gerontologist, 60*(3), 368–375. https://doi.org/10.1093/geront/gnaa006

Wahl, H. W., Iwarsson, S., & Oswald, F. (2012). Aging well and the environment: Toward an integrative model and research agenda for the future. *The Gerontologist, 52*(3), 306–316. https://doi.org/10.1093/geront/gnr154

Wahl, H. W., & Oswald, F. (2016). Theories of environmental gerontology: Old and new avenues for person-environmental views of aging. In V. L. Bengston & R. A. Settersen (Eds.), *Handbook of theories of aging* (pp. 621–641). Springer.

Wiles, J. L., Leibing, A., Guberman, N., Reeve, J., & Allen, R. E. S. (2012). The meaning of "aging in place" to older people. *The Gerontologist, 52*(3), 357–366. https://doi.org/10.1093/geront/gnr098

Williams, J. (2005). Designing neighbourhoods for social interaction: The case of cohousing. *Journal of Urban Design, 10*(2), 195–227. https://doi.org/10.1080/13574800500086998

Fermina Rojo-Pérez, PhD, is Research Scientist at the Institute of Economics, Geography and Demography (IEGD), in the Spanish National Research Council (CSIC), Madrid, Spain. She carries out her work in the Research Group on Ageing (GIE-CSIC) and in the Spanish Research Group on Quality of Life and Ageing. The domains related to age-friendly settings and living conditions, as facilitators or barriers for ageing well with quality of life, are among its priority research topics. She has coordinated several research projects and networks on active ageing and participates in projects on her subject of interest. Recently, she was the Principal Researcher of the scientific project on "The Construction of Active Ageing in Spain: Self-assessment, Determinants and Expectations of Quality of Later Life" (MINECO-FEDER, ref. CSO2015–64115-R). Currently she is involved in the study of conditions and behaviour of old people in long-term care settings during the COVID-19 pandemic (CSIC, ref. 202010E158). She belongs to several scientific societies, as ISQOLS (member of the board of directors 2012–2016), AGE (member of the board of directors 2013–2017), ALAP and SEGG. More information: http://iegd.csic.es/es/personal/fermina.rojo.

Diego Sánchez-González, PhD in Geography and MSc in Social Gerontology from the University of Granada, and associate professor in the Department of Geography at the National Distance Education University (UNED) (Spain). He has been a visiting professor at Spanish and Mexican universities, and a researcher at the Spanish Agency for International Development Cooperation (AECID). He has participated as responsible and evaluator of research projects in Spain, Holland, Switzerland, Israel, Mexico, Colombia, Chile and Uruguay. He has been an advisor to governments on ageing, urban planning, housing and climate change. Research lines: geography of ageing, environmental gerontology, environment, health and quality of life of the elderly; age-friendly cities and communities; health and social services planning; housing; processes of urbanisation and rural depopulation; social development and city; social vulnerability and adaptation to climate change.

Vicente Rodríguez-Rodríguez holds a PhD in Geography from the Universidad Complutense, Madrid and currently he is Research Professor at the Centre for Humanities and Social Sciences, Spanish National Research Council. As a member of the Research Group on Ageing (GIE-CSIC), his research focuses on various aspects of ageing of the Spanish population and its social consequences (residential strategies, health and dependency, active ageing, quality of life, qualitative analysis). The latest publications (peer-reviewed papers, books and book chapters) have focused mainly on active ageing and some other aspects of the social approach to ageing. Recently, Dr. Rodríguez is leading two research initiatives on older people care homes during the pandemic as a response of the CSIC calls on COVID-19. A general view of his CV can be accessed at: www.iegd.csic.es/es/personal/vicente.rodriguez and https://scholar.google.com/citations?user=VXSJA68AAAAJ.

Dr. Gloria Fernández-Mayoralas PhD in Human Geography by the Complutense University of Madrid, Dr. Gloria Fernández-Mayoralas is a Research Scientist in the Institute of Economics, Geography and Demography (IEGD), Centre for Human and Social Sciences (CCHS), at the Spanish National Research Council (CSIC). She works in the field of population studies, focusing on living conditions and well-being in old age, particularly on health conditions and physical and social environments for an active ageing with quality of life. Dr. Fernández-Mayoralas belongs to the Research Group on Ageing (GIE-CSIC) and she is also involved in the Spanish Research Group on Quality of Life and Ageing, a multidisciplinary network between CSIC and the Institute of Health Carlos III (ISCIII) aimed the study of the quality of life of older adults living in different residential contexts. From 2016 she coordinates the R & D Programme on Active Ageing, Quality of Life and Gender: promoting a positive image of ageing and old age (ENCAGE-CM and ENCAGEn-CM), a research consortium forms by public and private research institutions, as well as civil society organisations (http://encage-cm.es/).

Chapter 11
An Evidenced-Based Approach to Optimise Age-Care Facility Design for People with Dementia

Neveen Hamza, Stuart Franklin, and Colin Elkington

Abstract People with dementia (PwD) suffer from syndromes that progressively deteriorate their physical and cognitive abilities, such as 'sundowning syndrome', 'wandering' and in most cases an inability to manage their own thermal environments. Designing and building purpose-built facilities that educate, re-habilitate and provide thermal and visual therapeutic environments are a need for both patients and their carers. Literature suggests that well-designed environments influence PwD's behaviour and attitude and aid in retention of physical abilities and reduction of agitation levels. Given such suggested benefits, there is a need to provide evidence-based designs based on environmental psychology theories while quantitatively predicting building environmental performance at design stage. This in turn leads to the use of building performance and computational fluid dynamics (CFD) tools to assess wind flow patterns, daylight availability and thermal conditions at conceptual design stage and to test iteration of architectural concepts to provide sensory and sustainable environments.

Keywords Designing for people with dementia · Architecture design pedagogy · Environmental design

11.1 Introduction

The World Health Organisation (2020) defines 'dementia' as 'a progressive syndrome leading to a set of symptoms that lead to deterioration in memory, difficulties with thinking, problem-solving, verbal communication and the ability to perform

N. Hamza (✉) · C. Elkington
Newcastle University, Newcastle upon Tyne, UK
e-mail: neveen.hamza@newcastle.ac.uk

S. Franklin
JDDK Architects, Newcastle upon Tyne, UK

E. Pozo Menéndez, E. Higueras García (eds.), *Urban Design and Planning for Age-Friendly Environments Across Europe: North and South*, Future City 19,
https://doi.org/10.1007/978-3-030-93875-8_11

daily tasks'. These changes are often incremental at the beginning, progressing to a stage where patients need to be hospitalised or taken into care homes. The mid to late onset of the syndrome, coupled with an inability to express personal discontent with their immediate environmental conditions, in a socially acceptable manner, necessitates a building design that can provide environmental comfort (thermal, visual and acoustical) compensating for sensory, cognitive and physical impairments. PwD also present the 'wandering syndrome' which adds to the need for a building that integrates well with its outdoor environments to provide wandering safe paths. Coupled with the 'sundowning syndrome', climate change and national legislative frameworks for carbon reduction, the prediction of environmental conditions at design stage is therefore a paramount output to align architectural propositions of the building programme with surrounding outdoor environments and climate.

The prevalence of dementia is alarming with 47.5 million people living with dementia worldwide, positioning it as a pressing societal issue. The rates of this syndrome are expected to continually rise, with estimates at 75.6 million cases by 2030 and an expected rise to 135.5 million by 2050 (WHO, 2015). The Alzheimer's Society UK (2015) estimates that there were about 850,000 people in the UK suffering from dementia, with an estimated growth to 2 million by 2051. It mainly affects people over the age of 65, where one in 14 people in this age group has dementia. The proportion of people diagnosed with dementia doubles for every 5 years in age, and one in six people over the age of 80 will be diagnosed with the disease. Unfortunately, it also affects a younger age group, 42,000 people in the UK under 65 are diagnosed with dementia. Globally dementia is believed to have cost approximately $818 billion in 2015. These costs include unpaid care by family members, social care and medical costs. Costs associated with direct medical care themselves account for around 20% of these global costs (WHO, 2015). It is estimated that the cost of care for dementia patients in the UK amounts to £26.2 billion annually (enough to pay the energy bills of every household in the country).

Medical, environmental psychology, social studies in gerontology, urban planning and indoor environmental studies have contributed to a better understanding of dementia and the development of interventions targeting dementia symptoms. As the symptoms of dementia progress from mild to severe, with mirrored efforts of the community to specialist care, there is a surprising lack of published research on how to design dementia user-centric buildings, where experienced environments lead to a slowing progression of the disease, ensure quality of life and reduction of agitation in PwD (HBN0802; van Hoof, 2013).

This chapter aims to add to a scarce body of published initiatives integrating building performance simulation in teaching in the design studio pedagogy (Hand, 1993; Hand & Crawley, 1997; Batty & Swann, 1997; Strand & Fisher, 2000; Soebrato, 2005; Norford, 2006; Hamza & Horne, 2007; Charles & Thomas, 2009; Kumaraswamy & de Wilde, 2015; AlSaadani & De Souza, 2019). However, this initiative deviates from previous attempts, as it is concerned with the introduction of building and urban performance environmental modelling tools into the graduation project for the Part I of a three-part architectural course in the UK. It also deviates from introducing the students to a vulnerable user of PwD, who would usually spend 6–9 months in assessment wards in hospitalised settings. The main idea was to draw out the design process from the micro to the macro, the micro being the views from the patients' pillow.

The aim of this chapter is to show how insights from environmental psychology and advances in building and urban performance simulation modelling can inform a dementia user-centric yet energy-efficient and sustainable architectural design approach in the design studio pedagogy and in practice. The challenge facing the architectural design under this brief is to facilitate buildings that provide meaningful multisensory experiences, social engagement and environmentally comfortable yet energy-efficient environment that reduces agitation behaviour and slows the progression of the disease.

The School of Architecture, Planning and Landscape, Newcastle University, UK, has piloted a pedagogy of integrating a design-centric approach for PwD and sustainability approaches in a graduation project theme in its Part I undergraduate course, since 2019. The aim is to educate a distinct graduate, capable of relational and critical thinking combining architectural design vision with targeted and specialised knowledge, to create environments that promote wellness for vulnerable users.

What started as a progressive pedagogy thinking approach integrating building simulation in the undergraduate design studio to assess the sustainability of students' design approaches is, interestingly, now being considered for its wider application in other UK architecture schools. Both, professional accrediting bodies in the UK, the Royal Institute of British Architects (RIBA) and the Architects Registration Board (ARB, 2021), are calling for evidence of inclusive design approaches, sustainability and climate change teaching starting September 2021.

However, research and accrediting bodies still address climate change and energy-conscious architecture based on parameters for healthy individuals; yet, an older generation with dementia and associated sensitivities to their environments adds another layer of the necessity to predict building performance and adjacent urban conditions at design stage.

11.2 Methodology and Methods

The need to increase the user-centric dementia care homes, with climate change concerns in mind, integrates the understanding of two disparate knowledge domains into architectural conceptualisation. The need to assess building carbon dioxide emissions and internal and external habitable conditions in thermal, visual and acoustic environments raises the need to use building and CFD wind flow urban simulations to quantify the impact of building on its users.

Critical to an increasing population with dementia is a need to change the social perceptions of mental health facilities, commonly confused in the popular imagination with Victorian asylums and prisons. The availability of facilities that educate, re-habilitate and provide therapeutic and healing environments is a need for both patients and their carers. Literature suggests that well-designed environments for a small group of patients in a home-like environment influences dementia patients' behavioural attitude and aid in retention of physical abilities to move within the spaces and reduces agitation levels. The term 'well-designed environment' is used here to refer to the physical attributes of a building and its sensory thermal, visual and acoustic environments (Lawton, 2001; HBN 08-02, 2015; Feddersen & Ludtke, 2014; Nagari & Hamza, 2016; Rodriguez & Hamza, 2016). Thermally comfortable and well day-lit indoor and outdoor environments alleviate symptoms of depression, sleep-disturbances,[1] and agitation, which are commonly treated with antipsychotic drugs. Studies show that antipsychotic treatments of elderly persons living with dementia are only modestly effective and produce a variety of serious side effects (Singleton, 2013).

This research shares experiences of setting the design brief and teaching a graduation project with an integrational and experiential architectural studio pedagogy. This is a live project that is to be built on the Newcastle University site for the Centre of Ageing and Vitality, located in the heart of the city. The teaching team aimed to:

- Create a design brief where the building and its surroundings offer a delicate balance between memorable sensory experience, while reducing information overload for PwDs.
- Integrating resolutions to an architectural design programme, with the environmental psychology theories of 'affordance' and the 'salutogenesis', the latter delineating the psychology of human transactions with their environment (Antonovsky, 1979).
- To amplify the intellectual and analytical capabilities of students to visualise and work with the invisible environmental energies that influence occupants' comfort and can be reiterated into early architectural decision-making processes.
- An understanding that one software will not deliver the required analysis. Instead, a process of building the students' capabilities for using two industry standard software (IESVE and ENVI-met) that will require time and dedication from the students added to their usual commitments in the studio.
- As the outputs of simulation tools are tables and graphs, students need to transfer this data into graphical visualisation images to facilitate design decisions and communication with non-experts in the design team (Hamza & De Wilde, 2014).

11.2.1 Choosing Building and Urban Performance Tools

Acknowledging the learning curve associated with the use of building simulation tools, a literature review of previous published precedents to integrate building simulation tools in the architectural design studio, suggests two main methodologies

[1] Sleep Disturbances. More information: https://www.sciencedirect.com/topics/computer-science/sleep-disturbance

and roles that the teaching team fulfil in an architectural studio design setting (Alsaadani & De Souza, 2019; Charles & Thomas, 2009). The first methodology tends to be for shorter projects where visual and simplified tools such as ECOTECT are used (He & Passe, 2015; Delbin et al., 2006; Roberts & Marsh, 2001; Soebarto et al., 2005) or controlled introductory massing assessments (Tarabieh et al., 2018). The second methodology, which we adopted, chooses validated industry standard software such as IESVE in the design studio to prepare graduates with tools that are used in practice (Kumaraswamy & De Wilde, 2015; Hamza & Horne, 2007). The latter choice of software tends to be introduced to students in architectural design studios over an academic year to allow for time and background knowledge attainment by the students. It is also noted that the school supported the high costs of the required computational infrastructure. The choice of software was also based on an integrated graphical user interphase that offered a level of interoperability with CAD tools.

PwDs suffer from a 'wandering syndrome' and there is a need to provide circular and passively surveyed wandering paths indoors and outdoors. The latter necessitated the assessment of the outdoor wind flow levels and shading. Urban wind flow simulation (ENVI-Met) is used to assess the effects of building massing, changing vegetation and surface materials on courtyard climatic conditions. This assessment leads to extending outdoor space use and provides indoor-sufficient natural ventilation and sheltered entrances to the building from the harsh northerly winds that hits Newcastle in the winter.

11.2.2 Forming the Teaching Team

An approach of co-production of knowledge between staff, professionals and students underpinned the methodology of teaching (explained later in the paper). The teaching team consisted of five different yet complementary roles to deliver the pedagogical framework:

- The first author as the studio leader has an experience of over 25 years as an academic teaching architecture and environmental design. She has implemented various pedagogical approaches integrating passive design principles using building and CFD wind flow urban simulation in the design studio. The studio benefited from first-hand findings from a Welcome Trust award to the first author to study dementia wards.
- The second author was appointed as a leading practitioner in user-centric approaches to architectural design for hospices, cancer respite centres and dementia housing. He is a keen practitioner of designing for vulnerable users, while designing energy-efficient and sustainable buildings.
- Two graduate tutors graduated from the MSc Sustainable Buildings and Environments, Newcastle University, who studied under a similar framework of design and environmental simulation. They were proficient and worked with the

first author to generate clear student guides of the definition of each variable of the data entry and how to analyse and visualise data outputs from the software.
- The third author was chosen as a student 'simulation champion'. His role was to support his colleagues in maintaining a dialogue of how to use the simulation tools and visualising simulation results through architectural presentation tools. He was showing progressive levels of engagement with the integration of the simulation results into the design process.
- Medical professionals and lead clinicians, working on dementia wards in the North East of England hospitals, engaged in design tutorials and gave an introductory lecture explaining the medical nature of the syndrome and their observations of working on the wards.

11.3 Evidence-Based Psycho-Spatial Effects on People with Dementia

The brief was constructed based on evidence-based research on psycho-spatial effects of space design on the vulnerable and frail in an ageing community.

The studio trajectory focused on engaging with targeted user requirements and the complexity of surrounding socio-economic neighbourhood, local climate, sustainability and climate change to deliver an integrating interpretation of these driving factors into a holistic, situated in place architecture.

Established dementia design guidelines (HBN 08-02, 2015 and New Zealand's Secure Dementia Unit Design Reference Guide, 2016) attempt to highlight the need to address contrasting design brief requirements when designing dementia facilities, contrasting a feeling of 'home away from home' while providing a hospital secure and easily sterilised accommodation. Other contrasting requirements include provision of outdoor spaces but where external environments reduce the effects of high wind speed and provide sunny areas, with reduced visual contrast in shaded areas, to extend the free use of outdoor space for the 'wandering syndrome', while easily being monitored by staff and carers.

11.3.1 Thermal Comfort for PwD

People suffering from age-related mental illnesses, such as dementia, experience physical limitations in interacting with the control infrastructures of a heating system, expressing thermal preferences to others. In the advanced stages of dementia, PwDs find it difficult to respond to the sensation of thermal and visual discomfort and are unable to attribute it to their surrounding environment (Walker et al., 2016). Instead, their thermal discomfort must often be inferred from patterns of behaviour interpreted by relatives and carers (van Hoof et al., 2010). Unfortunately, in most hospital wards reviewed, there is little, if any, local control on the central mechanical heating and ventilation systems which place greater limits on the adaptations that can be made. There are no thermal comfort design guidelines for PwD. Currently, thermal comfort for PwD is mapped from guidelines for the healthy and abled body.

There is a consensus among researchers and medical practitioners that the provision of a naturally well-lit and naturally ventilated environment will reduce agitation behaviour and simultaneously the need for pharmacological treatment. Literature suggests a narrower comfort range than younger groups and that the usual 18–27 °C range is observed to be too cold and too hot for PwDs (Walker et al., 2016; Mendes et al., 2013). For the sake of facilitating thermal assessment of spaces and its implication on energy use, enclosures' operative temperature was to be kept between 20 °C and 22 °C. Architectural layering and massing had to passively reduce dependency on heating and mechanical ventilation systems and provide 50% of the time where thermal and visual ranges for comfort are met without recourse to electrical or mechanical systems (free-floating conditions). Building performance simulations are used to assess a new design concept against the performance of a base case. The students compared their own design concepts against thermal and daylight levels of a typical double-sided corridor ward, a 1960s configuration in the NE of England. Interestingly overheating was recorded in about 9 months of the year in the existing wards (Fig. 11.1). The visualisation of thermal conditions allowed students to avoid this through testing their own design concepts and providing design iterations in terms of changing natural ventilation pathways and changing apertures' design for better daylight distribution.

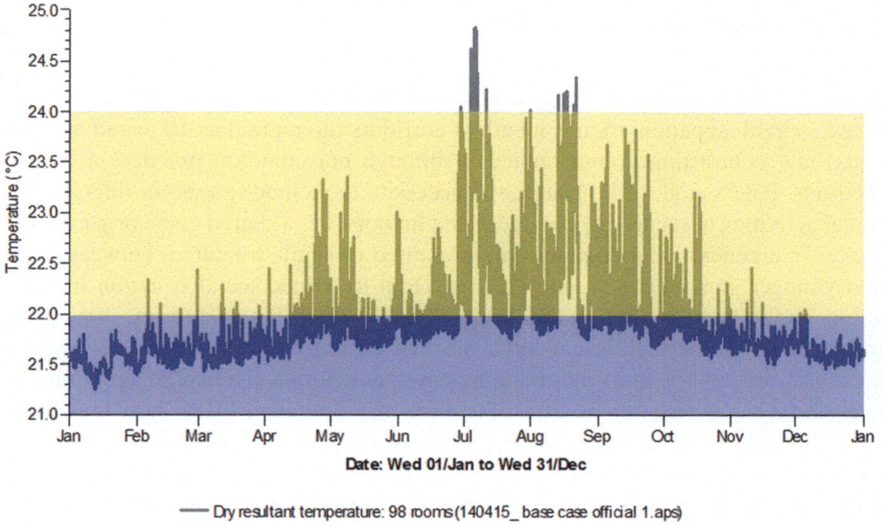

Fig. 11.1 Records of overheating in existing wards in NE of England. Colours code: Blue, temperatures between 20 °C and 22 °C; yellow, slight overheating; white, thermal stress. (Source: Authors' own)

11.3.2 Daylight for PwD

Daylighting studies quantifying required levels of indoor lighting levels for PwD are more advanced than thermal studies. Providing adequate levels of daylight for PwDs to improve visibility of their surroundings also links to medical studies on common vision impairment related to ageing, such as macular degeneration and glaucoma, stating that light levels required are substantially higher than people without visual impairment (van Hoof, 2013).

Research suggests that there is a need to provide appropriate day-lit indoor and outdoor environments, where 30 minutes per day of sun exposure is recommended for maintaining the circadian rhythm and improving older adults' sleep quality (Duzgun & Durmaz Akyol, 2017; Rodriguez & Hamza 2016; Lin & Liao, 2016). Alarmingly, Shochat et al. (2000) found that older adults in nursing homes only spent 10.5 minutes a day with lighting levels higher than 1000 lux; the limited time exposure to daylight decreased older adults' sleep quality.

Energy-saving lighting fixtures with daylight colour rendering are used to compensate for lack of access to day-lit corridors and residential spaces in care homes. However, Konis (2018) concluded that using artificial lighting can also be viewed as an indicator of the poor performance of existing indoor environments[2] to provide sufficient natural light stimulus to residents. Reliance on specialised electrical lighting fixtures or areas can also restrict the mobility of residents in other areas and add a significant financial cost.

11.3.3 Social Interaction Spaces

Students complied with design guidelines (HBN 0802) on medical service facility areas, individual patient rooms, length of corridors (no more than 10 m and no dead ends) and contrasting colour choices. Although not common practice in British facilities, Peters et al. (2021) suggest a necessity to include spaces for intergenerational activities in secured open spaces and indoors, i.e. a shared game or gardening space. Intergenerational engagement, structured or semi-structured, between PwD and younger generations is emerging as a tool to reduce social isolation in older adults and to benefit children and adults alike. In environmental terms, this means the provision of thermal, visual and acoustically comfortable spaces that allows for socialisation, yet not an overstimulating environment that can lead to agitation and aggressive behaviours by the PwD. Stephens and Breheny (2018) show how these spaces enable residents to continue to engage in activities that have meaning for them, like dining out with family and friends. Living in an environment that supports daily activities and values is likely to contribute to PwD's overall well-being and slow the progression of the disease. Social isolation is associated with an

---

[2] Indoor environments. More information: https://www.sciencedirect.com/topics/earth-and-planetary-sciences/indoor-environment

increased risk of adverse health outcomes, including functional decline, cognitive decline and dementia.

However, the brief emphasised the integration of environmental psychology, i.e. 'overstimulation' and required separate quiet rooms. The thermal and visual environmental analysis of these spaces is needed as they maintain a close link between indoor and outdoor environments and need to be passively designed to maximise the use of natural local climate potentials.

11.3.4 The Wandering Path

PwDs suffer from progressive decline in cognitive function, an individuals' basic and functional abilities. The wandering syndrome is observed, in which loss of wayfinding abilities leads to agitation. Although a wide range of non-pharmacological treatments are applied to reduce agitation behaviours, such as music therapy, exercise, games and psychological therapy, wandering needs to be encouraged as a sign of good health and activity. However, the design of the wandering path is critical as it needs to follow rules to prevent residents from wandering off the site. Increasingly frequent wandering episodes, in part, account for why caregivers of community-residing PwD make the decision to move their loved ones to long-term care.

Algase et al. (2007) define wandering as a 'locomotion behaviour' having a frequent, repetitive, temporally disordered and/or spatially disorientated nature that is manifested in lapping, random pacing patterns, attempts to wander off-site or getting lost unless accompanied. Attempts to design appropriate wandering paths are also challenging in institutions. Research conducted by Gaffney (1986) in an institutional setting found that over a 15-hour period, a population of 28 wanderers attempted to leave the unit 457 times and attempted to use an exit 274 times.

The design of the wandering path is a crucial element to keep wanderers able to wander in passively surveyed environments, by staff, to ensure PwDs safety from trips and falls.

Criteria for the design of the outdoor and indoor pathways in the brief were:

1. A simple looped path where entrance and exit doors are the same, leading a PwD to the same place.
2. Daylight availability and uniformity simulations to provide shadowed cooler areas in the summer and maximise exposure to the sun in the winter.
3. Provision of benches and sheltered areas for use by the able and less-abled bodies/users from staff, family and PwDs.
4. Provision of games areas that can be used for intergenerational activities between PwDs and children.
5. Sheltered areas of refuge outdoors where PwD can still be observed remotely.
6. Raised planters to allow easy access to planting and at the height of a wheelchair to allow PwD to experience the smell and textures of plants, with emphasis on those when brushed to release strong fragrances such as herbs and lavender.

7. Avoid using poisonous plants that might be mistakenly tried by PwDs.
8. The material finish of the pathway should be smooth and low in glare.
9. A shallow gradient in ground levels, if gradients cannot be avoided.
10. Paths wide enough for a wheelchair and a carer.
11. Upward bevelled edged walkways that prevent wheelchairs rolling into the grassed areas.
12. Avoiding light contrasts that might be cast by the construction surrounding the courtyard or by using patterns with sharp colour/texture contrasts as these are perceived as holes in the ground (AAO, 2019).
13. Fences or high green hedges to provide visual connection to the neighbourhood but prevent PwD from wandering outside the site.
14. A cloister arrangement when possible, for a shelter outdoor walkway.
15. A wandering path inside the building, providing corridors with easy single-sided access to rooms while the other side has a view to indoor external spaces and places of rest.
16. Circular pathways inside the building without intersecting the staff areas and common rooms.

11.4 Mapping Contrasting Design Brief Requirements to an Operational Design Studio Framework

Considering time limitations and an aspirational studio brief, the teaching team took the view that supporting cognitive and experiential thinking requires a pedagogical framework. The framework offers a structured pedagogical experience that involve the learner as an active participant in the process and co-generator of knowledge and its application into individual design propositions. The aspiration was that design projects can incorporate designed environments that require deeper levels of relational thinking of existing knowledge domains, i.e. building thermal and visual simulation and urban CFD wind flow modelling and environmental psychology theories.

The Structure of the Observed Learning Outcomes (SOLO) taxonomy proposed by Biggs and Collis (1982) identifies five levels of competence and knowledge acquisition for students to reach an expert-like level of competence. The SOLO taxonomy offers a framework for mapping project design activities to assess the hypothesis that simulation models offer a platform for hypothesis testing, while considering diverse assumptions of architectural design, building technologies and environmental psychology for a specific vulnerable user. Mapping the student project deliverables offered an opportunity for students to move towards more complex levels of understanding, multi-structural, relational and optimally to level that emulates working in architectural practices as graduate). Figure 11.2 presents a structural pedagogical framework mapped to the 'Structure of the Learning Outcome Theory' the SOLO taxonomy (*x*-axis), and the expected development in students design outputs (*y*-axis).

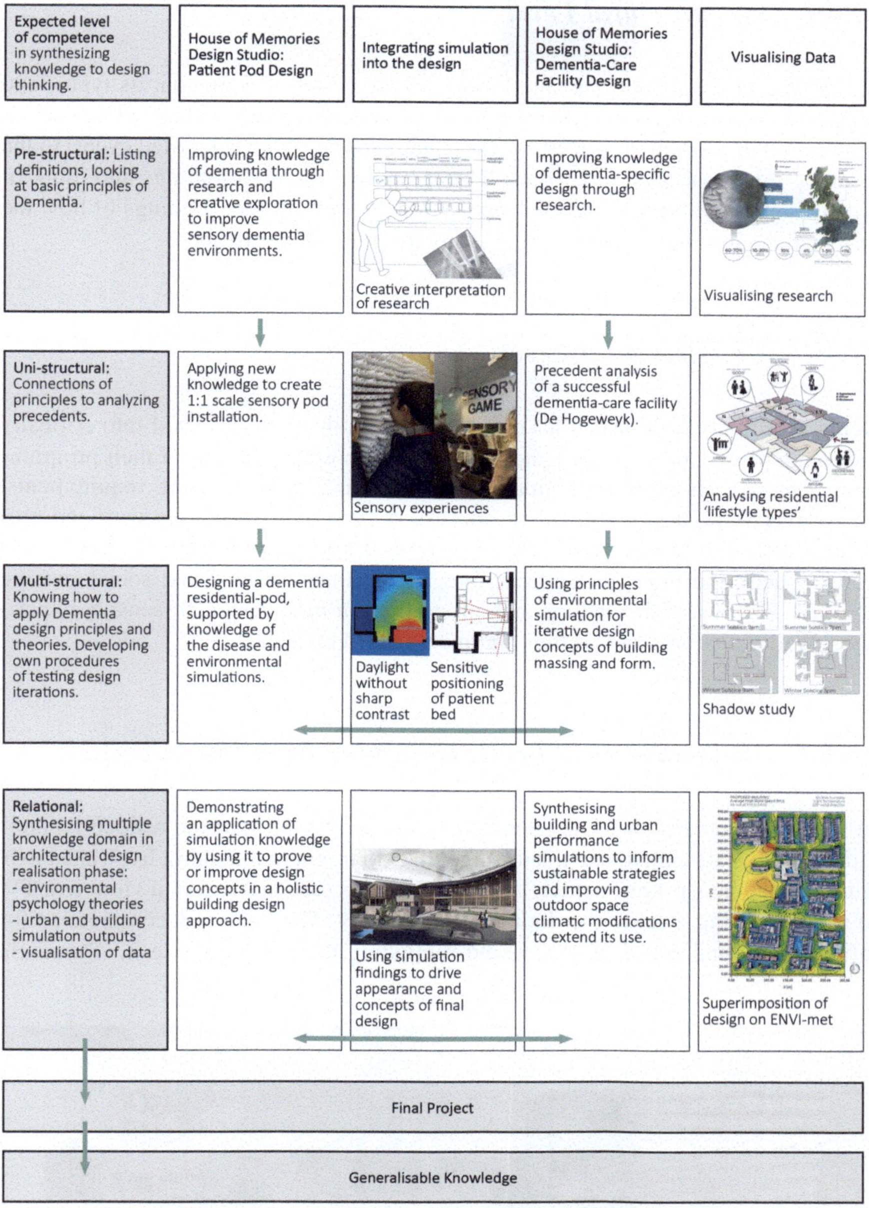

Fig. 11.2 Presents in rows the stages of the structural pedagogical framework and students' presentations at each stage. (Source: Authors' own)

11.4.1 Pre-structural Level

Being able to list definitions, and basic understanding of dementia, its types, and progression of the syndrome, sensory materials and technologies for PwD. Students were asked to work in groups at this stage and present (in a student-led event) to the teaching team and each other their bibliographical research findings and think of creative interpretations and data visualisations to present their findings of how the designed environment can positively delay the onset of dementia.

11.4.2 Uni-structural Level

Being able to connect principles understood from the previous level into critically analysing precedents of built dementia-specific facilities, in terms of their programmatic interpretation of individual, communal and medical staff requirements. Precedents were analysed for their success in providing spaces for care and also what could be improved. This stage also looked into the provision of intergenerational spaces and how the facilities integrate to their climatic and social-specific contexts. This was followed by each group presenting a designed 'sensory experience' wall of textures and edible fragrances (Fig. 11.3).

11.4.3 Multi-structural Level: Designing from Inside-Outside

In this stage, the design of the 'dementia pod', in Fig. 11.4, as the residential individual room is designed with a late-stage PwD in mind, maintaining five visual axis from the patient's pillow, namely a direct visual access to an external view, to a TV, to a carer seating area, to the door and to the toilet. The latter is recommended for more dignity and speed of access and reminding the person to use it. 'Dementia

Fig. 11.3 Sensory wall explorations: exhibition (students). (Source: Authors' own)

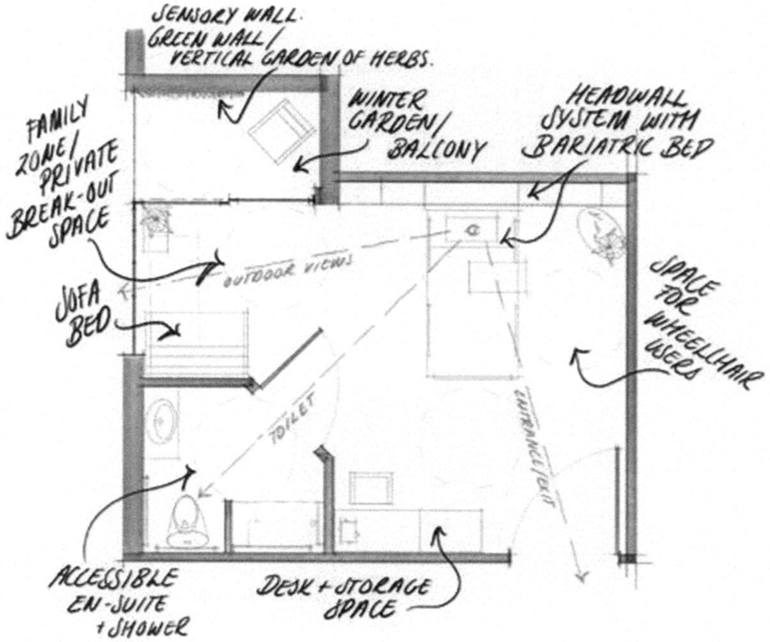

Fig. 11.4 The five visual axis in a PwD pod. (Source: Katie Hughes, student 2021)

pods' were tested in peak seasonal conditions over a day to test the varying levels of daylight and their impact on patient vision and perception of space. Design iterations were undertaken to reduce sharp contrasting shadows on floors that could be perceived by PwDs as a hole in the ground.

In this level, IESVE was used to test year-round iterations of the project massing, for shadowing and external environmental conditions, to provide solar access to all 'dementia pods' and to ensure that photovoltaics (PV) positioning on the roof will not be overshadowed by the massing.

CFD wind flow urban performance simulation (ENVI-met), in Fig. 11.5, was also used to test areas of higher wind velocities to place building entrances away from wind tunnels and test positioning of trees as natural elements to reduce wind speed around the building and inside the courtyard. Figure 11.5 illustrates wind assessment discrepancies, using a comparative study of wind prevalence and speed at conceptual design stage, when applying climatic rules of thumb (city wind rose) and a more local assessment by urban performance simulation. However, ENVI-met presents buildings as 'black masses' (middle image). The final analysis stage visualised the individual designs of the building with wind analysis superimposed on Google Maps of the neighbourhood (right).

Fig. 11.5 Comparative practices of studying prevailing wind and speeds using rules of thumb (left) and ENVI-met. (Source: Colin Elkington, 2021)

11.4.4 Relational Level

It is the final stage of demonstrating the application of building and urban environmental simulation to test the final design proposal meets set standards for thermal and visual comfort and its response to the 'salutogenic model'. In this stage, the synthesis of simulation outputs is tested for the provision of comfortable thermal, visual, natural ventilation indoors and in the wandering paths.

Designing for dementia deals with contrasting requirements, such as the need to provide sunlit spaces but without casting shadows on the floor that may be perceived as holes, wandering paths that are not close-ended corridors, etc. The need to extend the use of outdoor spaces and movement in and around the building is aided by building performance simulation.

Local climatic analysis, as seen in Fig. 11.6, leads to massing deciduous trees, sheltering entrances from higher wind speeds, changing entrances where the prevailing wind is increased in speed around building corners to calmer wind areas and how the positioning of the patient should be.

11.5 Conclusions

A structured pedagogy framework to advance architectural education is presented in this chapter. However, this methodology can also be used in practice to optimise the environmental performance of buildings for PwD. In this methodology a design from the inside to outside is proposed, which introduces environmental assessment tools at conceptual stage. The design of buildings and outdoor environments can have a long-term and significant impact on PWD behaviours and the progression of the disease. This chapter presents a pedagogical methodology based on the SOLO educational model in the design studio to educate a new generation of architects on how to design for vulnerable building users.

Designing for dementia requires the integration of different knowledge domains, linking environmental psychology theories of 'space perceptions', 'space affordance' and 'salutogenesis' to building and urban environmental performance modelling, in the architectural design process. Visualising the differences in environmental and space perception between a cognitively sound person and a person with dementia (PwD) helps reduce agitation for PwD. Design iterations, based on environmental simulations results, provide easier environments for care provision and work towards a better integration with local climatic conditions to maximise renewable energy integration.

Reflecting on our experience, we thought there is no better way than including students' feedback and letting them write our conclusion. Initially, as tutors, we were pleasantly surprised by the level of enthusiasm and continuous appetite to work with the integration methodology and the cogeneration of knowledge in our studio. We asked the students why they decided to choose our studio and to give us

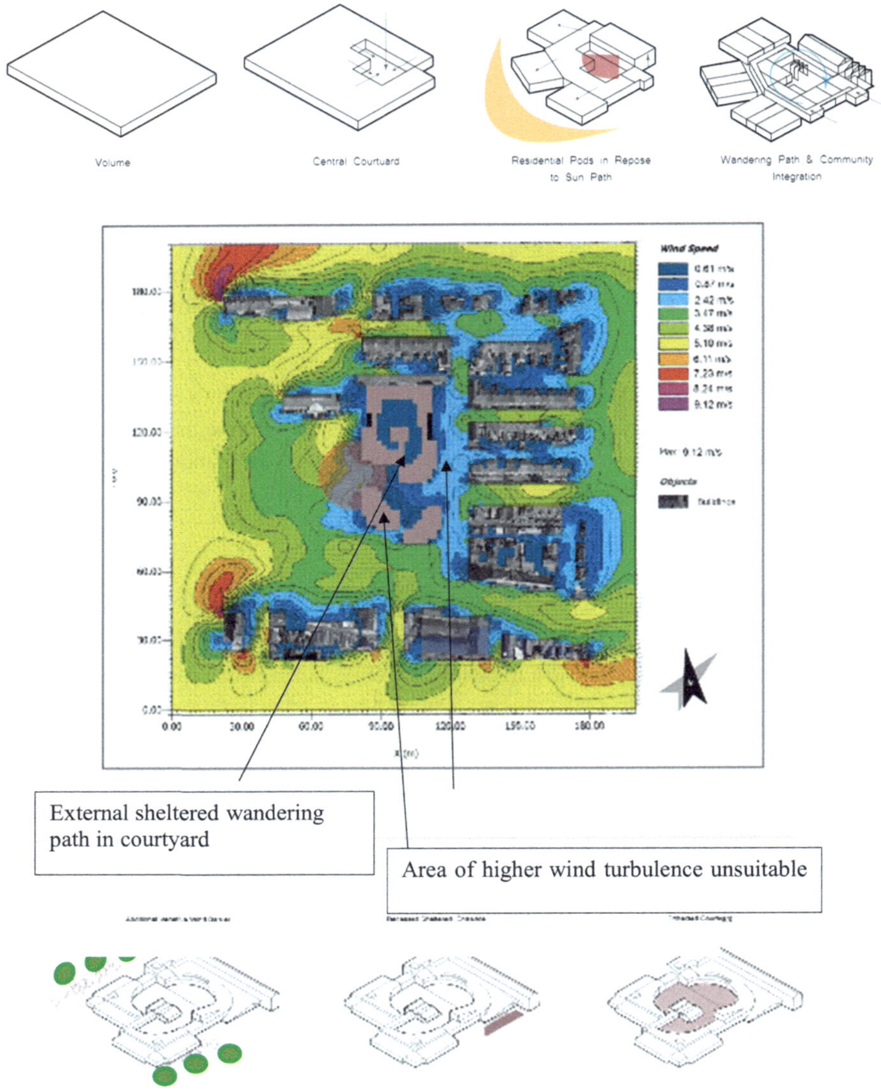

Fig. 11.6 Urban performance simulation to inform massing of building and landscape treatments of the site to maximise external environments use. (Source: Lewis Baylin, student 2021)

feedback on the process. All 36 students concurred that it was a steep learning curve on how to use the two software. The external help of the two graduate tutors and the student champion were acknowledged. However, feedback suggests that it was an enjoyable experience although at times stressful and challenging when the software crashed. Feedback suggests that they were happy to prove how their designs could work to quantitatively comply with design standards rather than just speculations.

'During my second year of studying Architecture at university, I found myself becoming more and more disconnected from my studies. I was immediately won over by the 'House of Memories' brief. I was excited at the prospect of working on a project that felt 'rooted in reality'; dealing with a live site, official regulations and scientific research, and with vulnerable end-users who form a significant part of our society – users for who empathetic design can be instrumental to their well-being and understanding of the world around them. In addition, the studio's intent on using environmental simulation to guide the project felt invaluable, as we prepared to enter the working world – a world where climate change and the environment are finally becoming of utmost importance across all industries and disciplines.'

'Using IESVE and EnviMET demanded a basic understanding of light, wind and other environmental factors, while the outputs were purely factual and informative, once again providing a refreshing difference to large proportions of architectural education which seem driven by aesthetics. It was interesting to learn not only how to produce the right diagrams or images, but to gain an understanding as to why certain results occur, and how changes can be made so that more optimal results can be achieved.'

Acknowledgements Special thanks to our graduate tutors Mrs. Paula Barrientos and Mrs. Maritina Markopoulou (MSc. Sustainable Buildings and Environments, Newcastle University). Thanks to our Medical collaborators, Dr. Jonathan Richardson and the 'Castleside' and 'Roker and Mowbray' wards' staff.

References

AAO. (2019). *American Academy of Ophthalmology. Alzheimer's disease, dementia and the eye*. Retrieved from: https://www.aao.org/eye-health/diseases/alzheimers-disease-dementia-eye

Algase, D. L., Helen Moore, D., Vandeweerd, C., & Gavin-Dreschnack, D. J. (2007). Mapping the maze of terms and definitions in dementia-related wandering. *Aging & Mental Health, 11*(6), 686–698. https://doi.org/10.1080/13607860701366434

Alsaadani, S., & Souza, D. (2019). *Teaching BPS to architects: A closer look at the building performance simulation 'consumer' and 'performer' training paradigms*. BSCAIRO 2019. 28–30 November.

Antonovsky, A. (1979). *Health, stress and coping*. Jossey-Bass Publishers.

ARB. (2021). *Architecture school consultation meeting on zoom, 8 January 2021*.

Batty, W., & Swann, B. (1997). *Integration of computer modelling and an interdisciplinary approach to building design in post graduate education. In: Proceedings of Building Simulation, 8–10. Prague: IPBSA, 29–34*.

Biggs, J., & Collis, K. (1982). *Evaluating the quality of learning: Structure of the observed learning outcome taxonomy*. Academic Press.

Charles, P., & Thomas, C. R. (2009) *Building performance simulation in undergraduate multidisciplinary education: Learning from an architecture and engineering collaboration. In: Proceedings of Building Simulation*.

Delbin, S., Gomes da Silva, V., Kowaltowksi, D., & Labaki, L. (2006). *Implementing building energy simulation into the design process: A teaching experience in Brazil*. Proceedings of PLEA 2006, pp. 755–760.

Duzgun, G., & Durmaz Akyol, A. (2017). Effect of natural sunlight on sleep problems and sleep quality of the elderly staying in the nursing home. *Holistic Nursing Practice, 31*(5), 295–302.

Feddersen, E., & Ludtke, I. (2014). *Architecture and dementia. Lost in Space*. Birkhauser.

Gaffney, J. (1986). Toward a less restrictive environment. *Journal of Geriatric Nursing, 7*, 94–95.

HBN 08-02 (2015). *Dementia-friendly health and social care environments*. Department of Health and Social Care. Government of United Kingdom. Retrieved from: https://www.gov.uk/government/publications/dementia-friendly-health-and-social-care-environments-hbn-08-02

Hamza, N., & De Wilde, P. (2014). Building simulation visualization for the boardroom: An exploratory study. *Journal of Building Performance Simulation, 7*(1), 52–67.

Hamza, N., & Horne, M. (2007). Educating the designer: An operational model for visualizing low-energy architecture. Building and Environment Hamza N, Horne M. Educating the designer: An operational model for visualizing lowenergy architecture. *Building and Environment, 42*(11), 3841–3847.

Hand, J. W. (1993). *How to train users of simulation based thermal performance analysis tools*. In: Proceedings of Building Simulation, 16–18. Adelaide: IBPSA, 39–45.

Hand, J., & Crawley, D. B. (1997). *Forget the tool when training new simulation users. In: Proceedings of Building Simulation, 8–10. Prague: IPBSA, 186–192.*

He, S. & Passe, U. (2015). *Architetctural students' attitude towards building energy modelling: A pilot study to improve integrated design education. Proceedings of BS2015, pp. 231–238.*

Konis, K. (2018). Field evaluation of the circadian stimulus potential of daylit and non-daylit spaces in dementia care facilities. *Building and Environment, 135, 112–123.*

Kumaraswamy, S., & de Wilde, P. (2015). *Simulation in education: Application in architectural technology design projects, Proceedings of BS2015 Conference, pp. 2773–2779.*

Lawton, M. (2001). The physical environment of the person with Alzheimer's disease. *Aging and Mental Health*, 56–64.

Lin, T. C., & Liao, Y. C. (2016). The impact of sunlight exposure on the health of older adults. *The Journal of Nursing, 63(4), 116–122.*

Mendes, A., Periera, C., Mendes, D., Aguiar, L., Neves, P., Silva, S., Batterman, S., & Teixeira, J. (2013). Indoor air quality and thermal comfort—Results of a pilot study in elderly care centers in Portugal. *Journal of Toxicology and Environmental Health Part A, 76*(4–5), 333–344.

New Zealand's Ministry of Health. (2016). *Secure dementia unit design reference guide: A person-centred perspective*.

Nagari, K., & Hamza, N. (2016). *Assessment of daylight in relation to the agitation levels of people with dementia*. Building Simulation and Optimization, Third IBPSA-England.

Norford, L. K. (2006). *Using simulation tools in a university laboratory course: assessing the performance of a health care center in Lusaka, Zambia*. In: Proceedings of SimBuild, 2–4. MIT: IPBSA-USA.

Peters, R., et al. (2021). Intergenerational programmes bringing together community dwelling non-familial older adults and children: A systematic review. *Archives of Gerontology and Geriatrics*.

Roberts, A., & Marsh, A. (2001). *ECOTECT: Environmental prediction in architectural education (pp. 342–347). Proceedings of the 19th eCAADe Conference.*

Rodriguez, M., & Hamza, N. (2016). Assessment of indoor visual environments using dementia friendly design criteria in day care centres. In N. Hamza & C. Underwood (Eds.), *Building simulation and optimization*. Newcastle University. 12–14 September.

Shochat, T., Martin, J., Marler, M., & Ancoli-Israel, S. (2000). Illumination levels in nursing home patients: Effects on sleep and activity rhythms. *Journal of Sleep Research, 9*(4), 373–379.

Singleton, M. (2013). *Understanding Alzheimer's disease*. A Review of Medical Advancements and Efforts to Address the Societal, Economic, and Personal Toll of an Impending Public Health Crisis California Senate Office of Research.

Soebarto, V. I., et al. (2005). *Teaching an energy simulation program in an architecture school: Lessons learned*. International IBPSA Conference.

Strand, R. K., & Fisher, D. E. (2000). *Developing thermal Intuition using building simulation in the classroom*. In: Proceedings of the association of collegiate school of architecture technology Conference, 14–17. MIT.

Stephens, C., & Breheny, M. (2018). *Healthy ageing: A capability approach to inclusive policy and practice*. Routledge.

Tarabieh, K., Aly, M., & Kazame, M. (2018). Teaching energy simulation in the architectural design studio: An experimental approach. *Engineering of Science and Military Technologies, 3, 59–64.*

van Hoof, J., & Verkerk, M. (2013). Developing an integrated design model incorporating technology philosophy for the design of healthcare environments. *Technology in Society, 35*, 1–13.

van Hoof, J., Kort, H. S. M., Hensen, J. L. M., Duijnstee, M. S. H., & Rutten, P. G. S. (2010). Thermal comfort and the integrated design of homes for older people with dementia. *Building and Environment, 45*(2), 358–370.

Walker, G., Brown, S., & Neven, L. (2016). Thermal comfort in care homes: Vulnerability, responsibility and 'thermal care'. *Building Research and Information, 44*(2), 135–146.

WHO. (2015). *Dementia.* World Health Organization. Retrieved from: http://www.who.int/mediacentre/factsheets/fs362/en/ Last access: February 2021.

Neveen Hamza looks at architecture as an interface between healthy and sustainable indoors and outdoors environments. With a background in architecture and urban design, she integrates her research into teaching in the design studio and environmental design lectures at both undergraduate and postgraduate levels. She is currently interested in the impact of the built environments on memory and well-being and the implementation of Salutogenic theories in design. She has supervised 10 PhDs to completion and examined PhDs at over 15 universities nationally and internationally. Neveen has over 100 original research journal articles, book chapters and refereed conference papers that reflect her main teaching and research, highlighting the relationships between Architectural design, building and urban performance simulation and energy consumption, human well-being and sustainability from a technical and environmental psychology perspective. You can find her publications and citations on https://scholar.google.co.uk/citations?user=aY6mHuUAAAAJ

Stuart Franklin is an Associate Director at JDDK Architects. He has been involved with numerous healthcare projects with the practice over the past 22 years, focussing particularly on hospice design throughout the UK since 1999. Locally built health care schemes include St Oswald's Hospice and Marie Curie Newcastle and Edinburgh. Stuart applies his knowledge of dementia design throughout his healthcare projects and is an external tutor at Newcastle University, running a Stage 3 Design Studio with Neveen Hamza, focussing on dementia and sustainable design.

Colin Elkington is student of the School of Architecture, Planning and Landscape at Newcastle University. In 2021, he was awarded the first prize in the national Architecture for Heath student award competition with a proposal for a dementia ward and a village for mid to advanced stage people with Dementia.

Chapter 12
Thermal Comfort for Older Population in Spain: Diagnosis and Strategies for a Climate Change Scenario

Emilia Román López, Rafael Córdoba Hernández, and Teresa Eiroa Escalada

Abstract Current cities face multiple challenges to combat problems such as pollution, shortage of green areas, invasion of vehicles, lack of complexity or urban heat island, which will be negatively influenced by the effects of climate change. In fact, extreme weather events are occurring with increasing frequency in Spain and the scenarios forecast more complex climatic changes. This situation affects the quality of life and health of citizens, increasing the risks of cardiovascular and respiratory diseases, obesity or mental health problems, aggravated by a progressively ageing population.

The ageing population will be especially hard hit by climate change and the associated rise in temperatures. Therefore, this chapter, based on the compilation of measures proposed by different guides for the case of Spain and linked to the issues of greatest impact on health in cities, raises a series of recommendations for urban intervention, taking into account different design and planning strategies to mitigate and adapt to climate change and their relationship with active ageing, within a framework of a design that encompasses health.

All this in order to achieve networks of naturalised public spaces and green areas in the vicinity, which improve the quality of life and facilitate activity, relationships and coexistence for the elderly in the best possible environmental conditions.

Keywords Healthy cities · Climate change · Extreme temperatures · Active ageing · Urban design

E. Román López (✉) · R. Córdoba Hernández · T. Eiroa Escalada
Department of Urban and Regional Planning, Escuela Técnica Superior de Arquitectura, Universidad Politécnica de Madrid, Madrid, Spain
e-mail: emilia.roman@upm.es; rafael.cordoba@upm.es; teresa.eiroa@upm.es

E. Pozo Menéndez, E. Higueras García (eds.), *Urban Design and Planning for Age-Friendly Environments Across Europe: North and South*, Future City 19,
https://doi.org/10.1007/978-3-030-93875-8_12

12.1 The Environmental Challenges of an Increasingly Urban Planet

12.1.1 Urban Development in Times of Uncertainty

In an increasingly urban planet, and subject to a multidimensional crisis, the ecological transition and design of cities take on great importance. Therefore, a substantial change is needed in the methodologies and analysis and intervention tools, which facilitate the understanding and approach to the problems we face, not only as technicians but also as citizens: "The scale of environmental problems, in a strongly interrelated society such as that of the 21st century, makes it necessary to take into account environmental problems through the eyes of the planet's citizens, so that the understanding of the problems in global terms allows people to understand fundamental issues such as the finitude of resources, the limits of the Earth's system…" (Novo Vilaverde, 2017, p. 203).

Of the different aspects of this planetary crisis (economic, political, health, social, environmental, urban, etc.), the ones to be dealt with in this chapter are those of an environmental and urban nature, related to design measures to improve thermal comfort in the public spaces of cities. This does not mean that these dimensions are the most important or that they should be treated independently of the others, as they are totally interrelated. In fact, a holistic approach is needed, which addresses the problems through a multivariate and multiscale understanding of these issues, linking analytical and sectoral approaches with comprehensive approaches, which allow a global vision of such complex and dynamic environments as are cities.

With regard to the environmental dimension of the crisis, there are three key issues that must be taken into account: climate change, depletion of resources and the deterioration of ecosystems (Córdoba Hernández, 2021b). Tackling climate change is one of the greatest challenges facing humanity and, from the disciplinary approach of urban planning, it implies changing the objectives and tools of urban and territorial planning. The importance of the local urban scale in addressing the environmental crisis should be highlighted. This challenge is inseparable from the other two key issues, depletion of resources and the deterioration of ecosystems, and it is, therefore, essential to look at all three issues together.

The urban crisis entails not only a crisis of the current city models but also of how they are experienced, thought about, designed and managed. In this sense, much more flexible forms of planning are needed in the face of uncertain times, since a historical turning point has been reached in which changes are occurring rapidly, chaotically and unexpectedly at planetary level: "The consideration of uncertainty can also be a determining factor in order to adopt adaptive management approaches, so that responses to the problems arising from climate change are implemented progressively…" (Feliu Torres et al., 2015, p. 12).

Faced with these circumstances, urban development models should be oriented towards a more ecologically balanced and socially equitable future, taking into account the integration of the existing context (history, identity, cultural and natural heritage, etc.), urban metabolism (saving energy and materials, closing cycles), citizens' quality of life (health, comfort and well-being) and social participation. The latter is essential to identify the needs, desires and satisfiers of the population. Decisions about the cities we want must be made with the citizens themselves, without imposing uses and designs.

12.1.2 Resilient Cities, Healthy Cities

The definition of the health concept has evolved throughout history. The WHO, in its formulation of the goals for the *Health for All in the 21st Century* Strategy (WHO, 1999), defined health as "that which enables all people to work productively and participate actively in the social life of the community in which they live." In this sense, cities, complex communities, play a fundamental role, as they are the places on the planet that are home to more and more people, something that will continue to increase in the coming years.

Added to this is a longer life expectancy and, therefore, an increase in the population's ageing rate. This circumstance is accentuated by the low birth rate in many European countries such as Spain, which has taken place over decades.[1] In fact, according to data from the Fundación General CSIC, in 2050 people over the age of 65 will account for more than 30% of the total Spanish population (some 13 million people) and there will be more than four million octogenarians (Fernández et al., 2010). The United Nations Organisation places Spain, for the same period, as the most aged country in the world, with 40% of the population being over 60 years old (Fig. 12.1).

The increasing ageing process of the country demands plans and strategic lines to intervene in urban spaces that are oriented at achieving more resilient, comfortable and healthy cities, improving the quality of life of their inhabitants. Currently, the dominant pattern in countries with higher levels of economic development is characterised by the dominance of non-communicable diseases (NCDs), also called chronic degenerative diseases, many of which cannot be "cured" but can be prevented and cared for, which will allow higher levels of "quality of life" or "healthy" living (Garcia Salinero et al., 2015) and where urban environments play a key role.

Another major challenge is to combat the effects of climate change, an increasingly daily reality, which enhances not only environmental problems but also those related to health: "The global climate is changing. This affects human health, well-being and life on earth. In the near future, it will lead to an amplification of current health problems, as well as new risks and pressures for the environment and the social and economic determinants of health. In the WHO European Region, health effects have already been observed from more frequent and intense extreme weather

[1] In the 1970s the average number of children was almost children per woman of childbearing age. At present this figure is 1.18 children per woman. Source: National Statistics Institute (INE)

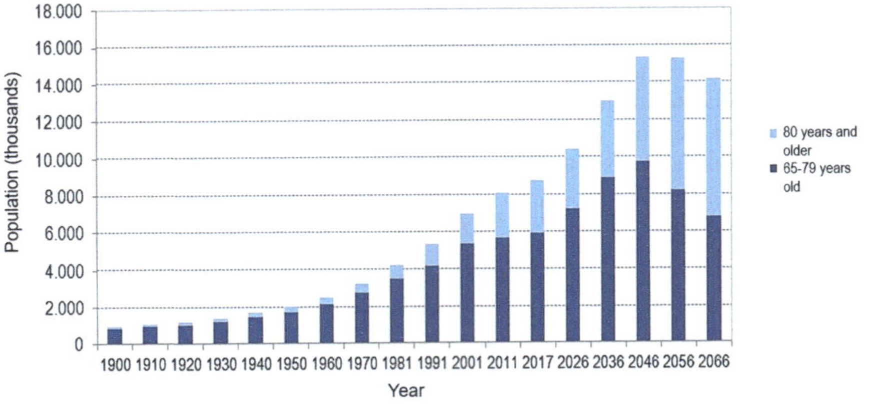

* From 1990 to 2006 the data are actual; from 2026 to 2066 they are projections.
Source: INE: INEBASE:1900-2011: Population and Housing Census. 2016: Continuous Register Statistics at 1 January 2017. Accessed January 2018; 2026-2066: Population projections. Accessed in January 2018

Fig. 12.1 Evolution of the elderly population in Spain, 1900–266. (Source: Spanish National Research Council (CSIC in Spanish), Social and Human Sciences Centre (CCHS in Spanish) and Envejecimiento en red)

events as well as changes in the geographic range of some infectious disease vectors" (WHO, n.d.).

In recent years, extreme weather phenomena are occurring more frequently in Spain, and some of the effects that are already being experienced, which are related to temperature, are heat waves and episodes of intense cold. Although historically the consequences of these thermal extremes in cities were already known, "it was not until the heat wave that hit Europe in 2003 that the health authorities incorporated the temperature variables into the public health acquis: in a first phase, heat and, in a second phase, cold" (Linares Gil et al., 2017). This is an emerging threat to the health of citizens and a challenge in order to protect the population, especially the most vulnerable, including the elderly. In the words of the *Global Compact Spanish Network* "the fight against climate change is the greatest challenge for global health in *the 21st century*" (Pacto Mundial. Red Española, 2021).

Finally, it is important to note that health in cities is not only a competence of urban planning: "It has a certain capacity to make possible environments that are favourable for a healthy city or, on the contrary, to make them impossible. But there are many other fields that have a capacity to influence this. By means of urban design, it is possible to achieve friendly streets that invite people to take strolls or green areas in the vicinity that make it possible, for example, for neighbours to get to know each other. However, education or public awareness campaigns can have as much influence as planning" (Higueras et al., 2021).

12.2 Recommendations and Strategies for Urban Intervention in the Twenty-First Century: Reference Documents and Methodology

In Spain, a multitude of guides, strategies, plans and documents have been drawn up aimed at adapting and improving the environmental quality of cities and towns when faced with the effects of climate change, as well as others focused on a design that encompasses health. Knowledge of the local climate is essential before undertaking different measures, especially in countries with a wide variety of climates, as is the case here. If the geographic situation of the peninsula already influences the presence of extreme conditions during prolonged periods of time, such as droughts that can last several years, climate change will contribute to making them more intense or frequent (IPCC, 2014).

Two reference texts will be used in this chapter, due to their relevance and specificity for the Spanish case. The first is focused on urban design and planning measures to mitigate and adapt to climate change and the second on a design that encompasses health, with special emphasis on active ageing.

A brief summary of the contents and structure of both will follow, because from them, and as will be explained later in the methodology, a comparative analysis of the most relevant urban variables regarding thermal comfort and its influence on the health of the elderly will be established, with a view to obtaining strategies for intervention in an urban scenario of climate change.

12.2.1 Measures to Mitigate and Adapt to Climate Change in Urban Planning: Methodological Guide

The first of the goals of *Measures to mitigate and adapt to climate change in urban planning*: *Methodological guide*[2] (Verdaguer Viana-Cárdenas et al., 2015) is to offer an operational tool to technicians, municipal managers and agents involved in urban and environmental processes in order to prepare coherent policies in the fight against climate change, from the perspective of urban and territorial planning. The second goal is aimed at improving the level of public awareness of the phenomenon of climate change and understanding its direct link with all aspects that characterise the urban life of towns. That is, to highlight the clear improvement of the quality of life from the broad framework of sustainability, emissions reduction, etc., but on a more day-to-day basis for citizens. Finally, the guide highlights the importance of the

[2] Verdaguer Viana-Cárdenas, C., Fariña Tojo, J., de Luxán García de Diego, M., Gómez Muñoz, G., Román López, M. E., Velázquez Valoria, I. and Sanz Alduán, A. (2015). Prepared for the Spanish Network of Cities for Climate, Department of the Spanish Federation of Municipalities and Provinces, with the collaboration of the Spanish Office for Climate Change of the Ministry of Agriculture, Food and Environment.

urban context as a scenario to tackle the environmental crisis, as well as the fundamental role of the local scale, complementary to the essential fight on a global scale.

It has a multidimensional and heterogeneous approach, addressing several instruments of analysis. Firstly, it establishes eight reference variables, which are applied as categories to each of the specific measures proposed and which respond to the multiple issues from which climate change can be addressed:

- The effects of climate change to be mitigated by adaptation measures
- Main causes of climate change to be addressed through mitigation measures.
- General sustainability goals to be met indirectly through adaptation and mitigation measures (comfort, health and quality of life/resource savings/cohesion and social welfare)
- Climatic regions of Spain
- Types of municipalities in which it can be applied, depending on their size and dominant activity
- Planning instruments from which it can be applied, depending on the scale of intervention: from general planning to the architectural project
- State-level climate change documents where the measure has been found
- Cost-benefit ratio, which in general terms provides indications of the relationship between the resources required for each measure and the level of benefits obtained from the climate change perspective

Twelve thematic areas are also proposed, which aim to cover the multidimensional scope of the urban phenomenon, taking into account different components and scales. Each of them contains a basic guideline and a series of specific measures to achieve the established goals:

- Thematic area 1: Relationship with the surrounding ecosystems
- Thematic area 2: Land use patterns
- Thematic area 3: Spatial distribution of land uses
- Thematic area 4: Urban density
- Thematic area 5: Metabolism/energy
- Thematic area 6: Metabolism/water
- Thematic area 7: Metabolism/materials, waste and emissions
- Thematic area 8: Mobility and accessibility
- Thematic area 9: Urban regeneration
- Thematic area 10: Building and urban form
- Thematic area 11: Public spaces
- Thematic area 12: Urban greenery

As will be seen below, this chapter will focus on the thematic areas "11. Public spaces" (with 19 specific measures) and "12. Urban Greenery" (with 14 specific measures), directly related to outdoor spaces in cities.

In addition, four framework measures are established, of a transversal nature and which transcend the field of urban planning, but are considered essential from the perspective of the fight against climate change, which are analysis of risks and impacts; planning, development, regulations and management; communication and awareness; and citizen participation.

12.2.2 City, Urbanism and Health

The research collected in *City, urbanism and health* (Fariña Tojo et al., 2019)[3] aims to promote healthy local environments, sharing urban criteria that contribute to promoting health and helping professionals and technicians in the design of healthier cities. It analyses the measures needed to ensure that urban environments enable people of all ages to lead healthy, active lives, with a special focus on active ageing.

The document is structured in two parts: the first part sets out the inspiring principles to change the planning model of cities, focusing on key urban elements such as streets, green spaces, facilities, social networks and coordination systems so that effective participation of the population can be achieved. The second describes a series of specific actions that can exemplify reference tasks and projects, without forgetting that these recommendations must necessarily be adapted to local contexts.

The proposed change of urban model focuses on three specific issues:

- A city designed for walking, where the pedestrian acquires the status of a basic piece in the use of urban space, being able to walk to the different functions offered by the city. As this is a form of active transport, which involves physical activity, it has a great health benefit.
- Bringing nature into the city. Urban green spaces also facilitate physical activity, reduce stress, provide a refuge from noise and offer spaces for social interaction and recreational use. At the same time, it is important to highlight their contribution to the reduction of urban pollution and the moderation of temperatures.
- To build spaces of coexistence in order to achieve cities with more solidarity and less inequality, trying to break the bubbles of loneliness and isolation that systematically confront or isolate citizens from each other. This is a key issue in the elderly.

These issues will also be taken into account in the following methodological stages.

[3] Fariña Tojo, J., Higueras García, E. and Román López, E. (2019). Prepared within the framework of the Agreement for the promotion of the Spanish Network of Healthy Cities and the local implementation of the Strategy for Health Promotion and Prevention, signed by the Ministry of Health, Consumption and Social Welfare and the Spanish Federation of Municipalities and Provinces.

12.2.3 Selection of Strategies to Adapt to Climate Change so as to Cope with High Temperatures in the Spanish Context

In order to analyse the strategies applicable to reduce the effects derived from climate change in Spain, we start with the compilation of adaptation measures proposed by different documents, specific to the Spanish case, which are collected by Verdaguer Viana-Cárdenas et al. (2015).

In accordance with the chapter's goal, the analysis has focused on those impacts derived from climate change with a direct effect on urban thermal comfort (Table 12.1).

Subsequently, the relationship of the thematic areas with the impacts of climate change that have the direct effect of increasing temperatures has been analysed in order to identify in which areas the greatest number of measures are grouped (Table 12.2).

In relation to the increase in temperature, the thematic areas that group the most measures are *10. Building and urban form, 11. Public areas and 12. Urban Greenery*. The first of the thematic areas has been dismissed, since they include measures related to the design of buildings, which are not the subject of this text, and to the development of new urban fabrics, which are not usually the urban areas where the elderly mostly live.

The article, therefore, focuses on the thematic areas *11. Public areas and 12. Urban greenery* with their measures being valid both to adapt existing urban spaces and for those newly created, giving priority to bioclimatic, habitability and comfort criteria.

In order to delve into and understand the implications of the measures included in these two areas with respect to thermal comfort conditions and health, we have proceeded to review them in detail and to evaluate the interrelation with the three main goals set out in Fariña Tojo et al. (2019) as well as the specific impacts to climate change that they contribute to mitigate, which are related to temperatures, as can be seen in the following methodological framework (Fig. 12.2).

Table 12.1 Climate change effects related to rising temperatures to be mitigated by adaptation measures

Causes	Code	Impacts on urban ecosystems
Rise in temperatures	T1	Increase of the "heat island" effect in urban areas
	T2	Greater need for shade in the middle of the day during the summer
	T3	Increased irrigation needs for urban greenery
	T4	Significant effects on human health
	T5	Increased evaporation of water from ponds, pools and reservoirs
	T6	Longer periods of thermal inversion
	T7	More pollution due to less ventilation with thermal inversion

Source: Compiled by author based on Verdaguer Viana-Cárdenas et al. (2015)

Table 12.2 Number of specific measures to mitigate and adapt to climate change identified in the FEMP Guide

Specific measures to mitigate and adapt to climate change		Thematic areas											
		1. Ecosystems of the environment	2. Land use	3. Distribution of uses	4. Urban density	5. Metabolism: Energy	6. Metabolism: Water	7. Metabolism: Materials, waste and emissions	8. Mobility and accessibility	9. Urban regeneration	10. Building and urban form	11. Public spaces	12. Urban greenery
Impacts on urban ecosystems	Increase of "heat island" in urban areas	1	1	1	1	3		2	5	8	10	13	10
	Increased needs for shade in summer										6	10	6
	Increased needs for irrigation of urban greenery						7				6	10	
	Significant effects on human health.										6	10	
	Increased evaporation in water bodies						3				6	10	

(continued)

Table 12.2 (continued)

Specific measures to mitigate and adapt to climate change		Thematic areas											
		1. Ecosystems of the environment	2. Land use	3. Distribution of uses	4. Urban density	5. Metabolism: Energy	6. Metabolism: Water	7. Metabolism: Materials, waste and emissions	8. Mobility and accessibility	9. Urban regeneration	10. Building and urban form	11. Public spaces	12. Urban greenery
	Longer periods of thermal inversion	1		1							6	10	8
	Pollution due to less ventilation with thermal inversion	1									6	10	8

Source: Compiled by author based on Verdaguer Viana-Cárdenas et al. (2015) and Fariña Tojo et al. (2019)

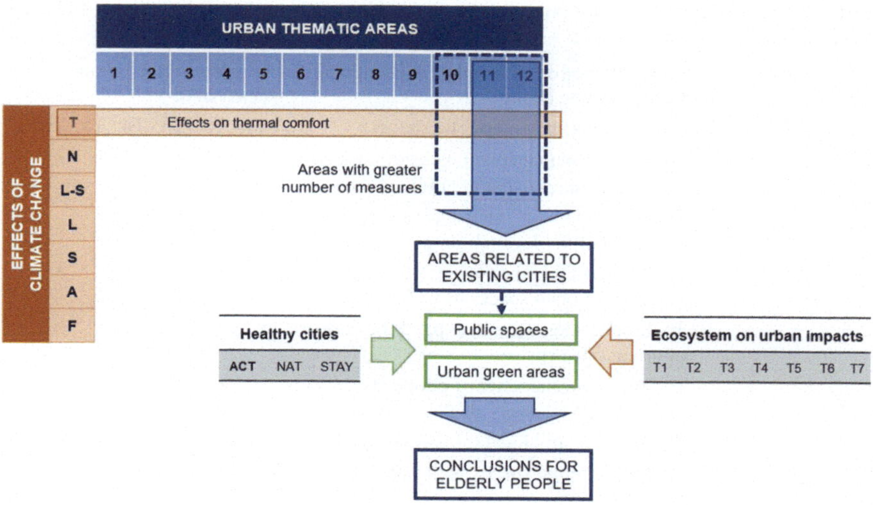

Fig. 12.2 Methodological scheme. T, Rise of temperatures; N, rise in sea levels; L-S, torrential rain/drought; L, torrential rain; S, drought; A, alteration and extinction of species; F, forest fires. (Source: Compiled by author based on Verdaguer Viana-Cárdenas et al. (2015) and Fariña Tojo et al. (2019))

12.3 Adaptation Measures, Taking into Account Thermal Comfort for the Elderly, in Order to Create Healthy Cities that Are Resilient to Climate Change

12.3.1 Public Spaces as a Vertebral Element

Open public spaces are a key urban element, where bioclimatic actions improve the population's quality of life and, at the same time, favour adaptation to climate change. These issues have an impact on all age groups but are particularly important for the elderly, who spend a large part of their time in these spaces and use them as places to socialise. They are areas that facilitate community life but also essential places to have access to the city's basic services and to a large number of urban activities.

Based on the methodology described above, a total of 13 specific measures to mitigate and adapt to climate change are identified, which can be applied to improve thermal comfort in these areas. The aim is to improve and enhance public spaces as the city's backbone, leaving behind the concept of urban development around road networks, so characteristic in the consolidated city as in current new developments, improving the safety, stay and enjoyment of these spaces for all age groups but, above all, for the elderly.

As can be seen in Table 12.3, the measures related to the three identified health-related issues in cities seek to mitigate the impacts on urban ecosystems, heat island

Table 12.3 Specific measures for public spaces in healthy neighbourhoods to mitigate and adapt to climate change

Healthy cities			Thermal comfort in healthy, resilient cities facing climate change	Impacts on the urban ecosystem						
ACT	NAT	STAY	Specific measures on public spaces	T1	T2	T3	T4	T5	T6	T7
x	x	x	To promote management measures aimed at optimising and vitalising public spaces	•	•	•	•	•	•	•
x	x	x	To adapt the use of open spaces to the evolution of the local climate	•						
	x		To consider and take advantage of the potential of open spaces as CO_2 sinks	•	•	•	•	•	•	•
x	x	x	To design, integrate and expand the network of urban open spaces	•	•	•	•	•	•	•
x	x	x	To favour the character of public spaces as a place to take leisurely strolls	•	•	•	•	•	•	•
x		x	To eliminate architectural barriers	•	•	•	•	•	•	•
x		x	To design public spaces with safe urban design criteria	•	•	•	•	•	•	•
x			To create and consolidate networks of functional, safe and attractive pedestrian and cycle routes	•	•	•	•	•	•	•
x			To dedicate no more than 25% of public roads to automobile traffic	•	•	•	•	•	•	•
	x		To reduce the impenetrability of public spaces and encourage vegetation	•						
x		x	To use high albedo materials and cold materials for floors and façades	•						
x			To establish gradients of relationships between public and private spaces	•	•	•	•	•	•	•
		x	To jointly assess the variables of the building and the attached public space	•	•	•	•	•	•	•

Source: Compiled by the author based on Verdaguer Viana-Cárdenas et al. (2015) and Fariña Tojo et al. (2019)

ACT activity, *NAT* nature, *STAY* stay

T1, increase of the "heat island" effect in urban centres; T2, increased need for shade in the middle of the day during the summer; T3, increase in irrigation needs for urban greenery; T4, significant effects on human health; T5, increased evaporation of water from ponds, pools and reservoirs; T6, longer periods of thermal inversion; T7, more pollution due to less ventilation with thermal inversion

effects, the need for shade and ventilation as well as evaporation or the need for irrigation, the negative effects of which are greater in the elderly. These different measures presented below can be strengthened if local administrations take the decision to *promote management measures aimed at optimising and vitalising public spaces*. These actions also contribute to the symbolic and functional appropriation of these spaces by citizens, establishing new community links or creating support networks among the most vulnerable (Córdoba Hernández & Garciá-Burgos Pérez, 2020; HIC-AL/Grupo e trabajo de PSH, 2017).

Given the variety of climates in Spain, another important measure is *to adapt the use of free space to the evolution of the local climate,* so as to correct or absorb the heat island effects generated as far as possible (Gunawardena et al., 2017; Shishegar, 2014), mainly in large cities and in overheated periods of the year.

One of the main actions, which can be considered common to many of the following is *to consider and take advantage of the potential of open spaces as* CO_2 sinks through plant photosynthesis. To increase green spaces in these areas boosts biodiversity and improves climate regulation capacity (Bowler et al., 2010; Demuzere et al., 2014).

Rising temperatures are becoming an increasingly serious threat to outdoor activities, especially for the elderly, who are more vulnerable to the effects of heatstroke and suffer from dehydration more easily. Thus, by *designing, integrating and expanding the network of urban open spaces,* coordinating a system capable of correcting and moderating environmental conditions through the introduction of bioclimatic and environmental quality criteria (thermal comfort, noise, pollution, etc.), the capacity of urban areas to adapt to climate change will be improved by providing strategies such as cooling through evaporation and rainwater infiltration (Carter, 2018).

Also, *to favour the character of public spaces as a place to rest and take leisurely strolls* in anticipation of more extreme temperature conditions in the overheated months, by planting of tall vegetation, which provides shade, and the increase of evapotranspiration through the increase of more natural, permeable soil, which favours the absorption, retention and presence of rainwater. In this way, this strategy is not only focused on adaptation, because by promoting the pedestrian use of these spaces, CO_2 emissions and energy costs associated with motorised mobility would be reduced (Nikolopoulou & Steemers, 2003; Pigliautile et al., 2020). Thus, possible actions such as making the road section more complex (Córdoba Hernández et al., 2020; Rueda Palenzuela, 2011), the use of asymmetrical pavements, the introduction of different uses for these, making them greener or the concept of a central boulevard with well-oriented street axes, depending on sunlight and natural ventilation, can provide important benefits for citizens, issues that will undoubtedly help to make these spaces safer and more accessible, facilitating intergenerational interaction, encouraging various activities to be carried out, promoting physical exercise and providing spaces for leisure and relaxation in the city.

Among the measures that support urban pedestrian mobility policies and benefit the creation of spaces for coexistence related to health issues, improving adaptability to climate change and favouring the physical activity of inhabitants, are the removal of physical obstacles that prevent certain population groups from reaching, accessing or moving through an urban space and their projection with safety criteria. By *eliminating architectural barriers*, this facilitates the autonomy of dependent groups or those with reduced mobility while promoting urban vitality (Kroll et al., 2006; Rimmer et al., 2004). This type of measure helps to guarantee universal accessibility to public facilities and collective equipment for all citizens through active ways that do not consume fossil fuels. The aim is to optimise the use of public spaces, encouraging people to stay in them and to socialise in all age groups, but especially the elderly population, which is particularly affected by the effects of extreme temperatures (Astrom et al., 2011; Diaz et al., 2002; McGeehin & Mirabelli, 2001).

Another of the strategies identified is *to design public spaces with safe urban design criteria* (Cozens, 2002; Parnaby, 2006). Different tools such as the *crime prevention through environmental design* (CPTED) method or the preparation of victimisation surveys can favour daily and frequent use by target and subjectively most vulnerable sectors of the population, especially the elderly population, who are at risk of reclusion and exclusion due to fear of public spaces. This type of issue again contributes to the optimisation of public spaces and to reducing all types of expenses related to the loss of autonomy (health expenses due to physical and mental pathologies, energy expenses associated with forced confinement, trips associated with unnecessary home care, etc.).

As we have seen, urban greenery is associated with a wide range of health benefits, including those related to local cooling (Nieuwenhuijsen et al., 2017; van den Bosch & Ode Sang, 2017). In this sense, actions aimed, for example, at improving the albedo of street surfaces are welcome. Therefore, *creating and consolidating functional, safe, attractive* and interconnected *networks of pedestrian and cycle routes* or *dedicating a percentage of no more than 25% of public roads to car traffic* has significant effects on reducing the heat island effect associated with asphalted surfaces, motorisation and the emission of greenhouse gases and pollutants by favouring other types of travel. Specific actions to achieve these effects are varied and can range from widening pavements in commercial areas to encourage walking, creating pedestrianised and traffic calming areas or treating local roads with a width of less than 6.0 metres as roads where road and pedestrian traffic can coexist (Kleerekoper et al., 2012; van den Bosch & Ode Sang, 2017).

The specific measure related to *reducing the impenetrability of public spaces and promoting their vegetation,* linked to the previous ones, but promoting the introduction of nature, is focused along the same lines. Actions aimed at this are the planting of vegetation, the penetrability of the land or the introduction of bodies of water that can help to improve thermal comfort. At the same time, the biodiversity of these spaces is increased, and more beneficial microclimates are achieved in the surroundings of the building. In the city, these measures should be aimed at reducing the impenetrability of the road surface and open spaces, promoting the existence of more vegetation and favouring the continuity of the water cycle, increasing the potential of urban land as a CO_2 sink and contributing to the reduction of the heat island effect (Fariña Tojo et al., 2013; Zimmermann et al., 2016).

The influence of the urban configuration on outdoor climatic conditions and the energy balance of buildings and on the diffusion of pollutants are very significant (Gros et al., 2014; Wheeler et al., 2019). In this way, *using high albedo (light) and cool materials for pavements and façades* or *establishing gradients of relationships between public and private spaces* in residential blocks can be of great importance. Within this set of actions, the treatment of the sequential access from the street to a dwelling should be emphasised, following traditional local patterns (doorways, vestibules, courtyards, etc.). This type of issue, in addition to establishing an

Fig. 12.3 Fountains in public spaces used to combat the heat in Cordoba (Spain). (Source: Eiroa, T. July 2021)

acclimatisation area, should avoid architectural typologies that encourage the spatial segregation of certain urban areas by forming closed communities, in order to favour urban vitality and the more efficient use of public spaces throughout the day.

The last measure considered would be *to jointly value the variables of the building and the surrounding public spaces*, so as to favour the optimal orientation of the buildings and their integration with the environment. The location and design of open spaces between buildings, so as to encourage the creation of microclimates that improve thermal comfort throughout the year in residential environments, is another crucial factor for the health and habitability of the population (Fig. 12.3).

12.3.2 The Potential of Urban Green Spaces for the Thermal Comfort of the Elderly

Nature is the original framework in which the vital processes of each ecosystem take place. The change of the natural balance of the territory due to the effects of urbanisation negatively affects the ecological functioning of the city. This fact has a particular impact on citizens' health, both physically and psychologically, an effect that is exacerbated in the case of the most vulnerable and the elderly.

According to the methodology used, a total of 11 specific measures have been identified to adapt to climate change referring to the intervention on green areas and aimed at improving the thermal comfort of the elderly. The integrated development of these measures, articulating a green infrastructure comparable in scope and connectivity to the city's road infrastructures, makes it possible to prioritise and optimise the services that this new ecosystem can provide to the city and its inhabitants.

As Table 12.4 shows, the measures related to the increase and improvement of green areas in the city have a direct influence on the identified issues related to health in cities.

Firstly, *increasing the existing green areas in the municipality and increasing the density of trees on public roads* contributes to the general reduction of the ambient temperature in public spaces, which has a great impact on the health of the elderly, who have a lower capacity for body thermoregulation.

In addition to increasing the amount of vegetation, paying attention to the introduction of diverse uses in the design of green areas in order *to establish criteria that allow a multifunctional and multipurpose character of green areas and public spaces* facilitate and encourage outdoor activities, the psychological and physical enjoyment of a friendly, comfortable environment in the vicinity, as well as establishing places to socialise, suitable for different age groups.

Regarding the planning of new green areas or the improvement of existing ones to *introduce bioclimatic and environmental quality criteria in the design of green areas and public spaces* allows the time in which these spaces are usable in thermal comfort conditions by the population to be extended, as well as a greater enjoyment of them. A correct design that establishes areas with good visibility and orientation elements favours the elderly staying in the parks reducing the possible sensation of insecurity and increasing their autonomy.

The positive effect of design measures aimed at maintaining cooler surfaces and spaces, such as *limiting impenetrable urban surfaces and maintaining natural terrain with occasional paving*, should not make us forget issues of universal accessibility. Thus, special attention should be paid to ensure sound, safe pavements without protrusions, to avoid the risk of falls for people with mobility problems.

On the other hand, the interventions aimed at giving continuity and connection to green areas by forming networks, such as *establishing ecological corridors between urban green areas and peri-urban natural spaces* and *increasing street trees appropriate to local climatic conditions and forecasts)*, allow the extension of accessible routes, protected from the effects of excessive heat thanks to the shade they offer, which favours physical activity associated with activities such as walking, cycling or exercising in these areas.

Moderate-intensity physical activity, such as walking, is essential for maintaining good physical and mental health. The World Health Organisation has on numerous occasions outlined how the benefits of this activity outweigh the disadvantages associated with the risk of accidents and exposure to pollution. In addition, accessibility requirements in public space for people with reduced mobility are similar to those for pedestrians in general. Improving the pedestrian environment can therefore make a significant contribution to addressing the challenges of climate change, air pollution and health (ITF, 2012).

Table 12.4 Specific measures in green areas to mitigate and adapt to climate change

Healthy cities			Thermal comfort in healthy, resilient cities facing climate change	Impacts on the urban ecosystem						
ACT	NAT	STAY	Specific measures on green areas	T1	T2	T3	T4	T5	T6	T7
x	x	x	To increase the existing green areas in the municipality and increase the density of trees on public roads	•	•				•	•
x	x	x	To establish criteria that allow a multifunctional and multipurpose character of green areas and public spaces	•	•				•	•
x	x	x	To introduce bioclimatic and environmental quality criteria in the design of green areas and public spaces, creating microclimates in the surroundings of the building	•	•				•	•
x		x	To limit the penetrable urban surface and maintain natural terrain with occasional paving improvements	•						
x	x		To establish ecological corridors between urban green areas and peri-urban natural spaces	•					•	•
x	x	x	To increase street trees appropriate to local climatic conditions and forecasts	•	•				•	•
x	x		To promote productive urban and peri-urban agriculture by inserting it into the network of urban green spaces	•						
	x		To use local species adapted to the climate and resistant to the aggressive urban environment, and with a high CO_2 retention capacity	•	•				•	•
	x		To conserve and increase the surface with CO_2 retention capacity, with adapted vegetation. To consider the CO_2 sink potential of undeveloped land	•			•		•	•
	x		To insert urban greenery into the building itself through patios, façades and green roofs	•	•				•	•
x	x	x	Bioclimatic systems to create microclimates of environmental quality in the building environment	•						

Source: Compiled by the author based on Verdaguer Viana-Cárdenas et al. (2015) and Fariña Tojo et al. (2019)

ACT activity, *NAT* nature, *STAY* stay

T1, increase of the "heat island" effect in urban centres; T2, increased need for shade in the middle of the day during the summer; T3, increase in irrigation needs for urban greenery; T4, significant effects on human health; T5, increased evaporation of water from ponds, pools and reservoirs; T6, longer periods of thermal inversion; T7, more pollution due to less ventilation with thermal inversion

In a scenario of extreme heat events, it is key to ensure that people, especially those residing in buildings most affected by high temperatures, can access public spaces and facilities with adequate thermal conditions. On the other hand, older people have greater mobility difficulties, which require them to reduce the distances between spaces without sun protection in order to reach these "climatic shelters".

Related to the previous actions, it is also advisable *to promote productive urban and peri-urban agriculture by inserting it into the network of urban green spaces* to create a sufficiently large network of parks, urban gardens and green areas connected to the peri-urban and rural environment through ecological corridors and inserting urban greenery into the built fabric through courtyards, façades and green roofs (Govindarajulu, 2014; Nerini et al., 2019). It is interesting to take advantage of the potential that the experience and the availability of time of many elderly people can have for the development and maintenance of urban orchards and gardens, being spaces for socialisation, reinforcement and generational exchange.

Measures to adopt criteria for the adequacy of plant species and the establishment of different scales of green areas, such as *to use local species adapted to the climate and resistant to the aggressive urban environment, and with high CO_2 sequestration capacity, to conserve and increase the surface with CO_2 retention capacity, with adapted vegetation and to consider the potential as CO_2 sinks of non-developable land*, are aimed at reducing the hygrothermal stress of plants, improving their resilience and optimising the services they provide (protection from radiation, CO_2 absorption, capture of suspended particles, improvement of ambient humidity, aesthetics, etc.), while reducing the need for phytosanitary treatments. Increasing green areas would also help to halt the loss of biodiversity and the degradation of ecosystem services (European Commission, 2011) and to make visible the role of ecosystem services in territorial planning and the need for their protection (Córdoba Hernández, 2021a; European Environment Agency, 2014) given the benefits for human well-being that they provide.

Finally, the improvement of conditions of the environment near the buildings, especially residential buildings *(inserting urban greenery in the building itself through patios, façades and green roofs and bioclimatic systems to create microclimates of environmental quality in the building environment)*, has a positive impact on the indoor conditions of buildings through the main passive strategies to improve comfort in public spaces: natural ventilation, lowering the temperature and increasing humidity. These strategies, without involving any investment cost in the buildings, are an improvement in the case of dwellings that are very exposed to overheating. In addition, reducing the distances between residential buildings and facilities (healthcare, commercial, etc.) without adequate thermal comfort conditions facilitates the independence of the elderly in carrying out daily activities at all times of the year (Fig. 12.4).

Fig. 12.4 Spaces of coexistence and shade in Cordoba (Spain). (Source: Eiroa, T., July 2021)

12.4 Conclusions

As has been seen during the development of this text, there is a close link between public spaces, green areas and the health of the population, especially the elderly. The independent consideration in Spanish planning of "local road infrastructures" and "green areas" hinders a necessary integrated approach and design, especially in a scenario of climate change, where urban greenery and its connection with the surrounding ecosystems play a fundamental role for the comfort and thermal conditioning of the outdoor spaces of our cities.

A change in city models is also necessary, especially in terms of mobility and transport, promoting cities and neighbourhoods for walking and reducing motorised traffic and, consequently, their negative effects on the health of the population. In the case of the elderly, walking and strolling will be the most common mode of transport, as opposed to other modes of transport that are also healthy but are aimed more at other age groups, such as cycling, for example. Walking can prevent numerous diseases, improve quality of life and increase psychological well-being.

The presence of squares and green spaces in the vicinity is also another essential element to combat a sedentary lifestyle and isolation that many elderly people suffer, especially in big cities. These spaces of relationship and coexistence must have certain characteristics so that they can be enjoyed, not only by the elderly but also by other age groups. In other words, in addition to being comfortable and accessible spaces for everyone, they should be places to socialise and for there to be intergenerational coexistence. To achieve this goal, in a scenario of climate change, it will be essential to have an adequate treatment of surfaces, as well as the presence of vegetation, especially tall vegetation, capable of offering shaded areas, which also regulate the temperature and humidity.

Taking into account the episodes of high temperatures associated with the climate change scenario in Spain, it is key to include hydration and refreshment elements in the design of public spaces and green areas. Such as drinking fountains, ornamental water features and water sheets to reduce both ambient and body temperature.

The elderly can be affected by many limitations, such as mobility difficulties, the need for companionship or reduced vision and hearing. The design of our cities must be approached from a participatory model where the elderly have a relevant role and with a focus oriented to their needs. These issues would also help to solve problems that benefit other age groups. If we think of the city, its design and functions to meet the needs of this group, we will be creating quality spaces for the entire population, favouring a lifestyle that keeps citizens more active, independent and healthy for more years.

References

Astrom, D. O., Forsberg, B., & Rocklov, J. (2011). Heat wave impact on morbidity and mortality in the elderly population: A review of recent studies. *Maturitas, 69*(2), 99–105. https://doi.org/10.1016/j.maturitas.2011.03.008

Bowler, D. E., Buyung-Ali, L., Knight, T. M., & Pullin, A. S. (2010). Urban greening to cool towns and cities: A systematic review of the empirical evidence. *Landscape and Urban Planning, 97*(3), 147–155. https://doi.org/10.1016/j.landurbplan.2010.05.006

Carter, J. G. (2018). Urban climate change adaptation: Exploring the implications of future land cover scenarios. *Cities, 77*, 73–80. https://doi.org/10.1016/j.cities.2018.01.014

Córdoba Hernández, R. (2021a). La vulnerabilidad ecosistémica según el planeamiento en la protección del suelo de la Comunidad de Madrid. *Urbano, 24*(43), 18–29. https://doi.org/10.22320/07183607.2021.24.43.02

Córdoba Hernández, R. (2021b). Estructura territorial resiliente: análisis y formalización a través del planeamiento urbanístico [Universidad Politécnica de Madrid]. https://doi.org/10.20868/UPM.thesis.69364

Córdoba Hernández, R. and García-Burgos Pérez, A. (2020). Urbanización inclusiva y resiliente en asentamientos informales. Ejemplificación en Latinoamérica y Caribe. In *Bitacora Urbano Territorial* (30, 2, pp. 61–74). Universidad Nacional de Colombia. https://doi.org/10.15446/BITACORA.V30N2.81767

Córdoba Hernández, R., Fernández Ramírez, C., Hernández Aja, A., Sánchez-Toscano Salgado, G., & Gómez Giménez, J. M. (2020). Áreas Urbanas frente a Barrios. Análisis de las características urbanas ante el reto de la ciudad post-COVID19: el caso de Madrid. *Ciudad y Territorio Estudios Territoriales, 52*(205), 665–684. https://doi.org/10.37230/CyTET.2020.205.13.2

Cozens, P. M. (2002). Sustainable urban development and crime prevention through environmental design for the British city. Towards an effective urban environmentalism for the 21st century. *Cities, 19*(2), 129–137. https://doi.org/10.1016/S0264-2751(02)00008-2

Demuzere, M., Orru, K., Heidrich, O., Olazabal, E., Geneletti, D., Orru, H., Bhave, A. G., Mittal, N., Feliu, E., & Faehnle, M. (2014). Mitigating and adapting to climate change: Multi-functional and multi-scale assessment of green urban infrastructure. *Journal of Environmental Management, 146*, 107–115. https://doi.org/10.1016/j.jenvman.2014.07.025

Diaz, J., Jordan, A., Garcia, R., Lopez, C., Alberdi, J. C., Hernandez, E., & Otero, A. (2002). Heat waves in Madrid 1986-1997: Effects on the health of the elderly. *International Archives of Occupational and Environmental Health, 75*(3), 163–170. https://doi.org/10.1007/s00420-001-0290-4

European Commission. (2011). The EU Biodiversity Strategy to 2020. https://doi.org/10.2779/39229

European Environment Agency. (2014). Spatial analysis of green infrastructure in Europe. In *EEA Technical Report* (Issue 2). Retrieved from: http://www.eea.europa.eu/publications/spatial-analysis-of-green-infrastructure

Fariña Tojo, J., Fernández Áñez, M. V., Gálvez Huerta, M. Á., Hernández Aja, A., & Urrutia del Campo, N. (2013). In A. Hernández Aja, J. Fariña Tojo, M. Á. Gálvez Huerta, M. V. Fernández Áñez, & A. Gonçalves (Eds.), *Manual de diseño bioclimático : Manual de recomendaciones para la elaboración de normativas urbanísticas*. Instituto Politécnico de Bragança. Retrieved from: http://oa.upm.es/15813/

Fariña Tojo, J., Higueras García, E., & Román López, E. (2019). Ciudad, Urbanismo y Salud. Documento Técnico de criterios generales sobre parámetros de diseño urbano para alcanzar los objetivos de una ciudad saludable con especial énfasis en el envejecimientos activo. Retrieved from: https://www.mscbs.gob.es/profesionales/saludPublica/prevPromocion/Estrategia/docs/ImplementacionLocal/Ciudad_urbanismo_y_salud.pdf

Feliu Torres, E., García Blanco, G., Gutierrez García, L., Abajo Alda, B., Mendizabal Zubeldia, M., Tapia García, C., & Alonso, J. A. (2015). *Guía para la elaboración de Planes Locales de Adaptación al Cambio Climático* (Ministerio). Oficina Española de Cambio Climático. Ministerio de Agricultura, Alimentación y Medio Ambiente. Retrieved from: https://www.miteco.gob.es/es/cambio-climatico/publicaciones/publicaciones/guia_local_para_adaptacion_cambio_climatico_en_municipios_espanoles_tcm30-178446.pdf

Fernández, J. L., Parapar, C., & Ruíz, M. (2010). El envejecimiento de la población. Revista Lychnos. Retrieved from: http://www.fgcsic.es/lychnos/es_es/articulos/envejecimiento_poblacion

Garcia Salinero, J., Cifuentes Ochoa, M., Eiroa Escalada, T., & Sanz Alduán, A. (2015). *Los beneficios de andar para la salud*. Retrieved from: https://asociacionapie.org/wp-content/uploads/2020/01/Activamadrid_documento-salud-web.pdf

Govindarajulu, D. (2014). Urban green space planning for climate adaptation in Indian cities. *Urban Climate, 10*(P1), 35–41. https://doi.org/10.1016/j.uclim.2014.09.006

Gros, A., Bozonnet, E., & Inard, C. (2014). Cool materials impact at district scale—Coupling building energy and microclimate models. *Sustainable Cities and Society, 13*, 254–266. https://doi.org/10.1016/j.scs.2014.02.002

Gunawardena, K. R., Wells, M. J., & Kershaw, T. (2017). Utilising green and bluespace to mitigate urban heat island intensity. *Science of the Total Environment, 584–585*, 1040–1055. https://doi.org/10.1016/j.scitotenv.2017.01.158

HIC-AL/Grupo de trabajo de PSH (2017). Utopías en construcción. Experiencias latinoamericanas de producción social del hábitat. HIC-AL. Retrieved from: http://autogestao.unmp.org.br/artigos-e-teses/utopias-en-construccion-experiencias-latinoamericanas-de-produccion-social-del-habitat/

Higueras, E., Román, E., & Fariña, J. (2021). Guidelines for healthier public spaces for the elderly population: Recommendations in the Spanish context. In J. Martinez, C. A. Mikkelsen, & R. Phillips (Eds.), *Handbook of quality of life and sustainability* (pp. 35–51). Handbook of Quality of Life and Sustainability. https://doi.org/10.1007/978-3-030-50540-0_3

IPCC. (2014). Climate change 2014 impacts, adaptation, and vulnerability Part B: Regional aspects: Working group ii contribution to the fifth assessment report of the intergovernmental panel on climate change. In *Climate change 2014: Impacts, adaptation and vulnerability: Part*

B: Regional aspects: Working Group II Contribution to the Fifth Assessment Report of the Intergovernmental Panel on Climate Change (Intergover). Cambridge University Press. https://doi.org/10.1017/CBO9781107415386

ITF. (2012). Pedestrian safety, urban space and health. *International Transport Forum*, 93–97. https://doi.org/10.1787/9789282103654-en

Kleerekoper, L., Van Esch, M., & Salcedo, T. B. (2012). How to make a city climate-proof, addressing the urban heat island effect. *Resources, Conservation and Recycling, 64*. https://doi.org/10.1016/j.resconrec.2011.06.004

Kroll, T., Jones, G. C., Kehn, M., & Neri, M. T. (2006). Barriers and strategies affecting the utilisation of primary preventive services for people with physical disabilities: A qualitative inquiry. *Health & Social Care in the Community, 14*(4), 284–293. https://doi.org/10.1111/j.1365-2524.2006.00613.x

Linares Gil, C., Carmona Alférez, R., Ortiz Burgos, C., & Díaz Jiménez, J. (2017). Temperaturas extremas y salud. *Cómo nos afectan las olas de calor y de frío*. Los libros de la Catarata.

McGeehin, M. A., & Mirabelli, M. (2001). The potential impacts of climate variability and change on temperature-related morbidity and mortality in the United States. *Environmental Health Perspectives, 109*(2), 185–189. https://doi.org/10.2307/3435008

Nerini, F. F., Slob, A., Engström, R. E., & Trutnevyte, E. (2019). A research and innovation agenda for zero-emission European cities. *Sustainability (Switzerland), 11*(6). https://doi.org/10.3390/su11061692

Nieuwenhuijsen, M. J., Khreis, H., Triguero-Mas, M., Gascon, M., & Dadvand, P. (2017). Fifty shades of green. *Epidemiology, 28*(1), 63–71. https://doi.org/10.1097/EDE.0000000000000549

Nikolopoulou, M., & Steemers, K. (2003). Thermal comfort and psychological adaptation as a guide for designing urban spaces. *Energy and Buildings, 35*(1), 95–101. https://doi.org/10.1016/S0378-7788(02)00084-1

Novo Vilaverde, M. (2017). *La educación ambiental. Bases éticas, conceptuales y metodológicas (Universita)*. Universitas.

Pacto Mundial. Red Española. (2021). *La lucha contra el cambio climático, el mayor reto para la salud mundial del siglo XXI*. Retrieved from: https://www.pactomundial.org/2021/05/la-lucha-contra-el-cambio-climatico-el-mayor-reto-para-la-salud-mundial-del-siglo-xxi/

Parnaby, P. F. (2006). Crime prevention through environmental design: Discourses of risk, social control, and a neo-liberal context. *Canadian Journal of Criminology and Criminal Justice, 48*(1), 1–29. https://doi.org/10.1353/ccj.2006.0011

Pigliautile, I., Marseglia, G., & Pisello, A. L. (2020). Investigation of CO2 variation and mapping through wearable sensing techniques for measuring pedestrians' exposure in urban areas. *Sustainability (Switzerland), 12*(9). https://doi.org/10.3390/su12093936

Rimmer, J. H., Riley, B., Wang, E., Rauworth, A., & Jurkowski, J. (2004). Physical activity participation among persons with disabilities – barriers and facilitators. *American Journal of Preventive Medicine, 26*(5), 419–425. https://doi.org/10.1016/j.amepre.2004.02.002

Rueda Palenzuela, S. (2011). El urbanismo ecológico: un nuevo urbanismo para abordar los retos de la sociedad actual. Territorio, Urbanismo, Sostenibilidad, Paisaje, Diseño Urbano, pp.1–34.

Shishegar, N. (2014). The impact of green areas on mitigating urban heat island effect: A review. *International Journal of Environmental Sustainability, 9*(1), 119–130. https://doi.org/10.18848/2325-1077/CGP/v09i01/55081

van den Bosch, M., & Ode Sang, Å. (2017). Urban natural environments as nature-based solutions for improved public health – A systematic review of reviews. *Environmental Research, 158*, 373–384. https://doi.org/10.1016/j.envres.2017.05.040

Verdaguer Viana-Cárdenas, C., Fariña Tojo, J., de Luxán García de Diego, M., Gómez Muñoz, G., Román López, M. E., Velázquez Valoria, I., & Sanz Alduán, A. (2015). *Medidas para la mitigación y la adaptación al cambio climático en el planeamiento urbano. Guía metodológica* (Federación Española de Municipios y Provincias (ed.))

Wheeler, S. M., Abunnasr, Y., Dialesandro, J., Assaf, E., Agopian, S., & Gamberini, V. C. (2019). Mitigating urban heating in dryland cities: A literature review. *Journal of Planning Literature, 34*(4), 434–446. https://doi.org/10.1177/0885412219855779

WHO. (1999). *Health for all in the 21st century*. The Health for All Policy Framework for the WHO European Region. World Health Organisation.

WHO. (n.d.). *Health topics. Environment and health. Climate change.* Retrieved from: https://www.euro.who.int/en/health-topics/environment-and-health/Climate-change. Accessed 12 July 2021.

Zimmermann, E., Bracalenti, L., Piacentini, R., & Inostroza, L. (2016). Urban flood risk reduction by increasing green areas for adaptation to climate change. In M. Drusa, I. Yilmaz, M. Marschalko, E. Coisson, & A. Segalini (Eds.), *World multidisciplinary civil engineering-architecture-urban planning symposium 2016* (Vol. 161, pp. 2241–2246). Wmcaus 2016. https://doi.org/10.1016/j.proeng.2016.08.822

Emilia Román López is PhD in Architecture at Universidad Politécnica de Madrid (UPM), december 2014. Architect, formed at the Universidad Politécnica de Madrid (UPM) since 1999. Foundational partner cc60 Architectural Studio. Currently working as Associate Professor in the Department of Urban and Regional Planning (DUyOT) at Escuela Técnica Superior de Arquitectura de Madrid, Universidad Politécnica de Madrid. She is member of Grupo de Investigación en Arquitectura, Urbanismo y Sostenibilidad (Architecture, Urbanism and Sustainability Research Group, GIAU+S) at UPM. Involved in several studies, investigations, and projects regarding bioclimatic and environmentally integrated architecture, healthy urban design and urban regeneration, at GIAU+S to which belongs since 2009. Taken part in many publications and lectured in numerous professional associations, besides national and international universities.

Rafael Córdoba Hernández is PhD in Arquitecture at Universidad Politécnica de Madrid (UPM), december 2021. He is member of GIAU+S (Research group on Architecture, Urbanism and Sustainability), AETU (Spanish Urban Planners Association, which is integrated in the European Council of Spatial Planners, CPT-CEU) and of Planners for Climate Action from UN-HABITAT Climate Change Planning Unit (P4AC). Since 2012, he has been teaching as an associate lecturer at the Department of Urban and Regional Planning (Universidad Politécnica de Madrid). His professional work is focused on the development of urban projects at different territorial scales in various communities, drafting normative and regulatory documents for Land Management, General and Development Planning. Also, he has carried out different national and international works of urban advisory in Projects and Reports of Technical and Legal Feasibility of projects. For his experience in the subject, he writes expert opinions of urban nature in contentious-administrative processes, and drafts technical reports. Furthermore, he has been called as an expert in environmental and sustainability issues in the Technical Committees for review of the General Plan of Madrid and in the Environmental Sectoral Working Group of the review of the Land Law of the Community of Madrid.

Teresa Eiroa Escalada, MSc Architecture and Urban Planning. Founding partner of CC60 Estudio de Arquitectura SLP since 1999, she develops building projects and technical assistance in urban planning in relation to various administrations from the perspective of sustainable development and the viability of actions. She is teaching as an associate lecturer at the Department of Urban and Regional Planning (Universidad Politécnica de Madrid), is professor of the Master's Degree in Environment and Bioclimatic Architecture (MAYAB-UPM) in Comprehensive Urban Rehabilitation and Environmental Impact and, previously at the Pontifical University of Salamanca, in Spatial Planning, Urbanism and Environmental Management, and in many postgraduate courses in architecture professional associations and national and international universities.

Chapter 13
Use of Public Spaces by Older Adults. Comparison Among Madrid and Newcastle Upon Tyne

María Teresa Baquero Larriva

Abstract This study aims to identify what factors influence the use of public spaces by older adults in two European cities, Madrid (Spain) and Newcastle upon Tyne (United Kingdom), through a mixed methodology based on environmental measurements, surveys and direct observation in open public spaces. There are significant differences between the two case studies in terms of permanence, frequency of use and activities. In Madrid, older adults correspond to 22.33% of the users; most of them visit those spaces daily and perform more physical activities, while in Newcastle, they represent 37% of the users; most of them perform passive activities and stay longer than in Madrid. It was found that some environmental factors such as air temperature, humidity, noise level and wind speed, as well as some design elements and the subjective perception of acoustic and thermal comfort, have a statistical influence on the permanence of older people. Although these are cities with different characteristics in size, climate, culture and urban structure, in both cases, older adults correspond to a high percentage of the users of open public spaces. An adequate design that promotes their health and quality of life is important, taking into account their needs and preferences to support active ageing.

Keywords Use of public spaces · Older adults · Open public spaces · Active ageing · Healthy ageing · Population ageing

13.1 Introduction

Population ageing is a challenge for urban planners for twenty-first century cities. International policies have focused on developing supportive urban environments for ageing societies, like the 'WHO Global Network for Age-friendly Cities and

M. T. Baquero Larriva (✉)
Universidad Politécnica de Madrid, Madrid, Spain

E. Pozo Menéndez, E. Higueras García (eds.), *Urban Design and Planning for Age-Friendly Environments Across Europe: North and South*, Future City 19,
https://doi.org/10.1007/978-3-030-93875-8_13

Communities' of which Madrid and Newcastle upon Tyne are involved, whose main objective is to facilitate active and healthy ageing (World Health Organization, 2020).

The first environmental gerontology studies appeared around 1970, analysing the influence of interactions between the elderly and their surroundings (Lawton & Nahemow, 1973; Lawton, 1985). It has been demonstrated that green and urban public spaces affect mental and physical health (Kaplan, 1995; Shanahan et al., 2015), and use of those spaces could contribute to keeping older people more independent, promoting active and healthy ageing in place (Sugihara & Evans, 2000; Vidal & Pol, 2005; Maas et al., 2009; Sánchez-González et al., 2018).

The main causes of morbidity among the elderly are related to cardiovascular diseases being sedentarism and stress some of the risk factors. Some authors have proven that infrastructure in neighbourhoods to do physical activities is associated with a 38% reduction of type two diabetes (Auchincloss et al., 2009). Lack of neighbourhood security affects stress, leading to the elderly restricting their movements and self-isolation (Beard et al., 2009). Loneliness is one of the main social problems in the developed countries; this particularly affects the elderly because of 'ageism'. For some authors, this is a multidimensional process and has a spatial manifestation; generally in the neighbourhoods, it can be economic, social, cultural, labour, participation, etc. (Scharf et al., 2001; Burchardt et al., 2002; Allen et al., 2012). For many people, public spaces play an important role in overcoming loneliness (Arup et al., 2015) being places for social interaction, participation and diverse activities.

Some studies have used observation through photographs, videos and notes to describe the use and behaviour in public spaces (Whyte, 1980); other authors have added measurements of environmental parameters and surveys to analyse the use of public spaces (Urrutia del Campo, 2001; Nikolopoulou & Lykoudis, 2007; Lai et al., 2014). Nikolopoulou and Lykoudis (2007) studied two public spaces in Athens, where 15% of the users were elderly people, and found a strong relationship between the microclimate and the use of these spaces. Likewise, in the study of Urrutia del Campo (2001), older people represented 11% of users of three squares in Madrid, finding a relationship between physical and environmental conditions and the way they use. Similarly, Lai et al. (2014) analysed a square and a park in Wuhan, China, where the elderly represented the highest percentage of users; they found that thermal comfort was the most influential factor in the use of these spaces, and the authors emphasised the need of designing urban spaces considering their needs.

The city of Madrid, capital of Spain, is located in the centre of the country (40°26′N 3°41′W), at 667 metres above sea level; its climate is a hot dry summer Mediterranean (Csa) according to Köppen's climate classification (mild wild winters and hot, dry summers). This city has a population of 3,266,126 inhabitants of which 7.4% are over 80 years old, while people over 65 represent 20.39% (Ayuntamiento de Madrid, 2019). While Newcastle upon Tyne is a city in the north-east of England (54°58′40″N 1°36′48″ W), at 30 metres above sea level, its climate, corresponding to 'Cfb' humid temperate oceanic climate, is characterised by cool summers, with abundant and well-distributed rainfall throughout the year, according to Köppen's climate classification. It is the eighth most populated city in England with 314,366 inhabitants as of July 1, 2019. In Newcastle upon Tyne, the

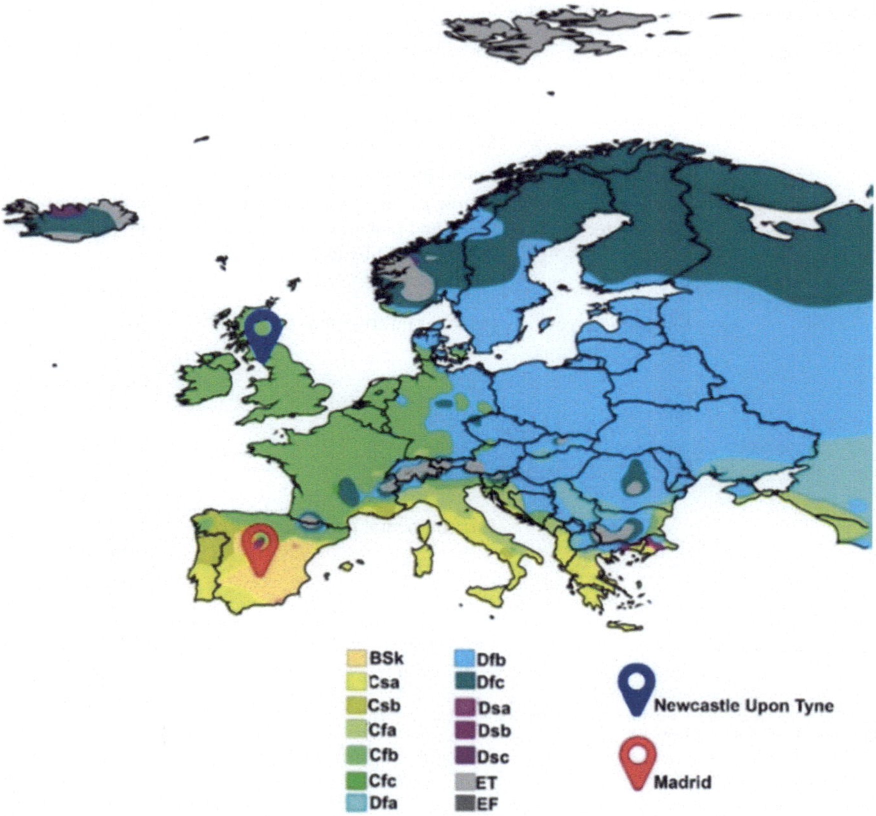

Fig. 13.1 Location of Madrid and Newcastle on Köppen's climate classification map of Europe. (Source: Own elaboration based on Ali Zifan (2016))

percentage of people over 60 years old was 19.4% in 2018 (World Health Organization, 2019) and is estimated to reach 44.2% by 2039 (Newcastle City Council, 2019). Figure 13.1 shows the location of these two cities within the Köppen's climate classification map of Europe.

Although those are cities with different characteristics in size, climate, culture and urban structure, this study aims to assess and compare the use of outdoor public spaces by older adults and identify which factors influence it in those European cities.

13.2 Methodology

A mixed methodology that includes measurements of environmental parameters, direct observations and surveys was applied in three open spaces in Madrid and two in Newcastle upon Tyne (Table 13.1). Fieldwork was conducted once a month, from

Table 13.1 Methodological framework

Methodology	Measurements of environmental parameters in situ	Air temperature (Ta)	°C
		Relative humidity (RH)	%
		Wind speed	m/s
		Sky view factor (SVF)	
		Sound pressure level (SPL)	LAeq (dB)
	Direct observation and photographic recording	People passing through the public space and people who stayed, differentiated by age group	Children
			Teenagers
			Adults
			Older adults
	Questionnaire survey	Personal information	Age
			Sex
			Frequency and time of the visit
			Activity
			Level of clothing
		Environmental perceptions	Thermal comfort
			Acoustic comfort
	Statistical analyses	Statistical test with significance 5%	T-student
			Kruskal-Wallis
			Chi-square

Source: Self elaboration

September to December 2018 in Madrid and from September to December 2019 in Newcastle, at three different points within each public space. Environmental parameters such as relative humidity (RH), air temperature (Ta), wind speed, sky view factor (SVF) and sound pressure level (SPL) were measured for 15 min at each point, twice a day between 10 h00 and 18 h00 as these are the central hours of the day, as well as being the moments of greatest sunshine and occupation.

Direct observation and photographic recording were applied to identify the use and stay of older adults (Katzschner et al., 2006; Martinelli et al., 2015; Nikolopoulou & Lykoudis, 2007; Whyte, 1980); older people who passed by and those who stayed at each measuring point by age group (older adults, adults, teenagers and children) were counted.

Data analysis aimed to identify statistical differences between the two case studies about the relationships between the use and the personal, environmental and design variables. All the collected data were processed in IBM SPSS Statistics software and analysed with a significance of 5%. The T-student test was applied for environmental variables; Kruskal-Wallis for the level of clothing (clo), age and visit time; and chi-square for gender and subjective comfort perception.

The questionnaire survey consisted of two parts. The first part concerned the personal characteristics of the interviewees (age, sex, frequency and time of the visit, activities carried out in the space and the level of clothing insulation 'clo'). In the second part, questions were about the subjective environmental perceptions of older people in each space, including thermal and acoustic comfort. To assess thermal comfort, the seven-point scale (ASHRAE, 1966) (TSV (thermal sensation vote)) (−3, cold; −2, cool; −1, slightly cold; 0, neutral; 1, slightly warm; 2, warm; 3, hot) was used for thermal sensation, and the McIntyre three-point scale (McIntyre, 1973) (1, warmer; 2, would not change; 3, colder) was used to identify thermal preference. For SPL, humidity and wind speed perceptions, a four-point Likert scale (very pleasant-very unpleasant) was applied; and for noise annoyance, a 'yes' or 'no' question was applied.

13.2.1 Case Studies

Two very different European cities were selected, Madrid and Newcastle upon Tyne. The first one is a metropolitan city, the capital of Spain, in which the Arapiles neighbourhood taken as a case study is a residential area in the consolidated city within one of the districts with the highest population density and the highest percentage of the ageing population, where three public spaces were chosen because of their different physical and functional characteristics, being close to health centres, care centres, nursing homes, schools, etc., and the presence of older people: Conde del Valle de Súchill Square (PCVS), José Luis Sampedro Park (PTG) and Vallehermoso Street (CVH). Regarding the park and the square, there are facilities such as exercises for the elderly, a playground, an ornamental water fountain and a dog area, while the street is a residential street with vehicular traffic and some commercial and leisure premises (Fig.13.2).

On the other hand, the city of Newcastle upon Tyne in the United Kingdom is a university city, in which the selected public spaces are located within the city centre, in the commercial area, as it was the only outdoor public place where older people were observed during the months of the research, identified after a tour by some neighbourhoods in the city. However, this neighbourhood has a very low-density population, and the student population is predominant in terms of residence. The selected spaces are different from each other; on the one hand, a square with vegetation, gardens and benches is located next to a high-traffic street and bus stops and surrounded by commercial premises. While the selected street is a pedestrian and constitutes one of the main commercial areas of the city and the region, it is equipped with urban furniture such as benches and little vegetation (Fig.13.3).

The samples in both cases were similar, with 70 people in Madrid and 73 in Newcastle upon Tyne. Table 13.2 shows the main characteristics of each assessed public space in each city including the number of older people interviewed, equipment nearby, functions, main sound sources and physical elements of age-friendly design.

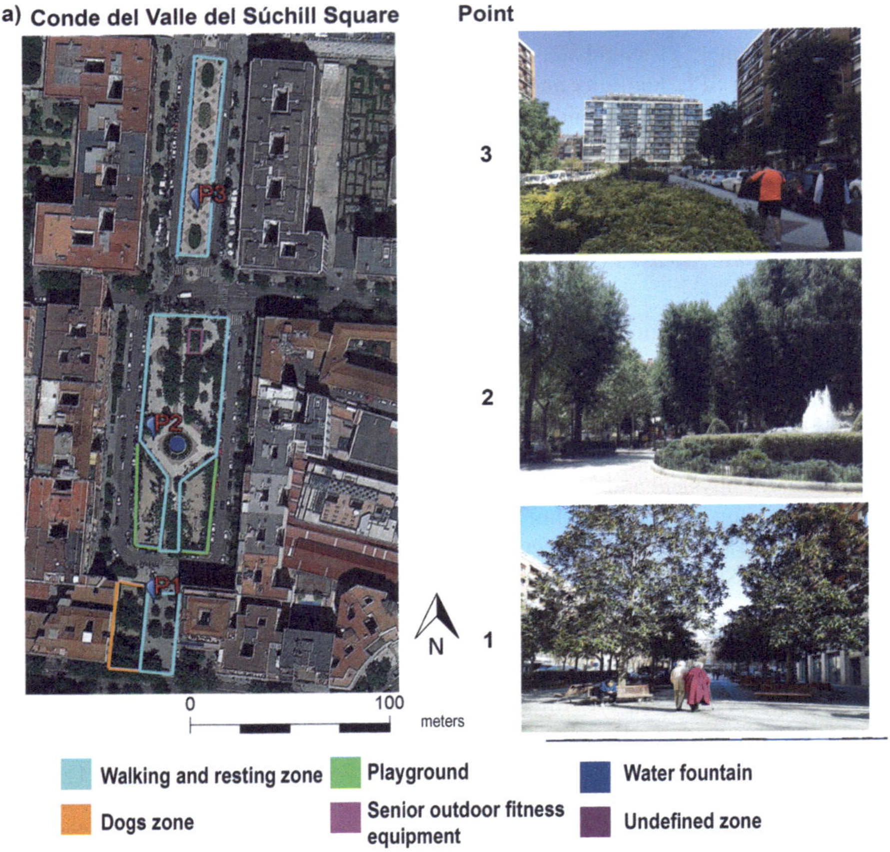

Fig. 13.2 Location, zoning and photographs of the measurement points of each public space in Arapiles neighbourhood, Madrid, Spain: (**a**) Conde del Valle de Súchill Square (PCVS), (**b**) José Luis Sampedro Park (PTG) and (**c**) Vallehermoso Street (CVH). (Source: Self elaboration, 2018)

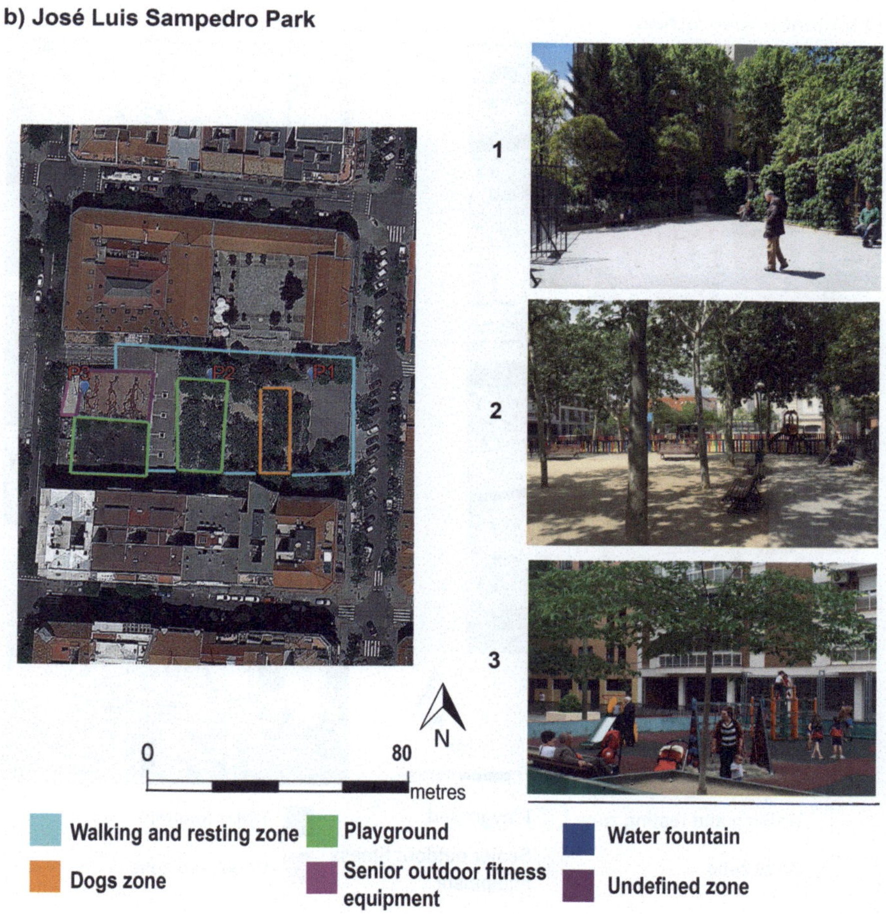

Fig. 13.2 (continued)

Fig. 13.2 (continued)

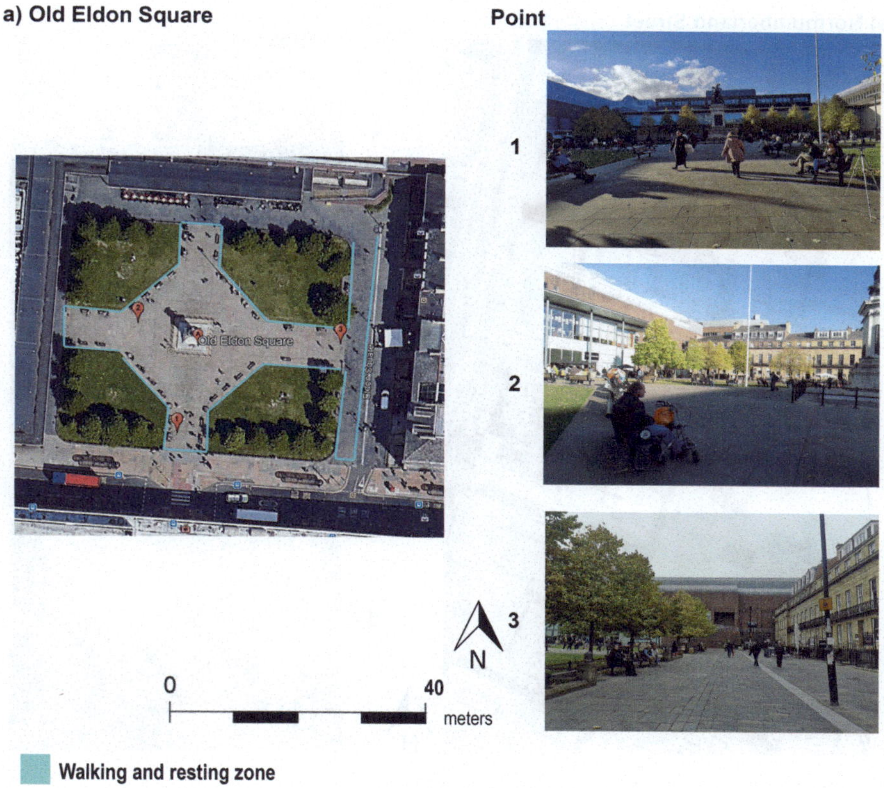

Fig. 13.3 Location and photographs of the measurement points of each public space in Newcastle upon Tyne: (**a**) Old Eldon Square and (**b**) Northumberland Street. (Source: Self elaboration, 2019)

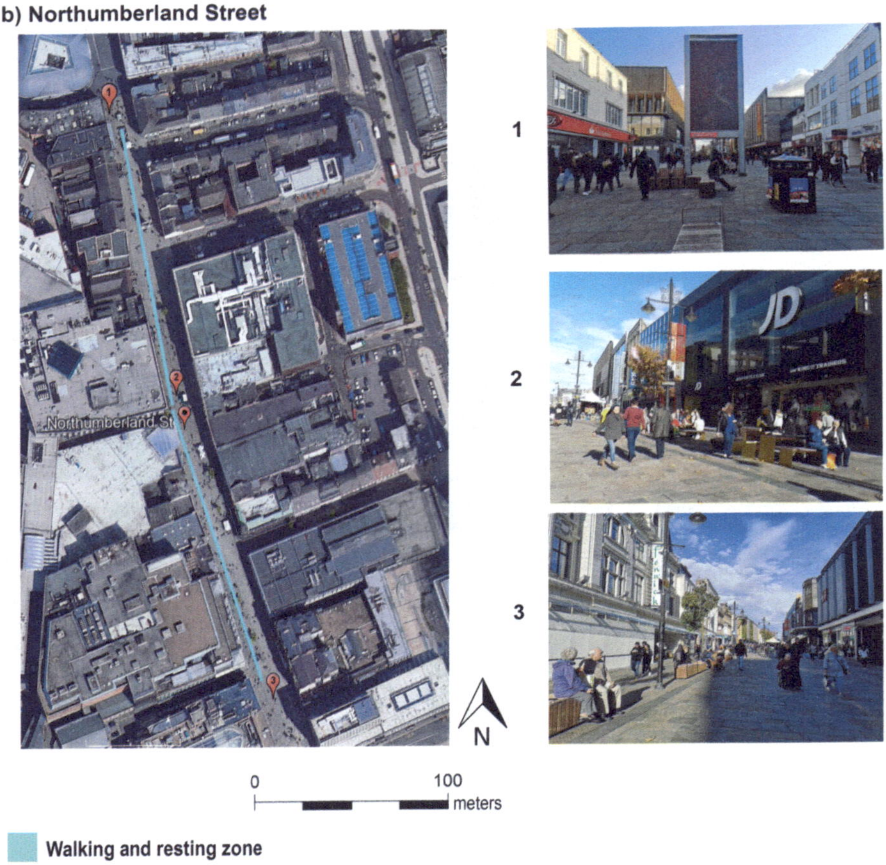

Fig. 13.3 (continued)

Table 13.2 Main characteristics of each assessed public space in each city

City	Public space	NT	NW	NM	% OPS	Size (Ha)	Building high (storey)	Nearby equipment	Function	Sound sources	Veg. (u)	GSM	Age-friendly design
Madrid	PCVS	26	13	13	25	0.68	5–10	School, hospital	Residential, playground	Traffic, water fountain, dogs barking, children playing, people	97 u	76% stone, concrete, 24% sand	Exercise equipment, playground, water fountain, dog area, benches, vegetation, shadow, accessibility, signage, security, good maintenance
	CVH	27	17	10	16	0.33	5–9	School, elderly Centre, theatre	Residential, playground	Traffic, works, dogs barking, children playing, people, construction works	42 u	65% stone, concrete, 35% sand	Benches, shadow, vegetation, accessibility
	PTG	17	9	8	16	0.072	2–10	School, market	Residential, shops, leisure	Traffic, people, leisure activities	35 u	100% concrete and hinged cement block	Playground, dog area, benches, vegetation, shadow, accessibility
Total		70	39	31	22.3								

(continued)

Table 13.2 (continued)

City	Public space	NT	NW	NM	% OPS	Size (Ha)	Building high (storey)	Nearby equipment	Function	Sound sources	Veg. (u)	GSM	Age-friendly design
Newcastle Upon Tyne	OE	43	19	24	40	0.42	4	Medical Centre, pharmacy, bus station, church	Historical, commercial	Busses, traffic, music, people, birds	30 u	28.5% stone, 71.42% grass	Benches, shadow, vegetation, accessibility
	NB	30	15	15	33	0.99	10	Pharmacy, municipal library, banks, church	Commercial	Live music, music from shops, sailors, people	5 u	100% stone	Benches, accessibility
Total		73	34	39	37								

PCVS Conde del Valle de Súchill Square, *PTG* José Luis Sampedro Park, *CVH* Vallehermoso Street, *OE* Old Eldon Square, *NB* Northumberland Street, *NT* total sample size, *NW* number of women interviewed, *NM* number of men interviewed, *OPS* older people's stay, *Veg* vegetation trees, *u* units of trees, *GSM* ground surface materiality
Source: Self elaboration, 2020

13.3 Results and Discussion

Table 13.3 presents a summary of the environmental data measured from autumn 2018 in Madrid and autumn 2019 in Newcastle. Regarding the average air temperature, there was a difference of 4.74 °C, with Newcastle being the coldest. However, Madrid presented lower minimum and higher maximum temperatures, which means there was a greater daily thermal amplitude in Madrid. The average radiant temperature was higher in Madrid too. On the other hand, relative humidity and wind speed were higher in Newcastle, as well as the noise levels, which could be attributed to its centrality and commercial function, different to that of the spaces studied in Madrid.

To identify statistical differences between the two case studies, the T-student test for environmental variables; Kruskal-Wallis for the level of clothing (clo), age and visit time; and chi-square for gender and subjective comfort perception were applied (Table 13.4). Results show that there are significant differences ($p < 0.005$) between all the environmental variables measured except for the sky view factor. Additionally, according to the subjective perception of thermal and acoustic comfort variables, there are differences except in the perception of humidity, in thermal preference and in thermal acceptability. And regarding the use of those spaces, there are significant differences in older people's stay, activity, time and frequency of visit. Figure 13.4 presents a summary of the analysis of these differences between the two cities.

Concerning thermal and acoustic comfort (Fig.13.4a), Madrid has a higher percentage of people in thermal comfort, while the subjective perception of ventilation is more pleasant in Madrid, coinciding with the fact that public spaces in Newcastle had a higher maximum wind speed (Table 13.3).

On the other hand, even though higher noise levels were found in public spaces in Newcastle (Table 13.3), acoustic comfort was higher in Newcastle, similar to the study by Ge and Hokao (2005) that found that the quality perception of the soundscape was lower in the busy streets than in the quiet ones; in both places, people were annoyed by traffic noise; however, it was more unpleasant for people present in the quiet street with natural and cultural elements than in the noisy and busy streets, which is attributed to expectation. Although in both cases they exceed the limits that are harmful to health (>55 dB) (World Health Organization Europe, 2005) and those allowed by local and European legislation (>65 LAeq dB day afternoon in residential areas) (European Parliament and Council, 2002), some authors associate this with physiological changes that lead to hearing loss (Miedema & Vos, 1999; Yu & Kang, 2008), and others relate it to greater tolerance in older adults (Miedema & Vos, 1999; Alain et al., 2006; Liu et al., 2019). However, although these high noise levels are not uncomfortable, they can affect health in unnoticed ways, physical and psychological (World Health Organization Europe, 2005).

In terms of the frequency of visit to each public space, in Madrid, most older people visit these spaces daily, while in Newcastle, they visit them several times a week, and men visit those spaces more often (Fig.13.4b). While regarding visiting time, it can be seen that in Madrid, most of the interviewees were 'just passing by',

Table 13.3 Characterisation of fieldwork environmental measurements

City	Public space	Air temperature (°C)			Relative humidity (%)			Wind speed (m/s)			Tmrt (°C)	SVF	LAeq (dB)
		Avg	Min	Max	Avg	Min	Max	Avg	Min	Max			
Madrid	PCVS	20.92	6.64	30.72	38.26	18.27	78.79	1.47	.80	2.54	34.96	.039	55.44
	CVH	17.53	7.22	28.12	49.97	25.45	76.41	1.13	.60	2.18	28.25	.023	62.60
	PTG	19.45	7.60	28.95	50.46	31.21	75.22	1.25	.80	1.72	30.52	.027	63.73
	Avg	19.30	7.15	29.26	46.23	24.98	76.81	1.28	0.73	2.15	31.24	.030	60.59
Newcastle upon Tyne	OE	13.27	12.57	15.07	60.81	54.08	64.04	1.61	1.01	2.51	22.69	.009	75.05
	NB	15.85	14.77	17.96	54.70	53.65	60.50	1.87	.92	3.04	20.29	.017	74.57
	Avg	14.56	13.67	16.52	57.76	53.87	62.27	1.74	0.97	2.78	21.49	.013	74.81
Differences		4.74	−6.52	12.75	−11.53	−28.89	14.54	−0.46	−0.23	−0.63	9.75	.017	−14.22

Source: Self elaboration, 2020

PCVS Conde del Valle de Súchill Square, *PTG* José Luis Sampedro Park, *CVH* Vallehermoso Street, *OE* Old Eldon Square, *NB* Northumberland Street, *Avg* average, *min* minimum, *max* maximum, *Tmrt* mean radiant temperature, *SVF* sky view factor, *LAeq* equivalent continuous sound level

Table 13.4 Statistical differences between variables between Madrid and Newcastle case studies

Variables		Statistical test		
		Chi-square	Kruskal-Wallis	T-student
		Sig (bilateral)	Sig (bilateral)	Sig (bilateral)
Environmental	Ta mean			0.000**
	Ta max			0.000**
	Ta min			0.000**
	RH mean			0.000**
	RH min			0.000**
	RH max			0.000**
	LAeq dB (A) mean			0.000**
	LAeq dB (A) mín			0.000**
	LAeq dB (A) max			0.000**
	Wind speed mean			0.000**
	Wind speed min			0.022*
	Wind speed max			0.000**
	SVF			0.080
	Trmt			0.000**
Personal	Clo		0.611	
	Age		0.415	
	Gender	0.274		
	Time of visit		0.001**	
	Activity	0.000**		
	Frequency of visit	0.000**		
Perception	Pleasant factors	0.006**		
	Unpleasant factors	0.000**		
	Pleasant sounds	0.000**		
	Unpleasant sounds	0.000**		
	Humidity perception	0.239		
	Ventilation perception	0.000**		
	Thermal sensation (neutral)	0.023*		
	Thermal preference (would no change)	0.172		
	UTCI 'no thermal stress'	0.0246*		
	PET 'no thermal stress'	0.007**		
	Thermal acceptability	0.061		
	Acoustic comfort	0.000**		
	No noise annoyance	0.022*		
Others	Older people's stay	0.000**		

Source: Self elaboration, 2020
*the correlation is significant at the level 0.05 (two tails). **the correlation is significant at the level 0.01 (two tails)

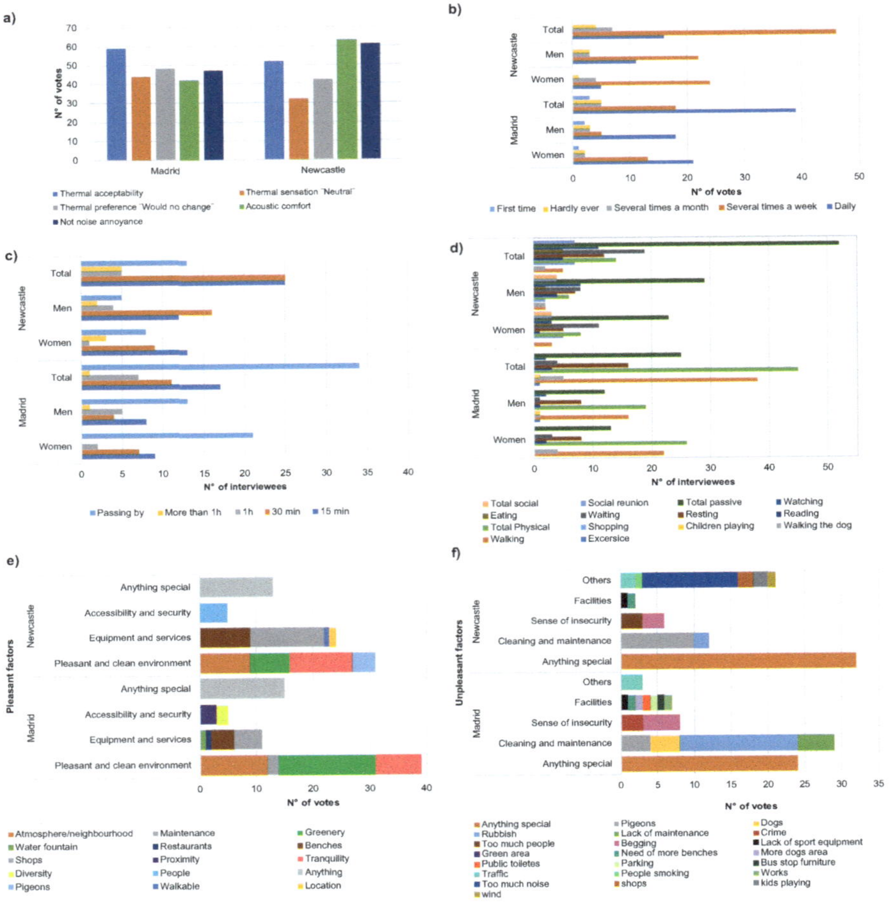

Fig. 13.4 Main differences between the uses by older people of open public spaces assessed in Madrid and Newcastle. (**a**) Thermal and acoustic comfort by city, (**b**) frequency of visit by city, (**c**) time of visit by city, (**d**) activity by city and gender, (**e**) pleasant factors by city and (**f**) unpleasant factors by city. (Source: Self elaboration)

while in Newcastle, most of them stay in the public space for around 15–30 min. In Madrid, men stay more time in those places than women (Fig.13.4c). Concerning the activities carried out in the public spaces studied in the two cities, it can be seen that in Madrid, older people interviewed performed more physical activities (walk, walk the dog, exercise), being women the most active, they walk and walk their dogs more often than men. While in Newcastle, most of the interviewees were engaged in passive activities such as waiting, resting and observing; this city was also more socially active (Fig.13.4d). This could be attributed in part to the physical characteristics of the public spaces assessed; in Newcastle, there were no equipment or spaces for this purpose; neither of the two public spaces studied has exercise machines for older people or walking trails. Those are only spaces for rest, being shopping the main activity in them. More physical activities could be encouraged in these spaces in a simple way such as the adaptation of existing benches with pedals.

Older people were asked about their general perception of pleasant (Fig.13.4e) or unpleasant factors (Fig.13.4f) in each public space. For a better analysis of these results, the answers were grouped into seven groups based on the bibliographic analysis of quality indicators of a public space adapted for older people (World Health Organization, 2007; Newton and I'DGO, 2012; Fariña Tojo & Merino, 2013; Ayuntamiento de Madrid, 2015; Fariña et al., 2019). The general subjective perception both in Madrid and in Newcastle was positive, as most of them do not dislike anything. In the first case, the indicator with the greatest satisfaction was 'pleasant and clean environment' which includes factors such as vegetation and tranquillity. In terms of gender differences, in general, women were more satisfied with vegetation (24.3% compared to 19.5% of men), while men have expressed greater satisfaction with tranquillity (21.5% compared to 9.2% of women). Similarly, in Newcastle, the most pleasant factors for the older people interviewed were the shops and those related to the 'pleasant and clean environment', the vegetation and the environment in general. Among the unpleasant factors, the noise was selected by 21.9%. In terms of gender, although there is no significant statistical relationship, it was found that women find certain elements more pleasant than men, such as pigeons.

Regarding the permanence of older people in those public spaces, significant differences were found between the two cities (Table 13.5). The percentage is higher in Newcastle (concerning the other age groups) being 37% of the users, while in Madrid, they correspond to 22.33%. Since the average of older people's stay was three people, this variable was categorised into none (no person), low (1–2 people), medium (3–5 people) and high (more than 5 people). The statistical analysis with a 5% degree of significance between older people's stay in each public space in each city and environmental, personal, design and other variables is presented in Table 13.5.

In the case of Madrid, concerning environmental variables, a significant relationship was found with temperature, humidity and noise levels; those influence the use that older people make of those public spaces, as well as the hours of daily sunshine. In a multivariate statistical model, it was found that the mean radiant temperature, the average, and maximum air temperature, as well as the maximum noise level, were the most influential (R = 11%):

$$\begin{aligned} \textit{older people stay} = 3.459 + (0.152\,\mathrm{Tmrt}) + (-0.053\,\mathrm{LAeq\,dB\,max}) \\ + (-0.586\,\mathrm{Ta}) + (0.477\,\mathrm{Ta\,max}), \end{aligned}$$

while in the case of Newcastle, only the wind speed influences this prediction (R2 = 7%): ***older people's stay* = 10.79 – 1.73 (*Ws max m/s*)**.

These results coincide with some authors which have found a strong relationship between the microclimate and the use of open spaces (Urrutia del Campo, 2001; Nikolopoulou & Lykoudis, 2007).

Table 13.5 Statistical analysis of the relationship between stay and other variables in each city

Variable		Statistical test							
		Spearman Rho correlation		Pearson		Mann-Whitney U		Kruskal-Wallis	
		Madrid	Newcastle	Madrid	Newcastle	Madrid	Newcastle	Madrid	Newcastle
		Coefficient correlation		Coefficient correlation		Sig (bilateral)		Sig (bilateral)	
Environmental	Ta			0.14**	0.128				
	RH			−0.13**	0.177				
	LAeq dB			−0.14**	0.036				
	Wind speed			0.10*	−0.252*				
	SVF			−0.03	−0.06				
	Trmt			0.24**	0.176				
	Humidity perception	−0.08	−0.38**						
	Ventilation perception	−0.12*	−0.31**						
	Sound level perception	−0.17**	−0.016						
	Thermal sensation	−0.03	0.36**						
	Thermal preference	−0.05	0.30**						
	Noise annoyance	0.19**	−0.37**						
	Pleasant sounds	0.05	−0.87						
	Unpleasant sounds	0.00	0.002						

Physical design elements	Benches					0.000**			
	Water fountain					0.000**			
	Trees					0.64			
	Shadow					0.483			
	Permeable surface					0.000**	0.033*		
	Signage					0.000**	0.026*		
	Security					0.000**			
	Maintenance					0.379			
	Exercises equipment					0.061			
	Dogs' area					0.000**			
	Playground					0.723			
	Terrace bar					0.000**			
	School					0.001**			
	Bus stop					0.000**	0.856		
Others	Sunlight hours	0.27**	0.031*						
	Morning/afternoon					0.015*	0.771		
	Public space							0.00**	0.026*
	Point							0.00**	0.00**

Source: Self elaboration, 2020

*The correlation is significant at the level 0.05 (two tails). **The correlation is significant at the level 0.01 (two tails)

Ta air temperature, *RH* relative humidity, *LAeq dB* equivalent continuous sound pressure level in a period of time, *SVF* sky view factor, and *Trmt* mean radiant temperature

In Newcastle, most of the older people stay in those spaces during the day and when the mean air temperature was between 10.30 °C and 19.07 °C. While in Madrid, most of the older people stay in the afternoon and when the mean air temperature was between 17.69 °C and 29.81 °C. Besides, it was found a relationship between the stay of older adults and the subjective perception of acoustic comfort in Madrid and thermal comfort in Newcastle (Table 13.5).

In both cities, most of the 'high-stay' situations were in the squares, especially in the areas with the greatest possibility of activities and variety of users, like point 2 in the Conde del Valle de Súchill Square; this was the point with the 'highest stay' and coincides with the location of children's areas and exercises for the elderly, which are intergenerational areas. In Madrid, the Conde del Valle de Súchill Square (PCVS) was also the public space where they perceived the greatest percentage of thermal and acoustic comfort, while the street is where the least amount of time older people spent, coinciding with the highest noise levels and the least perception of thermal and acoustic comfort.

While in Newcastle, it was evident in points 2 and 3 of Old Eldon Square and Northumberland Street, respectively, which are where the higher stays were registered and coincide with the highest percentages of people in thermal and acoustic comfort too. Finally, it was found that some design elements of the public space are statistically determinant in older people's stay, such as benches, water fountains and shade, among others (Table 13.5), being the areas with most of those elements where most of the 'high-stay' situations were registered in those public spaces.

13.4 Conclusions

This study assessed the use of public spaces by older adults in two very different cities of Europe, Madrid in Spain and Newcastle upon Tyne in the United Kingdom, based on environmental measurements, surveys and direct observation in open public spaces during autumns 2018 and 2019, respectively.

In both cases, older adults represented an important percentage of the users of the public spaces analysed, 22% in Madrid and 37% in Newcastle. In the first case, older people were more active, and they visited those places more frequently, while in the second case, they stay longer. The fact that older people in Newcastle performed more passive activities is attributed to the lack of facilities for exercise or walk in the public spaces studied. More physical activities could be encouraged in these spaces in a simple way such as the adaptation of existing benches with pedals. This is important because sedentariness is one of the risk factors associated with many of the non-communicable diseases prevalent in older people; public spaces that encourage more physical activities can help older people maintain their cognitive skills and independence for longer.

The use and stay of older people in public spaces in both cases were related to environmental factors like temperature and noise level in Madrid and wind speed in Newcastle. Additionally, there is a relationship with the subjective perception of thermal and acoustic comfort with the 'high-stay' situations in both cities, being the public spaces with more facilities, variety of users and activities where more older people were registered. While for most of the interviewees the most pleasant factors were vegetation, the cleanliness, and in the case of Newcastle the shops, these were statistically related to staying.

It should be emphasised that although in both cities, noise levels were above the 55 dB recommended limit for health, older people have expressed satisfaction with the noise levels, indicating a possible risk of these high noise levels on the health of older people, although they do not perceive it as such. These levels are higher in the public spaces studied in the city of Newcastle as well as the percentage of acoustic comfort. Nevertheless, the most unpleasant factor for most of the interviewees was the traffic noise in Madrid; this matches that the street was the less crowded place by older people. This fact can be attributed to the expectation.

Therefore, this study found some environmental and design important factors to take into account to encourage greater use and activity of older people in public spaces, which could contribute to achieving the objectives of active ageing and improving the quality of life and health of older people in their neighbourhoods.

Acknowledgements This work was supported by the *Secretaría de Educación Superior, Ciencia, Tecnología e Innovación del Ecuador* SENESCYT under Grant [109-2017 (SENESCYT-SDFC-DSEFC-2017-4040-O)].

References

Alain, C., Dyson, B. J., & Snyder, J. S. (2006). Aging and the perceptual Organization of Sounds: A change of scene ? In *Handbook of model human aging*, pp. 759–770.

Allen, J., Cars, G., & Madanipour, A. (2012). *Social exclusion in European cities: Processes, experiences and responses*. Routledge.

Arup., Help Age International., Intel., & Systematica. (2015). *Shaping ageing cities: 10 European case studies*.

ASHRAE. (1966). *Thermal comfort conditions. ASHRAE Standar 55–1966*.

Auchincloss, A. H., et al. (2009). Neighborhood resources for physical activity and healthy foods and incidence of type 2 diabetes mellitus. *Archives of Internal Medicine, 169*, 1698. https://doi.org/10.1001/archinternmed.2009.302

Ayuntamiento de Madrid. (2019). Demografía y población. In *Estadística*. Retrieved from: https://www.madrid.es/portales/munimadrid/es/Inicio/El-Ayuntamiento/Estadistica/Areas-de-informacion-estadistica/Demografia-y-poblacion/Poblacion-extranjera/Poblacion-extranjera-en-la-ciudad-de-Madrid/?vgnextfmt=default&vgnextoid=c289d54944580510VgnVCM200. Last Accessed 14 June 2019.

Ayuntamiento de Madrid. (2015). *Madrid Ciudad Amigable con las Personas Mayores*.

Beard, A. J. R., Cerdá, M., Blaney, S., et al. (2009). Neighborhood characteristics and change in depressive symptoms among older residents of New York City. *American Journal of Public Health, 99*, 1308–1314. https://doi.org/10.2105/AJPH.2007.125104

Burchardt, T., Le Grand, J., & Piachaud, D. (2002). Degrees of exclusion: Developing a dynamic, multidimensional measure. In J. Hills & J. P. D. Le Grand (Eds.), *Understanding social exclusion* (pp. 30–43). Oxford University Press.

European Parliament and Council (2002) Directive 2002/49/EC relating to the assessment and management of environmental noise. .

Fariña, J., Higueras, E., & Román, E. (2019). *Ciudad Urbanismo y Salud*. Documento Técnico de criterios generales sobre parámetros de diseño urbano para alcanzar los objetivos de una ciudad saludable con especial énfasis en el envejecimiento activo.

Fariña Tojo, J., & Merino, B. (2013). *Cuadernos del OSE (Observatorio de Salud de Europa) sobre políticas de salud en la UE*. Número 7. Urban y Salud Pública.

Ge, J., & Hokao, K. (2005). Applying the methods of image evaluation and spatial analysis to study the sound environment of urban street areas. *Journal of Environmental Psychology, 25*, 455–466. https://doi.org/10.1016/j.jenvp.2005.10.003

Kaplan, S. (1995). The restorative benefits of nature. *Journal of Environmental Psychology, 15*, 169–182. https://doi.org/10.1016/0272-4944(95)90001-2

Katzschner L., Bosch U., & Röttgen M. (2006). Behaviour of people in open spaces in dependence of thermal comfort conditions. Proceedings of 23rd Conf Passiv low energy Archit I–505.

Lai, D., Zhou, C., Huang, J., & Jiang, Y. (2014). Outdoor space quality : A field study in an urban residential community in Central China. *Energy and Buildings, 68*, 713–720. https://doi.org/10.1016/j.enbuild.2013.02.051

Lawton, M. P. (1985). The elderly in context. Perspectives from environmental psychology and gerontology. *Envrionment and Behavior, 17*, 501–519. https://doi.org/10.1177/0013916585174005

Lawton, M. P., & Nahemow, L. (1973). Ecology and the aging process. In C. Eisdorfer & M. P. Lawton (Eds.), *The psychology of adult development and aging* (pp. 132–160).

Liu, J., Wang, Y., Zimmer, C., et al. (2019). Factors associated with soundscape experiences in urban green spaces: A case study in Rostock, Germany. *Urban Forestry & Urban Greening, 37*, 135–146. https://doi.org/10.1016/j.ufug.2017.11.003

Maas, J., Verheij, R. A., De, V. S., et al. (2009). Morbidity is related to a green living environment. *Journal of Epidemiology Community Health*, 967–973. https://doi.org/10.1136/jech.2008.079038

Martinelli, L., Lin, T. P., & Matzarakis, A. (2015). Assessment of the influence of daily shadings pattern on human thermal comfort and attendance in Rome during summer period. *Building and Environment, 92*, 30–38. https://doi.org/10.1016/j.buildenv.2015.04.013

Mclntyre, D. (1973). A guide to thermal comfort. *Applied Ergonomics, 4*(2), 66–72.

Miedema, H. M. E., & Vos, H. (1999). Demographic and attitudinal factors that modify annoyance from transportation noise. *The Journal of the Acoustical Society of America, 105*, 3336–3344. https://doi.org/10.1121/1.424662

Newcastle City Council. (2019). Newcastle City Council. In *Equal. Stat. Res. Inf.* Retrieved from: https://www.newcastle.gov.uk/your-council-and-democracy/statistics-and-census-information/equality-statistics-research-and-information#population. Last Accessed 14 Mar 2019.

Newton R., & I'DGO. (2012). *The design of streets with older people in mind. Design guide 002: Bus stops.* United Kingdom.

Nikolopoulou, M., & Lykoudis, S. (2007). Use of outdoor spaces and microclimate in a Mediterranean urban area. *Building and Environment, 42*, 3691–3707. https://doi.org/10.1016/j.buildenv.2006.09.008

Sánchez-González, D., Rivera Adame, L. M., & Rodríguez-Rodríguez, V. (2018). Paisaje natural y envejecimiento saludable en el lugar: el caso del Parque Nacional Cumbres de Monterrey (México) Natural landscape and healthy ageing in place: The case of the Cumbres de Monterrery National Park in Mexico. *Boletín la Asoc Geógrafos Españoles, 76*, 20–51. https://doi.org/10.21138/bage.2514

Scharf, T., Phillipson, C., Kingston, P., & Smith, A. E. (2001). Social exclusion and older people : Exploring the connections. *Education Ageing, 16*, 303–320.

Shanahan, D. F., Lin, B. B., Bush, R., et al. (2015). Pathways to health benefits from urban nature. *American Journal of Public Health, 105*, 470–477.

Sugihara, S., & Evans, G. (2000). Place attachment and social support at continuing care retirement communities. *Environment and Behavior, 32*, 400–409.
Urrutia del Campo, N. (2001). Clima, diseño y diversidad urbana en el uso de tres plazas de Madrid. *Rev Territorios en Formación, 1*, 161–178.
Vidal, T., & Pol, E. (2005). The appropriation of space: A theorical proposal to understand the bond between the persons and the places. *Anu Psicol, 36*, 281–297.
Whyte W. H. (1980) The social life of small urban spaces. .
World Health Organization Europe (2005) Experts consultation on methods of quantifying burden of disease related to environmental noise. .
World Health Organization. (2020). Age-friendly world. In *What is an age-friendly world?* Retrieved from: https://extranet.who.int/agefriendlyworld/about-us/. Last accessed 17 June 2020.
World Health Organization. (2019). Newcastle – A city adapting to demographic change. In *Age friendly Word. Newcastle upon Tyne*. Retrieved from: https://extranet.who.int/agefriendlyworld/network/newcastle-upon-tyne/. Last Accessed 14 Jan 2019.
World Health Organization. (2007). *Global age-friendly cities: A guide*. France.
Yu, L., & Kang, J. (2008). Effects of social, demographical and behavioral factors on the sound level evaluation in urban open spaces. *The Journal of the Acoustical Society of America, 123*, 772–783. https://doi.org/10.1121/1.2821955

María Teresa Baquero Larriva is a doctor in urbanism (cum laude, International PhD) in the Universidad Politécnica de Madrid (UPM). She is an architect and magister in constructions in the Universidad de Cuenca, Ecuador. She is also a researcher in urban regeneration, sustainability, energy efficiency and age-friendly cities.

Chapter 14
Greenery Urban Design for Good Mental Health. Analysis of a Vulnerable District of Madrid

Elisa Pozo Menéndez

Abstract Mental health problems have become one of the major public health concerns in many countries affecting all ages and sectors of the population. From depression to dementia or anxiety, mental health in urban environments is a consequence of a stressful lifestyle as well as many other factors, where the environment where people live is directly related. In this context, green and natural spaces play a major role to increase quality of life and well-being. The relation between care homes or mental health institutions has been historically related directly with green spaces. In this research is presented the relation between green infrastructure and the centres for older people in the district of Usera in the city of Madrid, as well as the urban environment where they are located. In total, 14 centres of the district of Usera in the south of Madrid have been analysed. The aim was to evaluate the current situation of this kind of municipal equipment and to understand the current challenges and opportunities of the existing built environment to create friendly healthier and more inclusive environments. The results show the lack of connectivity and accessibility with green spaces and the challenges of zoning in a district of the consolidated city. Most of the centres that have been studied include a green space inside its limits; however, these spaces are disconnected from the surroundings and are usually kept as decorative elements. The discussion highlights the need of a more multidisciplinary approach to plan urban spaces and the cities, integrating in a coherent strategy the concepts of sustainability, green infrastructure and age-friendliness.

Keywords Mental health · Green spaces · Outdoor · Care homes

E. Pozo Menéndez (✉)
ABIO Research Group, Universidad Politécnica de Madrid, Madrid, Spain
e-mail: e.pozo@upm.es

E. Pozo Menéndez, E. Higueras García (eds.), *Urban Design and Planning for Age-Friendly Environments Across Europe: North and South*, Future City 19,
https://doi.org/10.1007/978-3-030-93875-8_14

14.1 Introduction: Preventing Mental Health Problems as a Major Public Concern

Mental health problems have become one of the major public health concerns in many countries. Depression has been identified as the primary driver of disability in 26 countries, which has been also recognized as the predominant mental health problem all over the world, followed by anxiety and stress, schizophrenia and bipolar disorders (Vos et al., 2013). The World Health Organization (WHO) estimates that around 35–50% of people with severe mental health problems in developed countries receive no treatment (Demyttenaere et al., 2004). Mental health problems have become one of the leading causes of disability in Europe (Mental Health Foundation, 2016), it is associated with the major risk of heart disease on mortality and disability and it directly impacts on the length and quality of life (Ferrari et al., 2010). Specially, older adults with intellectual disabilities are at particular risk, as they are often marginalized and do not receive the minimal support; they would need in order to play an active role in their communities (WHO, 2000). There is also a difference in how mental health affects differently by gender, with a major prevalence of all types of common mental health problems in women than in men (Mental Health Foundation, 2016). Among all the mental health problems, preventing dementia has been also recognized by WHO as a major public health issue. There is no adequate cure for this disease, but there is a growing knowledge of actions that can support its prevention as well as the care management to reduce the prevalence and to improve the quality of life when symptoms start to appear. Many of them can be related with the living environment and lifestyles (Smith & Yaffe, 2014). Moreover, the annual cost for the care and support for people living with dementia worldwide is estimated to be over €515 billion in 2020, which is another fact that is challenging the sustainability of the social well-being system (Ritchie et al., 2015).

Even though there is a minor risk of certain mental health problems and better access to services in cities in comparison with rural areas (WHO, 2021), as urban population is increasing continuously and it is expected that by 2050 68% of the world's population will live in urban areas (UN, 2018), mental health has become an urgent challenge for urban planners and policymakers.

14.2 Background

14.2.1 The Impact of the Built Environment on Mental Health and the Role of Nature

Cities have become unhealthy places in many senses, both for physical and mental health, specially great metropolitan areas. Focusing on mental issues, there is almost 40% higher risk of depression in urban contexts, over 20% of higher levels of anxiety and a double risk of schizophrenia, as well as other public health issues such as loneliness, stress and isolation (Peen et al., 2010; Vassos et al., 2012; Gruebner et al., 2017). There are several reasons established by the Centre for Urban Design and Mental Health (McCay et al., 2017) that explain this relation, but the ones which are related with the built environment are mainly two: the overload of stimuli and the weakening of protective factors. People who live in urban areas experience increased stimulus levels, sometimes related to unhealthy parameters, such as crowding, noise, air and visual pollution, as well as other factors such as density and high intensity of inputs like smells or sights than might cause overload. In addition, accessibility to protective factors such as access to nature, daily exercise and strong social bonds might increase the feeling of unsafety or anxiety.

In this sense, the lack of nature is described as an important problem with many health-related issues, while accessible green spaces in neighbourhoods and in the daily routines are likely to improve general mental well-being, improve social and cognitive functioning and ADHD, reduce depression and stress and improve the mood and reduce aggression for people with dementia (Ulrich et al., 1991; Chiesura, 2004; Croucher et al., 2007; Shanahan et al., 2015, 2016). Another important factor which affects mental health is an active lifestyle. Walkability and possibility to exercise in the daily routine, as regular exercise is recognized to reduce general levels of anxiety and stress and to improve well-being, reduce some symptoms associated with ADHD, schizophrenia and dementia. Reducing the risk of cardiovascular disease and diabetes through exercise is as well a preventive action for cognitive deterioration and the apparition of certain forms of dementia (Bedimo-Rung et al., 2005; Alty et al., 2020). Even in the higher stages of dementia, enabling exercise reduces the aggressive behaviours or agitation (Alzheimer's Society, 2019), and it is also related with better sleep patterns (Kodama et al., 2016). In addition, exercising and walking in the outdoor space enable the sunlight exposition, which improves the sleeping patterns (Konis et al., 2018) as well as the calcium absorption and thus prevents bone fragility disease in all ages, but especially in the late elderly (Larrosa et al., 2008). Finally, the relation with others and opportunities to meet and interact with the community is one of the major necessities for human well-being to create relationships and develop the sense of feeling part of a community (Fariña Tojo et al., 2019). Indeed, social interaction is related with improving health and well-being (Berkman et al., 2014) and reduction of mortality (Holt-Lunstad et al., 2010). Even though there is a direct relation with other socio-economical factors, it is also important to consider urban and public space design (Whyte, 1980).

In this context, green areas play a key role in healthy urban design and planning. Indeed, multiple benefits from ecosystems have been acknowledged by international entities such as FAO (Food and Agriculture Organization) (2021). Natural environments assure carbon sequestration and oxygen supply, reduce the urban heat island and improve thermal comfort, and its presence in urban neighbourhoods is related with reduced levels of stress or anxiety and decreased risk of overweight or diabetes, among other diseases. There is also a link between the spatial scale in territories and the visit recurrence in relation with social activities (Rodrigue et al., 2013). Research acknowledges the need for a gender perspective planning in transport, as it has a lot to do with daily and care activities (de Madariaga & Zucchini,

2019). In this sense, also for older adults, it is important to carefully consider their perspective and limitations, in order to correctly locate and design land use distribution, services and activities for them, so they can remain independent as long as possible; at the same time, they exercise their autonomy with daily routines and walks, as part of a healthy lifestyle. The 15-minute city concept (Moreno et al., 2021) for older people and healthy ageing is an idea that has relevance nowadays. Thus, accessibility to green areas, how they are integrated in the urban landscape and how they are related with meaningful activities is one of the main strategies for healthier environments for all (Fig. 14.1).

Understanding the direct link between healthy outcomes of green spaces on human health, especially psychology, it is evident that more green areas and nature-based solutions should be implemented in the consolidated city.

14.2.2 Therapeutic and Restorative Spaces

Gardens and architecture have been closely related during the whole history of humanity, and the presence of gardens – as controlled and anthropized nature – has been usually related with well-being in many civilizations. Indeed, the art of gardening dates 3000 BC in Ancient Egypt for multiple purposes and uses. In Persia, there were already found examples of medical gardens, becoming the first hospitals in Arabic civilization, the *bîmâristâns*. In the twelfth century, the king Parkama Bâhu of Ceylan founded many hospitals surrounded by alignments of trees of different species (Le Mandat, 2009). In Europe, the Greek and Roman civilizations used plants for therapeutic purposes, including medicines as well as ointments. During Middle Ages, medical gardens were associated with monasteries (Hortus Medicus), as the knowledge of botanical properties of plants remained to the only state of society with access to books and libraries and so predominantly religious. However, botanical research became progressively more scientific in the sixteenth century thanks to Paracelse, who considered nature as an authentic laboratory of empirical knowledge. By the end of this century, the major challenge was due to the augmentation of deprived populations and increase in health inequalities. It is by the end of the eighteenth century when the major renovation of design of hospitals occurred, in a context of extreme precarity of hospices as well as new demands of medicine due to advances of the field. Thus, after the fire of the Hôtel de Dieu in Paris (1772), new hospitals were conceived as *machines à guérir* for patients (Foucault et al., 1979). In this sense, health policies included places for gardening and production as spaces for healing, especially for patients with mental health problems.

The evolution of medicine and care services has also influenced the changes in the built environment. Indeed, at the beginning of the twentieth century, isolation was the main treatment for mental health patients. Asylums were conceived integrated in a major landscape, with big extensions of ornamental gardens, for walking

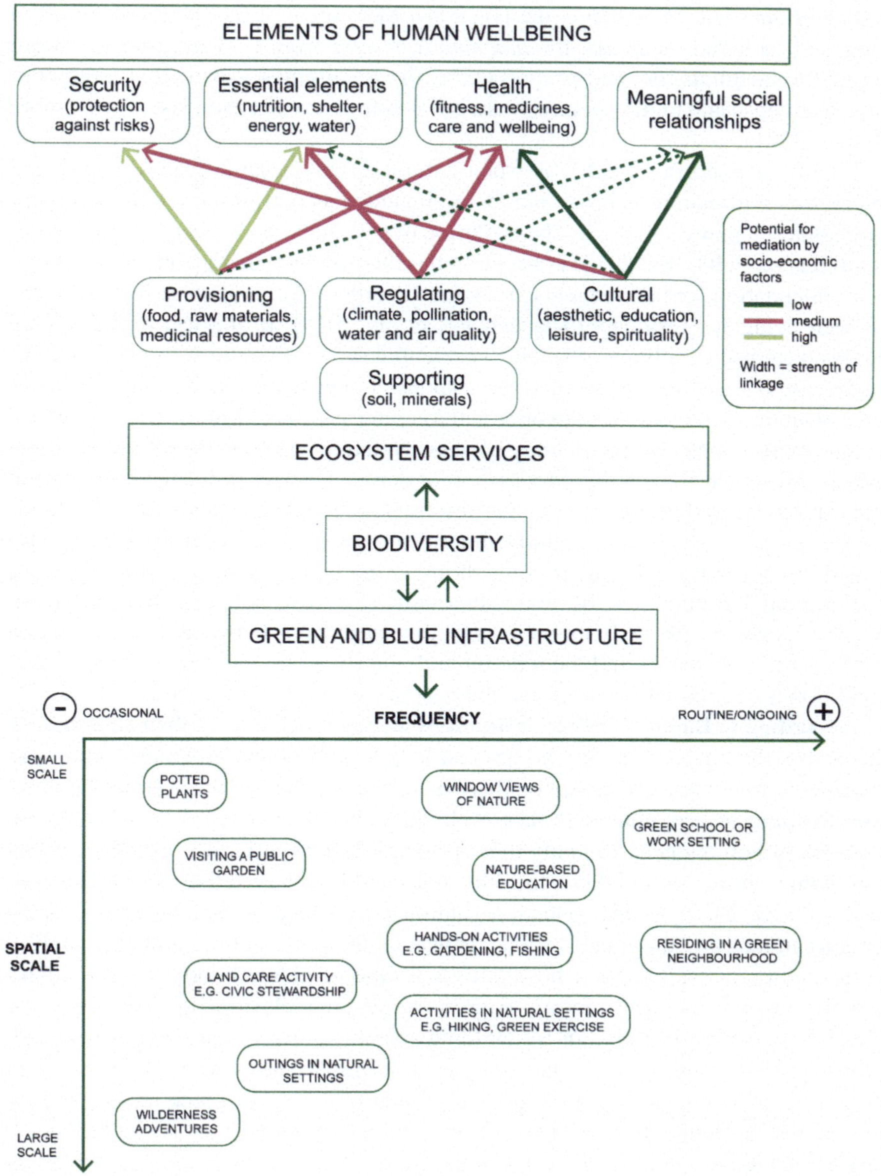

Fig. 14.1 Links between human health and green and blue infrastructure in territories. (Source: Author's own adapted from Morand and Pipien (2013) and Jędrzejczyk (2021))

and entertainment. Meanwhile, the Hygienist Movement (1920) and new trends in urbanism included ideas like the *machine à habiter* from Le Corbusier that were applied to hospitals too, and progressively, these buildings – as well as were conceived as dense and high structures – are isolated from nature, even visually (Cambra Rufino, 2021).

Finally, it is during the 1970s when natural spaces became places to stay and interaction with nature is promoted, being included as part of the requirements for care homes in many countries. The evolution of psychiatry and new needs of person-centred approach (HIN, 2014) conception has incorporated therapeutic gardens as a recommendation for nursing and care homes in order to promote physical and mental well-being for residents (Thaneshwari et al., 2018). In the last decades, care homes have been evolving to smaller residential or pavilion units, to prevent communicable disease and to improve the response to epidemics. In addition, the benefits of maintaining social networks and keeping patients close to their relatives, social environments and community have evolved into spaces in the middle of urban tissues. However, this evolution varies a lot across Europe, influenced by cultural and traditional backgrounds, as well as different approaches to landscape and building design. For instance, while in Denmark care services for older people are integrated in normalized living lifetime homes, in Spain, there are still big-scale institutional buildings as the main alternative for care services in elderlyhood (García Lantarón, 2015). In the same way, the interactions between the outdoor and indoor, common spaces and the relation with the street are important elements that condition the social interactions and the person's experience (Câmara et al., 2015).

According to Barenys (1992), institutions are the reflections of their societies. In this sense, the analysis of care homes can help to understand how older adults are considered in our society. In Spain, there is not a consistent framework for building specifications of care homes. In fact, each region has its own regulations, and there is a lack of homogeneity to apply these principles. In Madrid, it is established that care home centres should be in "healthy areas and be integrated or near urban centres" (SEGG, 2020). In this context, regulations regarding outdoor surroundings are missing in many of the specifications. However, the urban setting and accessibility to the centre, as well as their hospitality with the neighbourhood, the community and the visitors, are key factors to promote social inclusion. In this sense, the Flemish regulation in Belgium for building specifications for care homes (Flemish Ministry, 2016) highlights the necessity of assuring accessibility to the centre and its direct surroundings, and it requires a minimum of 3 m^2 of green areas (outdoor spaces) per resident to be used by residents, caregivers and visitors. Also, it is necessary to include a secure bike shelter in the limits. Another interesting example can be found in the urban planning in Weesp (Amsterdam, the Netherlands), where the well-known De Hogeweyk residence or "dementia village" is located. The city council of Weesp has established in the local strategy for 2013–2030 that the relationships with this singular care centre should be strengthened, creating new links and synergies with the city as another social equipment open to the community (Brilman & Reimus, 2013).

14.3 Methodology

To explore the relationship between green spaces and the built environment, this study takes one of the districts of Madrid identified as one of the most vulnerable ones: Usera. Usera is in the south of Madrid, on the other side of River Manzanares and Madrid Rio Linear Park. As many other neighbourhoods in the periphery of the municipality limits, it is characterized by a massive increase in population due to the migration from rural areas from the 1950s onwards. The result was rapid urban growth, with a huge development of social housing and public infrastructures, following the tendency of open blocks as building typology. Although there is one of the biggest public parks in the city, this district lacks green spaces, and access to green areas is still a challenge (Pozo Menéndez et al., 2019). Considering the benefits of green spaces on human health, the aim of this study is to understand the link between the care homes and daycare centres of the district and the availability of outdoor and green spaces in the immediate surroundings. First, the specialized centres for older population, which includes nursing homes, day care centres and municipal social centres have been identified and selected. Each one is examined by a surrounding analysis of the urban characteristics in a ratio of 500 m, where the principles of therapeutic gardens (Thaneshwari et al., 2018) are evaluated on those centres that have outdoor green spaces.

14.3.1 Vulnerability and Environmental Problems in the District of Usera

The district of Usera is characterized by a high urban complexity. The connectivity between neighbourhoods and different areas of the district is one of the main challenges. Both the district and the neighbourhoods are separated by main roads and grey infrastructures that not only contribute to the problems of accessibility but also deteriorate the environment, with important peaks of bad air quality and noise. The presence of the cars, the high density of the district and the morphotypology of the urban tissue contribute to the vulnerability of the population of the district. Moreover, among the 29 neighbourhoods of Madrid that are included in the urban heat island, 6 of them are in Usera, corroborating the direct link between climate change and social vulnerability (Neila González et al., 2020; Ayuntamiento de Madrid, 2016). Social vulnerability in Usera is produced by economic precarity, being the second district in Madrid with the highest rate of social exclusion, evictions and unemployment, among others. On the other hand, the high rate of migratory population hides the high presence of older people in the district. Indeed, the district has one of the highest shares of population under 14 years old, whereas all the neighbourhoods overpass the share of over-ageing with people older than 80 years old with a percentage of 41% and more, while the average in the whole municipality is around 36% (Cabello et al., 2019).

14.3.2 Green Infrastructure and the Built Environment

In 2018, the City Council of Madrid published the Green Infrastructure and Biodiversity Plan in which the current situation of the existing green and blue infrastructure was analysed and quantified. The district of Usera counts with one of the biggest city parks of the capital: Pradolongo. However, the continuity with other natural elements in the neighbourhoods is not yet met. There are very heterogeneous urban morphotypologies in Usera that also determine the specific problems in each neighbourhood. In the north and central areas of the district, the urban blocks are compacted and dense, streets are narrow and they are organized in an orthogonal grid. There are important accessibility problems, and it is the area of the district that is affected the most by the heat island effect. The south part of the district corresponds to residential developments with a typology of open blocks and open interblock spaces. There is a huge difference between the maintenance of these open spaces between Orcasitas neighbourhood and Orcasur, and it is also related with the vulnerability, being Orcasur the most affected by lack of good quality spaces, as well as services.

Finally, the east part of the district presents a very fragmented typology, as many of the blocks are tertiary services, including the Doce de Octubre Hospital, which is one of the main hospitals in the city in Madrid and so has an important influence area in the surroundings.

The district limits with Madrid Rio Linear Park in the north and east, where the River Manzanares has been recovered creating a natural high-quality environment both for health promotion and for supporting biodiversity (Fig. 14.2).

14.3.3 Care Facilities and the Relation Between Outdoor and Indoor Spaces

The population in Usera is 142,894 habitants, where 17% are older than 65 years old (23,853 habitants). This percentage is lower in comparison with the average of the city (21%) due to the high percentage of migrant and young population (24%). The over-ageing rate is 39%, with 13,629 citizens older than 75 years old (Ayuntamiento de Madrid, 2020). As a general offer for older adults in Spain, in the district of Usera, different services are found: care homes, daycare centres and municipal social centres. There are two big-scale care homes, located in an equipment area, with a high number of residents, and managed by private companies. The other two care homes are in the south of the district, in Orcasur neighbourhood, and they consist of smaller facilities. Daycare centres are sometimes combined with these care homes; the rest of the centres are in the streets of the densest part of the district. Finally, the networks of municipal social centres for older people are distributed in each neighbourhood; they offer as well day care services, organizing also multiple activities and engaging older people in the management and organization.

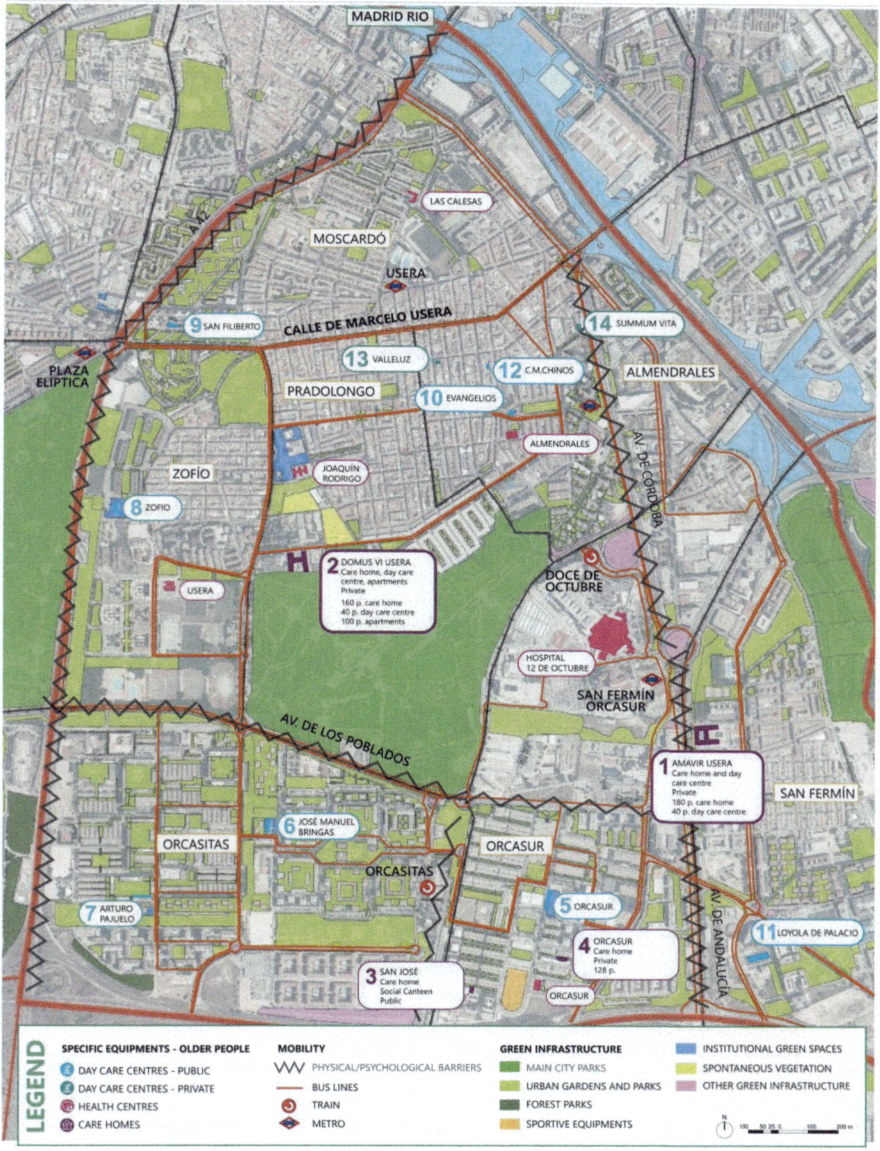

Fig. 14.2 Green infrastructure of the district of Usera. Big city parks dominate the southern districts of Madrid. However, this main green infrastructure is not always connected, and it presents a lack of continuity in the urban tissue of many neighbourhoods. Additionally, the southern districts are usually surrounded by high-speed roads, creating important physical and psychological barriers. Consequently, the district of Usera is also one of the most polluted districts in Madrid. In dark pink, the main health services are also included, as they are important services also related with health care as they are usually frequented by older adults. (Source: Author's own edition of the Plan of Green Infrastructure and Biodiversity (Ayuntamiento de Madrid, 2018))

The evaluation of the design of the care homes and centres themselves is out of the scope of this research, although it is evident that they are not considering the principles of a person-centred approach design, where small units and normalized living have also a specific translation into architectonic design and configuration. Figure 14.3 presents all the centres identified in the district, analysing the immediate surroundings and Table 14.1 shows a global overview of all the elements that have been analyzed.

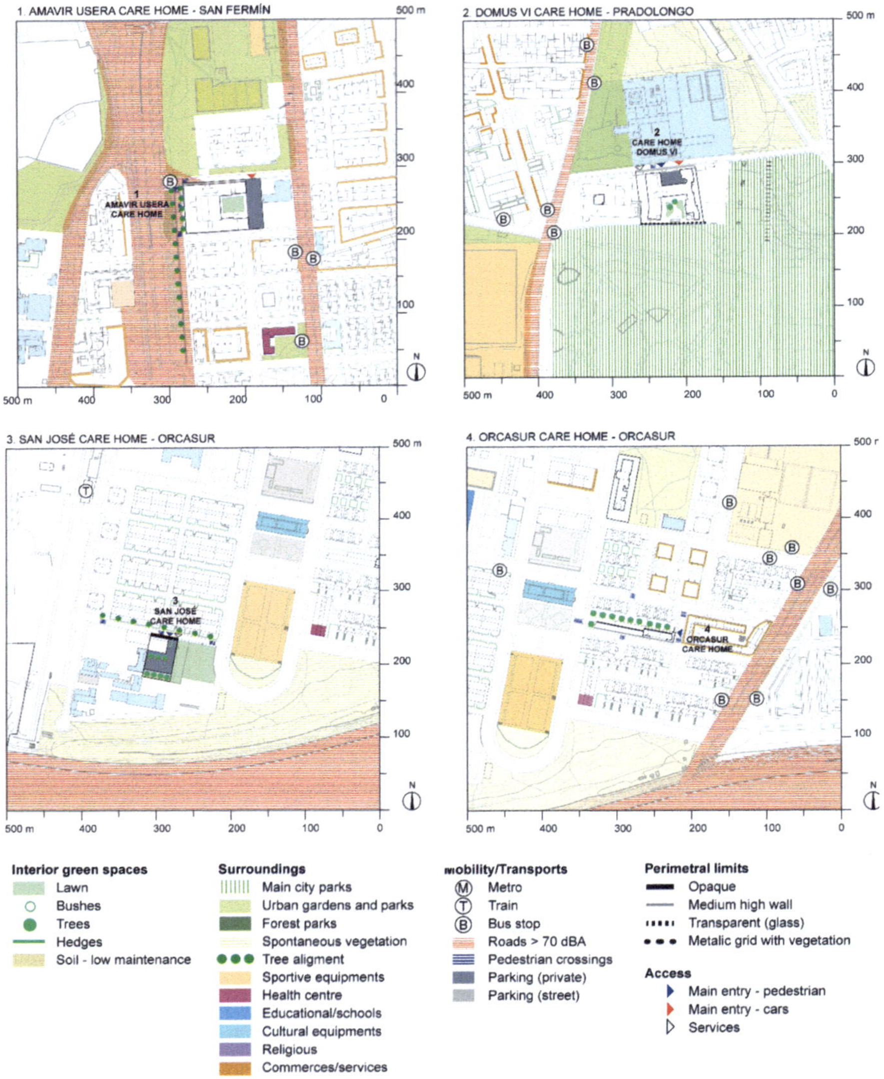

Fig. 14.3 Urban analysis of the surroundings of the centres for older people in the district of Usera. (Source: Author's own)

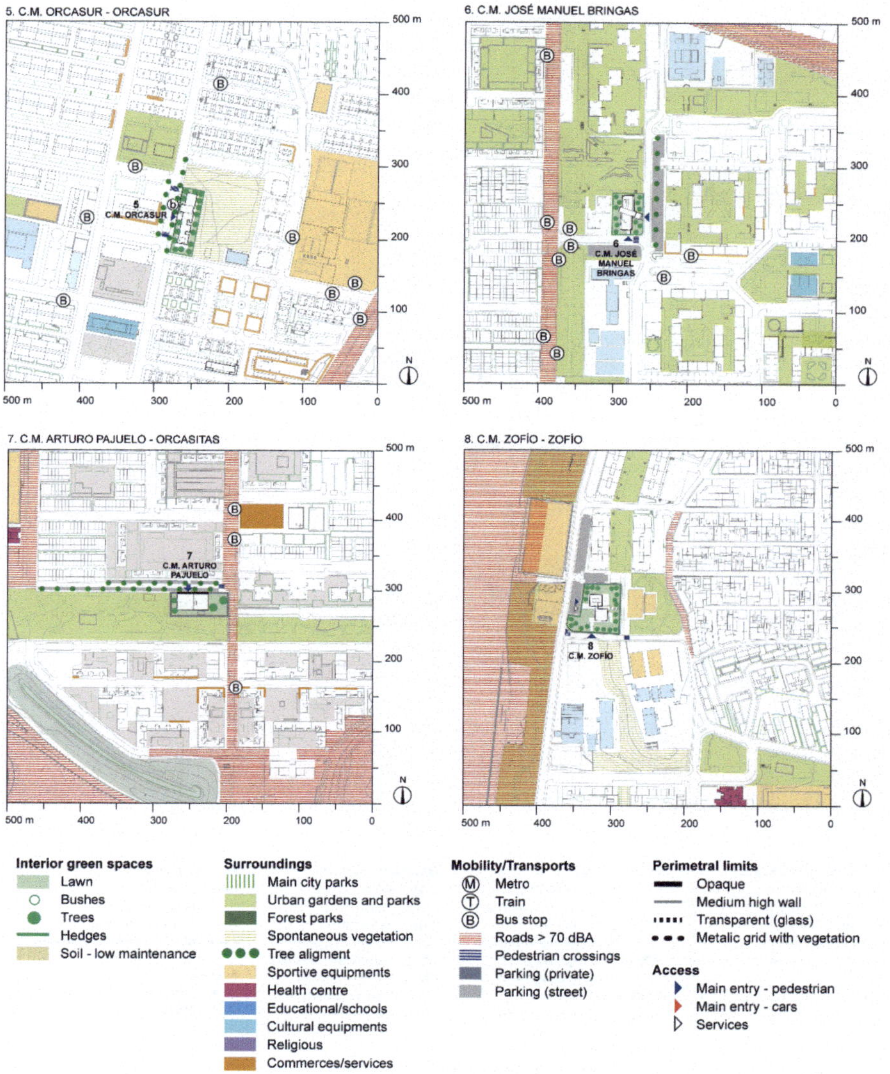

Fig. 14.3 (continued)

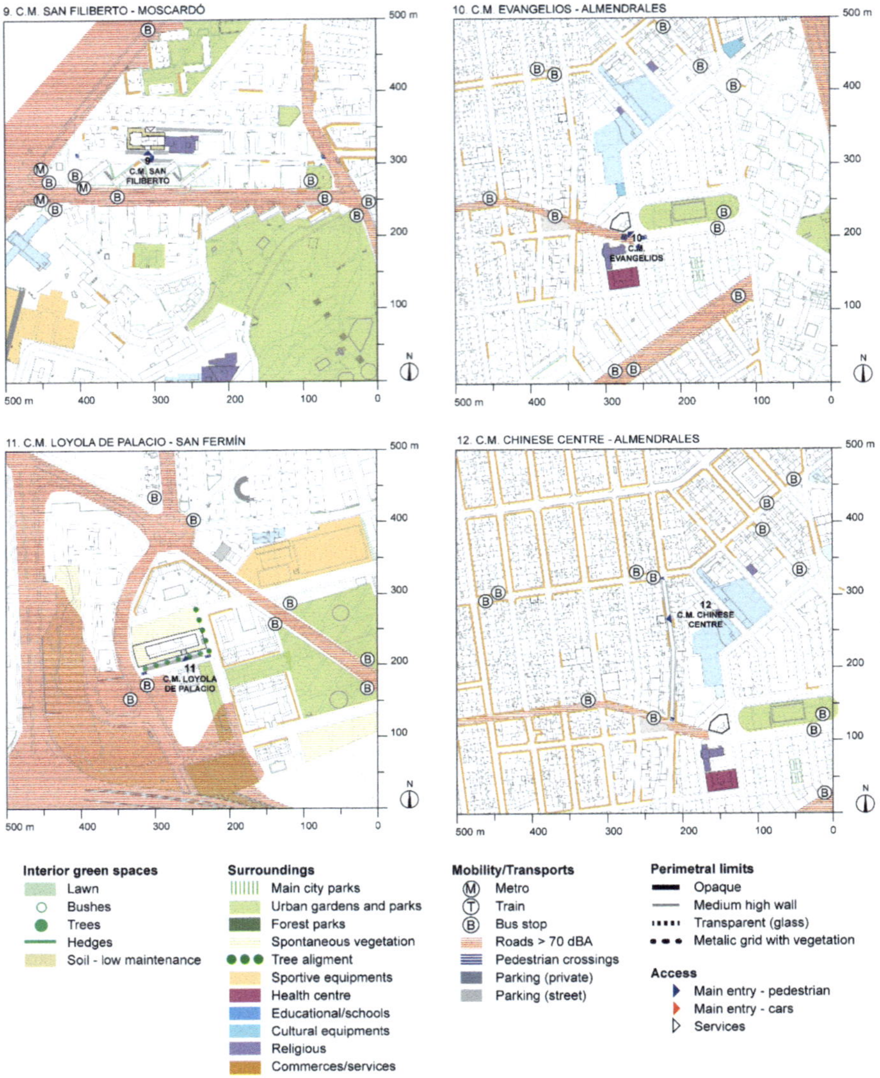

Fig. 14.3 (continued)

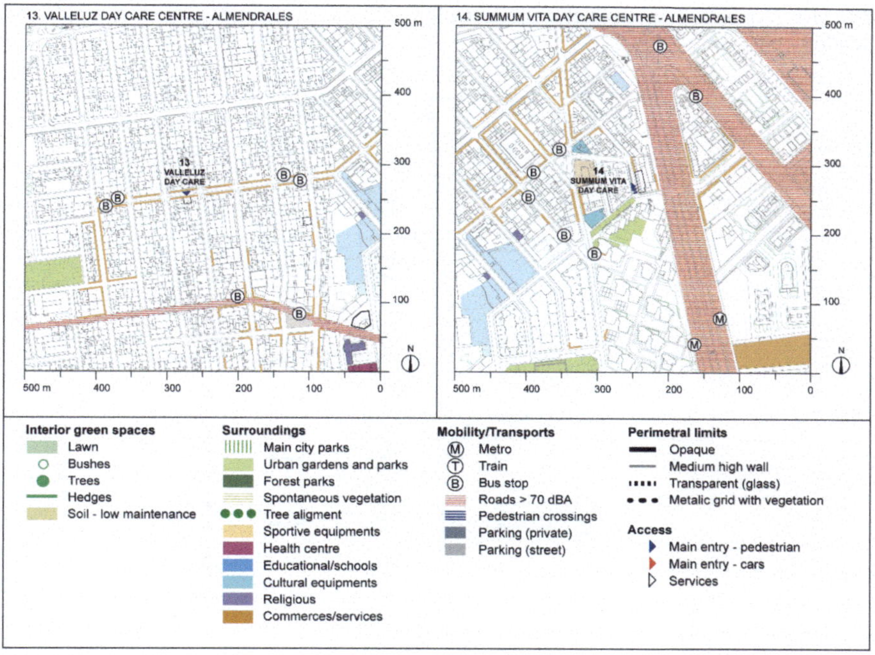

Fig. 14.3 (continued)

14.4 Discussion: Urban Green Spaces for Healthy Ageing and Well-Being

The analysis of the built environment in the surroundings of the centres for older adults in the district of Usera shows a lack of planning with natural elements. The centres are either integrated in the urban tissue (57%) in continuity with residential blocks, either in equipment areas (42%) as a consequence of the zoning. The scale and typology of the centres depend also on this location. The centres that are located in equipment areas present multiple situations, but it is proposed to regulate compatible uses in order to avoid situations such as a recycling centre or fuel stations in the immediate surroundings. Also, it results in a physical proximity to other equipment like sports facilities or educational centres; however, the size and treatment of the perimeters do not relate visual spaces; neither does it promote a direct interaction.

Centres 1 and 2 correspond to public care homes with private management. Both consist of big buildings, with a high number of residents, and combine home care with daycare services. Even though they occupy a big area, and there is a presence of big green spaces in the immediate surroundings, the connectivity is low. Centres 3 and 4 are other typologies of care homes, of smaller scale, more integrated in the neighbourhood and near a variety of equipment. However, the connectivity with green spaces is low. A noisy road passes by at the south of the neighbourhood,

Table 14.1 Synthesis of the main characteristics of the older adults' centres in the district of Usera in Madrid

	Identification and main characteristics of the centre					Connectivity and accessibility			Relation with other activities and surroundings			Relation with nature	
	Equipment/ facility	Total surface [m²]	Max occupancy [m²]	Nº residents	Address	Parking [m²]	Public transport	Connectivity/ accessibility	Other activities r<400 m	Perimetral limits	Green areas	Accessibility/ connectivity with surrounding green areas	Hierarchy and variety of green areas
1	Care home + day care centre Amavir Usera	7,001	4,667	180 p. 40 p.	Av. de la Perla	[P] 1,305 (P)	(B) 50 m (M) 261 m	(F)	primary school, residential, sports centre, services	Opaque Transparent	● lawn 590 m² ○○○	[L]	[H]
2	Care home + day care centre + apartments	6,421	2,545*	160 p. 40 p. 50 p.	C. Cristo de la Victoria 247	[P] 747	(B) 220 m	(f)	recycling centre, primary school, residential, city park	Transparent	● bushes 988 m²	[L]	[H]
3	Care home + social centre	2,201	412,4*	no data	C. Trévelez 3	[P] 1,244 (P)	(B) 400 m (T) 300 m	(F)	2 primary schools, kindergarten, sports courts, health centre	Opaque	● trees 430 m² ○○○	[L]	[L]
4	Care home	551	435*	128 p.	C. Salobreña 2	(P)	(B) 170 m	(F)	health centre, day care centre, sport courts, library, cultural centre, commerces (low), residential	Opaque	○ ○○○	[L]	[L]
5	Daycare centre Orcasur	2,678	953*	30 p.	Av. Orcasur 60	(P)	(B) 99 m	(f) (b)	care home, health centre, sport courts, library, cultural centre, residential	Opaque Transparent	● trees 1,145 m² ○○○	[M]	[L]
6	Daycare centre José Manuel Bringas	2,271	787	30 p.	C. del Camino Viejo de Villaverde, 26	(P)	(B) 150 m	(F)	primary school, secondary school, residential	Opaque	● trees 1,344 m²	[H]	[M]
7	Daycare centre Arturo Pajuelo	2,573	980	no data	C. Segura 3	(P)	(B) 100 m	(F)	parks, squares, services (high), residential	Opaque Transparent	● trees 1,593 m²	[H]	[L]
8	Daycare centre Zofio	4,032	561	no data	C. Ricardo Beltran y Rozpide 31	(P)	(B) 400 m	(f)	parks, primary school, secondary school, sports centre, petrol station, residential, commerces (low)	Transparent	● trees 2,677 m²	[H]	[H]
9	Daycare centre San Filiberto	1,336	674*	no data	C. San Filiberto 7	(P)	(B) 210 m (M) 200 m	(F)	public social services, church, residential, commerces, private educational centre, park	Opaque Transparent	● trees 603 m² ○	[L]	[M]
10	Daycare centre Evangelios	698	698	no data	C. Porta Coeli	(P)	(B) 110 m (M) 350 m	(F)	park, health centre, squares, church, residential, citizen association, primary schools	Opaque	○ ○○○	[M]	[L]
11	Daycare centre Loyola de Palacio	2,876	1,188	no data	C. Silvina 10	(P)	(B) 110 m	(F)	squares, sports courts, petrol stations, offices, residential (low), terciary activities	Opaque	● 1,095 m²	[M]	[L]
12	Chinese daycare centre	117	117	no data	C. Olvido 64	(P)	(B) 150 m	(f)	park, health centre, squares, church, residential, citizen association, primary schools	Opaque	-	[L]	[L]
13	Daycare centre Valleluz	110	110	no data	C. Antonio Salvador 79	(P)	(B) 160 m	(f)	square, commerces, residential, citizen association, primary school	Transparent	-	[L]	[L]
14	Daycare centre Summum vita	311*	311*	no data	Av. Córdoba 10	(P)	(B) 160 m (M) 270 m	(f)	sociocultural centre, drama theatre, residential, terciary activities, commerces	Opaque Transparent	-	[L]	[L]

[P] own parking (B) bus (F) on foot - high accessibility ▬ Opaque ● own garden [L] low
(P) exterior parking (M) metro (f) on foot - low accessibility ▭ Transparent ○ exterior garden [M] medium
(T) train (b) by bike ○○○ trees in street [H] high

Source: Author's own from Ayuntamiento de Madrid (2021) and Consorcio Regional de Transportes de Madrid (2020) and methodology adapted from Ajuntament de Barcelona (1989)

separated by a fringe of spontaneous vegetation in very low maintenance state. On the other hand, municipal social centres (n = 8, 57% from the total) are distributed in each of the neighbourhoods of the district. Centres 5, 6, 7, 8 and 11 have a surrounding own green space, usually with trees and lawn and high level of maintenance, which represents a 63% of the municipal centres and 36% of the whole specific equipment for older adults. However, most of them are just decorative spaces instead of integrating other uses for outdoor activities. Centres 8 and 10 have gymnastic features for older adults in the park of the immediate surroundings. On the other side, it has been identified several situations of open spaces nearby the centres (3, 5 and 11) with no maintenance or urban development planned. This presents a unique opportunity to expand the green areas of the centres, intergenerational equipment or even new complementary facilities, as alternative housing of small scale. Finally, daycare centres 12, 13 and 14 are integrated in the urban tissue of Almendrales neighbourhood, and they are all small scale, run by private management. However, the immediate surroundings are very low quality, with any accessibility or connectivity to green spaces, but also accessibility in terms of mobility is a challenge, because of the narrow width of the sidewalks and absence of any facility to encourage autonomy in the streets.

All the green spaces that are found in the centres are mainly decorative elements, without a proper design to use these spaces or low possibility to organize meaningful activities. Regarding the proximity to other green spaces and complementary equipment or services, all the centres are in the urban tissue, according to the regulation of Madrid for care homes. However, the good accessibility of the centres and the connectivity with the neighbourhood and green areas do not meet age-friendly recommendations (WHO, 2007). For instance, the pedestrian crosses are not located near the main entrance, connectivity with public transport stops is not always near (3 and 8), there are not many benches or supports in the perimeters and there are not spaces for stay in the intermediate environment between indoor and outdoor of these centres. Fourteen percent of the residences have a bus stop in the immediate surroundings (<100 m), 71% have a bus stop in the proximity area (100 m–300 m) and 14% have a bus stop equal to or further than 400 m.

Another interesting element is the perimetral treatment, which is usually an opaque wall; either it is the building wall and either it consists of an exterior limit. In the best cases, this wall is medium-high, and it is completed with a perimetral hedge. This need of separation answers to the requirements of security and privacy but also is part of the Mediterranean culture and traditional architecture and morphotypology of urban blocks. However, the connectivity with the public space and the community also requires a friendlier design of the access and the visual relation with the street.

This research relied only on an urban analysis due to the circumstances of COVID-19. All the municipal social centres have been closed since the beginning of the pandemics. Also, care home centres have also restricted all the visits and many of the activities. Further analysis will include interviews to the carers and managers of the centres in order to identify potential activities that could be done outdoors and how to design these spaces and to older people to include their perception about the importance of green spaces in the daily routines and activities. It is also important to identify other community spaces that might welcome older people, such as urban gardens in the neighbourhood or in schools.

14.5 Conclusions

COVID-19 lockdown in the beginning of 2020 has questioned the sustainability of our societies from many different perspectives. In cities, including big metropolitan areas, one outcome has been the impact – both positive and negative – on the surrounding environments and neighbourhoods. Restrictions to mobility had made many people realize the importance of having a good network of proximity services, commerces and spaces for recreation as well as good quality of streets to walk around and have access to open spaces to distress and meet other people. On the other hand, combining this general perception with the impact of the pandemic on the care sector, which has been one of the most affected ones, it is possible to see that normalizing care into daily life is becoming an urgent policy across Europe.

From the perspective of healthy ageing, there is a need for distress spaces in the community as part of the daily activity; this would enable the inclusion of care and mental health in planning with a more sensitive approach. Spaces of proximity where a person can feel safe, calm and connected with nature and with others need a specific approach to address the challenge of creating this kind of spaces in daily life in urban environments. This kind of spaces should be welcoming to all the neighbours in a common space, so intergenerational meetings or intercultural ones can take place. This feeling of belonging to a community as well as to a place probably is the major challenge for urban planning.

Also, it is necessary to highlight the need to adapt Mediterranean cities to the climate change scenario. Green infrastructure needs to be integrated in cities; however, this requires a multiscale approach that can be translated in a smaller scale to the citizen's respect for green spaces and their commitment to be engaged with nature-based solutions and contributing to community spaces. Many of the guidelines and published resources for creating therapeutic gardens are based on more humid climates and higher latitudes (i.e. UK, France). Also, the culture of public-private limits of the urban blocks conditions the relation between indoor and outdoor spaces and the activities that can be organized in both. Mediterranean cities have also a specific typology of gardens and green spaces, not always meaning a green lawn coverture, as this would suppose a major need of water for its maintenance. Further research would be necessary in order to understand the relationship between older people, intergenerational and meaningful activities related to gardening and the need for specific spaces for it. Also the commitment of managers of the centres in order to invest in green solutions to design therapeutic gardens and community green spaces.

References

Ajuntament de Barcelona. (1989). *Ordenació Urbanística els Equipaments Hospitalaris de Barcelona*. Ambit d'Urbanisme i Serveis Municipals. Àmbit de Salut Pública.

Alzheimer's Society. (2019). Physical activity and exercise. Factsheet 529LP. Retrieved from: https://www.alzheimers.org.uk/sites/default/files/2019-09/529LP%20Exercise%20and%20physical%20activity.pdf

Ayuntamiento de Madrid. (2016). Mapa de áreas preferentes de impulso a la regeneración urbana (APIRU). Área de Gobierno de Desarrollo Urbano Sostenible. Dirección General de Estrategia de Regeneración Urbana. Retrieved from: https://www.madrid.es/UnidadesDescentralizadas/UrbanismoyVivienda/Urbanismo/Destacamos/ficheros/MapaAreasPreferentes.pdf

Ayuntamiento de Madrid. (2018). Plan de Infraestructura Verde y Biodiversidad. Retrieved from: https://www.madrid.es/portales/munimadrid/es/Inicio/Medio-ambiente/Parques-y-jardines/Plan-de-Infraestructura-Verde-y-Biodiversidad/?vgnextoid=5fdec0f221714610VgnVCM2000001f4a900aRCRD&vgnextchannel=2ba279ed268fe410VgnVCM1000000b205a0aRCRD

Ayuntamiento de Madrid. (2020). Distritos en cifras (Información de barrios). Last update: 26/11/2021. Retrieved from: https://www.madrid.es/portales/munimadrid/es/Inicio/El-Ayuntamiento/Estadistica/Distritos-en-cifras/Distritos-en-cifras-Informacion-de-Barrios-/?vgnextoid=0e9bcc2419cdd410VgnVCM2000000c205a0aRCRD&vgnextchannel=27002d05cb71b310VgnVCM1000000b205a0aRCRD

Ayuntamiento de Madrid. (2021). Sistema de Información Geográfica Urbanismo. Área Desarrollo Urbano. Retrieved from: http://www-2.munimadrid.es/urbanismo_inter/visualizador/index_inter.jsp

Alty, J., Farrow, M., & Lawler, K. (2020). 'Exercise and dementia prevention', practical neurology. *BMJ Publishing Group Ltd, 20*(3), 234–240. https://doi.org/10.1136/practneurol-2019-002335

Barenys, M. P. (1992). Las residencias de ancianos y su significado sociológico. *Papers: Revista de Sociología, 40*, 121–135. Retrieved from: https://dialnet.unirioja.es/servlet/articulo?codigo=5206450

Bedimo-Rung, A. L., Mowen, J., & Cohen, D. A. (2005). The significance of parks to physical activity and public health: A conceptual model. *American Journal of Preventive Medicine, 28*(2–2), 159–168. https://doi.org/10.1016/j.amepre.2004.10.024

Berkman, L. F., Kawachi, M., & Glymour, M. (2014). *Social epidemiology*. Oxford University Press. ISBN 0199395330.

Brilman, J., & Reimus, S. (2013). *Structuurvisie Wees 2013–2030 "Weesp vertelt haar verhaal"*. Retrieved from: https://www.weesp.nl/bestuur-organisatie/projecten_42384/item/structuurvisie-weesp_80246.html

Cabello, M. N., Guilló, R. P. and Mairal, M. (2019). Líneas estratégicas para el desarrollo del Distrito de Usera. Ayuntamiento de Madrid. Retrieved from: https://www.madrid.es/UnidadWeb/UGBBDD/Actividades/Distritos/Usera/DiagnosticoUsera/LineasEstrategicas.pdf

Câmara, A. N. D., et al. (2015). Cuando el edificio encuentra la calle: las relaciones entre la normativa y la construcción del espacio urbano. In *VII International Seminar of Research in Urbanism, Barcelona-Montevideo*. DUOT. https://doi.org/10.5821/siiu.6105

Cambra Rufino, L. (2021). *Evaluación de los factores funcionales y ambientales en el diseño de hospitales de agudos españoles*. PhD Thesis. Universidad Politécnica de Madrid.

Consorcio Regional de Transportes de Madrid. (2020). Transporte Público de Madrid. Retrieved from: https://www.crtm.es/media/147671/serie_3_madrid_plano_es.pdf

Chiesura, A. (2004). The role of urban parks for the sustainable city. *Landscape and Urban Planning, 68*(1), 129–138. https://doi.org/10.1016/j.landurbplan.2003.08.003

Croucher, K., Myers, L., & Bretherton, J. (2007). Transforming urban spaces. The links between greenspace and health: A critical literature review. *Greenspace Scotland Research Report*, 1–61.

Demyttenaere, K., Bruffaerts, R., & Posada-villa, J. (2004). Prevalence, severity, and unmet need for treatment of mental disorders in the World Health Organization World Mental Health Surveys The WHO World Mental Health Survey Consortium. *The Journal of the American Medical Association, 291*(21), 2581–2590. https://doi.org/10.1001/jama.291.21.2581

FAO. (2021). *Ecosystem services & biodiversity*. Retrieved from: http://www.fao.org/ecosystem-services-biodiversity/background/regulating-services/en/

Fariña Tojo, J., Higueras García, E., & Román López, E. (2019). Ciudad, Urbanismo y Salud. Documento Técnico de criterios generales sobre parámetros de diseño urbano para alcanzar los objetivos de una ciudad saludable con especial énfasis en el envejecimientos activo. Retrieved from: https://www.mscbs.gob.es/profesionales/saludPublica/prevPromocion/Estrategia/docs/ImplementacionLocal/Ciudad_urbanismo_y_salud.pdf

Ferrari, A. J., et al. (2010). Burden of depressive disorders by country, sex, age, and year: Findings from the global burden of disease study 2010. *PLoS Medicine, 10*(11). https://doi.org/10.1371/journal.pmed.1001547

Flemish Ministry. (2016). Act of the residential care modification of annex XII of the act of the Flemish government of 24 July 2009. Ministry of Wealth, Health and Family. Retrieved from: https://www.departementwvg.be/

Foucault, M. et al. (1979) Les Machines à guérir: aux origines de l'hôpital moderne. .

García Lantarón, H. (2015) *Vivienda para un envejecimiento activo. El Paradigma Danés*. PhD Thesis. Universidad Politécnica de Madrid. Retrieved from: http://oa.upm.es/40528/1/HEITOR_GARCIA_LANTARON.pdf

Gruebner, O., et al. (2017). Cities and mental health. *Deutsches Arzteblatt International, 114*(8), 121–127. https://doi.org/10.3238/arztebl.2017.0121

HIN, H. I. N. S. L. (2014). *What is person-centred care and why is it important?* London. Retrieved from: https://healthinnovationnetwork.com/resources/what-is-person-centred-care/

Holt-Lunstad, J., Smith, T. B., & Layton, J. B. (2010). Social relationships and mortality risk: A meta-analytic review. *PLoS Medicine, 7*(7). https://doi.org/10.1371/journal.pmed.1000316

Kodama, A., et al. (2016). Deriving the reference value from the circadian motor active patterns in the "non-dementia" population, compared to the "dementia" population: What is the amount of physical activity conducive to the good circadian rhythm. *Chronobiology International. Taylor & Francis, 33*(8), 1056–1063. https://doi.org/10.1080/07420528.2016.1196696

Konis, K., Mack, W., & Schneider, E. (2018). Pilot study to examine the effects of indoor daylight exposure on depression and other neuropsychiatric symptoms in people living with dementia in long-term care communities. *Clinical Interventions in Aging, 13*, 1071–1077.

Larrosa, M., et al. (2008). Vitamin D deficiency and related factors in patients with osteoporotic hip fracture. *Medicina clinica. Spain, 130*(1), 6–9. https://doi.org/10.1157/13114538

Le Mandat, M. (2009). Les plantes médicinales et les Hori medici dans les hôpitaux médievaux. *Revue de la Société française d'histoire des hôpitaux, 133–134*, 3–6. Retrieved from: https://www.bnds.fr/revue/rsfhh/sfhh-133.html

Jędrzejczyk, T. (2021). *Intro to urban development planning. MSc in urbanism. Module: Urban planning 26 April 2021(class material)*. University Polytechnic of Gdansk.

de Madariaga, I. S., & Zucchini, E. (2019). Measuring Mobilities of care, a challenge for transport agendas. In C. Scholten & T. Joelsson (Eds.), *Integrating gender into transport planning*. Palgrave Macmillan. https://doi.org/10.1007/978-3-030-05042-9_7

McCay, L., et al. (2017). Urban Design and mental health. In N. Okkels, C. Blanner Kristiansen, & P. Munk-Jorgensen (Eds.), *Mental health and illness in the City* (pp. 421–444). Springer. https://doi.org/10.1007/978-981-10-2327-9_12

Mental Health Foundation. (2016). *Fundamental facts about mental health*. London. Retrieved from: https://www.mentalhealth.org.uk/publications/fundamental-facts-about-mental-health-2016

Morand, S., & Pipien, G. (Eds.). (2013). *Notre Santé et la biodiversité*. Chastel.

Moreno, C., Didier Chabaud, Z. A., Gall, C., & Pratlong, F. (2021). Introducing the "15-Minute City": Sustainability, resilience and place identity in future post-pandemic cities. *Smart Cities, 4*(1), 93–111. https://doi.org/10.3390/smartcities4010006

Neila González, F. J., Sánchez, G. C., Núñez, P. M., Román, L. E., López, H., Gómez, G., & de Luxán, M. (2020). *Urban Heat Island, urban climate and building energy consumption. MODIFICA project*. García-Maroto Editores S.L.

Peen, J., et al. (2010). The current status of urban-rural differences in psychiatric disorders. *Acta Psychiatrica Scandinavica, 121*(2), 84–93. https://doi.org/10.1111/j.1600-0447.2009.01438.x

Pozo Menéndez, E., Gallego, G. C., Román, L. E., & García, H. (2019). UNI-Health, Programa Europeo de Innovación Educativa para la Salud Urbana. In D. García Escudero & B. Bardí Milà (Eds.), *VII Jornadas sobre Innovación Docente en Arquitectura (JIDA'19), Escuela Técnica Superior de Arquitectura de Madrid, 14 y 15 de Noviembre de 2019* (pp. 740–753). UPC IDP; GILDA. https://doi.org/10.5821/jida.2019.8406

Ritchie, C. W., et al. (2015). Development of interventions for the secondary prevention of Alzheimer's dementia: the European Prevention of Alzheimer's Dementia (EPAD) project. *The Lancet, 3*(2), 179–186. https://doi.org/10.1016/S2215-0366(15)00454-X; https://www.thelancet.com/journals/lanpsy/article/PIIS2215-0366(15)00454-X/fulltext

Rodrigue, J. P., Comtois, C., & Slack, B. (2013). *The geography of transport systems*. Routledge. Retrieved from: https://transportgeography.org/media/e-book/

Shanahan, D. F., et al. (2015). The health benefits of urban nature: How much do we need? *Bioscience, 65*(5), 476–485. https://doi.org/10.1093/biosci/biv032

Shanahan, D. F., et al. (2016). 'Health benefits from nature experiences depend on dose', scientific reports. *Nature Publishing Group, 6*(June), 1–10. https://doi.org/10.1038/srep28551

Smith, A. D., & Yaffe, K. (2014). Dementia (including Alzheimer's disease) can be prevented: Statement supported by international experts. *Journal of Alzheimer's Disease, 38*(4), 699–703. https://doi.org/10.3233/JAD-132372

Sociedad Española de Geriatría y Gerontología. (2020). Resumen Normativa de residencias. Documento de trabajo. Retrieved from: https://www.segg.es/media/descargas/Cuadro_resumen_SEGG_Normativa_Residencias_Rev.17junio2020.pdf

Thaneshwari, et al. (2018). Therapeutic gardens in healthcare: A review. *Annals of Biology, 34*(2), 162–166.

Ulrich, R. S., et al. (1991). Stress recovery during exposure to natural and urban environments. *Journal of Environmental Psychology, 11*(3), 201–230. https://doi.org/10.1016/S0272-4944(05)80184-7

United Nations. (2018). *Revision of World Urbanization Prospects*. Retrieved from: https://population.un.org/wup/

Vassos, E., et al. (2012). Meta-analysis of the Association of Urbanicity with Schizophrenia. *Schizophrenia Bulletin, 38*, 1118–1123. Available at: https://academic.oup.com/schizophreniabulletin/article/38/6/1118/1860155

Vos, T., et al. (2013). Global, regional, and national incidence, prevalence, and years lived with disability for 301 acute and chronic diseases and injuries in 188 countries, 1990-2013: A systematic analysis for the global burden of disease study 2013. *The Lancet, 386*(9995), 743–800. https://doi.org/10.1016/S0140-6736(15)60692-4

WHO (2000) Healthy ageing – adults with intellectual disabilities summative report. Retrieved from: https://www.who.int/mental_health/media/en/20.pdf

WHO. (2007). Global age-friendly cities: A guide. In *WHO library cataloguing-in-publication data*. Retrieved from: https://www.who.int/ageing/publications/Global_age_friendly_cities_Guide_English.pdf

WHO. (2021). *Mental health in cities*. Retrieved from: https://www.who.int/health-topics/urban%2D%2D-health/cities-spotlight/mental-health-in-cities

Whyte, W. H. (1980). *The social life of small urban spaces*. The Conservation Foundation.

Elisa Pozo Menéndez is an architect and urban planner and an MsC in *Villes durables* (UPEM-IFU, 2014) and MsC in the *Environment and Bioclimatic Architecture* at UPM (MAYAB, 2019). She is currently a researcher in ABIO-UPM Group and PhD candidate on active ageing and urbanism, coordinating European projects at the Universidad Politécnica de Madrid on innovative education and urban health. She collaborates as an architect and consultant with different studios and engineering teams in urban studies, mobility plans, architecture projects and competitions. She teaches workflows and digital tools in the Istituto Europeo di Design. Since 2006, she has collaborated in multiple NGOs and associations, working with vulnerable collectives. In 2020, she received the third prize solidarity award of Young Talent of the Region of Madrid.

Chapter 15
Dementia-Friendly Communities and Challenges from Built Environment Design: The Belgian Case

Veerle Baert, Elise Cornelis, and Patricia DeVriendt

Abstract Dementia-friendly communities make living with dementia possible. After all, a large number of people are affected by dementia in one way or another. In a dementia-friendly community, people continue to participate in life by being addressed on their abilities rather than their limitations. Designing dementia-friendly environments is one of the key elements. In this chapter, the ten design principles of Alzheimer's Disease International are discussed and illustrated with Belgian examples.

Keywords Dementia · Dementia-friendly community · Built environment · In- and outdoors

---

V. Baert (✉)
Department of Social Work, Artevelde University of Applied Sciences, Ghent, Belgium
e-mail: veerle.baert@arteveldehs.be

E. Cornelis
Department of Occupational Therapy, Research and Development Centres Health & Care, Artevelde University of Applied Sciences, Ghent, Belgium

P. DeVriendt
Faculty of Medicine and Health Sciences, Department of Rehabilitation Sciences, Ghent University, Ghent, Belgium

Department of Occupational Therapy, Artevelde University of Applied Sciences, Ghent, Belgium

Mental Health Research group, Frailty in Ageing Research Group, Vrije Universiteit Brussel, Brussel, Belgium

E. Pozo Menéndez, E. Higueras García (eds.), *Urban Design and Planning for Age-Friendly Environments Across Europe: North and South*, Future City 19,
https://doi.org/10.1007/978-3-030-93875-8_15

15.1 Introduction

15.1.1 *Dementia: A Worldwide Challenge*

Every year, there are worldwide nearly 10 million new cases of dementia (WHO, Dementia, 2020). The World Health Organization (WHO) projected the total number of people with dementia to reach 82 million in 2030 and 152 million in 2050. Dementia is declared as 'one of the major causes of disability and dependency among older people worldwide' (WHO, Dementia 2020), but also as one of the major financial challenges (Oba et al., 2021; Junfang et al., 2017). In Europe, in 2018, the number of people living with dementia was estimated at about 9.8 million. By 2050, the forecast is that the total number of people with dementia will almost double to 18.8 million (Alzheimer Europe, 2020). In Belgium in 2020, according to an estimate based on the EuroCode (a European epidemiological project led by the Alzheimer Europe) risk model, approximately 220,104 people are living with dementia (Steyaert, 2020).

15.1.2 *What Is Dementia?*

Dementia is not a single disease; it is a broad term that covers more than 50 diseases since there are many different underlying causes of dementia. The most common forms are Alzheimer's disease, frontotemporal dementia, vascular dementia and dementia associated with Parkinson's disease. In older persons, it is common to have mixed dementia, a combination of two or more types of dementia.

15.1.3 *Cognitive Impairment*

Dementia is a progressive clinical syndrome. Depending on the types of brain changes that may be taking place, cognitive abilities are primarily affected. The brain is no longer able to process information properly, characterised by impairment of at least two brain functions, such as memory loss and judgement. Dementia might also lead to different orientation problems and declines in navigation skills, including spatial disorientation. We speak of spatial disorientation when someone no longer recognises where he lives or when he does not know where to find the toilet in his house or loses his way in the neighbourhood where he lives. The person does not recognise the place or does not remember how to get from one place to another. Besides, also visual-related changes in visual space and visual receptor processing might manifest as navigation problems and occur early in the disease process, especially in Alzheimer's disease. Initially, navigation in unfamiliar locations is impaired, but in the later stages of the disease, it is even difficult for people to navigate in familiar environments (Rowe et al., 2015). Given the fact that dementia is driven by deterioration of brain functions in addition to the fact that many people with dementia will have other age-related health problems such as mobility, visual or hearing difficulties, one can say that its impacts are tremendously the responses in the design of the environment (Barrett et al., 2019).

15.1.4 Loss of Independence

The cognitive impairment in dementia significantly interferes – in a negative way – with someone's abilities to perform activities of daily living (ADL). The process of functional decline shows a typical and distinctive progression from mild forms of cognitive impairment to more severe forms. In mild forms of cognitive impairment (such as mild cognitive impairment, MCI), individuals remain autonomous although subtle problems may already occur in complex activities such as leisure, hobbies, etc. The household activities such as cooking, shopping and managing medication will become to some extent limited when MCI evolves to a diagnosed dementia requiring more and more support when people are evolving to more severe dementia. However, the self-care activities, the so-called personal hygiene, such as dressing and eating, remain stable the longest. This means that – at some point in the progression of the disease – people are not able anymore to live independently in a safe way and that support will be needed at home or that moving to a residential care home will become inevitable.

15.1.5 Challenging Behaviour

Dementia encompasses not only impairment in cognitive and daily functioning but also problems in other domains such as personality, mood and behaviour. These neuropsychiatric disturbances such as agitation, aggression, depression and apathy are called 'behavioural and psychological symptoms of dementia' (BPSD). BPSD affects up to 97% of community-dwelling people during the trajectory of dementia. BPSD has a serious negative impact on prognosis and institutionalisation and is associated with a poorer quality of life of both the person with dementia and his caregiver (Cloak & Al Khalili, 2020).

15.1.6 Interventions and Approaches

To date, there is no cure to solve or stop the process of dementia. Therefore, good care should focus on rehabilitation, information provision and practical support of both the person with dementia and their caregiver. Ideally, interventions should be initiated at the earliest stages of dementia. Therefore, early and timely diagnosis is of primordial value. Subsequently, interventions, both pharmacological and psychosocial, targeting the consequences of the disease and the quality of life and care should be initiated. In literature, a variety of evidence-based interventions and approaches are available, of which is the concept of 'a dementia-friendly environment' (Shih-Yin, 2017). This is an important concept since we all live in and interact with our own environment. The way our environment is adapted to our identity, capacities and wishes influence not only our autonomy but also our quality of life (Huang et al., 2019). This is also the case in dementia (Whear et al., 2014). An environment can have a significant effect on a person with dementia, both supportive and hindering. An 'adapted' living environment, both material/physical and social – where sufficient care and support are available – increases the well-being of people with dementia but also of the carers and can give utmost independence. About 70% of people with dementia live at home and usually want to stay at home for as long as possible. This requires a dementia-friendly home but also a dementia-friendly neighbourhood and city. And since a considerable part of them needs to go into residential care homes at some point, because their own homes are not designed to enable them to live independently or hard to adapt to meet their needs, also the residential care homes should be adapted to their needs and capacities.

15.2 Universal Design, Age-Friendly Communities and Dementia-Friendly Communities: What Is in a Name?

A variety of concepts has been launched in the last decennia, all targeting better living environments for people: for all people and for vulnerable people. Universal design, age or dementia-friendly communities, etc. are often used interchangeably or are interdependent and need some explanation how they all fit together in the broader concept of 'a dementia-friendly environment'.

15.2.1 Universal Design

To begin with, designing for dementia goes hand in hand with the benefits and principles of universal design. Universal design is the design and composition of the environment so that everyone can access, understand and use it, regardless of age, size, ability or disability. An environment (or buildings, products or services in that environment) should be designed to meet the needs of all people who wish to use it. This is not a special requirement only for the benefit of a small population. This is the starting point of good design. If the environment is accessible, usable, easy and pleasant to use, everyone benefits (National Disability Authority, 2020). This approach will allow people to live at their own home and their familiar community for as long as possible.

15.2.2 Age-Friendly Communities

Secondly, the concept of 'age-friendly community' had been put forward by the WHO in 2007 (WHO, 2007). The WHO has identified eight domains of age-friendly cities: the built environment, transport, housing, social participation, respect and social inclusion, civic participation and employment, communication and community support and health services (WHO, 2021a, b, c). Age-friendly communities are places that enable people of all ages 'to actively participate in community activities and treat everyone with respect, regardless of their age'. It is a place that makes it easy for older people to stay connected to people that are important to them. And it helps people stay healthy and active even at the old stages and provides appropriate support to those who can no longer look after themselves (WHO, 2021a, b, c). The concept of age-friendly communities can be a starting point to help cities, towns or villages to become more dementia-friendly, but they lack a deeper understanding of the needs of people living with dementia. Therefore, more emphasis has been put on the concept of dementia-friendly communities.

15.2.3 Dementia-Friendly Communities: Living Well with Dementia

A dementia-friendly community (DFC) is a city, town or village where people with dementia are understood, respected and supported (Alzheimer's Society, 2021). In a DFC, citizens will be aware of and understand dementia. This is vital in helping people live well with – or despite – dementia and remain a citizen and a part of their community. People with dementia can continue to live in the way they want to and in the community they choose for as long as possible with a good quality of life.

As the ability to access and utilise the in- and outdoor environment is essential for the successful performance of meaningful daily activities and participation in society of people with dementia, this has major implications for the interior and exterior environmental design. This requires improvements in social and material environments and puts challenges on the built environment. This is paramount since there is absolute evidence that optimising pace for those living with dementia, in combination with a medical and social approach, can improve their quality of life (Barrett et al., 2019).

Nevertheless, the concept of 'dementia-friendly community' cannot be seen separately from the concept of 'age-friendly community' as put forward by the World Health Organization (Age-friendly environments, 2021a).

15.3 Designing for Dementia

Very recently, in 2020, the Alzheimer's Disease International published the World Alzheimer Report about dementia-related design and the built environment (Fleming et al., 2020; World Alzheimer Report, 2020: Design Dignity Dementia: dementia-related design and the built environment, 2020). In this report, the authors put forward the following ten design principles to keep in mind when designing for people with dementia:

- Unobtrusively reducing risks
- Providing a human scale
- Allowing people to see and be seen
- Reducing unhelpful stimulation
- Optimising helpful stimulation
- Supporting movement and engagement
- Creating a familiar place
- Providing opportunities to be alone or with others
- Linking to the community
- Designing in response to vision for way of life

In this text, the design principles have been used to structure key elements of designing for people with dementia in different environments. They are illustrated with examples from Belgium.

The principles described throughout this chapter are just that principles. They are not a checklist of items to be ticked off one by one. They are not rules to be applied in the same way in every situation. Instead, they are principles to be explored and interpreted in each setting to determine how they can be applied. Of course, most important of all in a dementia-friendly community is assisting people to 'live well with dementia', as far as possible.

15.3.1 Unobtrusively Reducing Risks

For all older people, falling is a major risk, a complex problem that is often caused by multiple factors. Dementia superimposed on older age increases this risk. Moreover, the type of dementia seems to influence the risk of falling. For example, people with Lewy body dementia are more likely to fall than people with Alzheimer's disease (Imamura et al., 2000). Therefore, people with dementia need an environment that is safe, secure and accessible to move around. Potential risks must be reduced indoor as well as outdoor. Ground floor thresholds need to be levelled. Steep stairs and slipping or curling rugs create unsafe situations in and around the house and should be removed. The provision of handrails and good (street) lighting can reduce the risk of falling. In the context of dementia, they can also act as a

Fig. 15.1 Stairs and an elevator to facilitate the connection between the parking lot and the hospital entrance – AZ Maria Middelares in Ghent (Belgium). (Source: Copyright Bart Vermandere (2020))

way-finding device and be orientating. A handrail that contrasts visually with the background will help to it more clearly (Grey et al., 2015).

External environmental modifications such as the elimination of street clutter or creating flat and clearly demarked footways, wide enough to pass other pedestrians, are suggested in order to create a safe dementia-friendly neighbourhood. The circulation can be made safe and accessible by designing ramps, stairs and access gates (see Fig. 15.1).

Clear pavements are also required with bin storage and storage shelters for mobility vehicles. Also frequently positioned outdoor seating, where people can take a rest, and good accesses to public transport are needed.

However, we also need to keep in mind that obvious safety features and barriers (such as fences and locked doors) can lead to frustration for persons with dementia (Fleming & Bennet, 2017). Therefore, agitation and anger and so potential risks need to be reduced unobtrusively.

15.3.2 Providing a Human Scale

The scale of a building can affect the behaviour and feelings of a person with dementia. The experience of scale is influenced by the number of people that the person encounters. Therefore, experts suggest that accommodations in residential care for people with dementia should host between 8 and 12 residents to keep a domestic character. Creating small units of people within larger buildings can be an alternative to multiple small separate units and is focusing on a more homelike, holistic and person-centred approach. Unfortunately, large institutions that have the look and feel of a hotel are still built. Research has shown that the physical environment of small-scale living facilities for people with dementia has more potential to be beneficial for residents' daily life than the physical environment of traditional large-scale nursing homes (de Boer et al., 2018). In De Wingerd in Leuven (Belgium),[1] among others, residents with dementia live together in a homely, familiar and recognisable environment. They retain maximum control over their own lives. Small-scale living is never separate from the immediate residential and living environment. This residential care concept is architecturally integrated into the surrounding district or municipality, and it is neighbourhood-oriented and therefore also integrated into the community. This principle offers professional support, guidance and care for people with dementia, their family members, informal carers and volunteers.

15.3.3 Allowing People to See and Be Seen

The idea of this principle is that cognitive and sensory deficits need to be compensated. The environment must be easy to understand for people with dementia: domestic and homelike. It will help them to minimise confusion. The creation of distinct spaces for different domestic activities within residential care is important (Grey et al., 2015). A room needs to be furnished and decorated in a way that one the purpose and function can be easily identified. Good visual access opens up opportunities for engagement and gives the persons with dementia the confidence to explore their environment. Therefore, elements need to be clearly legible. This way, people with dementia can orientate themselves independently much longer. A living room is a living room and a kitchen is undeniably a kitchen. That may seem obvious, but in most residential care homes, it is not so clear.

Good lines of sight will help people living with dementia to see the key places they want to go to, and it will engage them to be active. A coloured toilet seating might be helpful to find or recognise the toilet (see Fig. 15.2). A clear vision and light levels influence the independence of people with dementia (see Fig. 15.3) (Siegelaar et al., 2020).

Sometimes it is tricky. What is good for one person can be disliked by another person. Street lighting should be bright enough for adults with dementia, but it must not cause light pollution for other persons. Besides this, aspects or objects that are relevant to the person with dementia should be highlighted, and what is not relevant should be removed.

[1] WZC De Wingerd. Retrieved from https://wingerd.info/

Fig. 15.2 Contrasting toilet seat will enable the person with dementia to make a distinction between the toilet and the surroundings (Belgium). (Source: Copyright Veerle Baert (2021))

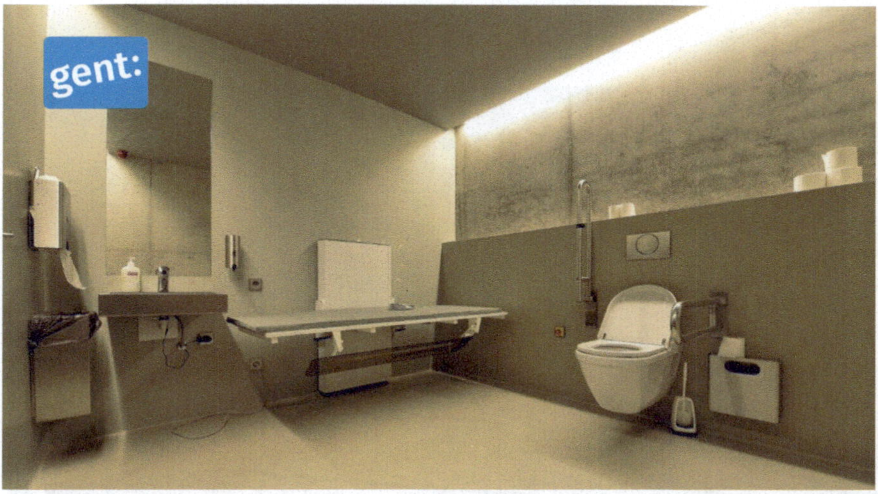

Fig. 15.3 Public toilet with a changing table for adults and adequate lighting in the De Krook, public library in Ghent (Belgium). (Source: Copyright Bart Vermandere (2020))

15.3.4 Reducing Unhelpful Stimulation

People with dementia have a short attention span and are easily distracted. People living with dementia can often face difficulties with sensory overstimulation which can increase agitation and confusion. Large amounts of stimulation can be stressful and can create unnecessary fear. Sensorial overstimulation comes in many different forms, too much noise, rooms that are too bright or too dark, etc., and in this way, attention should be paid to reduce them all.

First, close attention needs to be paid to the acoustic disturbances. The provision of quiet areas is important as people with dementia can be hypersensitive to noise. Planting or fencing can reduce street and background noise. Also, in residential care, it is important to reduce background noise. Among others, Brown, Rutherford, and Crawford (2015) describe the role of sound in clinical environments and the damaging effects of sound, with emphasis on mental health care. Excess unwanted noise can be harmful to health and impede recovery. In buildings, it is advised to use specific materials to reduce noise with regard to floors and ceiling, but also, for example, for curtains (Devos et al., 2020) (see Fig. 15.4).

Secondly, since many people with dementia have visual impairments or are facing problems with interpreting what they see due to the cognitive impairments, therefore, they need more natural light or stronger artificial light to compensate. Visual disturbances come in different ways. Shiny floors or dark tiles in the floor can cause anxiety as dementia reduces the ability to focus. Good lighting will reduce the effects of glare and enhance task visibility. Even more it can play an important role in reducing the opportunities for crime and the fear of crime (Grey et al., 2015).

Fig. 15.4 Noise-reducing panels on the ceiling and walls in a living room in a nursing home (Belgium). (Source: Copyright Acusticare (2020))

The full range of senses must be considered. Too much visual stimulation is as stressful as too much auditory stimulation can be (Fleming & Bennet, 2017).

15.3.5 *Optimising Helpful Stimulation*

Enabling the persons with dementia to see and hear can help to minimise their confusion and optimise their autonomy. Together with the growing evidence of non-pharmacological interventions to cope with BPSD, there is an increasing interest in adapting the environment, in order to create 'healing environments'. Healing environments can be defined as places that produce positive health outcomes within the physical environment (Jonas & Chez, 2004; Ulrich et al., 2008). Healing environments focus on all senses, such as smells, colours, light, green nature, etc.

Specifically, the sound environment, which is already a topic in many investigations in various settings such as schools, restaurants, parks and hospitals, is far less studied in dementia (Joosse, 2012; To & Chung, 2014; Wilson et al., 2016) but offers opportunities for dementia care (Remington, 2002). It is recognised that sound as a part of the environment has an impact on people's behaviour and their quality of life (Andringa & Lanser, 2013). Quiet and pleasant sounds promote health and nuisance noises limit health. Soundscapes – defined as acoustic environment as perceived or experienced and/or understood by a person or people, in context – are increasingly used to reduce adverse consequences and improve positive effects (Devos et al., 2018). Soundscapes usually encompass different sounds that occur simultaneously or consequently (Axelsson et al., 2010). To date, there are little to no studies on the effect of soundscapes on BPSD except the project in Belgium, Canada and Groningen in collaboration with the Belgian researchers. Though it is not wrong to think that soundscapes might have a positive influence on the BPSD, since BPSD have a neurological basis which makes the person with dementia more vulnerable to environmental, physical and psychological factors (Benoit, 2005; Liperoti et al., 2008). The soundscape research project in nursing homes in Belgium 'AcustiCare' (Acusticare, 2021) tries to put people and their experiences at the centre of the process of evaluating an acoustic environment. Personalised soundscapes were developed, encompassing natural, homelike, recognisable sounds. Soundscapes were subsequently displayed in living rooms, hallways and private rooms of the residents with dementia. The preliminary findings are promising; the number of BPSD, the frequency and severity and also the moral distress of caregivers significantly decreased during and after the soundscapes. In this project, 11 nursing homes were actively involved in the living labs which were experimented with soundscapes, while numerous others were engaged in the workshops afterwards.

Also for depression, superimposed on dementia, healing environments can be beneficial. About 20–30% of people diagnosed with dementia suffer from depression. Exposure to sufficient sunlight is a potentially effective non-medicated therapy for depression. A preliminary study by Konis et al. (2018) showed that interventions

Fig. 15.5 Black button on white background to call up the elevator. The text at the bottom reads 'Call lift'. (Source: Strobbe Mobility Solutions)

that increase the amount of sunlight in the morning (8 am–10 pm) can be used to reduce depression in patients with dementia.

A decreased vision is part of ageing. Vision requires the eyes and the brain to work together. Dementia can cause a person to have difficulty in processing what he/she sees. This is not caused by an eye problem, but by a problem in the brain. For example, it is less able to estimate distances and see depth. For people with dementia, it is often difficult to distinguish between different shades of colour or to recognise certain images or colours. Knowing that older people with or without dementia can see their surroundings less well is a good starting point for designing environments. Good design can greatly simplify the process of safely interpreting and navigating a building or environment, and the use of colour and contrast can help in many ways (see Fig. 15.5). Contrast is key in comprehending the environment and understanding the features and elements of a building that people use in their daily lives (Dementia Services Development Centre, 2012).

15.3.6 Supporting Movement and Engagement

The numerous health benefits of physical activity for people with dementia are well described (Nuzum et al., 2020). Physical activity improves both physical and mental health, increases the risk of developing depression, facilitates daily activities, stimulates independence and enhances the quality of life and well-being for people with dementia. Also social benefits can be important: for example, exercising in a small group can increase social networks and reduce loneliness, which is a problem for many people with dementia (Bowes et al., 2013). The type of exercise is also important. For example, walking outside can help reconnect with nature and local communities; dancing can bring joy and happiness. Exercise can help maintain a normal lifestyle and reduce the risk of other diseases.

A study of Clark et al. (2015) stresses the importance of considering urban and intelligent design for the cognitive health of older adults who are ageing in place. Residents in a neighbourhood with community resources, proximity to public transit and public spaces in good condition were associated with slower rates of cognitive decline, possibly by increasing opportunities for social and physical activities or access to destinations that facilitate engagement in activities (Clarke et al., 2015). This was confirmed in another study that found out that people with limited activity space and living in a deprived neighbourhood are particularly at risk of dementia and should be targeted for prevention (Letellier et al., 2019) (Fig. 15.6).

Encouraging people with dementia to stay physically active is incredibly important. Everyone living with dementia can enjoy the benefits of walking or even biking. The 'Kwiek beweegroute' (Kortrijk, 2021) exercise route is a unique concept. Pavement tiles with simple instructions are meant to get (senior) citizens moving by a well-defined pathway, free of obstacles and complex decision points. The exercises make use of street furniture, such as stairs or a bench, to perform simple actions. Through the exercises, balance, strength and endurance are trained (Fig. 15.7).

Fig. 15.6 Outdoor sport facilities in a public park in Ghent (Belgium). (Source: Copyright Stad Gent (2020))

Fig. 15.7 Tile and plan from the 'Kwiek beweegroute' – Kortrijk (Belgium). (Source: Kwiek beweegroute)

Another example of how older adults with dementia can be engaged in physical activity is the concept of 'Cycling Without Age' (Cycling without age, 2021). The founder of the project, Ole Kassow, wanted to help the elderly get back on their bicycle, but he had to find a solution to their limited mobility. The answer was a trishaw and he started offering free bike rides to the local nursing home residents.

Fig. 15.8 Tribikes can be rented by the citizens of Ghent (Belgium). (Source: Copyright De Fietsambassade (2020))

Trishaws, a hybrid of a bicycle and a rickshaw, seat two passengers in the front and a pilot in the rear (Fig. 15.8).

Many senior citizens have often cycled in the past and lack that experience: enjoying nature and the outdoors, getting a breath of fresh air, the wind that plays through their hair. With the trishaw, they get a little further and their world becomes just that little bit bigger. Vulnerable, home-sitting seniors come into contact with their peers and have the opportunity to talk to the volunteer who transports them along the way. The older people make friendships and feel safer in their community.

Cycling Without Age is now an intergenerational project which brings together younger and older people through shared trishaw bike rides. Their narrative is supported with measurable impact, illustrating a trishaw ride as a successful intervention that can combat social isolation and thereby promote health and well-being for both age groups. As known, loneliness and social isolation are as deadly as smoking and by some reports more deadly than obesity. Currently, the Cycling Without Age movement has spread nationally and globally and has already pedal edits way into 41 countries. In the city of Ghent, citizens can rent or use various adapted bicycles, such as duo-bikes, tandems or trishaw bikes. Projects like this require an adapted environment such as wide bicycle lanes preferably on flat ground (Fig. 15.9).

Fig. 15.9 Wide bicycle lanes in Gentbrugge. Benches (oriented to the river) for resting are provided along the way (Belgium). (Source: Copyright Stad Gent (2020))

15.3.7 Creating a Familiar Place

People with dementia rely on familiar, regularly used routes and environments (Mitchell & Burton, 2006). Clarity of use and function of outdoor environments will help people with dementia to understand and to learn (new) places, buildings and streets. The familiarity of the surroundings plays an important role in preventing and mitigating spatial disorientation, confusion and diverse memory problems (Mitchell et al., 2003). To decrease spatial disorientation in people with dementia, circulation systems should be simple (Marquardt et al., 2011; Marquardt & Schmieg, 2009). The easiest floor plans though are straight circulation systems, with no changes in direction.

In order to support the reality orientation of people with dementia, simple, structured and familiar environments may support wayfinding.

Moreover, the concept of familiarity refers to predictability and continuity; it is also important for creating a sense of being at home for persons with dementia in addition to supporting the maintenance of their social relationships, identity, autonomy and privacy (Førsund et al., 2018).

15.3.8 Providing Opportunities to Be Alone or with Others

The diagnosis of dementia should not be a reason to stop social relationships and meaningful daily activities. A rich and varied social life is one of the best weapons against the disease (Fleming et al., 2020).

Therefore, in a dementia-friendly community, people with dementia must be able to decide whether they want to be alone or with others. This means that the environment must offer a variety of places where one can retreat and places where one can meet with others. Frequent interaction with friends and relatives can help to maintain that identity, and visitors should be able to drop in easily and enjoy being in places that encourage interaction.

15.3.9 Linking to the Community

Almost half of the people with dementia say that they are afraid to participate in the community, 28% is scared to leave the house and more than 60% are afraid to get lost. Only one-third sees opportunities to continue participating in the community (Baert, 2015). Striving for a dementia-friendly community means that policymakers need to create opportunities for people with dementia to participate in society for as long as possible.

In Belgium, there are some nice initiatives that promote dementia friendliness. Like in the city of Aalbeke, the reminiscence walk 'levend geheugen' (living memory) was developed in cooperation with the local history society – in the vicinity of the residential care facility De Weister. This signposted walk via so-called slow roads has the residential care centre as its starting and finishing point and is accessible to every tourist. Reminiscence walks give a sense of recognition and relive stories from the past. This reminiscence reinforces the sense of self-worth of people with dementia. In a world that is rapidly alienating them, slow roads offer an experience very similar to the past: there is less noise from cars and more sounds of nature. These triggers from the past allow people to flourish.

The Huis Perrekes in Geel (Belgium) is a non-profit organisation that cares for people with dementia. The starting point is not a medical vision of care but a human approach. The residents live in an ordinary house in a normal street in a small village.

They start from the presence of the person with dementia in his/her experience of his/her environment (material and immaterial) and look at what is and not only at what is gone or damaged. Not only in the way how they care but also in the architecture of their homes, they link up with previous phases of life in order to form a continuum with what 'home' and 'dignity' mean. As one resident's husband said, 'it was as if Huis Perrekes had become an extension of our home'. Huis Perrekes recognises the need for embedding in a larger whole interwoven with the structure of the village. Their large garden is accessible to the public.

15.3.10 Designing in Response to Vision for a Way of Life

There are many different approaches to dealing with people with dementia. There is no evidence as to which approach is the best, but for sure, the familiarity of a residential care facility is beneficial for people with dementia.

Nursing home 'Heiveld' (Ghent, Belgium) is a colourful residential care facility in Ghent (Belgium). They installed beautiful furniture that the residents could choose themselves. It is surrounded by gardens laid out so that they have a homely environment. The staff is trained within the framework of 'active ageing' (WHO, 2002). This is a concept from gerontology that looks at the possibilities rather than the limitations of ageing. The site looks like a small village: there is a small café, a cultural centre, a playground, a market square and a garden with a lot of animals including a pig that is allowed to go in the nursing home. The residents can work in the greenhouse and the vegetable garden and process the fruits and vegetables from the garden. Even the elderly who do not actively participate in other activities can enjoy the various kitchen events with the own harvest: they sit at the table, and they experience the smells, the tastes and the social interaction.

The cafeteria is converted into a 1950s café. This has a particular reason. The average age of the residents is 85. So they were 20 in the 1950s. In the café, people are able to choose their own music on a jukebox, so they can receive their families in an environment that is stimulating (see Fig. 15.10). They even developed their own beer based on the tastes of the residents. The beer has become a symbol of

Fig. 15.10 A 1950 look-a-like cafeteria in nursing home 'Het Heiveld' in Ghent (Belgium). (Source: Copyright Geert Roggeman (2020))

homeliness and cosiness, and it makes a nice gift to give to visitors. The residents designed the packaging themselves (*Woonzorgcentrum*, 2016).

15.4 Challenges

In the text above, the ten design principles to keep in mind when designing for people with dementia, as put forward in 2020 by the Alzheimer's Disease International (Fleming et al., 2020), are described and illustrated with examples from Belgium. The main aim would be to combat loneliness and enable participation in society. Too many people affected by dementia feel that society fails to understand the condition they live with, its impact or how to interact with them. That is why people with dementia sometimes feel they need to withdraw from their community as the condition progresses. In fact, over a third of people with dementia reported that they have felt lonely recently (Victor et al., 2020). This is regrettably since people affected by dementia still have an incredible amount to offer to their community. If appropriately supported, they can continue to play an active and valuable role even years after diagnosis. This even applies to the informal caregivers as only 32.0% have as much social contact as they would like and thus the majority suffers from loneliness (NHS, 2016). In Belgium, many good practices are already developed, though we have to confess that in Belgium we also know other practices. What we can learn from studying them is that a variety of challenges need to be met to fully realise the ideal dementia-friendly community.

Firstly, and most importantly, the involvement of people with dementia in the design process is warranted as it will lead to better proposals because designers can take the wishes and needs but also dislikes of persons with dementia into account. This will result in fewer problems and unwanted features afterwards (Hendriks & Wilkinson, 2020). Dementia-friendly communities should be shaped according to the needs and expectations of people living with dementia and their loved ones. After all, each neighbourhood consists of different communities and has demographic differences. For instance, the needs of citizens living in rural areas will differ from those of citizens living in urban areas (Baert, 2015). Therefore, a participatory design approach where people with dementia and their carers are involved must be encouraged.

But caregivers and people living with dementia are not the only persons who are responsible for the road to a dementia-friendly community; this also belongs to the government who can facilitate or hamper. Creating a dementia-friendly society requires policy choices at different levels. In Belgium, this is a challenging task since Belgium is a unitarian state made in a complex three-level structure (Belgium, a federal state, 2021). Fortunately, the development of dementia-friendly communities goes hand in hand with the tendency of socialisation of care and community-based care. The municipalities – the level of government closest to the citizen – have a good insight into the needs of its citizens and have important tools to enhance dementia-friendly communities. Also, the Flemish Government – which is the leading authority for Flanders, the Dutch-speaking part of Belgium – urges the municipalities by launching 'the Dementia Plan 2016–2019' (Vandeurzen, 2016) and the 'Flemish Elderly Policy Plan 2015–2020' (Vandeurzen, 2015) to strive for an age and dementia-friendly approach. Very recently, in spring 2021, the Flemish government announced to invest in 'caring neighbourhoods' in the coming years. The minister of welfare, Beke, stated that caring neighbourhoods should be the future model for the organisation of support, assistance and care in Flanders. This vision is required in order to keep assistance and care accessible, available and affordable for everyone. This model offers opportunities to increase the quality of life and reduce the costs of care and support. It aims for a coherent and neighbourhood-oriented approach to living, care and welfare (Schuermans, 2020). Based on the governmental vision, the local governments may consider it their responsibility to include a dementia-friendly approach in their services, built environment and contacts with citizens. The local government should act as a facilitator and director in the creation, by, among others, raising awareness in the community. Citizens need to understand dementia before they can take dementia-friendly actions. It is important to ensure the level of information is appropriate to the audience. Raising awareness is different to attending an accredited training course. Also, local governments can share good practices or form learning communities.

However, the main challenge for addressing urban design from the perspective of ageing and dementia is for sure the financial implication or better 'to convince governments that this shouldn't be a barrier'. For sure, for architects and urban designers, there is a role to play in combating the effect of the disease and helping people with dementia to live a long life in dignity. A dementia-friendly environment, which allows people to stay at home longer and improves their quality of life, also means that the financial costs are kept manageable. The costs of dementia care are divided into three main sectors: health care (hospitalisation of patients with dementia, which can be attributed to various reasons), social care (costs for services such as nursing homes, home care and respite care) and informal care (family and friends who provide unpaid care to people with dementia) (Prince, 2014). Most of the costs are borne by informal carers. The total cost of illness of dementia disorders in EU 27 countries in 2008 was estimated to be 160 billion euros of which 56% were costs of informal care. The cost per person with dementia in the EU was about 22,000 euros per year (Alzheimer Europe, 2020). Investing in a dementia-friendly community does not have to cost a lot of money. On the contrary, British research shows that by investing in a dementia-friendly community, a local authority in the UK can save up to 13,000 euros per person, allowing a person with dementia to live independently, avoid residential care and maintain a higher quality of life. A study by Mittelman showed that an admission to a care home could be postponed by 557 days if the person with dementia and their carer were sufficiently supported and guided (Mittelman et al., 2006).

We also see a task for education. Young people should come into contact with older people (with dementia) as much as possible. In training courses for architecture, social work, engineering, design, etc., there is a need for a better understanding of the ageing and dementia. Another room for improvement can be seen at the level of research. There is little to no research focusing on this topic in a comprehensive and interdisciplinary way.

15.5 Conclusion

A dementia-friendly community will allow people with dementia to stay at home longer and improve their quality of life. A good living environment where sufficient care and support are available increases the well-being of people with dementia and their carers. It ensures that people with dementia still count and makes maintaining their social identity a matter of course. The ten principles put forward in 2020 by the Alzheimer's Disease International can guide everyone in developing the dementia-friendly environment. However, it is the warmth, involvement and humanity of everyone within the municipality that can realise various dementia-friendly actions. Besides this, if you get an area right for people living with dementia, you get it right the whole community: older adults, people with a disability, families with small children and ultimately everyone.

References

Acusticare. (2021). *Soundscapes en hun benutting*. Retrieved from: https://www.acusticare.be

Alzheimer Europe. (2020). *European Dementia Monitor 2020*. Comparing and benchmarking national dementia strategies and policies.

Alzheimer's Society (2021). *What is a dementia-friendly community?* Retrieved from: https://www.alzheimers.org.uk/get-involved/dementia-friendly-communities/what-dementia-friendly-community#:~:text=A%20dementia%2Dfriendly%20community%20is,in%20the%20community%20they%20choose

Andringa, T., & Lanser, J. (2013). How pleasant sounds promote and annoying sounds impede health: A cognitive approach. *International Journal of Environmental Research and Public Health*, 1439–1461.

Axelsson, Ö., Nilsson, M., & Berglund, B. (2010). A principal components model of soundscape perception. *Journal of the Acoustical Society of America*, 2836–2846.

Baert, V. (2015). *Handleiding dementievriendelijke gemeenten*. Politeia.

Barrett, P., Sharma, M., & Zeisel, J. (2019). Optimal spaces for those living with dementia: Principles and evidence. *Building Research and Information*, 734–746.

Belgium, a Federal State. (2021). Retrieved from: Belgium.be. https://www.belgium.be/en/about_belgium/government/federale_staat

Benoit, M. B. P. (2005). Behavioral and psychological symptoms in dementia: Description and management. *Revue Neurologique (Paris)*, 357–366.

Bowes, A., Dawson, A., Jepson, R., & McCabe, L. (2013). Physical activity for people with dementia: A scoping study. *BMC Geriatrics, 13*, 129.

Brown, B., Rutherford, P., & Crawford, P. (2015). The role of noise in clinical environments with particular reference to mental health care: A narrative review. *International Journal of Nursing Studies*, 1514–1524.

Clarke, P., Weuve, J., Barnes, L., Evans, D., & Mendes de Leon, C. (2015). Cognitive decline and the neighborhood environment. *Annals of Epidemiology*, 849–854.

Cloak, N., & Al Khalili, Y. (2020). *Behavioral and psychological symptoms in dementia*. StatPearls Publishing.
Cycling Without Age. (2021). Retrieved from: https://cyclingwithoutage.org/
de Boer, B., Beerens, H., Katterbach, M., Viduka, M., Willemse, B., & Verbeek, H. (2018). The physical environment of nursing homes for people with dementia: Traditional nursing homes, small-scale living facilities, and green care farms. *Healthcare*, 137–149.
Dementia Services Development Centre. (2012). *The importance of colour and contrast*. Retrieved from: https://dementia.stir.ac.uk/design/virtual-environments/importance-colour-and-contrast
Devos, P., Aletta, F., Thomas, P., Filipan, K., Petrovic, M., Botteldooren, D., ... de Vriendt, P. (2018). Soundscape design for management of behavioral disorders: A pilot study among nursing home residents with dementia.
Devos, P., Aletta, F., Thomas, P., Vander Mynsbrugge, T., Petrovic, M., Van de Velde, D., ... Botteldooren, D. (2020). Application of a prediction model for ambient noise levels and acoustical capacity for living rooms in nursing homes hosting older people with dementia. *Applied Sciences*, 1–13.
Fleming, R., & Bennet, K. (2017). *Applying the key design principles in environments for people with dementia*. Opgehaald van. https://dta.com.au/resources/applying-the-keydesign-principles/; https://www.enablingenvironments.com.au/uploads/5/0/4/5/50459523/dta_resource2_digital.pdf
Fleming, R., Zeidel, J., & Bennet, K. (2020). *World Alzheimer report 2020: Design dignity dementia: Dementia-related design and the built environment*. Alzheimer's Disease International.
Førsund, L., Grov, E., Helvik, A., Juvet, L., Skovdahl, K., & S, E. (2018). The experience of lived space in persons with dementia: A systematic meta-synthesis. *BMC Geriatrics*.
Grey, T., Pierce, M., Chahill, S., & Dyer, M. (2015). *Universal design guidelines. Dementia friendly dwellings for people with dementia, their families and carers*. National Disability Authority of Ireland.
Hendriks, N., & Wilkinson, A. (2020). *The involvement of people with dementia in the design process: A (political) choice to make*. Alzheimer's Disease International.
Huang, N., Chu, C., Kung, S., & Hu, S. (2019). Association of the built environments and health-related quality of life in community-dwelling older adults: A cross-sectional study. *Quality of Life Research*, 2393–2407.
Imamura, T., Hirono, N., Hashimoto, M., Kazui, H., Tanimukai, S., Hanihara, T., ... Mori, E. (2000). Fall-related injuries in dementia with Lewy bodies (DLB) and Alzheimer's disease. *European Journal of Neurology*, 77–79.
Joosse, L. (2012). Do sound levels and space contribute to agitation in nursing home residents with dementia? *Research in Gerontological Nursing*, 174–184.
Jonas, W., & Chez, R. (2004). Toward optimal healing environments in health care. *Journal of Alternative and Complementary Medicine*, S1–S6.
Junfang, X., Yuqian, Z., Chengxuan, Q., & Feng, C. (2017). Global and regional economic costs of dementia: A systematic review. *The Lancet*.
Konis, K., Mack, W., & Schneider, E. (2018). Pilot study to examine the effects of indoor daylight exposure on depression and other neuropsychiatric symptoms in people living with dementia in long-term care communities. *Clinical Interventions in Aging*, 1071–1077.
Kortrijk, S. (2021). *Kwiek beweegroute Buda's Hart*. Retrieved from https://www.kortrijk.be/kwiekroute
Letellier, N., Carrière, I., Gutierrez, L.-A., Gabelle, A., Dartigues, J.-F., Dufouil, C., ... Berr, C. (2019). Influence of activity space on the association between neighborhood characteristics and dementia risk: Results from the 3-City study cohort. *BMC Geriatrics*.
Liperoti, R., Pedone, C., & Corsonello, A. (2008). Antipsychotics for the treatment of behavioral and psychological symptoms of dementia (BPSD). *Current Neuropharmacology*, 117–124.
Marquardt, G., & Schmieg, P. (2009). Dementia-friendly architecture: Environments that facilitate wayfinding in nursing homes. *American Journal of Alzheimer's Disease and Other Dementias*, 333–340.

Marquardt, G., Johnston, D., Black, B., Morrison, A., Rosenblatt, A., & Lyketsos, C. (2011). A descriptive study of home modifications for people with dementia and barriers to implementation. *Journal of Housing for the Elderly*, 258–273.

Mitchell, L., & Burton, E. (2006). *Neighbourhoods for life: Designing dementia-friendly outdoor environments* (pp. 26–33). Quality in Ageing.

Mitchell, L., Burton, E., & Raman, S. (2003). Making the outside world dementia-friendly: Design issues and considerations. *Environment and Planning. B, Planning & Design*, 605–632.

Mittelman, M., Haley, W., Clay, O., & Roth, D. (2006). Improving caregiver well-being delays nursing home placement of patients with Alzheimer disease. *Neurology*, 1592–1599.

National Disability Authority. (2020). *What is universal design?* Retrieved from http://universaldesign.ie/

NHS. (2016). *Impact on carers*. Personal social services survey of adult carers in England. Retrieved from: https://www.dementiastatistics.org/statistics/impact-on-carers/

Nuzum, H., Stickel, A., Corona, M., Zeller, M., Melrose, R. J., & Wilkins, S. S. (2020). Potential benefits of physical activity in MCI and dementia. *Behavioural Neurology*.

Oba, H., Kadoya, Y., Okamoto, H., Matsuoka, T., Abe, Y., Shibata, K., & Narumoto, J. (2021). *The economic burden of dementia: Evidence from a survey of households of people with dementia and their caregivers* (p. 2717). International journal of environmental research and public health.

Prince, M. E. (2014). Dementia UK: Update second edition report

Remington, R. (2002). Calming music and hand massage with agitated elderly. *Nursing Research*, 317–323.

Rowe, M., Houston, A., Molinari, V., Bulat, T., Bowen, M. E., Spring, H., … McKenzie, B. (2015). The concept of missing incidents in persons with dementia. *Healthcare*, 1121–1132.

Schuermans, G. (2020). Zorgzame buurt moet bovenal oog hebben voor haar kwetsbare bewoners. *Sociaal.net, 1*.

Shih-Yin, L. (2017). 'Dementia-friendly communities' and being dementia friendly in healthcare settings. *Current Opinion in Psychiatry*, 145–150.

Siegelaar, A., Boersma, F. Z., & Mobach, M. (2020). *Controlling the stimulation of senses in design for dementia*. Companion Proceedings of the European Facility Management International Conference 2020.

Steyaert, J. (2020). *Prevalentie Dementie*. Retrieved from www.dementie.be: https://www.dementie.be/home/sample-page/prevalentie/

To, W., & Chung, A. (2014). Noise in restaurants: Levels and mathematical model. *Noise & Health*, 368–373.

Ulrich, R., Zimring, C., Barch, X., Dubose, J., Seo, H., & Choi, Y. (2008). A review of the research literature on evidence-based healthcare design. *HERD*, 61–125.

Vandeurzen, J. (2015). *Vlaams ouderenbeleidsplan 2015–2020*. Retrieved from: http://docs.vlaamsparlement.be/pfile?id=1223247

Vandeurzen, J. (2016). Retrieved from Samen verder bouwen aan een dementievriendelijk Vlaanderen: Geactualiseerd Dementieplan Vlaanderen 2016–2019. https://www.dementie.be/wp-content/uploads/2016/07/bis-dementieplan-mededeling-vlareg-def.pdf

Victor, C., Rippon, I., Nelis, S., Martyr, A., Litherland, R., Pickett, J., … Clare, L. (2020). Prevalence and determinants of loneliness in people living with dementia: Findings from the IDEAL programme. *International Journal of Geriatric Psychiatry*, 851–858.

Whear, R., Coon, J., Bethel, A., Abbott, R., Stein, K., & Garside, R. (2014). What is the impact of using outdoor spaces such as gardens on the physical and mental well-being of those with dementia? A systematic review of quantitative and qualitative evidence. *Journal of the American Medical Directors Association*, 697–705.

WHO. (2002). *Active ageing: A policy framework*. World Health Organization.

WHO. (2007). *Global age-friendly cities: A guide*. World Health Organization.

WHO. (2020). *Dementia*. Factsheets. Retrieved from: https://www.who.int/news-room/fact-sheets/detail/dementia. Accessed Sept 2021.

WHO. (2021a). *Age-friendly environments*. Retrieved from www.who.int: https://www.who.int/ageing/age-friendly-environments/en/

WHO. (2021b). *Ageing and life-course*. Retrieved from https://www.who.int/ageing/projects/age-friendly-cities-communities/en/#:~:text=The%20WHO%20age%2Dfriendly%20cities,participation%20and%20employment%2C%20communication%2C%20and

WHO. (2021c). *Towards an age-friendly world*. Retrieved from https://www.who.int/ageing/age-friendly-world/en/

Wilson, J., McGinnis, N., Latkova, P., Tierney, P., & Yoshino, A. (2016). Urban park soundscapes: Association of noise and danger with perceived restoration. *Journal of Park and Recreation Administration*, 16–35.

Woonzorgcentrum. (2016). *Het Heiveld behandelt dementie vanuit 'Active Aging' concept*. Zorgmagazine. 19 October 2016.

Veerle Baert is a social worker (Bach) and a gerontologist (MSc, PhD). She is a former staff member at the Association of Flemish Cities and Municipalities (VVSG) where she supported dementia-friendly and age-friendly municipalities in Flanders (Belgium). Since 2018, she works as a lecturer and researcher at the Artevelde University of Applied Sciences in Ghent (Belgium).

She contributed to several research projects on physical activity in older adults and dementia-friendly and age-friendly communities.

Elise Cornelis is an occupational therapist (Bach) and gerontologist (MSc, PhD). For almost ten years, she has been working in a geriatric day hospital and memory clinic for community-dwelling persons with dementia in a university hospital. Currently, she works as a lecturer and researcher at the Artevelde University of Applied Sciences in Ghent (Belgium). She contributed to several research projects concerning activities of daily living in older persons with neurocognitive disorders, active ageing, shared decision-making in dementia and meaningful activities in nursing homes.

Patricia DeVriendt is an occupational therapist (Bach), gerontologist (MSc, PhD) and professor at the Vrije Universiteit Brussel. She combines the job with a position as senior researcher at the Artevelde University of Applied Sciences and as visiting professor at Ghent University (Belgium) and SOMT university college (the Netherlands). For her work, she received several scientific prices and grants. As expert in biopsychosocial gerontological research with main topics such as active ageing, participation and meaningful activities in old age, frailty and mild cognitive impairment, she published several full articles in international peer-reviewed journals, several national publications and (chapters in) books: she had many presentations in peer-reviewed, internationally established conferences as invited speaker. She can look back to a very broad clinical carrier in which she worked with older persons in a variety of settings in different positions (Mental Health, Nursing Home, Department of Geriatrics of University Hospitals).

Chapter 16
The Covered Market Between Crossroads and Narration of Memories. A Brief Glimpse of an Italian Market and the Imagery That Runs Through It Between Past and Present

Caterina Gallerani

Abstract This chapter draws attention to the threads of memory that bind some elderly people to the idea of the market, using as a reference the covered market in Via Crespi. This market is currently at the centre of a well-known redevelopment project. The chapter analyses the recent situation of it as a place in evolution and change and at the same time tries to revive the memory of those who have spent part of their lives at the market, evoking their memories. The interest on the dimension of the market starts in April 2019, when the author conducted an Erasmus research project in Madrid, for the cultural association PEC (Puesto en Construccion), a collective of architects who deals with dynamics such as gentrification and environmental sustainability directly in the field. In fact, through an open tender, PEC was able to access a space within the San Fernando market, where it could promote cultural initiatives and debates on the subject. For this association, the author conducted, within 4 months, a small ethnographic research to analyse the relationship between the covered market, the neighbourhood and the new dynamics of gentrification and touristification, present in recent years in the surrounding neighbourhood, Lavapiés (Madrid). The covered market in Via Crespi is a complex reality with meeting points with San Fernando market in Lavapiés and differences; but both are located in neighbourhoods with the media spotlight shining on for the redevelopment and gentrification processes. In this chapter, a timid parallelism between the two realities is drawn, while always bearing in mind that they are localised realities and therefore with influences, characteristics and narratives only partially coinciding. The language, as well as the structure of the chapter itself, can only at times be considered academic, because the choice, made consciously, is to allow anyone to be able to read the text fluently, whereas the narrative form, the story, is easier to

C. Gallerani (✉)
Università degli Studi di Milano-Bicocca, Milan, Italy

E. Pozo Menéndez, E. Higueras García (eds.), *Urban Design and Planning for Age-Friendly Environments Across Europe: North and South*, Future City 19,
https://doi.org/10.1007/978-3-030-93875-8_16

enjoy than the form of the essay or scientific article, especially for people quite far away from academic world or young researchers or practitioners.

Keywords Gentrification · Market · Social memories · Narratives

16.1 Introduction

It is from the memory of my father that I want to introduce my analysis, a memory that starts in the neighbourhood and gradually reaches the market:

> The neighbourhood we lived in was very diverse – he told me one evening – there were the very rich and the very poor. In front of our house there was an apartment where you could sometimes catch a glimpse of gala dinners in the evening. There were waiters serving in ceremonial waistcoats, and sumptuous chandeliers whose pendants reflected their glow as far as our window. On Sundays at the market in Via Benedetto Marcello you could see them all, rich and poor alike, going to buy cheese from Giovanni or olives from Checco. On Sundays at the market you could see them all together.

This brief anecdote describes well the district of Milan, which stretches between Loreto, Caiazzo and Lima – three abstract points coinciding with the main stops of the green and red subway – which draw the imaginary boundaries of a district that is still extremely varied. On the one hand, the centre of Senegalese and Nigerian immigration – together with the nearby Porta Venezia district – and at the same time the residential home of important wealthy families who have been rooted in the area for decades.

This anecdote also traces another aspect: the importance of the market as a crossroad of realities that live in the territory. In this article, I will try to examine in depth this aspect, but first, it will be necessary to specify and investigate the Italian context in its specificity. I was able to take a closer look at some of the characteristics of the Milanese markets thanks to a comparison with the Spanish reality, specifically with the covered market of San Fernando in Madrid, where I worked between April and July 2019. The Italian covered market that I focused on when I came back to Milan was the market in Via Crespi, the focal point of the neighbourhood that for the last 4–5 years has been called "NoLo" North of Loreto, an alternative and avant-garde district. Regarding what it means to live and work in NoLo, the role and destiny of the market and those who live in it, I will try to reflect later. For now, I would like to concentrate on the description of the Italian markets so that the reader can use his/her imagination to construct a potentially similar representation to the scenario in question.

Listening to my father's words, a first relevant aspect is that the market my father is referring to is not a covered market. In fact, the quintessential market in Italy is the open-air market, and often when you hear someone say "I have been to the market", they are not referring to the covered market unless it is specified ("I have been to the municipal/covered/regional Giambellino market").

In Italy, particularly in Milan, open-air markets are the weekly events where you can taste specialities from all over Italy, Sicilian oranges, Sardinian pecorino cheese, Bronte pistachios and Calabrian 'nduja, which market vendors offer you to taste while you wait in line. It is the place of smells announcing a familiar flavour and the place where all the Italian dialects meet amidst the shouts and cries of vendors who want to sell their products. A bit like circus workers, open-air market vendors retain a fascination for travelling and discovering valuable products, which they present to their customers with wonder. The market is a place of offers and bargains, where the second-hand clothes stall can hold surprises for those who look carefully.

The covered market, on the other hand, is less dynamic at first glance. First, because it is open all weeks, both during the day and in the evening, and the influx of people is more spread out over various times and days, unlike the 1 day a week when there is an open-air market, usually Saturday or Sunday morning. Second, because in recent decades, it has experienced real economic difficulties being undermined by the opening of large supermarkets and large retailers. Only recently – at least in some markets – something has been changing, as we shall see.

When I told Maurizio – a manager who runs a small wine bar in the covered market in Via Crespi – that I'd been in Spain for 4 months conducting an ethnography on various covered markets, he said to me:

> "Oh, my dear, but you're talking about Spain... there is a real market culture there, not like here!". A few days later Claudio – the butcher in Via Crespi – told me: "I had a Spanish girlfriend! We did a lot of tours of the covered markets and there I got inspiration for a few things that I would like to propose here in Milan. Spain is a leader in market culture".

Actually, they are right. It was in Spain that I realised how the covered market can be considered a total social reality: a place where the economic, political and social aspects inextricably linked to our society are manifested and reproduced.

In Madrid, for example, the San Antonio market is fully part of the political project of the touristification of the city centre. Its objective is to make the market not so much a place of proximity for the neighbourhood inhabitants but a brand for "marketing experiencial", a place where tourists *se sienten inmersos en sus forma de vida local y obtienen una imagen positiva del destino visitado.*[1] Different from San Antonio, the San Fernando market seemed to me to be in a liminal phase – of transition and in a sense therefore fragile – since its multiple identities have led to a lack of identification with both a gourmet market and the traditional one. By virtue of this, the identity of the market is being negotiated by the many social actors who live it: vendors, customers, political collectives and neighbourhood associations or institutions. The liminal phase between the old and the new world – which as Gramsci said is "late in coming" – seems to me to reflect very much the sociopolitical situation in which a large part of Western culture finds itself. The market in a broad sense is therefore a useful tool to talk about ourselves and the historical period we are living in.

---

[1] "They feel immersed in their local way of life and get a positive image of the visited destination".

In this article, I will try to carry out this operation in the humblest way possible, to frame some of the characteristics of the Italian markets starting from the voluntary and involuntary memories of those who frequent them and those who have frequented them, and I will try to carry out a reflection on the future of the Italian markets. The market that for me will be the point of reference for this analysis is the one in Via Crespi, in constant parallelism with the experience of San Fernando market.

16.2 Market of the Past and Market of the Future

The covered market in Via Crespi was created in 1933 following a large increase of population and commercial activities (Figs. 16.1 and 16.2). It is structured in a three-aisle space designed by engineers Massari and Secchi. "The location is very bright thanks to the large window that forms the arched façade and those that open above the roofs of the two side aisles. The large barrel vault in reinforced concrete is reminiscent of the metal roofs typical of railway stations in the 19th century," writes the Fondo per l'Ambiente Italiano (FAI) on its website.

I attended the market both to conduct interviews and to observe the flow of people passing through at different times of the day and evening. During the interviews,

Fig. 16.1 The covered market in Via Crespi. (Source: FAI. Fondo per l'Ambiente Italiano. Retrieved from: https://www.fondoambiente.it/luoghi/mercato-comunale-coperto-di-via-crespi)

Fig. 16.2 Ancient image of the Crespi market. (Source: Massari and Secchi, from FAI, Fondo per l'Ambiente Italiano. Retrieved from: https://www.fondoambiente.it/luoghi/mercato-comunale-coperto-di-via-crespi)

I could see that both in people's conversations and in my questions, there was a kind of comparison between the market that it was and the market that it is today. In order to try to reconstruct the history of the market and to direct the interlocutor more towards a discourse that touches on the memory of the place, I encouraged people to tell me anecdotes, situations and memories of a *market that once was*, indirectly creating a polarisation, and a decidedly marked antithesis between an uncertain present and a consolidated and ideal past:

> "What has changed in the last eight years since you came here?" I asked Maurizio – the manager of the Crespi market wine bar – "The ability to respect each other has changed" he answered me. "There is less solidarity and spirit of collaboration. The market is like a body and if the organs are not well connected you work badly. I organised events for 35 years in various markets in Milan, I was in charge of the technical planning of events and I was always very open and willing to collaborate. Now I see that all this is missing and so I am not as active as before. I do not even go to the vendors' meeting for example".

Claudio – the Crespi market butcher – told me:

> We worked better in the past because there was an interest in growing together... if you were doing well I was doing better too, but now it is like *morstua vita mea* because there is no work, no money and you count your neighbour's customers. People look at each other more out of boredom and distrust than to learn something.

Similarly, in the San Fernando market – where I interviewed more customers than vendors – Mr. Fernando, sitting on the bench in the square opposite to the market, told me:

> This market is named like me and it was a glory then. Now look at the state of it, I cannot go in there anymore, it makes me sad.

Two elderly gentlemen walking through the long neon-lit corridors of San Fernando told me similar things:

> El mercado me da pena, ¡mucha pena! Hace un tiempo era todo, era maravilloso y ¡ahora no es nada! ¡Mira ahora cómo está triste! Y es así para eso, porque ninguno se pregunta nada… seguimos sin hacernos preguntas sobre lo que pasa.[2]

But not everyone feels this way; the same people mentioned who expressed their sadness about this to me actually made me realise that they have strong future prospects. Claudio, for example, is planning to transform his standby introducing a new type of meat, typical of the US tradition, such as the Tomahawk steak, tasted at the counter.

Also, the brothers who run the Monza Green grocery, who have been at the market since 1975, told me that business has never been as good as it is now. When I asked them if they have had any crisis with COVID-19, they said no. On the contrary, people have rediscovered the pleasure of going there. People have rediscovered the pleasure of going to the market at the expense of the supermarket, to queue less and to be safer avoiding the risks of an enclosed and crowded place and gaining in quality.

So how are things going? How has the market changed in recent years?

To answer these questions, we need to leave the market for a moment – but we will return to it soon – and observe the surrounding neighbourhoods. The Crespi and the San Fernando markets share a common factor that makes them closer than they seem. Both are located in two neighbourhoods that are at the centre of a real redevelopment project.

In the case of Lavapiés – the neighbourhood where the San Fernando market is located in Madrid – the redevelopment has already been underway for a long time being a rather central neighbourhood as well, and the consequences, even negative ones, have already been the subject of debate for several years, also in the newspapers. The redevelopment of Lavapiés has led to a real gentrification process contested by several neighbourhood committees. On February 1, 2020, *El Itagnol*, an Italian online newspaper entirely dedicated to Spanish news, published an emblematic article about the gentrification of Lavapiés:

> MADRID. The transformation of the centre of Madrid, with the gentrification of neighbourhoods once considered 'problematic' or 'vulnerable' also affects symbolic places. In recent days Madrid newspapers have reported the news of the closure of the Senegalese restaurant Baobab, in the Lavapiés district. "Gentrification closes Baobab, the most iconic

[2] "The market makes me sad, very sad! Once it was everything, it was wonderful and now it is nothing! Now look how sad it is! And that's why it is like that, because no one asks any questions… we still do not ask ourselves questions about what is going on."

> Senegalese restaurant in the neighbourhood", headlines eldiario.es, which goes on to explain how the old neighbourhood in the centre of Madrid has been transformed into a "trendy neighbourhood" with the boom of Airbnb apts and room rentals for tourists, but also as a trendy neighbourhood for cultural activity and multiculturalism.
>
> Gentrification has led to an increase in the value of real estate, resulting in a rapid transformation of the neighbourhood. Businesses such as Baobab are being forced to close because of rising rents, linked to the revaluation of property. It is likely that a hotel will be built in the same building where the Senegalese restaurant was located.

Lavapiés is therefore a territory of gentrification and touristisation, phenomena that have been hitting Madrid hard in recent years. San Fernando market is at the centre of the storm of this process – between urban planning and social studies. The Via Crespi market, on the other hand, is at the centre of a neighbourhood that has only been undergoing a process of redevelopment for a few years. The neighbourhood has been renamed NoLo (North of Loreto), but before being so called, it had no name and was simply known as Via Padova and surroundings or Viale Monza and surroundings. Today, if we type "NoLo" into Google Maps, we have a map of a well-defined area recently invented (see Fig. 16.3).

In any case, the terminology that we find in *El Itagnol*'s article – rising rents, high cost of living, multiculturalism and also fight against neglect – is recurrent in the articles dedicated to NoLo as well. I tried to trace some of the keywords found in the articles on Lavapies to NoLo as well by simply Googling them.

Fig. 16.3 NoLo area registered in Google Maps. (Source: Own elaboration)

On the website of the *Il Milanese imbruttito*, a social channel popular among the very young, we can read:

> And then it was the turn of NoLo. Today NoLo is cool, because from a simple and completely anonymous suburb the area beyond Loreto has become more sparkling, more alive. Many bars and shops have opened, there are wonderful murals and the atmosphere among the people is very engaging.

There is also an official website, run by Radio NoLo, where you can read about the neighbourhood's most exciting events to attend, such as the NoLo Fringe Festival:

> NoLo Fringe Festival is about to start. Until Sunday 13 September our neighbourhood is tinged with magenta, the colour of the Fringe, and Radio NoLo will follow the festival through the voices of the protagonists. Interviews and curiosities will be the ingredients of these daily podcasts that will accompany you until Monday 14th with the voice of the winner of this second edition.

The Artists' Garden, the design Salumeria and the BienNolo – the NoLo biennial – are all parts of what *magazine.it* calls "the Brooklyn of Milanese artists". "From working-class neighbourhood to creative district" – headline of another tabloid – gives us this picture of the neighbourhood:

> The Egyptian baker, the Middle Eastern confectioner, the Chinese tailor and the kebab seller are being joined by new brands. There are those who, taking advantage of the low rents, have opened their own art gallery or showroom here, others who have transformed traditional shops into more modern and trendy "concept stores", where selling "sensory experiences" instead of clothes (Spazio Nolo 43) or flowers and bicycles in pairs, making you feel like a pioneer of a more measured and sustainable lifestyle (Bici e radici).

These are potentially captivating places for those who want to spend a weekend in Milan enjoying the city and savouring the bohemian spirit that the artist quarter exudes.

Davide Fassi is a person who knows NoLo very well. He is a lecturer at the Design Department of the Milan Polytechnic, and since 2016, he chose to develop projects with his students in this neighbourhood. He chose this area because of the need to bring students into contact with a dynamic urban reality that is "proactive" – as he defines it – "fertile and willing to interact, to do, to change" even before NoLo became famous. When they started working in the neighbourhood, the NoLo Social District had already been set up, a group of people from the neighbourhood and elsewhere whose aim is to redevelop the area through social networks. So Davide Fassi and his students immediately set about organising a "project community" with the social district, a design workshop for the neighbourhood mediated by the tool of design. They set up art and design exhibitions with various artists from the area, including at the covered market in Via Crespi, and quickly came into contact with this reality, on which they have focused their efforts since 2018 with more continuity:

> We already had a network in the area that we could reactivate, so we had the students work in the market. Why the market? Because the first year we worked on the neighbourhood, we did a sketch mob, a flash mob in which you sketch the space by moving from place to place. At the end of this flash mob, we set up the exhibition here with the sketches and we had a brainwave because in the end you are in a place with a lot of potential that is not being used.

Fassi and his students began to organise workshops inside the market, with an unexpected media resonance. Back in 2016, one of the first articles published about NoLo of the daily newspaper *la Repubblica* began like this:

> They are still spores, but they are already turning into sprouts. In the place of a junk shop, a mechanic's, a poultry shop and a body shop, new signs are opening. Between kebabs and Chinese clothes, they are opening vintage shops, coworking desks, art galleries, white walls and DJ sets, shop windows on Cinelli bicycles and vegan ice creams. It is happening now in Milan. In Via Padova. Yes, "that" via Padova: the "street of drug dealing", "of retired prostitutes", "of gangs", "of transsexuals", "the Bronx", according to recent newspaper articles. The district where the residents between 25 and 34 are more foreigners than Italians (see box at the bottom of the page), where a grocery shop was looking for "a Filipino or Sri Lankan salesman" and for this reason the Lega (right wing party) has spoken of "reverse racism". A third immigration has now arrived in this street. After workers and immigrants, young creative people are making their way in.

And they are indeed young creative people. Professors, researchers and university assignees like Davide can work in a renovated space thanks to funding from the Polytechnic University, a free allocated space by the municipality as an association of cultural interest. "How has the market changed with your projects?" I asked:

> It is no longer just a market of shops, it is also a market that has to bring sociality. If you want to be competitive in an urban context where you have a thousand shopping centres, home delivery, you can access consumption at any time and in any form, you actually ask yourself: why do these places still exist? neighbourhood shops are disappearing and it is strange that there are still places like markets, when with Amazon Prime you can get fresh food in two hours. How can we combat this? By maintaining customer loyalty, which does not only come from selling a product but also from having an experience. For example, Beppe advises you on how to cook meat, he recommends which piece to get; Hassan makes your roast at the moment, tying it in front of your eyes; the greengrocer prepares the right vegetables for your favourite seasonal soup. There is an extra service that you do not get in supermarkets.

The aims are different: to stimulate the recomposition of a participatory community that orbits around the market, upgrading it and making it an attractive place for people from other parts of the city as well; to support a process that leads vendors to set themselves up as a single entity, as happened for the Lagosta or Giambellino markets; and to encourage the presence of creative workshops, as the anthropological workshop Dynamoscopio has done – much cited by Fassi as a model to be inspired by and located in the Giambellino market – that makes the market a cultural centre and a place for aggregation of an intellectual nature as well:

> There's the LaNoLo group, for example, which tells you that there are 15 ladies knitting and looking for a place [...] or the reading group... they're also looking for a place; the aim is to try to be a bit flexible in the occupation of the public space to become a place of aggregation; and then there's the idea of bringing in a radio station – RadioNoLo – which will animate the market from 6 p.m. and develop the values of a social promotion association within the market; so another function is also created: they are not here to sell anything but to bring content.

I like this professor's enthusiasm and the confidence with which he looks at the project, so, after the interview, I left the market with a pleasant feeling. However, something does not sit right with me. Not so much for the speech in general, but for the feeling that a process like this, in addition to the positive aspects described, also conceals less visible dynamics, whose impact on the neighbourhood could also have other implications. When I got home, I checked out NoLo's Instagram page and found a photo of a graffiti that reads "NoLo or: the stupidity of the poor who, pretending to be rich, have raised their own prices". In the caption, the comment from NoLo Social District is certainly not flattering towards this graffiti. Among all the comments praising NoLo, I find one that strikes me:

> When this creative community becomes annoyed staying in Nolo and living in 120 sqm studios at 600 € a month and migrates to more popular areas of Milan, what will be left? Nothing, because the community you are talking about lives in super cool clubs and in few design studios. But for the community and the social composition nothing has changed at all. There are the same problems as before. You talk about Nolo but in reality, you are referring to the area from Via Venini to Piazza Morbegno and Via PopoliUniti. What has changed in recent years in Via Crespi, Via Pasteur, Via Arquà, Via Termopili, Via Clitunno or Via Padova? For me, having grown up between Viale Padova and Viale Monza, I can assure you that the only change there has been is in the price of houses.

Only towards the end of the interview with Davide Fassi, I was able to talk about gentrification, our time was running out and he had another interview to do. He told me that yes, gentrification is a problem, and we should think more about it, but on the other hand, the world is changing, and the urban dynamics that always affect neighbourhoods are in a constant state of dynamism, almost uncontrollably.

To get another point of view, I got in touch with Gaia Sartori Pallotta, an active member of a collective in NoLo that implements strategies of resistance against gentrification.

Gaia told me that her interest in gentrification dynamics stems from her university studies in sociology and a trip to the United States for her final thesis. The study of urban sociology, specifically the one of the Chicago school theory,[3] helped her to develop the analytical tools to design a project in New York City focused on the gentrification process due to the expansion plan of the Columbia University[4] in West Harlem, a historically African-American and Latino neighbourhood. Through interviews with Columbia professors and with the historical habitants of the neighbourhood, she came into contact with an anti-gentrification student group and some Columbia professors such as the famous sociologists Saskia Sassen and David Harvey, who strongly criticised the expansion plan.

Once back in Milan, Gaia continued her research on a personal level and began to get involved in the right to housing group in the north-eastern Milan neighbourhood of Lambrate. It was at that time that the Fuorisalone[5] event was being held in Lambrate and real estate agencies and landlords were starting to favour short-term

[3] First school of urban sociology in the United States founded in the early twentieth century.

[4] One of the eight most prestigious and elite private universities in the United States.

[5] Annual event in Milan as part of the Milan Design Week, together with the Salone del Mobile itis the world's most important event in the industrial design and furniture sector.

rentals on platforms such as Airbnb, so they stopped renewing long-term contracts with tenants who were being evicted. Gaia told me:

> The anti-eviction pickets allowed us to get closer to everything that was the housing issue in Milan on a legislative level and we immediately realised that housing policies were not going to protect the tenant who was about to be evicted. On the contrary the housing policies were established with the aim of activating the expulsion process which on the one hand favours the landlords' interests and on the other they have the foresight to change a historically popular neighbourhood to allocate space for other activities. From that moment on I began to find strong similarities with the dynamics observed in the United States, and I better understood what had already happened in Isola, where even the associations and clubs of the area were dismembered to allocate space for luxury buildings and small shops at prohibitive prices. In 2015 I started hearing about NoLo and this really struck me because I felt catapulted back to NYC and re-experienced the process of renaming pre-existing neighbourhoods like Soho, Noho etc which are now high fashion districts. Naming something or someone is not a banal thing. When I think of the act of naming I immediately think of the birth of a child, which is given a name as the act that begins life. Re-naming is therefore a deliberate giving of a new identity to something, marginalising if not cancelling its past history. Together with the others, we immediately asked ourselves how the decision to give it this name came about, whether through a petition, a collection of signatures or something else because it seemed to us to be more of an attempt at "branding" the area in order to attract new people who might be fascinated by a new urban identity.

Therefore, Gaia and others of the group began to investigate this new dynamic noting various urban changes such as the creation of fences around small parks and the increase in cameras and several small shops, bars and restaurants with higher prices. "The streets have been altered to allow cars to drive in front of the new shops, square shave been rebuilt, the facades of empty buildings have been repainted without people being placed inside, festivals and events have been organised in new spaces while the historical locations of neighbourhood associations have been ignored. And the promoters of the events were people from outside the neighbourhood".

Gaia and others began to propose initiatives, questionnaires to the inhabitants, screenings and open discussions in order to engage with the neighbourhood's historical residents and try to oppose the ongoing gentrification process. At this point, the group decided to use the pseudonym Martin Loreto, referring to the religious revolution made by Luther in 1500 with the idea of bringing a strong criticism to the direction in which the neighbourhood was going. They also appealed to some graffiti writers asking them not to participate in a redevelopment project proposed by the municipality for a bridge defined as "damaged".

> "In some articles it had become one of the most dangerous places in the city where rapes, assaults, muggings took place... What was described in this article was far beyond reality" she told me.

The group is still active today, and, according to Gaia, with COVID-19, there are also new challenges added to this dynamic that they are ready to face.

When I asked her about the market, she told me that in her opinion, it is a place that has been invested by gentrification, precisely because of its strategic position halfway between the Morbegno area and the more popular area (Viale Monza/Via Padova) just as the local markets have been in other areas of the city such as Isola,

Giambellino and C.so XXII Marzo. She does not like the fact that a public space like the market has at times become a private place where, for events such as the Fuorisalone or Fashion Week, access was not guaranteed to everyone since one could only enter with a specific pass.

When I asked her "Can there be a positive gentrification?" she said:

> In my experience there is no good gentrification. Inhabitants and residents have to be aware that they will not be able to enjoy what they are being told about, an idyllic life in a more beautiful, more sustainable, cleaner and safer neighbourhood as a result of the gentrification process. Moreover, talking about "redevelopment" means that what you have has no quality and need to be re-developed from the outside, from someone who can do better than you! Talking about "degradation" means that the place you are in is horrible. Words carry a lot of weight, and these words get through, let us say, into people.

Gentrification is an English little used term, and if we wanted to translate it, it objectively indicates a process of substitution of one population (poorer) for another (wealthier) and also implies the heterodirectionality of how people live in a given place: everything is decided and constructed from above in order to be able to better control, modify and refine it, while those who are invested in it are not at all considered and involved. Moreover, once this process has begun, it is difficult to interrupt it. While gentrification only responds to the interests of those who promote it, if we are talking about redevelopment, we need to understand why, what its purpose is and for whom this better living is created. This makes all the difference.

If we well look at it, this renovation has not seen the opening of libraries, toy libraries or spaces for socialising outside consumption, nor has it proposed green areas or places for free sport. On the contrary, services for the neighbourhood such as the local health centre in Via Don Orione and the one in Via Padova are rumoured to be closing. The same fate awaits the registry office in Via Ricordi. In Piazza Costantino, there is a public building where since years are located different historic associations such as the ANPI club, the Via Padova band, Legambiente, etc., and for 3 years, that building has been put out to tender to be used for other purposes. There are segments of the city neighbourhood earmarked for short-term rentals and to become hostels with 25 euro dormitory beds, as well as other areas of the same neighbourhood earmarked for specific activities.

In my opinion, living well and feeling good means to be able to choose one's own dreams, desires, relationships and needs and to move around the space freely and independently of people's economic situation.

For me, gentrification has not improved the life of the people who historically live in the neighbourhood. It certainly benefits the real estate agencies, the big investors, some merchants and maybe some landlords but certainly not the people who really need to improve their lives. Moving out of the neighbourhood because it has become inaccessible for many people means losing the social network that surrounds and supports them, changing schools or moving away from their workplace and starting a life all over again, not because they chose to but because those who have no purchasing power cannot choose anything.

Sometime after these interviews, I found myself talking about gentrification with my class, an eighth grade class where I teach geography. In their schoolbook, we read this sentence by Spike Lee:

> They arrive with their fancy dogs and think they own the city, with no respect for its history and traditions. When I was born there, they never collected the rubbish, you never saw a police car, there was no decent school. Today you see mothers pushing strollers at three o'clock in the morning...

After finishing reading aloud, Francesca asked me "Teacher, I do not understand, why do not people want their neighbourhood to become nicer?" I answered her that people would like to live in a nicer neighbourhood but those same people, who would be entitled to enjoy the announced improvements, very often find themselves forced to change neighbourhoods because of rising rents. In any case, that question – which is anything but stupid – immediately led to the next point: what other ways are there of living in a neighbourhood, and what parameters should be respected in order not to exclude anyone but to ensure that the market remains a crossroad of different histories, cultures and social classes that we mentioned at the beginning? Is there a way to avoid that every redevelopment hides processes of gentrification and therefore selects an elite in being able to afford access to certain places?

16.3 The Ghost of Oral History

When I worked at the San Fernando market, I noticed that the redevelopment of the market brought in new people and, at the same time, drove out others. This process of moving in and out is part of the normal dynamism of a place that is as static as it is flexible and mouldable according to the social and economic needs of the moment. Therefore, the brand "market" becomes a reminder of a chess game between the concept of innovation and tradition adapting in people's minds, discourses and reflections according to what they want to highlight in their arguments. There are those orbiting the market who regret its glorious past, but at the same time, they see innovation as the only way to rise like phoenixes. Others, on the other hand, see innovation as a problem of identity, not only for the market itself but for themselves, who are no longer able to reflect themselves and find their place in such a fast-changing world.

The game between innovation and tradition is therefore fought in the re-elaboration of memories and their selection according to the relationship that the subject weaves over time with the historical places of his life. Maurice Halbwachs, in his book the *Collective Memory*, effectively describes the nature of this relationship:

> If, between the houses, the streets and the groups of its inhabitants, there were only to be an accidental relationship and one of short duration, men could destroy their home, their neighbourhood, and their city and rebuild another, in the same place, following a different

> design; but should the stones let themselves be transported, it is not equally easy to modify the relations that are established between the stones and men. When a human group lives for a long time in a place adapted to its habits, not only its movements but even its thoughts are regulated by the succession of material images that external objects represent to it.

It was easy, for example, for Elvira – an anthropologist from Madrid who does ethnographic research on covered markets – to associate the idea of the market with the memory of her grandmother who used to say "voy a la plaza"[6] to go there. It was just as easy for Pablo – the architect of Pec (Puesto en Construccion), an architecture collective based in San Fernando – to remember how, as a child, he used to run back and forth like a spinning top through all the aisles of the market in a small town in Gibraltar where he was born and raised, believing he was in a magical place that had nothing to do with the banality of the world.

One day in San Fernando, I met two elderly men who had entered the market to see what it had become. With them, unlike some of the users who only went in to buy, it was very easy to bring back the memory of the place: they told me that near the market, there was a *descampado*[7] and all the parties in the neighbourhood were held there. There was dancing in there, and the other man, a little older, told me "the most beautiful parties I have ever seen! They do not happen anymore…". They also told me that behind the market, near where there is now a bookshop, there was a patio where they used to organise the zarzuela, a Spanish opera drama genre in which spoken scenes alternate with singing and dancing. I'm not sure if I got it right, but they told me that there was a *corala*[8] where they used to have the greatest zarzuela actors in Madrid perform. The actors played on the balconies, and the two old men seemed to remember them all: Nati Mistral, Marujita Diaz and Concha Velasco are just some of the names I managed to record in my haste to capture their every word. In Calle Caravaca, two blocks from the market, there were several places where they served *gallinecas*[9] and a tavern where they served *chupitos de madroño*.[10] In front of this tavern, there was a magnificent mosaic depicting the scene of the bear and the *madroño*. The mosaic no longer exists because it has been covered with a layer of plaster by the businessman who bought the place, but one of the two old men confessed to me that he once scratched away some of the plasters, and now a piece of the mosaic can be seen.

They also told me about a *gypsy*, a "traveller" guy, who grew up in the barrio and has become famous in Italy. I immediately begun by saying "Tonino Carotone!" whom I had met a few nights earlier at the Bodegas Lo Maximo, a popular club he owns where he performs every Wednesday, but they referred to the far more regal Joaquín Cortés.

[6] I'm going to the square.

[7] Empty land without buildings.

[8] A typical Madrid building.

[9] A typical Madrid food.

[10] Tree that is in the coat of arms of Madrid.

They told me other anecdotes that oscillate between the history of the neighbourhood and that of the market, and finally they vented to me saying that the market is suffering the same fate as the neighbourhood for them; it is being destroyed. They told me that they wanted to sell their houses because they are no longer interested in living there, where the time when people used to put their chairs outside in the evenings and take a cool drink together, talking to their neighbours, is now denied. Describing the architectural, political and socio-economic dynamics that influence their current life, it produces a comparison of places and life in the past, giving rise to a process of reconstruction that extends from a central place such as the market to an entire neighbourhood. Together with the image of San Fernando, in fact, they have managed to draw – also in my mind – the contours of an entire neighbourhood that has now completely changed. It is not a monologue but a multi-voiced, cooperative tale, in which the gaps of one are furrowed by the memory of the other and by my own zealous questioning. The final outburst highlights well what the anthropologist Alessandro Portelli says in his text Oral Stories: "the past and the present were so intertwined in the dialogue that talking about the events and problems of the past was clearly a way of evoking those of the present".

Portelli emphasises that remembering is a political as well as a cultural act, and I use this consideration to reflect on how easy it was for me to collect stories and memories in a territory like Spain which – as Maurizio reminded me – has a very strong culture in terms of covered markets. In addition to the cultural factor, however, there are also other reasons of a strictly methodological nature linked to the structuring of the research. In Madrid, I carried out a real ethnographic research, as an Erasmus student, spending everyday for 3 months inside the market. In Via Crespi, on the other hand, I concentrated more on targeted, occasional and punctual interviews and less on observing the dynamics of the place, an observation which therefore tended to be partial and which makes this article more of a narrative text than a scientific article. For this series of reasons, I was more easily able to trace the coordinates of the history of San Fernando, while I struggled and had less opportunity to collect and rework the oral history of the Via Crespi market. The COVID-19 situation furthermore contributed to limit and circumscribe my research although, as I will explain in the next chapter, it gave me the opportunity to get closer to people I might never have met otherwise.

16.4 Between Memories and Silences

In early March 2020, several cases of COVID-19 occurred in Lombardy, the region where Milan is the provincial county seat. After Wuhan, Lombardy was the first region in the world to be severely affected by the pandemic and became quickly a hotspot and testing ground for geopolitical intervention strategies. It was a very hard time for the population because while the TV was broadcasting messages from the VIPs about how nice it was to stay at home and enjoy the family environment during the lockdown, a part of the population was living more and more isolated and/or in a home environment that had a number of complexities that did not make it at all easy to share the home environment with other family members. I had already come back from Spain in September and was working in tourism, so I was one of the first to stay home because my category could not work. So I decided to do volunteer work in the *Milano Aiuta* project with the *Solidarity Brigades*, a project carried out in collaboration with social centres, associations such as Emergency and the Municipality of Milan which guaranteed the mobility of volunteers in the area for aid reasons. Although I did not like some of the structural dynamics of this project and found it to have profound limitations, it was also one of the few developed responses in the total uncertainty atmosphere of that period, an uncertainty engulfed by the very unpreparedness of the institutions. In any case, through the project, I had the opportunity to know several elderly people in the neighbourhood who relied on the volunteers to bring their grocery shopping home. I live in Precotto, a neighbourhood bordering NoLo, just two subway stops away from Turro, a border stop you can easily find on the map in Fig. 16.3. I, therefore, got in touch with some elderly people living in NoLo, but I have made deeper connections with the elderly people living closer to my home, so that I could be an on-call support and physically close to them for any emergency. The elderly people in question were all in their 70s or 60s, lucid people, far from dementia or any form of brain fatigue. During the period of the pandemic, I did not try to conduct any kind of structured interview, but I usually remained outside their door to talk for a few minutes and discuss with them about the national health situation and their emotional state. In particular, there was Mrs. Cirruzzi (fictional name) who was very distressed by the situation and often called me for some comfort. One day when she was particularly intolerant of the situation, in addition to my support, I gave her the number of a switchboard that deals with mental health, created precisely to calm down the anxieties of coronavirus and lockdown.

Then the day of the general reopening – May 4 – came. People were able to start moving again, and by summer, in the sweltering heat of June and July, the atmosphere was more relaxed. The ghost of the coronavirus was always present, but on some occasions, it was put to one side, like the anxieties and discomfort experienced at that time. I had already scheduled the three interviews with the three elderly people with whom I had the closest connection and was looking forward to hearing what they would tell me about the markets – specifically about the covered market in Gorla, the one closest to their homes which, unfortunately, has been closed for years. I had already interviewed the owner of the greengrocer's shop in Via Crespi, who told me that the elderly were the great absentees from the market after the pandemic: "The elderly have hardly been seen at the market in recent years. This is a market frequented more by families, middle-aged people and recently also by young people". This statement further convinced me that I would find romantic and full of emotion stories as they were driven by a certain nostalgia of not being able – for different reasons – to access a beloved place anymore. The fact that the Gorla market had been closed for years, I thought it was a cause for disdain for them. The surprise, however, was to find a certain underlying apathy and disinterest about that. I had announced to everyone the topic of my interview – the market and the memory

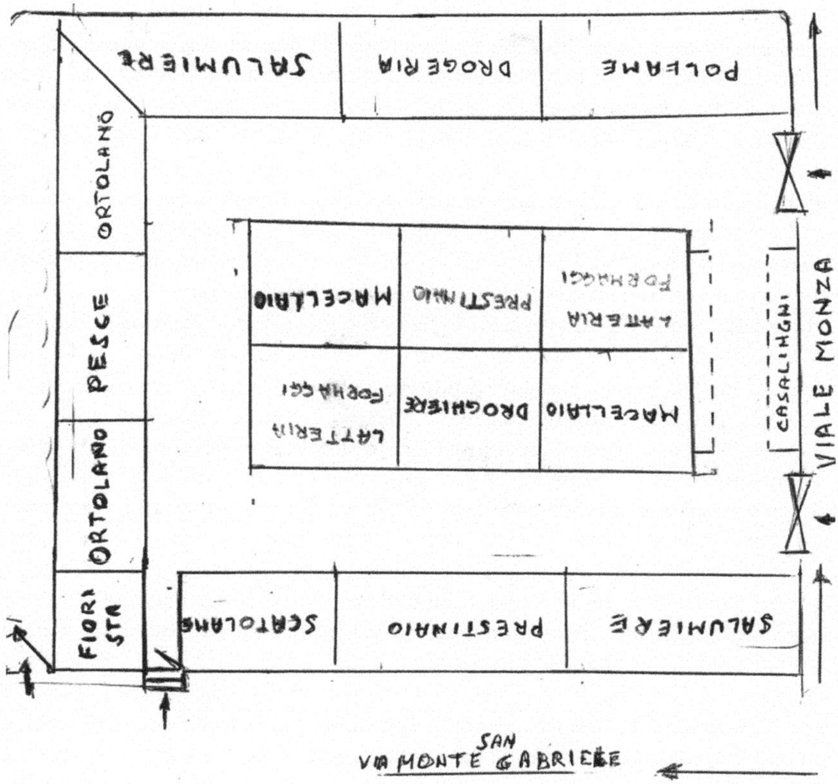

Fig. 16.4 Scheme of the market in Gorla, Viale Monza. (Source: Sketch by Mr. Moroni, 2019)

of places – and Mr. Moroni had grasped the spirit of the thing perfectly, bringing me a map of the Gorla market drawn by him, on which he had meticulously arranged every stand he remembered (Fig. 16.4).

I was very moved by this, but when I asked him what the market represented for him and whether he remembered any anecdotes to share, Mr. Moroni shrugged and told me that he did not remember anything exceptional in particular but a certain cordiality that contradicted the relationship with the vendors, a relationship based on complete trust in the quality of the product. Mr. Moroni was a great historical memory of the neighbourhood, giving me detailed information about the types of houses that were built and who lived in them. He also told me a lot about himself and about his career in the technology sector back in the day and told me that today, at the age of 80, he is proudly an apartment block administrator, pointing out to me the building wall decorations he designed. He told me he has always lived in this area and his wife never wanted to move away. I also found out that his wife is one of the survivors of the massacre of the Gorla martyrs – a massacre that took place in a school as a result of an American bombing during the Second World War – and I asked her a few questions about that focusing the interview on the emotions that were pervading me and that I felt were even stronger in her.

Mrs. Cirruzzi, the second interviewee, also welcomed me with great enthusiasm, happy to finally be able to offer me an ice cream on her balcony, since during the most agitated moments of this pandemic, I preferred not to step inside her house. She also told me a lot about herself, about her career in a large telephone company and about the painting course that changed her life defining herself today mainly as an all-round painter. She proudly showed me the house she has lived in all her life and all her paintings. However, she remembered little of the market except that she liked to go shopping for clothes but in open-air markets. I then tried to lead the interview towards the COVID-19 situation and how this virus has altered her habits and the consequences on her well-being. I asked her these questions because during the whole pandemic, Mrs. Cirruzzi was one of the most fragile persons I dealt with. She often called me asking when this situation would end and told me that she was suffering from anxiety and anguish; during that period, her sister died (not for COVID-19), and her suffering reached rather high levels, understandably. All I could do was listen to her and recommend a few numbers of psychological support. My astonishment was great when Mrs. Cirruzzi told me that after all, although the lockdown was a heavy moment, she felt she had faced it with tranquillity and serenity and that it was not such a terrible period as she had experienced in other time when she felt much more alone. I then continued with a few more questions on the same topic, and she reiterated that I should not worry about her. Since she has become elderly, she has always managed to get by herself as if in some way my questions were challenging her autonomy.

Mrs. Grechi, the last elderly person I interviewed in the area, smiled at me while we are drinking a coffee together and said that all these questions were a bit difficult for her and that she thought I should have asked young people about the market. She has not been there for quite a while, so her opinion on certain things did not count, given that she was old and she stopped walking out because of a leg illness that forces her to rest as much as possible.

Out of the three elderly people who have little inclination to talk about the covered market, there must be a hundred who are prepared to recount in detail how their own history has been intertwined with that public place, without mentioning that my inexperience as an interviewer may not have encouraged the opening of a deeper dialogue. I had thought about continuing to look for other people and keep conducting interviews until I finally found what I wanted to hear. But from a certain point of view, it was not right. Moreover, those talks with their uncertainties, digressions and denials highlighted to me extremely interesting and complex aspects of the reality I was trying to tell in an idyllic way. I thought about some common aspects of the three interviews: these three people, with their advancing age, hardly found interlocutors interested in their stories, so what they were interested in was to focus the story on aspects that were much more important for them than the market. It was evident that although all three had a desire to tell their stories, they also wanted to reiterate how their opinions were irrevocably linked to their seniority which in some ways diminished their importance. Their considerations turned to public space as a space that was no longer accessible to them, so the only safe space in which they could fully recognise themselves remained that of their own home, the focal point of their condition of old age and the memory of their life path, which – in old age – they were ready to rework coherently.

Maurizio's statement that in Italy – compared to Spain – there is "a culture of markets but in a much lesser way" could explain why these people hardly managed to collect memories around this place. However, the interviewees did not limit themselves to the simple observation that they had no specific memories of the market but linked this narrative incapacity not so much to the lack of relevance for them of the place itself but rather to their partial – and sometimes total – ontological incapacity to be in the world due to their new condition of seniority. The narrative inadequacy on the topic – which was totally absent on other subjects since they were able to fluidly and densely recount the most picturesque aspects of their lives and of certain places in the neighbourhood – was perceived as a natural consequence of the fact that, by now old, they had become inadequate. And not only passing through that public place but also in facing a discourse and a journey through their memories.

I have reflected on this common dynamic among the interviewees, and I believe that this link – between distancing oneself from public space and ontological inadequacy in reworking one's relationship with the world – is only partly attributable to an ageing body which restricts movement and isolates. Instead, I believe that inadequacy – transported in the context of the reconstruction of memories – is more strongly structured on a sense of fear. The fear of finding themselves living in a much-changed metropolis, enriched and impoverished by dynamics they often only know through the media and on a public discourse – meant as a collective metadiscourse – indirectly narrates the exploits of their constant exclusion from community life.

A collective discourse works as a device and as a network of discourses, actions, media information, practices and philosophical propositions which communicate – covertly or not – how the elderly, insofar as they are outside the relations of production, are irremediably placed outside the social community itself and the public space. The elderly are excluded both physically, with their forced placement in retirement homes for example, and intellectually, as they are not considered capable of keeping up with the speed of modernity. This discourse has become more radical in the period of the pandemic leading to the production of propositions, or rather tweets, such as that of the governor of Liguria, Giovanni Toti, who tweeted:

> As much as we are saddened by every single victim of COVID-19, we must take into account this fact: yesterday alone, among the 25 deaths in Liguria 22 were very elderly patients. People who are mostly retired, not essential to the productive effort of the country, but who must be protected.

The controversy which followed after Toti's tweet indicates that our "norm image" of the elderly is not only of this kind but is studded with varied and positive social representations of the inclusiveness and importance of these subjects in society. However, the coronavirus crisis is as if it had revealed moral and value propositions that were already being proposed in times of so-called normality. So, I think back to a university lesson that I took for a course in the Science of Education, in which we discussed about the elderly and their role. The dialogue I reported was between my professor and a student who worked as an educator in a retirement home:

Teacher:	"In the retirement home where you worked, were there any proposals, any speeches, concerning the role of the elderly in society?"
Student:	"No, not really"
Teacher:	"In your opinion, is this absence purely coincidental?"
Student:	"Well no, because they were no longer, effectively, in society so why delude them with an activity or a speech that dealt with this topic?"

During the same lesson, we discussed other discourses and ways of seeing the elderly, which are typical of our common imagination:

Student:	"Teacher, this happens because sometimes the elderly are not cooperative and we, as educators, have to take them by the hand, as we do with children, and make them understand what they have to do for their own good".
Teacher:	"What you have just said is very interesting for our discussion, because your statements are philosophical propositions which frame the elderly as children, as subjects who in our society notoriously need figures to educate and control them. Like the child, the elderly person is considered a figure to be guided and his relationship with the world is mediated by an operator, a nurse, an educator. The relationship with the world is not direct but mediated. The elderly person no longer has the power to have a relationship with the world".

I would like to reiterate that this kind of meta-discourse on the elderly is certainly not the only one being proposed. My aim is certainly not to simplify or propose a deterministic-mechanistic reflection. I think it is important to specify that this narrative is also part of other more varied public discourses. For example, the discourse on the valorisation of the elderly role as witnesses of Italian history is a much appreciated narrative addressed to the veterans of the resistance, especially as the date of April 25 approaches. However, in Italy, there is also a day dedicated to the *loneliness of the elderly* – November 15 – with the aim of raising public awareness of this issue. In my opinion, this choice is part of a more complex discourse that sees the placement of the image we have of the elderly outside the metropolis, in the bare life of Agamben for whom "In Western politics, bare life has the peculiar privilege of being that whose exclusion founds the city of men" (Agamben, 1995).

Why would this network of propositions – which induces us to look at the elderly person as an outsider – influence the self-perception's interviewee at a time when we are discussing a topic which is now past and linked to memory, rather than a current topic? For anthropologist Alessandro Portelli, the memory is the "way people place themselves in history. How they place their personal experiences in relation to their idea of the institutions laws, value systems and the culture of the societies in which they live". In this case, the positioning of one's own presence is oriented through the reconstruction in a new form and the constant reinterpretation of the past. An inability to narrate or a density of narrative can indirectly indicate priorities and needs of the present. "Once again, the most valuable knowledge consists precisely in the silences, the reticence, the deformations" (Portelli, 2007).

16.5 Conclusions

In this article, I tried to rework a bifocal narrative on two realities that are extremely linked – or so I thought ideally. The reality of the covered market and the one of the elderly are the ideal protagonists of a proximity market in which they can find their place in the community. Furthermore, I tried to find out whether this relationship was actually a social fact that is still relevant today. What I deduced was that, whether this link exists or has ever existed, it has been considerably weakened by a series of socio-political dynamics emphasised by the coronavirus emergency. What I have written may certainly be approximate and far from the large-scale values of statistical science, but regardless of whether you place this article among undeniable truths or pure fiction, it is certainly a starting point for re-imagining the social places that belong to our neighbourhoods. Between redevelopment projects and creation of imaginary communities, I believe there is a need to count the absentees: those whose memory is still so perfect to reproduce meticulously on paper the places of a 1960's market, but people considered at the same time unsuitable to stay in the common social space. The future absentees – those who will be forced to change neighbourhood when the gentrification process will be irreversible – and the absentees of the present, the entire multicultural substratum that dominates Via Padova but is excluded from the market, despite the presence of Hassan's halal butcher, shop in it.

Maybe we need to ask ourselves – in our process of imaginative reworking of community – if we want the market to become once again a crossroad for all the experiences and diversities that are currently among the minorities. Maybe we need to ask ourselves once again: What kind of possible world are we looking for, and what kind of market would we like to pass through? Concerning the market, my imagination continues to focus on the crossroad image that my father used to talk about and which for me continues to be the very essence of the market and to which I hope we will never renounce.

References

Agamben, G. (1995). *Homo Sacer. Il potere sovrano e la nuda vita*. Enaudi Editori.

Appadurai, A. (2012). *Modernity at large*. Edizioni Cortina Raffaello.

Bettoli, A. (2017). *Nolo: North of Loreto, la Brooklyn dei creativi di Milano*. unconventionaltour.net/nolo-north-of-loreto-la-brooklyn-dei-creativi-di-milano/. Accessed 15 June 2021.

El Itagnol. (2021). *Gentrificazione a Madrid, chiude il ristorante dei senegalesi a Lavapies*. itagnol.com/2020/02/gentrificazione-madrid-chiude-il-rsitorante-dei-senegalesi-a-lavapies. Accessed 21 June 2020.

Giacomantonio, A., & Luciano, E. (2018). Immagini d'infanzia e biopolitica. Congetture foucaultiane. In *Ricerche Pedagogiche*. EA ancia Editori.

Gledhill, J. (2000). *Power and its disguises: Anthropological perspectives on politics*. Pluto Pr Ediction.

Hannerz U. (1992) Esplorare la città. Antropologia della vita urbana [1980], .

Low, S. (2017). *Spatialising culture: The ethnography of space and place* (1st ed.). Routledge.
Meli, E., Franceschini, A., Go! 3 (2018). *Mondadori scuola FAI-Fondo per l'ambiente italiano*. https://www.fondoambiente.it/luoghi/mercato-comunale-coperto-di-via-crespi. Accessed 29 June 2021.
Ortner, S. (2006). *Anthropology and social theory. Culture, power, and the acting subject*. Duke University Press.
Poidimani, N. (2020). *Krisis Corpi confine e conflitto*. Catartica Edizioni.
Portelli, A. (2007). *Storie orali*. Donzelli Editore.
Rossi, U., & Vanolo, A. (2016). *Geografia politica urban*. Editori Laterza.
Samuele Tramontano. (2019). *Tutte le ragioni per cui il quartiere di NoLo sta diventando una città-stato*. imbruttito.com/2019/01/10/tutte-le-ragioni-per-cui-il-quartiere-di-nolo%2D%2Dnorth-of-loreto-sta-diventando-una-città-stato. Accessed 19 June 2020.
Semi, G. (2015). *Gentrification. Tutte le città come Disneyland?* Editore Il Mulino.
Vansina, J. (1976). *La tradizione orale. Saggio di metodologia storica*. Officina Editore.
Vollmer, L. (2018). *Strategies against gentrification*. Katakrak Editore.

Caterina Gallerani is an Italian anthropologist graduated with honour from the Università degli Studi di Milano-Bicocca, under Leonardo Piasere, professor of Social and Cultural Anthropology and director of the CREAa (Centre for Research in Ethnography and Applied Anthropology), which studies social, cultural, multi-ethnic and multireligious issues for application purposes. In her anthropological work, she discusses the importance and limits of participatory observation within anthropological analysis on themes such as dance, body and performance. In the last 2 years, thanks to an internship at Nada Colectivo in Madrid, she had focused on issues such as urbanisation, gentrification and urban regeneration.

Chapter 17
Inclusive Innovation for Age-Friendly Environments. The Role of Living Labs Dynamics in Territories

Luciana Castro Gonçalves

Abstract Aware of the economic and societal challenges concerning the ageing of the population in Europe, many territorial actors wished to participate actively in the silver economy. In this context, hybrid public-private structures dedicated to innovation, such as Living Labs, have multiplied in recent years to encourage age-friendly environments in territories. The purpose of this research is to expand knowledge about the development of inclusive innovation by exploring the particular roles played by three Living Labs in France dedicated to the ageing field. The analysis based on a qualitative research methodology shows three forms of value creation in this field related to the distinct characteristics of these structures: the dissemination of methodological knowledge to co-design with elderly; the encouragement of business, product and services strategies adapted; and the dissemination of the culture of innovation in the territory while involving the elderly in the decision-making process. The research also points to the challenges that still need to be addressed in order to converge the technological, economic and social dimensions in inclusive innovation activities.

Keywords Living Labs · Inclusive innovation · Elderly · Public-private-citizen stakeholders · Territories

17.1 Introduction

The concept of inclusive innovation is often associated with the development of products and services for low-income populations (Horton, 2008, Kaplinsky et al., 2009; Joseph et al., 2011; Foster & Heeks, 2013) or for other vulnerable populations (women, youth, the disabled and ethnic minorities) (Codagnone, 2009). According

L. Castro Gonçalves (✉)
ESIEE Paris, Université Gustave Eiffel, Paris, France
e-mail: luciana.castro@esiee.fr

E. Pozo Menéndez, E. Higueras García (eds.), *Urban Design and Planning for Age-Friendly Environments Across Europe: North and South*, Future City 19,
https://doi.org/10.1007/978-3-030-93875-8_17

to George et al. (2012), 'the boundaries of this construct are evolving and remain ill-defined in the management literature, and observers of the general phenomenon often colloquially consider inclusive innovation as primarily dealing with business model breakthroughs that enable participation in high-growth, high-profit ventures by previously disenfranchised poor people'. Yet each group of vulnerable people has specific characteristics and histories that need to be considered (Joseph, 2014), and other dimensions of innovation management can highlight the notion of inclusion of vulnerable populations.

In this line of work and in view of the growing interest of various companies to be part of the silver economy, the vulnerability of elderly has attracted the attention of many researchers in innovation management science to better understand their needs and the transformations of organisations associated with that context (Shutlz et al., 2016). The progressive loss of autonomy of this population throughout their years of life can make them vulnerable from an economic and social point of view (Donzelot, 1991) but also from a motivational, emotional, cognitive or psychological point of view (Brenkert, 1998) beyond their physical conditions. The medical field alone remains insufficient to meet the challenge of maintaining the independence of the elderly for as long as possible and prevent their exclusion and isolation from active society (Iwasaki, 2013). These issues are a real demographic, economic and social challenge to overcome, particularly in Europe, where the number of over-65 s will increase by almost 30% by 2028 (Eurostat, 2019). Innovation management researchers then try to propose original studies, among others, by focusing on multi-actor ecosystems dynamics that seek to better understand the vulnerability of this population and thus respond more adequately to their expectations and the local needs of the territory (Mostaghel, 2016). Among these ecosystems, Living Labs (LLs) are potentially interesting to analyse the collective co-construction of innovation for the elderly. Despite the growth in the number of living labs around the world, no research has focused specifically on their contribution to age-friendly environments in territories. By aiming to provide an answer to this theoretical gap, the objective of this research is to review the contours of inclusive innovation through the dynamics of Living Labs (LLs) that are dedicated to the development of new products and services for the elderly. It addresses particularly attention in the role of these structures in building a collective representation of inclusive innovation to other stakeholders and in disseminating new ways of understanding well-ageing in the territories.

Like innovation clusters, digital platforms, hubs, fab labs and other initiatives that have emerged in recent years in territories, LLs are part of the current trend towards encouraging creative, 'out of the box' and 'cross-fertilisation' innovation. According to the open innovation approach (Chesbrough, 2003; Dahlander & Gann, 2010) and the collaborative innovation approach (Davis & Eisenhardt, 2011), these mixed public/private/citizen structures bring together different types of stakeholders in order to provide ideas for innovation from outside the boundaries of organisations. They are defined as 'physical regions or virtual realities, or interaction spaces, in which stakeholders form public-private people partnerships (4Ps) of companies, public agencies, universities, users, and other stakeholders, all collaborating for creation, prototyping, validating, and testing of new technologies, services, products, and systems in real-life contexts' (Westerlund & Leminen, 2011). Through this systemic innovation approach, stakeholders are directly and collectively involved in the innovation process in multi-contextual, empirical real-world environments (Feurstein et al., 2008). As open spaces for collaborative innovation, LLs play an intermediation role between these stakeholders (Fasshauer & Zadra-Veil, 2020) through digital platforms (Dodgson et al., 2006) or in physical spaces (Bergvall-Kareborn et al., 2009). By offering real-use environments for experimentations (Eriksson et al., 2005), these structures provide knowledge-brokering activities and promote the collective creation and maturation of new innovation concepts and their adoption and the appropriation of associated transformations. Numerous innovation research studies show the diversity of activities, products and technologies that are developed in these territorialised structures (Almirall & Wareham, 2011; Leminen et al., 2012; Veeckman et al., 2013). Authors have focused on Living Labs in the health field (Picard & Noury, 2015), but the role that these structures play in the development of inclusive innovations has not been addressed in innovation literature.

By focusing on the elderly to explore the development of inclusive innovation, this research addresses two major issues in our society:

- The ageing of the population in Europe and the need to increase the autonomy of the elderly in the coming years in light of their wish to stay at home (Eurostat, 2019).
- The apprehension of ageing, with sustainable and non-stigmatising approaches for future generations of elderly people, while considering their well-being as a whole in addition to their physical health (Hesse & Cade, 2009).

The creation of LLs that are dedicated to innovations for the elderly in many cities is in line with these initiatives. We pay particular attention to the co-design of technological innovations that are developed in this context of the silver economy. Technological innovations are recognised for helping to increase the independence and well-being of the elderly (Iwasaki, 2013; Smither & Braun, 1994), but the study of their adoption remains contradictory (Mathur, 1999). The adoption of technologies by the elderly depends on certain specificities of the technologies and several factors (perception of needs, education, product complexity, familiarity with the product, etc.) (Czaja et al., 2006). In a context where new generations of the elderly are increasingly familiar with digital technology, we wish to focus on the one hand on how LLs take the vulnerability of the elderly into account in inclusive innovation co-design and on the other hand on how their managerial dynamics encourage the collaborative involvement of the many stakeholders in the territory in shaping inclusive innovation specificities for this population. The real-life environments that are mobilised to co-designing innovation in LLs and the central role given to a more proactive involvement of users and user communities offer a rich space to explore these two dimensions and make contributions to the literature about LLs and inclusive innovation.

Thus, the research question that motivates us is formulated as follows: To what extent does the dynamic of Living Labs that are dedicated to the silver economy encourage the development of inclusive innovation with and for the elderly within territories?

To address this issue, we have structured this chapter in three parts. In the first part, we begin with a more detailed presentation of the literature on LL and then on innovation processes. In the second part, we outline the qualitative research methodology used to study the dynamics of three LLs in France and the inclusive innovations related to urban spaces that are developed within them. Finally, in the third part, we present the results of the research.

17.2 Living Lab Approaches for Co-designing Inclusive Innovation

17.2.1 The Multi-stakeholder Dynamic of Living Labs: The Role of User Involvement

Despite the craze for the creation of LLs, their study is recent and has not yet reached a consensual definition of these collaborative territorialised innovation spaces (Almirall & Wareham 2008; Følstad, 2008). They are sometimes seen as physical spaces in regions or as virtual platforms (Westerlund & Leminen, 2011), as a systemic innovation approach of collaborative R&D (Feurstein et al., 2008) or even as platforms for open innovation (Leminen et al., 2017), but all the authors agree on the multi-stakeholder nature and the empirical real-life environments, as well as the central and more proactive role that is attributed to users and user communities in co-design innovation activities (Veeckman et al., 2013; Béjean & Moisdon, 2017; Picard, 2017). LLs tend to engage citizens in their activities and be located in urban districts (Capdevilla, 2017). These mixed public/private structures are full-fledged members of territorial innovation ecosystems (Leminen, 2013; Béjean et al., 2020).

Driven by companies (the top-down approach) or by users (the bottom-up approach), according to an out-in or in-out approach (Capdevilla, 2017), LLs play intermediary roles in the collaboration of the different stakeholders in the territory (Fasshauer & Zadra-Veil, 2020), involving them directly in the innovation process. Like in other collaborative spaces present on the territory, each stakeholder will have distinct expectations, values and operating methods (Castro Goncalves, 2015). According to the scope of application and their internal environment, LLs can have different degrees of openness. The challenge of LLs is thus to lead dynamic innovation networks and orchestrate organic connections between technology development and local communities of people (Leminen & Westerlund, 2019).

In this collective dynamic, the way of mobilising users may be different from one LL to another. Westerlund & Leminen (2011) identified four different mechanisms with which LLs coordinate public-private partnerships:

- The provider-driven mechanism aims at developing an innovation for participants or other stakeholders; it can also defend an educational purpose.
- In the enabler-driven mechanism, the activities focus on the development of the local social needs of a community or association.
- In the activities of the utilizer-driven mechanism, the participants contribute to developing or improving the product or service of a third party (the utilizer).
- In the user-driven mechanism, the participants collaborate to develop their own personal projects.

According to this typology, the dynamics of LLs can generate different outcomes in a privileged way. They can either promote the creation of knowledge in a determined field (utilizer-driven and provider-driven living labs), promote new development strategies (enabler-driven living labs) or encourage the deployment of innovative solutions for users' specific needs (user-driven living labs) (Leminen et al., 2012; Leminen, 2013).

These typologies are instructive, but the modalities of participation of each stakeholder are still poorly understood (Hakkarainen & Hyysalo, 2013), particularly concerning the inclusion of users in the collective dynamic. Kommonen and Botero (2013) call for a clearer distinction between user involvement and user-driven innovation in studying the degree of participation of users and user communities in the open and collaborative innovation processes of LLs. Capdevilla (2017) also highlights the importance of the motivation of each stakeholder concerned.

Today, many LLs encourage the development of innovation in different fields. Most of them are members of the European Network of Living Labs (ENoLL). Together with Spain, France is the country with the most ENoLL members (23 according to the website in August 2020), and many of them are more particularly associated with the Living Lab Health and Autonomy (LLSA) Forum that was created in France, where we can find innovations dedicated to the silver economy. The way in which the different stakeholders involved in innovation address the vulnerability of the elderly and represent the notion of inclusion has not been addressed in the literature on LLs. This topic can be explored by confronting the coordination mechanisms of public-private-citizen partnerships with the innovation processes and methodologies that are implemented within LLs and which promote the development of age-friendly environments in cities.

17.2.2 The Collaborative Innovation Process and Co-design Methodologies to Include the Elderly

The innovation process can be broken down into five generic phases: (1) recognising a problem or need from internal and external information sources, (2) the search for solutions with basic or applied research, (3) the development of a selected solution and (4) the commercialisation of the new product and (5) its wider diffusion (Rogers, 2003). This process is not necessarily linear, as there may be back and forth between the phases. Even though these phases are associated with distinct activities, they are interdependent (Chen & Guan, 2011; Rogers, 2003).

Much research has focused on innovation processes. Some authors show that in some of the collaborative spaces (e.g. FabLabs), innovation processes can be shortened through the use of addictive technologies (Rayna et al., 2015). The same acceleration effects are highlighted by authors who are interested in the participation of users and user communities in the innovation process (Von Hipel, 2007). Their participation helps to better understand their needs in the first stage of the innovation process, and they also contribute also to prototyping, developing and disseminating solutions according to their needs. In this sense, LLs provide real added value by offering laboratories that experiment with innovation in real-life situations with users and user communities during the innovation process. In their co-creation activities, participants must be selected and coached through the process (Kristensson et al., 2008), using an interactive and participatory approach with a specific set of tools (Veeckman et al., 2013). Thus, in these structures, users should have played more proactive roles in innovation processes by testing, evaluating and co-creating innovation. This would mean that other stakeholders would be ready to welcome both positive and negative assessments of innovation or even the transformation of the initial idea. However, Leminen et al. (2012) show that users can have four degrees of participation in the innovation process in LLs (informants, testers, contributors, co-creators) and this can produce different effects concerning the inclusion of the elderly.

In addition, it is important to question the continuity and sustainability of these approaches for the effective inclusion of these users. Most LLs depend on state or company funding that justifies the successful commercialisation of innovations that may impact the exploratory dimensions of certain innovations (Kommonen & Botero, 2013). In order to secure revenues, some LLs build business models where each phase in the innovation represents a service that is provided to a client company. This brick-by-brick service proposal does not necessarily allow for a continuum between the different phases of the innovation process and a long-term approach to the inclusion of users and user communities.

With regard to the elderly as users, the methodologies that are mobilised in the innovation process are even more central. They need to take their health vulnerability into account, but they also hide other vulnerabilities such as social exclusion, whether economic (Donzelot, 1991) motivational, emotional, cognitive or psychological (Brenkert, 1998). Our interest in the concept of inclusive innovation in this research addresses the dynamics that create or enhance opportunities to improve the well-being of the elderly in daily and active life in cities. According to a preventive approach, it makes it possible to mitigate these different types of vulnerability as much as possible and that aims at exploring the remaining capacities of this population. Even if certain care structures and associative and economic entities try to diffuse a non-stigmatising vision of this population, in many countries, ageing is still apprehended according to a physical health approach. This vision can lead to the devaluation or even exclusion of the elderly vis-à-vis the active population (Henrard, 2002), even though the question of developing their autonomy remains central. Thus, the involvement of the elderly in the process of co-design innovation can be an inherent part of the inclusive approach. This can constitute an 'identity renaissance' for those who are motivated to participate in the development of innovations that concern issues that they are passionate about (Schau et al., 2009). We would like to better understand inclusive innovation through the involvement of the elderly in LLs, so it is important to analyse how these participants are involved in the innovation process.

17.3 Methodology

17.3.1 The Design of the Research

The methodology adopted is based on explorative qualitative research (Miles & Huberman, 2003) through a multiple case study approach (Yin, 2003). Our objective is to explore a still-emerging phenomenon.

We selected three LLs that are dedicated to health or specifically to the silver economy, which are located in three different regions in France. Madopa is located in Champagne-Ardenne, I-Care in Rhône-Alpes and AutonomLab in Nouvelle-Aquitaine. The executive boards of these structures vary between 8 and 15 members and include a wide variety of actors (research and higher education, mutual insurance companies, territorial actors, health and prevention actors and manufacturers, user associations). They all simultaneously encompass societal and technological dimensions in business-citizen-government-university partnerships, but their structures and innovation dynamics differ.

In order to respond to the problem statement of this research, we structured our design research in three steps. The first step is dedicated to characterising the three LLs by confronting the typologies of Westerlund & Leminen (2011) and Leminen et al. (2012) with their own specificities. We particularly paid attention to how the emergent collaborative dynamics between different local stakeholders participated in the main purpose of LLs in terms of inclusive innovation. In the second step, we explore the ability of the LLs studied to capture users' involvement, the collaborative dynamics considering the phases of innovation process and methodologies used. We also focus on challenges concerning the co-design of innovations that are

Table 17.1 The prescribers for the analysis of inclusive innovation through Living Labs dynamics

Analysis of inclusive innovation	
Collaborative approach in the territory	Top-down approach (driven by companies) Bottom-up approach (driven by users)
Internal characteristics	Expectations Values Operation methods (offer services, participate in the collective R&D, bringing together actors) Business model
Coordination mechanisms of multi-actors	Provider-driven Enabler-driven Utilizer-driven User-driven
Methodologies involve the elderly in the co-design process	Experiments in LL premises Experiments in elderly home Experiments in other spaces
Implication on the elderly in the innovation co-design process	Recognition of a problem or needs (phase 1) Search for solutions (phase 2) Development of a selected solution (phase 3) Commercialisation of the new product (phase 4) Diffusion of innovation (phase 5)
Value creation	Promotion of new knowledge Promotion of new strategies Promotion of deployment of innovative solutions

Source: Own elaboration

dedicated to the elderly in urban spaces. Finally, we address an overview of inclusive innovation according to the analyses at the end of these first two steps (Table 17.1).

17.3.2 Data Collection and Analysis

We started to collect secondary data in the website content of the three LLs, through documentation published by them, in the mainstream and specialised press and in the websites of the projects, consortiums and companies that are concerned by the collaborative dynamic. This data allowed us to create a certain familiarity with the context of LLs and the environment in which they emerged. We were thus better prepared for the collection of primary data.

Primary data collection consisted in carrying out semi-structured interviews with the directors of each LL and the innovation project manager of one of them. The interview guide consisted of 15 leading questions structured around the 3 main topics that are presented in our research design (LL specificities, user involvement and innovation process and reactions about the notion of inclusive innovation). Each of the three interviews lasted between one and a half hours and two hours. The three interviews were re-transcribed to facilitate the structuring of the case studies. We

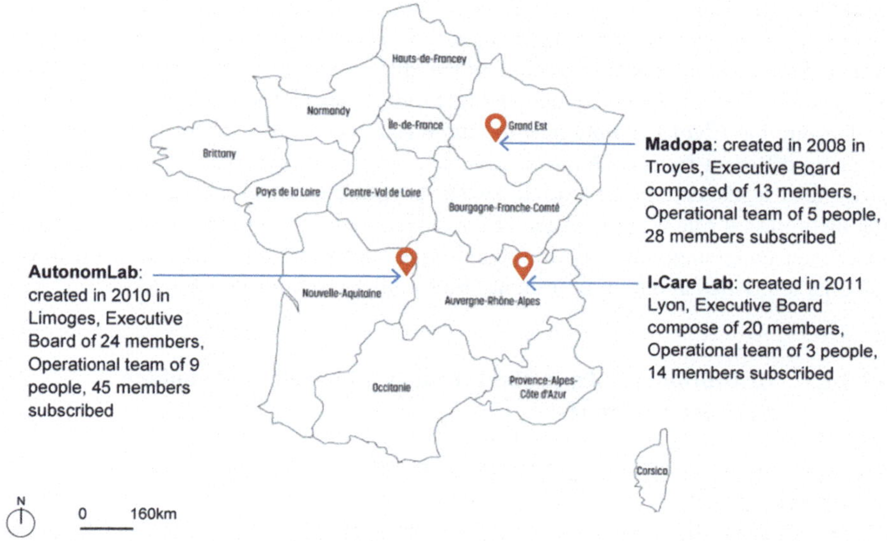

Fig. 17.1 Localisation and structure of the three Living Labs studied. (Source: Own elaboration)

coded the data in a theme dictionary according to the existing literature (Miles & Huberman, 2003). This analysis method allowed us to compare the variables in the three cases and identify emergent themes related to the research topic.

The purpose of the research is not only to examine particular features of inclusive innovations for the elderly but also to identify specific managerial mechanisms in Living Labs that lead the engagement of multiple stakeholders in developing the well-being of the elderly through innovation activities. We, therefore, wish to establish a new perspective about inclusive innovation through the roles that Living Labs play in promoting the development of age-friendly environments in cities.

The Fig. 17.1 shows the localisation in France of the three LLs studied and summarises their main characteristics.

17.4 Results

17.4.1 The Specificities of Living Labs That Are Dedicated to Innovation for the Elderly

The innovations that are emerging in LLSA cover a large spectrum of outcomes from medical to entertainment products and services, and they mobilise a great heterogeneity of public and private stakeholders (public hospitals, user associations, industrial and service companies, public institutions, researchers in various scientific fields, start-ups, etc.). To encourage the development of LLs in different countries, the European Union offered its financial support for creating 19 LLs called the 'European Network of Living Labs' (ENoLL) in 2006. The LLs that were selected are those that best respected all the characteristics of this type of structure. The network has grown since then and had more than 150 active members in 2018.

Despite the ENoLL guidelines, the three LLs that were studied in this research are very different from each other, and each have their own specificities. This heterogeneity leads us to go beyond the existing typologies of LLs to take into account the institutional role that they have progressively built up in the territory. This leads to a better understanding of their collaborative approach in the territory, their internal characteristics and their coordination of multi-actor's mechanisms.

17.4.1.1 Institutional Roles in the Territory and Multi-stakeholder Network Dynamics

The three LLs were created in 2010 through different local dynamics. They are all certified by ENoLL.

According to the typologies of Leminen et al. (2012), we can consider Madopa as an enable-driven LL. The Madopa structure was created without a clear positioning on ageing and relied on university support through the implementation of a first European project that was dedicated to the elderly. The contribution of other European projects has enabled the LL to gradually meet other potential partners until today forming an ecosystem of multiple stakeholders. Through this bottom-up approach, the mission that Madopa has set for itself is to disseminate the socio-anthropological methodology in a territory that is larger today, as it is the most suitable for putting the motivations of the elderly at the heart of the implementation of innovation projects. By doing so, the LL seeks to influence the collaborative R&D strategy by offering a societal vision that is more focused on health and well-being rather than insisting on the traditional vision that is focused on solutions to tackling pathologies. However, economic vision is not excluded from the collaborative dynamic to the detriment of the new societal vision. Madopa's income is based on the financing of projects by various stakeholders. The project-based approach is the privileged place to disseminate this societal vision in which each stakeholder plays its own role in reaching common objectives, without a prior phase of negotiation about the relevance of the methodology. The change in practice that is needed in order to adopt this methodology is sometimes more difficult with institutions at the national level. The inclusive innovation dynamics of LLs also tends to hinder collaborations with start-ups which have fewer resources or which are in the embryonic stages of their innovation in relation to the time to market.

I-Care Lab can be characterised as utiliser-driven in the health sector in general. This LL was created basing on an innovation cluster model with a mission to support companies in order to innovate and increase their competitiveness. As the economic dimension is privileged, the structure does not claim to be a full LL. I-Care Lab sees itself as a tool for interaction and interface between the stakeholders that are present in the territory to facilitate and guide the development of solutions in the health sector. Funded in part by the regional public institution, the LL favours local partnerships around two activities: the emergence of new problem statement in health to light through the organisation of hackathons and the deployment of in situ actions in the post-development of solutions with users. Beyond the role of supporting companies, I-Care Lab can also position itself as a co-financer of innovation projects. In this case, it participates in project management, in the search for additional budgets and in the downstream regulation of problems linked to delays that are too long in order to avoid the demobilisation of those involved, for example. If the company is alone in carrying out the project, there is sometimes a gap between the stakeholders who want to develop and those who want to deploy (e.g. health institutions). The company's goal is to succeed and to move up to a level of scale while the health institutions have a social and health mission. There is sometimes tension between these two worlds. The LL thus works through responses to calls for projects. For the analysis of behaviour users, I-Care Lab calls on external experts such as other LLs from the territory.

The generation of Autonom'Lab is based on the result of a study which states that the Limousin region is where people live longer than elsewhere. Interest was then focused on identifying projects and initiatives that promote this phenomenon. The objective was then to bring together these initiatives, which were very disparate, around innovation dynamics to create a real ecosystem. However, innovation was (and remains) unknown to certain stakeholders, particularly in the medico-social and health fields (hospitals). As a user-driven LL, Autonom'Lab has the objective of economic development of the territory but above all of social cohesion in order to develop service offers that are appropriate for the everyday lives of the elderly. For them, they implement a regional competence with training issues. Because of its proximity to the different stakeholders in the territory, the LL has succeeded in placing trust at the centre of the collaborative dynamic. This has an impact on the speed of operation of their activities (to obtain strategic information, to make decisions, to mobilize resources, etc.). This process of legitimation based on their experience has taken on a large scale within the territory. For this reason, Autonom'Lab has been invited by public policies to become a public geriatric hub with three missions: (1) the development of innovation (which Autonom'Lab already did with the economic dimension), (2) support for public and territorial policies and (3) training and development of professions and associating public and private research, LL-type operations and the inclusive dimension. To date, the elderly already participate in the governance of this structure. The LL's mission is gradually evolving towards a provider-driven approach. Initially based on a problem-solving approach through collaborative projects, Autonom'Lab also focuses on the mission of disseminating the culture of innovation to the various stakeholders in order to support new developments, including those who are not yet at ease in the field.

17.4.2 Forms of Value Creation

The process of inclusive innovation can be observed through a first level of analysis linked to its progressive institutionalisation in territories but also in relation with the different forms of value creation targeted by the LLs.

Madopa places a great emphasis on the role of knowledge creation through the dissemination of its enabler-driven methodological approach. Three books retracing the experience of the LL and the methodology it uses have been published between 2013 and 2018. This enhances the value of the experience that has been linked in the 60 innovation projects carried out within the LL. Madopa thus seeks to go beyond the scope of the projects carried out in order to raise awareness of the salutogenic societal vision concerning the elderly.

I-Care Lab is rather well positioned in the continuity of the dynamics of the health innovation cluster whose objective is to structure collaborative R&D consortiums between different stakeholders of the territory. The LL then mobilises their relational knowledge to encourage the setting up of new projects. It serves both as a translator of new strategic utilizer-driven orientations and as a mentor for the articulation of distinct or even convergent modes of operation. The social impact that this role brings to the different stakeholders is directly linked to the economic value it is supposed to generate in terms of developing new products and services and new businesses.

Autonom'Lab started from an approach that was based on valuing solutions to the everyday life problems of the elderly, even the simplest ones. Today, the LL combines this approach with the dissemination of a customised innovation culture to support the development of these operations by the stakeholders in the territory. The elderly are involved in the decision-making process and knowledge creation, ensuring the relevance of the new knowledge that is created and disseminated in the local ecosystem. The LL strengthens social effects in addition to economic value.

17.4.3 The Methodologies Used to Structure Inclusive Innovation Through Collaborative Approaches

17.4.3.1 The Methodologies to Involve the Elderly in Innovation Processes

One of the main specificities of LL is to be based on real-life environments to develop innovations that are as close as possible to the needs and expectations of users. The three LLs that were studied show a great diversity of activities that are connected to the elderly through the innovation process. They work in partnership with local users' associations and service and care providers to mobilise elderly people whose profiles correspond to the issues of innovation projects. The participation of the elderly is generally voluntary. Their recruitment seems to be easier when they feel concerned because they expect to contribute to identifying solutions.

The analysis of the two main methodologies that are used by the LLs points to different levels of the involvement of the elderly in the innovation process. The methodologies that greatly involve the elderly are based on an ethnographically based approach and aim at exploring recurrences and difficulties in the everyday lives of elderly people at home or in their work environments. These methodologies are used in the upstream phase of innovation projects to identify needs and search for solutions. They can be used a little later in the process to test the solutions that are developed or identify the best way to market the product or service. This exploratory approach is favoured by Madopa and is implemented by the LL team or by doctors of sociology and anthropology from their university partner. The implementation of the results according to the needs of the economic partner is necessary so that it can be taken into account in the development process. For I-Care Lab, the challenge is rather to get users to embrace the innovation process. For them, users are experienced contributors and testers. They call upon experts to apply these methodologies (independent professionals or researchers). Autonom'Lab can also mobilize experts when specific spaces or structures are needed for interaction with the elderly.

The methodologies that are used for the emergence of ideas such as hackathons are implemented to bring out problems and find possible solutions to existing problems. In this case, the elderly interact with each other and with other people concerned. The participants are involved for a shorter time to foster the early phases of the innovation process. I-Care Lab and Autonom'Lab use these methodologies to formulate promising themes in an inductive way or to raise the awareness of certain stakeholders about challenges or their implications for well-ageing.

Through these iterative methodologies, the LLs explore new ideas or test new products or services. However, the elderly are rarely involved over a long period in the innovation process. These methodologies sometimes disrupt the practices and culture of socio-economic stakeholders, but the fact that some of them have adopted them spontaneously in their operations shows a progressive acculturation of innovation around these methodologies. Some of them are also more open to accepting changes to their product or service ideas in the upstream phases of the project. The non-acceptability of feedback from the elderly by the project leaders can go as far as stopping the collaborative project with LLs. Co-design activities with the elderly are not present in any of the LLs.

17.4.3.2 Inclusive Innovation Projects in the Urban Field

Innovations which link well-ageing and urban spaces are present in the three LLs that were studied, but they are still few in number. In this research, we explore in an example of a project that was dedicated to the elderly in urban spaces in each LL that was studied. In all three cases, the projects encountered difficulties in reaching their objectives. The question of the involvement of private-public stakeholders takes on its full meaning in this context.

The projects that were developed by the LLs relied on the internet of things (IOT) to develop smart environments for the elderly to live better at home in the case of chronic illness, loss of autonomy, risk management and security in vulnerable situations. Different lessons can be learnt from these experiences. Firstly, city-wide technologies rarely work. This raises the problem of compatibility between different technologies and the management of big data (collection, processing, protection, etc.). Secondly, for the moment, public and parapublic institutions are resistant to changes in their practices. This is why LLs tend to consider it more plausible to use pre-established policies to adjust the offer of services. Finally, it is necessary to rely on the presence of solidarity networks that are present in the territory. These stakeholders are the driving forces behind changes in inclusive approaches.

In the I-Care urban project, the aim was to develop a meta-project linking different technologies with distinct but complementary functionalities (e.g. sensors to evaluate air pollution, IOT to link the elderly in their homes, etc.). The idea was to build a centralised service offer that was based on a preventative approach to health. The biggest challenge in the project was to work with the stakeholders in the real estate field. As this is a highly regulated sector, various constraints were highlighted in implementing the technologies in buildings.

The project underway at Autonom'Lab, which is currently being implemented in collaboration with the city, focuses on smart mobility. The objective is to explore new modes of transport in everyday life that include fragile populations and provide customised, sustainable and inclusive solutions. The new ways of consuming mobility risk are colliding with the practices of transport operators. But we can also consider that the new elderly are coming up with the digital age.

Thus, the greatest challenge concerning the design of inclusive innovation projects in cities does not seem to be related to the involvement of the elderly or the identification of their needs. It is rather a question of being confronted with both novelty through the development of technologies that are not yet stable and the weight of the past, in that their deployment may call into question the functioning of major stakeholders whose practices are historically rooted (transport, real estate, electricity, etc.) and go beyond perimeter of the city perimeter. Financial support is offered by regional, national and European public authorities to encourage the development of these projects within innovation ecosystems. However, it would appear that the transformations are deeper and that institutional flexibility and the culture of innovation can play a major role in bringing them about. At the local level, the dissemination of inclusive innovation by LLs can contribute to a progressive development in that field.

17.5 Discussion

The result of this research shows the ways of LLs in encouraging the development of inclusive innovation in territories. If the typologies of Leminen et al. (2012) help us to characterise LLs by how they develop inclusive innovation, the analyses push us to explore other dimensions beyond the factors that are mobilised. Whether through the governance of LLs, collaborative project management or the methodologies that are applied to involve users, LLs develop inclusive approaches to the elderly in different ways. The dissemination of what really concerns the elderly population through a bottom-up approach and ethnographic co-design methodologies helps to raise awareness of ecosystems about the perspectives of inclusive innovation. The users' contribution to the innovation process produces transformations beyond the minor adjustments made by a prototype test. As open innovation intermediaries (Ollila & Elmquist, 2011), the LLs contribute to educating and rallying local stakeholders through a collaborative approach.

Despite the strong potential to encourage collective innovation in the territories, of the 400 LLs created in 2006, 150 are active in 2018. These structures remain fragile and are faced with certain challenges. The analyses of the LLs dedicated to silver economy point to three main internal and external challenges in going further and cultivating age-friendly environments. The first one concerns the convergence of economical and inclusive approach in the innovation projects. The major industrial players remain attached to the first dimension, sometimes to the detriment of the contradictory or controversial information that the co-design methodologies for analysing usage behaviour provide in relation to the innovation they wish to develop. However, this convergence is not unattainable. Some countries in Europe have had positive experiences in this direction for cultural and economic reasons. Sweden is a good example. This country has the largest proportion of people with age of 80 or more (Swedish Institute, 2015) and is considered as the second most innovative country in Europe since 2007 (European Commission, 2015).

The second challenge concerns to addressing urban design from the perspective of ageing. Despite the advent of smart cities, this notion, which is currently fashionable, tends to be restricted to a very computational notion (management of a large volume of data to make cities intelligent), sometimes to the detriment of social and inclusive aspects (Picon, 2015). Although some LLs such as AutonomLab are constantly making progress in this direction, the combination of these different dimensions is still fragile and requires a profound transformation of institutions and entire professional sectors (such as the real estate sectors, for example, which are governed by strong regulations in France). The network of associations representing different categories of elderly people plays an essential role in the dynamics that LLs can play in the territory to overcome this difficulty. The easy access to elderly populations through these stakeholders allows LLs to identify present and future trends and to bring a deep understanding of the behaviour of the elderly, beyond the targeted work on innovations for which they are solicited.

The third challenge concerns the participation of the elderly in LL dynamics. The literature on LLs already points to this difficulty. Initiatives such as those of AutonomLab, which seek to integrate the elderly into the governance of the LL itself, and initiatives such as those of AutonomLab, which seek to integrate the elderly into the governance of the LL, are a way of putting their voice at the centre of the process of co-designing innovation and the social and societal issues it must address. Thus, the elderly can also assume a transformative role into value the inclusive environments for all ages in our societies.

17.6 Conclusion

The three cases that were studied in this ongoing research show how the dynamics of LLs participate differently in developing age-friendly environments through public-private-citizen coordination and the forms of value creation that are produced among these stakeholders in the territory. The results show more broadly the strategies that these structures implement to encourage inclusive innovation according to their particular dynamics. The approaches can be based as much on the dissemination of methodological knowledge to understand the consumption behaviour of the elderly, on the encouragement to develop strategies for the creation of products and services after the actors of the territory or on the dissemination of the culture of innovation in the territory while involving the elderly in the decision-making process.

The research also discusses the challenges concerning the convergence of economical and inclusive approaches, the development of a broader understanding of the smart city that goes beyond the digital approach to include more of the social dimension and finally the strengthening of the participation of the elderly in the co-design activities and decision-making process.

This research presents future perspectives about the understanding of inclusive innovation within LLs and territories. It would be relevant to compare the vision of the CEOs of these innovation spaces with other stakeholders involved and especially with the elderly. Furthermore, a more detailed analysis of innovation projects dedicated to cities and urban spaces would allow us to explore more closely the technological, social and political challenges that remain to be met to inclusive co-design environments. The French specificity of creating LLs dedicated to health may play a role in the territorial dynamics in this field. This hypothesis should be verified in future research in relation to other European countries.

This research offers finally a look at public policies to use LLs as a basis for structuring the initiatives of urban planning and the design of better ageing in cities. International feedback from countries in Europe further ahead than France on these issues could be useful to reinforce the results of future research.

References

Almirall, E., & Wareham, J. (2008). Living labs and open innovation: Roles and applicability. *eJOV: The Electronic Journal for Virtual Organization & Networks, 10*.

Almirall, E., & Wareham, J. (2011). Living labs: Arbiters of mid-and ground-level innovation. *Technology Analyses and Strategy Management, 23*(1), 87–102.

Béjean, M., & J.-C. Moisdon (2017). Recherche-intervention et Living Labs, une même lignée de « recherches immersives » ? *Annales des Mines - Réalités industrielles Mai, 2017*(2), 54–9.

Bejean, M., Picard, R., & Breda, G. (2020). «Living labs, innovation collaborative et écosystèmes: le cas de l'initiative 'concept maturitylevels' dans les medtech», Innovations. *Revue d'Economie et Management de l'innovation*.

Béjean, M., Picard, R., & Breda, G. (2021), «Living labs, innovation collaborative et écosystèmes: le cas de l'initiative 'concept maturitylevels' dans les medtech», Innovations, *Revue d'Economie et Management de l'innovation*, N° 65(2), 81–110.

Bergvall-Kåreborn, B., Eriksson, C. I., Ståhlbröst, A., & Svensson, J. (2009). A milieu for innovation: Defining living labs. In *ISPIM Innovation Symposium: 06-09/12/2009, USA*.

Brenkert, G. G. (1998). Marketing and the vulnerable. *Business Ethics Quarterly, Ruffin Series, 1*, 7–20.

Capdevila, I. (2017). Motivating participation in collaborative spaces. *Journal of Business Strategy, 40*(2), 14–21.

Castro Gonçalves, L. (2015). Strategizing across boundaries: Revisiting knowledge brokering activities in French innovation clusters. *Journal of Knowledge Management, 19*(5), 1048–1068.

Chen, K., & Guan, J. (2011). Mapping the innovation production process from accumulative advantage to economic outcomes: A path modeling approach. *Technovation, 31*(7), 336–346.

Chesbrough, H. (2003). *Open innovation: The new imperative for creating and profiting from technology*. Harvard Business School Press.

Codagnone, C. (Ed.). (2009). *Vienna study on inclusive innovation for growth and cohesion*. European Commission.

Czaja, S. J., Charness, N., Fisk, A. D., Hertzog, C., Nair, S. N., Rogers, W. A., & Sharit, J. (2006). Factors predicting the use of technology: Findings from the center for research and education on aging and technology enhancement. *Psychology and Aging, 21*(2), 333–352.

Dahlander, L., & Gann, D. (2010). How open is innovation? *Research Policy, 39*, 699–709.

Davis, J. P., & Eisenhardt, K. M. (2011). Rotating leadership and collaborative innovation: Recombination processes in symbiotic relationships. *Administrative Science Quarterly, 56*(2), 159–201.

Dodgson, M., Gann, D., & Salter, A. (2006). The role of technology in the shift towards open innovation: The case of Procter and Gamble. *RandD Management, 36*(3), 333–346.

Donzelot, J. (1991). *Face à l'exclusion, le modèle français* (pp. 137–168). Éditions Esprit.

Eriksson, M., Niitamo, V., & Kulkki, S. (2005). *State-of-the-art in utilizing living labs approach to user-centric ICT innovation: A European approach*. Center for Distance-Spanning, *1*(13), pp. 1–13.

European Commission. (2015). *European innovation scoreboards project*. Retrieved from http://ec.europa.eu/growth/industry/innovation/factsfigures/scoreboards/index_en.html

Eurostat. (2019). *Augmentation de la part de la population âgée de 65 ans et plus entre 2007 et 2017*. Retrieved from ec.Europa.eu/Eurostat. Last access Oct 2021.

Fasshauer, I., & Zadra-Veil, C. (2020). Le living lab, un intermédiaire d'innovation ouverte pour les territoires ruraux ou péri-urbains? *Innovations, 1*, 15–40.

Feurstein, K., Hesmer, A., Hribernik, K. A., Thoben, K. D., & Schumacher, J. (2008). Living labs: A new development strategy. In *European living labs – A new approach for human centric regional innovation* (pp. 1–14).

Følstad, A. (2008). Living labs for innovation and development of information and communication technology: A literature review. *The Electronic Journal for Virtual Organizations and Networks (Special Issue on Living Labs), 10*, 99–131.

Foster, C., & Heeks, R. (2013). Conceptualizing inclusive innovation: Modifying systems of innovation frameworks to understand diffusion of new technology to law-income consumers. *European Journal of Development Research, 25*(3), 333–355.
George, G., McGahan, A. M., & Prabhu, J. (2012). Innovation for inclusive growth: Towards a theoretical framework and a research agenda. *Journal of Management Studies, 49*(4), 661–683. Research Collection Lee Kong Chian School of Business.
Hakkarainen, L., & Hyysalo, S. (2013). How do we keep the living laboratory alive? Learning and conflicts in living lab collaboration. *Technology Innovation Management Review, 3*(12), 16–22.
Henrard, J. C. (2002). Le système français d'aide et de soins aux personnes âgées. *Santé, société et solidarité, 1*(2), 73–82.
Hesse, H., & Cade, A. (2009). *Éloge de la vieillesse*. Générale Française.
Horton, D. (2008). *Facilitating propoor market chain innovation*. Papa Andina.
Iwasaki, N. (2013). Usability of ICT applications for elderly people in disasterreduction. *Journal of E-Governance, 36*(2), 73–78.
Joseph, K. (2014). Exploring exclusion in innovation systems: Case of plantation agriculture in India. *Innovation and Development, 4*(1), 73–90.
Joseph, K. J., Chaminade, C., Dutrenit, G., Sutz, J., Muchie, M., Cozzen, S., & Turpin, T. (2011). Editorial. *Innovation and Development, 1*(1), 1–3.
Kaplinsky, R., Chataway, J., Clark, N., Hanlin, R., Kale, D., Muraguri, I., Papaionnou, T., Robbins, P., & Wamae, W. (2009). Below the radar: What does innovation in emerging economies have to offer other lowincome economies? *International Journal of Technology Management and Sustainable Development, 8*(3), 177–197.
Kommonen, K., & Botero, A. (2013). Are the users driving, and how open is open? Experiences from living lab and user-driven innovation projects. *The Journal of Community Informatics, 9*(3).
Kristensson, P., Matthing, J., & Johansson, N. (2008). Key strategies for the successful involvement of customers in the co-creation of new technology-based services. *International Journal of Service Industry Management*. https://doi.org/10.1108/09564230810891914
Leminen, S. (2013). *Coordination and participation in living lab networks, technology innovation management review*, November 5–14.
Leminen, S., & Westerlund, M. (2019). Living labs: From scattered initiatives to a global movement. *Creativity and Innovation Management, 28*(2), 250–264.
Leminen, S., Westerlund, M., & Nyström, A. G. (2012). Living labs as open-innovation networks. *Technology Innovation Management Review, 2*(9), 6–11.
Leminen, S., Rajahonka, M., & Westerlund, M. (2017). Towards third-generation living lab networks in cities. *Technology Innovation Management Review, 7*(11), 21–35.
Mathur, A. (1999). Adoption of technological innovations by the elderly: A consumer socialization perspective. *Journal of Marketing Management, 9*(3), 21–35.
Miles, M. B., & Huberman, A. M. (2003). *Analyse des données qualitatives*. De Boeck Supérieur.
Mostaghel, R. (2016). Innovation and technology for the elderly: Systematic literature review. *Journal of Business Research, 69*(11), 4896–4900.
Ollila, S., & Elmquist, M. (2011). Managing open innovation: Exploring challenges at the interfaces of an open innovation arena. *Creativity and Innovation Management, 20*(4), 273–283.
Picard, R. (2017). *La co-conception en Living Lab santé et autonomie 2: témoignage de terrain*. ISTE Editions, Londres.
Picard, R., & Noury, N. (2015). The development of the living lab approach in the health and autonomy sector. In *17th international conference on E-health networking, Application & Services (HealthCom)* (pp. 182–188).
Picon, A. (2015). *Smart cities: A spatialised intelligence*. Wiley.
Rayna, T., Striukova, L., & Darlington, J. (2015). Co-creation and user innovation: The role of online 3D printing platforms. *Journal of Engineering and Technology Management, 37*, 90–102.
Rogers, E. M. (2003). *Diffusion of innovations* (S. & S. Inc., Ed.). Free Press.

Schau, H. J., Gilly, M. C., & Wolfinbarger, M. (2009). Consumer identity renaissance: The resurgence of identity-inspired consumption in retirement. *Journal of Consumer Research, 36*(2), 255–276.

Schultz, J. S., Andre, B., & Sjøvold, E. (2016). Managing innovation in eldercare: A glimpse into what and how public organizations are planning to deliver healthcare services for their future elderly. *International Journal of Healthcare Management, 9*(3), 169–180.

Smither, J. A., & Braun, C. C. (1994). Technology and older adults: Factors affecting the adoption of automatic teller machines. *The Journal of General Psychology, 121*(4), 381–389.

Swedish Institute. (2015). *Elderly care in Sweden.* Retrieved from https://sweden.se/society/elderly-care-in-sweden/

Veeckman, C., Schuurman, D., Leminen, S., & Westerlund, M. (2013). Linking living lab characteristics and their outcomes: Towards a conceptual framework. *Technology Innovation Management Review, 3*(12 December), 6–15.

Von Hippel, E. (2007). Horizontal innovation networks – By and for users. *Industrial and Corporate Change, 16*(2), 293–315.

Westerlund, M., & Leminen, S. (2011). Managing the challenges of becoming an open innovation company: Experiences from living labs. *Technology Innovation Management Review, 1*(1), 19–25.

Yin, R. K. (2003). Designing case studies. In *Qualitative research methods* (pp. 359–386).

Luciana Castro Gonçalves holds a PhD in management science from the École Centrale Paris. She is Associate Professor at the ESIEE Paris/Gustave Eiffel University where she holds the responsibility of student entrepreneurship. As a researcher from the Management Research Institute, she is responsible for the RINNODI research group (www.rinnodi.com) and the standing track on the Open Innovation from the European Academy of Management. Her research focuses on collaborative innovation. She has been publishing in international journals on collective learning, innovation community dynamics, collaboration within innovation clusters and the development of innovations in the territories. She is currently leading a research project based on citizen science, which aims at co-constructing with citizens the inclusive city of the future from the perspective of ageing populations.

Chapter 18
Citizen Engagement: Inclusive Methodologies Towards Creating a City for all Ages

Barbara Douglas

Abstract 'The success of communities in becoming age-friendly will, to a large extent, depend on whether older people, including those facing social exclusion, will be involved as key actors in setting the agenda for future research and policies on age-friendly developments' (Buffel et al., Age-friendly cities and communities: a global perspective. Policy Press, Bristol, 2018, p. 282).

In this chapter, we will explore the experiences of an older people-led organisation, the Elders Council of Newcastle, in developing projects and partnerships to enable the participation of older people in shaping research, policy and practice. The approaches outlined in this chapter go beyond consultation and engagement and illustrate how older people can be co-creators and agents of change.

Older people's activism and the leadership of the Elders Council have made a critical contribution to raising awareness in the city of the need to address demographic change and to seizing the opportunity of participating in national and international initiatives such as the Better Government for Older People and the WHO Global Network for Age-friendly Cities and Communities. These initiatives provide a valuable platform for learning from others and contributing to citizen-led understanding of the issues and co-creation of solutions.

Older people are not a homogenous group, so engagement requires a range of methodologies to appeal to a wide audience and in particular to engage with people who through lack of confidence, knowledge, opportunity and resources find it difficult to participate. Online methods of engagement, whilst useful and essential during a pandemic, tend to exclude many, thereby exacerbating, pre-existing inequalities.

Strong partnerships have been critical to developing successful approaches. These range from partnerships with academics to share knowledge, support learning and ensure robustness of methodologies and with public, private and voluntary sec-

B. Douglas (✉)
Elders Council of Newcastle, Newcastle upon Tyne, UK
e-mail: barbara.douglas@elderscouncil.org.uk

E. Pozo Menéndez, E. Higueras García (eds.), *Urban Design and Planning for Age-Friendly Environments Across Europe: North and South*, Future City 19,
https://doi.org/10.1007/978-3-030-93875-8_18

tor partners to ensure relevance and opportunities to provide a platform to present insights to decision-makers and drive change.

Key ingredients of successful approaches to engagement include creativity and fun, trust, openness, mutuality, taking risks, valuing each other's expertise, iteration to build learning and insights over time and connections and resources.

Keywords Active ageing · Citizen participation · Social inclusion · Innovation

18.1 Elders Council of Newcastle

The Elders Council of Newcastle[1] is an older people-led organisation in Newcastle upon Tyne in the north-east of England. The work of the organisation is delivered by older activists, who are volunteers, with small amount of staff support. The aim of the organisation is to improve the quality of life of older people by:

- Providing a platform from which older people can have a voice.
- Increasing older people's knowledge and understanding of the issues that affect them.
- Sharing older people's insights with academia, policymakers and service providers.
- Promoting positive images of older people as contributors to society.

The origins of the Elders Council date back to 1995 and an initiative of the King's Fund to test out 'whole system working'[2] as an approach to tackling complex social issues such as urban regeneration, which are not within the power of an individual or single agency to 'fix'. Improving the quality of life of older people was the focus of their work in Newcastle upon Tyne. The approach was innovative in bringing key stakeholders together in a different way and, especially, in including older people as essential and equal contributors. Older people were seen as part of the solution, not the problem.

This trend towards recognising the importance of older people's voices was reinforced when Newcastle became one of the pilot sites for the Better Government for Older People Initiative (1998–2000)[3] which aimed to address the issues arising from the ageing demographic. This initiative placed significant emphasis on the involvement of older people in setting the agenda and marked a shift from thinking about older people primarily as recipients of health and care towards older people as contributors and change makers and putting in place the infrastructure that would enable this (Hayden & Boaz, 2000).

---

[1] Elders Council's website: www.elderscouncil.org.uk. Last access: 2 May 2021

[2] Working Whole Systems Putting theory into practice in organisations Julian Pratt, Pat Gordon, Diane Plamping 1999.

[3] Retrieved from https://hansard.parliament.uk/commons/1998-03-26/debates/6c3963a9-1f14-4080-9411-9559ae703550/BetterGovernment (Older People).

'The strategy that emerges from the evaluation of the 28 pilot projects clearly shows that public service improvements can be achieved by involving older people—people who are over 50—more directly by taking more careful note of their views and recognising the value of their contribution more clearly than in the past' (Mr. Archy Kirkwood (MP Roxburgh and Berwickshire) Hansard, 16 January 2001).

At the same time, Age Concern launched the Millennium Debate of the Age[4] (O'Sullivan et al., 1999) stimulating a national intergenerational debate on the impact of the ageing demographic.

Taken together, these welcomed national initiatives placed significant demands on older activists and older people-led groups. This led older people in Newcastle to form a citywide organisation which would provide a platform for older people to share their views, contribute their ideas and build their skills to engage more fully as equal partners. In 2001, the Elders Council of Newcastle was formally established.

Through its work, the Elders Council has developed a strong position in the city and makes a significant contribution at both strategic and grassroots levels. Its manifesto *The Way Ahead*, published in 2003 and updated in 2005, laid the foundation for 'Everyone's Tomorrow', The Strategy for Older People and an Ageing Population in Newcastle upon Tyne (2007). This comprehensive strategy brought a greater awareness that issues of importance to older people are not primarily health and social care. They include living in age-friendly environments which enable older people to be active contributors to society and to prepare for their later life. The economic crash of 2007–2008 followed by a period of austerity hampered the implementation of some of the ambitions of the strategy. The Elders Council has continued to be involved in strategic development in the city as a member of the Newcastle's Age Friendly City Group, which is part of the WHO Global Network for Age-friendly Cities and Communities,[5] which reinforces the more holistic approach to later life.

This chapter illustrates the different methodologies through which older people have contributed their insights and the approaches taken to maximise the participation of older people, particularly those whose voices are not often heard.

18.2 Older People as Researchers

The experience of 'whole systems working' and Better Government for Older People gave older people the confidence to see themselves as researchers in their own right and agents of change. The following examples illustrate how older people in Newcastle took practical action to ensure that their insights and experience would be taken into account. Their starting point is what matters to them in their daily

[4] Retrieved from https://publications.parliament.uk/pa/ld199899/ldhansrd/vo990310/text/90310-08.htm

[5] Retrieved from https://www.who.int/ageing/projects/age-friendly-cities-communities/en/

lives, and the examples illustrate how through their own stories together with their research and learning they have influenced local policy and practices. Some of the achievements may seem small, but gradually and incrementally they start to influence change.

18.2.1 House for Life (1999–2001)

Being able to live in a home which is suitable to your changing needs is critical to enabling people to live independently in later life. Older people in Newcastle decided to take a very practical approach to how this might be achieved and approached the Newcastle City Council with a request to work alongside their designers on the adaptation of a standard, semi-detached house to make it a home for life. Their request was accepted, a suitable property found and staff time allocated to the project.

A reference group of 15 older people drawn from across the city was set up to work on the project. The group participated in a series of workshops starting with working with an artist to think creatively about space, reflecting on how their use of space in the home had changed over their life course. This was followed by practical workshops learning about aids and adaptations and assistive technologies and how they might mitigate sensory and physical impairments. The workshops included listening to the advice of housing and care professionals, trying out equipment and deliberating on the most suitable adaptations for the house. The group also took part in regular site visits to monitor progress.

Representatives from the group negotiated with city council officers to ensure that older people's priorities were considered, whilst keeping the work within the standard budget for a refurbishment. Negotiation was not always easy as local authority officers had to learn that older people also have expertise and would stand their ground on key issues. For example, with the support of the occupational therapists who had worked with them, the group won key decisions: level access was provided to both the front and the back of the house, and the small, high windows were replaced with patio doors letting in more light and providing direct access to the garden.

The property was open as a show house for a limited period, giving other older people as well as health, housing and care professionals the opportunity to see what had been achieved and to learn. Older people were available to guide people around the house and explain the rationale for the adaptations. The decisions made by the group raised awareness amongst the professionals of what mattered to older people and influenced the city council's guidelines for housing adaptations:

> The House for Life project made a telling contribution to the education and awareness of personnel who can influence and develop City Council Policy and Standards. (Designer, Newcastle City Council)

Many of the older people involved in the project were unaware of the kinds of adaptations that could be made to an 'ordinary' house, and so the project became a learning experience. This led to a determination to ensure that older people are well informed about the options available to them, which continues to be a significant strand of the Elders Council's work today. It also set the stage for housing to be a key priority in the work of the Elders Council.

Older people committed over 500 hours of their time to the project, and importantly, they felt valued and enjoyed the process:

> We had a strong issue we were all committed to. We found a practical way of working on it, which involved all our skills and experience. We had tangible results – a well-adapted house and a good partnership team. We had fun! (participant)

At the start of the project, older people recognised that they had a lot to learn:

At first, we needed the experts as a shoulder
to lean on. Why can't more people have a say?

By the end, they recognised the value of their own expertise and the potential to continue the relationship into future projects:

So why can't more people have a say?
We could offer our shoulders for the experts.
to lean on. We could offer our feet,
hearts, minds and eyes.[6]

Relationships of trust and respect were established that are keys to collaborative working (Gilroy, 2003). Having established these foundations, the group went on to collaborate with the city council on a number of projects, and the Elders Council is still seen as a valued resource, most recently assessing the quality of customer service in the tenders for the housing adaptation service.

18.2.2 Older Person-Friendly City (2003–2013)

Forecasts of increasing longevity and evidence highlighting the importance of staying active in later life (WHO, 2017) prompted members of the Elders Council to undertake an audit to assess the 'older person friendliness' of Newcastle city centre. The city centre was chosen because it is a focus for the commercial and cultural life of the city and because older people were aware that new developments were being planned.

The starting point for the work was a simple questionnaire distributed to older people to find out what really mattered to them. Although not a scientific sample of opinion, the response to the survey showed a reasonable spread of responses from

[6] Extracts from a Group Poem written by members of the House for Life Group.

different parts of the city and across the 50+ age range. The survey results determined the priorities for the audit.

During the audit, members of the group visited key areas of the city centre assessing the quality and availability of public transport; signposting and information; the quality and cleanliness of pavements, as well as ease of circulation; the quality and opening hours of public toilets; street furniture; street entertainment; open spaces; and buildings of historical and architectural interest and public art. Following their in-depth audit, the group presented the report of their findings and recommendations to the Newcastle City Council and to the business improvement district (NE1). A particular concern of older people is the provision of public toilets. This led to the implementation of a 'Use our Loos' scheme where shops and cafes were encouraged to allow members of the public to use their toilet facilities without the requirement to purchase anything.

Inspired by this approach, the Elders Council also created the opportunity for older people to be creative and to have fun in exploring the city centre. Working with a photographer and a storyteller/creative writer, they produced a short video of 'Newcastle Past, Present and Future' complementing the rich, factual information from the audit with personal interpretations of what the city meant to them.

Having fun and being creative have their place alongside more rigorous methodologies to enable people to uncover and articulate what is important to them, whilst at the same time presenting issues in a way that touches hearts and minds. As a contribution to Architecture Week (2005), the Elders Council worked with Xsite Architecture to run a workshop inspired by William Heath Robinson, a trained draughtsman, who created a world of weird and wonderful contraptions to deal with some of the everyday challenges of living. A morning of fun, creativity and laughter produced a range of wonderful images. The lack of toilet facilities in public parks led the group to devise a 'floating loo' to be pulled down from the sky whenever and wherever it is needed. The solutions proposed were not realistic nor were they intended to be, but they did give indicators for the measures that could be taken by architects and planners. The images were turned into coasters which were widely distributed amongst planners and senior officers as a constant reminder of what matters to older people.

Several members of the group were also part of the Old Spice drama group, which created an amusing but pointed sketch on going out and about in the city centre. The Old Spice drama group was formed by members of the Elders Council. They created their own sketches and songs using humour to lampoon ageism and raise serious issues. As well as performing at conferences and events for older people and professionals, they used drama in the education of nursing and medical students (Fig. 18.1).

Fig. 18.1 The Old Spice drama group in their 'out and about in the city centre' kit. (Source: Elders Council of Newcastle (2011))

18.3 Older People as Partners in Research and Practice

The following two examples are illustrations of the Elders Council being involved as an equal partner in research and practice. The longevity of the Elders Council enabled it to establish strong partnerships and relationships across the city and the reputation of an organisation that delivers. The trust and mutuality which have built up with partners over many years have laid down foundations which enable partners to be opportunistic and take risks.

18.3.1 Growing Older in My Home and Neighbourhood (2014–Ongoing)

As we grow older, we tend to spend more time in our own homes and immediate neighbourhood (Horgas et al., 1998). The location of our home and the surrounding amenities become as important as the bricks and mortar of the house. Older people are often reluctant to move to what might be considered a more appropriate dwelling when this means moving to a different area with the consequent loss of relationships with friends, local shopkeepers and known acquaintances.

Following the financial crash of 2007–2008, in 2009, the UK Government announced that we were entering an 'age of austerity' resulting in significant reductions in the provision of public services and local public amenities such as libraries and community centres which are very important in the everyday life of older people. It was in this context that the Elders Council was invited, in 2014–2015, to work with researchers from Northumbria and Newcastle Universities to explore with older people what they do to support themselves and others and what it means to age well in their home and neighbourhood.

Representatives from the Elders Council's long-established health, housing and care interest group became part of the steering group for the research alongside colleagues from Northumbria and Newcastle Universities and a local arts organisation, Skimstone Arts.

The research team identified a neighbourhood in the west end of Newcastle as the location for the research and worked through networks of local organisations to recruit participants. Twenty-two older people took part reflecting the local socio-economic and ethnic mix.

We used the World Café[7] methodology inviting participants to take part in three workshops. The first two events were based around a set of questions devised by the researchers from Northumbria University drawn from their analysis of the ageing in place literature and agreed with the steering group. Participants were invited to sit in small groups at a table with a facilitator. Each table was allocated a question which participants discussed, writing their notes and comments on a tablecloth. After an allotted time, the facilitator moved to the next table with the tablecloth and repeated the exercise with a new group, layering up the responses and insights on the tablecloth. As the conversations were taking place, an illustrator captured the key issues in a series of bold cartoons which enlivened the conversations and contributed to the findings. In the third workshop, participants were invited to co-analyse and prioritise the issues to produce a set of statements of what older people need to enable them to age well in place. The Artistic Director of Skimstone Arts participated throughout, capturing older people's stories to inform a collaborative performance. She led the final part of the third workshop by giving participants a taster of the Arts Residency that was to come (Fig. 18.2).

The next stage of the project gave participants the opportunity to participate in an Arts Residency with Skimstone Arts to co-create a fictionalised performance based on the key issues arising from our conversations. The residency resulted in a short performance 'Doorbells of Delight' which was performed to a mixed audience of older and younger people and representatives from a range of public services. The performance was well received as a different and more engaging way of presenting research findings, alongside a formal report.

The creative and messy nature of the workshops and Arts Residency were enjoyable for the participants and gave them space to explore their own ideas, whilst at the same time enabling them to uncover and articulate key issues. The process had practical, tangible outcomes for older people who discovered their own individual resources but also found a community platform through which to share their knowledge. It led to practical action as some participants gained the confidence and knowledge from others to establish an activity group within their housing complex.

[7] Retrieved from www.theworldcafe.com. Last access: 21 April 2021.

Fig. 18.2 World Café workshop or example of illustration. (Source: Rose Gilroy)

The impact of this research contributed to further work by the Elders Council on the issues of home and neighbourhood. Focusing on the recommendation that we need to plan ahead for later years, including thinking about the suitability of our current homes, the Care and Repair England joined the partnership to continue to use performance as a way of engaging older people in thinking about their housing options. A new performance, 'Doorbells – Dreaming for the Future',[8] was produced, taking account particularly of the circumstances of people who are ageing without children and those who have primary responsibility for the care of another older adult. The performances are being used to stimulate conversation and debate amongst older people and service providers. The pandemic put a temporary stop to performances, but the passion of the partners to continue to work led to developing a shorter film 'Doorbells – COVID-19' reflecting life during the pandemic, which has been used with significant impact in four successful webinars. Using performance has enabled our mixed audiences of older people and professionals to engage both personally and professionally in the issues raised and enabled an intergenerational conversation around this often difficult and challenging question. The findings from these events are being analysed into key messages for policy makers and practitioners.

Bailey et al. (2018) sum up the power of creative processes to 'foster honest and messy public conversations that have the potential to shift social care policy and practice, and neighbourhood budgets, from focusing too narrowly on older persons' deficits or needs; listening instead to personal and collective stories of the creative use of local resources'.

[8] Retrieved from www.skimstone.org.uk/doorbells. Last access: 20 April 2021.

From an Elders Council's perspective, working in partnership with academic researchers and arts practitioners brings in the new skills and creativity, rigour and knowledge of the evidence which helps to bring more credibility to the work of older activists. Older people's motivation for engaging is because they want to make a difference, preferably in their own lifetimes, but also for future generations. Reed et al. (2006) describes this as a utilitarian view of research contrasting it with notions of research rigour, validity and veracity – 'a choice between useful but 'messy' research, which has no academic credibility or pristine research, which has no use and no credibility with users'. Goulding (2018, p.12) raises a similar question about whether more creative methods challenge notions of whose knowledge is valid – *In aiming to break down traditional notions of expert* versus *layperson and to value different kinds of experiential knowledge, how do we ensure that scientific knowledge is still valued and informs the dialogue?*

18.3.2 Future Homes (2016–Ongoing)

Future Homes emerged from the ambition of a range of actors in Newcastle to address the housing needs of an ageing population. Based on the quadruple helix model used by the Newcastle City Futures (Vallance et al., 2020), Professor Rose Gilroy of Newcastle University brought together partners from the public, private and community sectors to explore how collectively they might come up with an innovative response.

The longevity and quality of the partnership between Professor Gilroy and Newcastle University played an important part in ensuring that the Elders Council had a seat at the table from the outset. From the House for Life project in 1999, the Elders Council had continuously worked with Professor Rose Gilroy and colleagues on a range of projects, establishing a relationship of mutuality and trust.

Initially Future Homes set out quite simply to find new ways of addressing the issue of housing for an ageing demographic. Quite quickly, the group made three key decisions: that this would not be solely a research project but would result in building real houses; that although addressing issues of an ageing demographic, the houses would not be *for* older people but would be suitable for all ages, flexing around our changing needs; and that the housing would be affordable rented housing, placing innovation within the reach of those on lower incomes.

As a partner in this venture, the Elders Council took the responsibility for leading on the engagement with older people and younger families in the co-design process. Working with the architects and an experienced facilitator, the Elders Council led a series of iterative co-design workshops to inform the design of the individual dwellings and, importantly, the character of the overall scheme to foster a sense of community through good design. At the same time, the architects conducted a series of workshops with occupational therapists and health professionals to gain their insights. The findings from this work were brought together in a set of design principles which informed the design of Future Homes (Fig. 18.3).

Fig. 18.3 Future Homes workshop. (Source: Rose Gilroy)

In 2018, the Future Homes Alliance was established as a community interest company. The Elders Council is one of the partners in this collaboration, illustrating the value placed on the insights and contribution of older people.

The skills and commitment of the Future Homes quadruple helix partnership are well on the way to delivering the tangible outcomes that older people seek from their participation – 66 dwellings on the helix site, a prestigious new development in the heart of the centre of Newcastle upon Tyne.

18.3.3 Older People as Commissioners of Research

In 2020, the Elders Council decided to revisit its city centre audit of 2007, in the light of proposals for significant changes and pedestrianisation of the city centre. They took the opportunity to commission a consultancy with planning students from Newcastle University. Elders Council members worked alongside the module leader at Newcastle University to develop the brief. Five final year students signed up to undertake the consultancy. Because of COVID-19, the project had to be conducted via zoom. The Elders Council recruited ten older people to participate in the project. The project started with a provocation on what we mean by age-friendly environments presented by Professor Gilroy followed by a presentation from a lead officer and politician from the council about current policy and plans. The students then led a series of focus groups with older people to find out from them what was important about their city centre. The students undertook research to find evidence to reinforce or challenge the opinions of older people. Their findings were discussed with the group of older participants and refined in preparation for a final report and presentation to city council officers and a lead politician. The students produced a comprehensive report which the Elders Council is now using as evidence to share with the City Centre Transformation Team. COVID-19 prevented the in-depth collaboration between students and older people that was originally envisaged in the brief, for example, joint walkabouts. However, older people valued the opportunity to discuss their views alongside the students' findings, and the intergenerational dialogue is reflected in the title of the report which changed from an 'older person-friendly city centre' to an 'all age-friendly city centre'.

18.4 Older People Empowering Older People

During consultation and engagement processes, significant attention is now paid to ensuring that older people are enabled to participate through providing transport, interpretation, information in large font or easy to read versions and refreshments and often recognition of their time in the form of vouchers or modest payments. All these considerations are important and fundamental to enabling the participation of a wider cohort of people who might otherwise be excluded.

The Elders Council would add to this list that older people need to be informed in order to participate with confidence and build their knowledge and understanding. In response to this, the Elders Council dedicates a significant amount of its resources to keeping older people informed about issues and opportunities through a range of mechanisms. The content is all created by older people for their peers and includes the following:

- A fortnightly e-bulletin.
- A fortnightly radio show, *Older Voices*, on community radio (Nova Radio 102.5 FM).
- A quarterly magazine – *The Echo* – available online and in hard copy.
- Social media including Facebook and Twitter.
- As a member of the Later Life Audio and Radio Co-operative[9] contributing ideas for content such as the Housing Matters series of podcasts.

To sustain its activity, the Elders Council draws on the skills, knowledge and expertise of older activists. To attract a wider range of older people, the Elders Council decided to invest in building the confidence and capacity of older people to participate both in work of the organisation and in playing their part in creating sustainable, age-friendly communities. The idea for the project came from the experience of the Touchstone project in Galway, Ireland (see Scharf et al., 2016). The Elders Council entered into a partnership with the Workers Educational Association,

[9] LLARCooperative | Mixcloud. Retrieved from https://www.mixcloud.com/LLARC/stream/. Last access: 21 April 2021.

a national adult educational organisation with a branch in the north-east of England, to deliver this project with guidance from Professor Tom Scharf. It proved challenging to deliver Active Voices, in particular, to recruit participants, but although small scale, it resulted in positive change for those who took part. 'Active Voices has led to powerful examples of groups of older people driving their own agendas for action and engagement, all underpinned by the project's careful sensitivity to their diverse needs and priorities' (Bevan, 2020, p.4).

18.5 Conclusion

> Community empowerment goes much further than 'consulting', 'involving' or 'engaging' people. It implies a process of renegotiating power and building capacities to gain access, networks and/or a voice in order to gain more control over the decisions that shape communities (Buffel et al., 2018, p.280).

The work of the Elders Council offers examples which illustrate how an older people-led organisation has endeavoured, over many years, to renegotiate power for the benefit of older people in the here and now and for future generations.

Renegotiating power means recognising that there are different kinds of knowledge, skills and perspectives which can be brought to bear to address complex issues. The recognition of the implications of this for changing the way in which policymakers, service providers and academia work is articulated by Reed et al. (2006) when reflecting on working with older people as co-researchers:

> Working with older people as co-researchers was both exciting and challenging. While the academic researchers quickly came to realise that the co-researchers were an enthusiastic group of people who brought a wealth of skills, knowledge and personal experiences that could be harnessed to enhance the project, they were also aware that new research practices and different ways of working were needed to elicit the added value of working together.

In a society marked by significant inequalities (Marmot et al., 2020), there is still a challenge in ensuring that we take account of the lived experiences of all older people, particularly those who are excluded. As highlighted by Buffel et al. (2018), there is a need for investment in community development and greater experimentation in participatory and collaborative approaches to build the confidence and capacity of older people to participate and the skills of policymakers and service providers to listen.

'We have the voice. Do you have the ears?' is the question posed by Vera Bolter, a founder member of the Elders Council of Newcastle. The challenge for urban design is that there are many voices, sometimes with conflicting demands and expectations. Creating opportunities for citizens to engage in negotiating what makes an inclusive environment for all helps to build understanding amongst citizens as well as between citizens and professionals of what creates an inclusive environment and gives everyone a stake in making places the best they can be.

Most people want to age in place, especially as moving to a new place and building new relationships become more difficult as people grow older. To make ageing *in* place possible, the design of our homes and neighbourhoods and related services need to accommodate our changing needs across the life course. This challenges urban planners and designers to develop their skills in the facilitation of creative methods of walking alongside citizens to understand their lived experiences and to draw on that knowledge in their work. This has significant resource implications for the planning and design process. In the UK, the capacity of local authorities has been significantly reduced as a result of austerity. Over the last decade, Newcastle has lost the equivalent of £2500 per household. This loss of resource, coupled with the failure to prioritise age issues, is driven partly by competing high priorities and ageism and fuels a lack of understanding of the social and economic contribution of older people to a city and the value in creating ownership and pride in the places in which we live.

References

Bailey, C., Gilroy, R., Reynolds, J., Douglas, B., Webster, S. C., Nicholls, M., Warwick, L., & Gollan, M. (2018). Ageing in place: Creativity and resilience in neighbourhoods in resilience in ageing creativity, culture and community. In A. Goulding, B. Davenport, & A. Newman (Eds.), *Creative practice in the resilience of older people* (pp. 157–180). Policy Press.

Bevan, M. (2020). *The views and experiences of people and organisations using active voices.* An evaluation report *(note: This is not published but available from elders council or WEA).*

Buffel, T., Handler, S., & Phillipson, C. (2018). *Age-friendly cities and communities: A global perspective*. Policy Press.

Gilroy, R. (2003). Why can't more people have a say? Learning to work with older people. *Ageing and Society, 23*(5), 659–674.

Goulding, A., Davenport, B., & Newman, A. (2018). Resilience and ageing: Creativity, culture and community. In A. Goulding, B. Davenport, & A. Newman (Eds.), *Creative practice in the resilience of older people* (p. 12). Policy Press.

Hayden, C., & Boaz, A. (2000). *Making a difference: The better government for older people programme evaluation report*. University of Warwick, Local Government Centre.

Horgas, A. L., Wilms, H.-L., & Baltes, M. M. (1998). Daily life in very old age: Everyday activities as expressions of daily life. *The Gerontologist, 38*, 556–568.

Marmot, M., Allen, J., Boyce, T., Goldblatt, P., & Morrison, J. (2020). *Health equity in England: The Marmot review 10 years on*. Institute of Health Equity.

O'Sullivan, P., Hanso, J., Russell, L., Bamford, T., Banister, D., Broom, D., Greenaway, P., Kellaher, L., Lockton, S.L., Meikle, J., Morris, M., Nutt, B., Ongeri, S., Rowlands, M., & Syllas, J. (1999). *The millennium debate of the age: The future of the built environment as society ages. London,* Retrieved from: http://www.age2000.org.uk/be. Accessed 7 Oct 1999. (*Note: website no longer available*).

Pratt, J., Gordon, P., & Plamping, D. (1999). *Working whole systems: Putting theory into practice in Organisations*. Kings Fund.

Reed, J., Cook, G., Bolter, V., & Douglas, B. (2006). *Older people 'getting things done' involvement in policy and planning initiatives*. York.

Scharf, T., McDonald, B., & Atkins, A. M. (2016). *Promoting civic engagement in later life through the touchstone Programme: A resource and research guide.* Irish Centre for Social Gerontology, NUI Galway.

Vallance, P., Tewdwr-Jones, M., & Kempton, L. (2020). Building collaborative platforms for urban innovation: Newcastle City futures as a quadruple helix intermediary`. *European Urban and Regional Studies, 27*(4), 325–341.

WHO. (2017). *Global strategy and action plan on ageing and health.* Retrieved from: https://www.who.int/ageing/WHO-GSAP-2017.pdf?ua=1#:~:text=The%20Global%20strategy%20and%20action%20plan%20on%20ageing,experience%20both%20a%20long%20and%20healthy%20life.%20Extensive

Barbara Douglas is Executive Officer for the Elders Council of Newcastle, the older people's forum in Newcastle upon Tyne. She has spent over 25 years working in the field of ageing, with a particular interest in creating opportunities for older people to have a voice and in developing partnerships to deliver improvements in the quality of life of older people today and in the future. Her work started with the King's Fund Whole Systems programme in 1995, leading to a role in coordinating Newcastle's involvement in Better Government for Older People (1998) and currently in WHO Global Network for Age-friendly Cities and Communities (E-mail: barbara.douglas@elderscouncil.org.uk).

Part III
Study Cases, Future Lines of Research and Conclusions

Chapter 19
Eleven Study Cases Across Europe

Elisa Pozo Menéndez

Abstract The different perspectives and studies that have been presented in the two previous part are complemented by the presentation of the following 11 study cases. These projects have been the result of field visits and interviews with several agents across Europe. The result is a diverse practical perspective with the key points of innovation regarding ageing in place, care services, urban developments and housing solutions as an inspirational source of knowledge and experiences for finding new opportunities for a longer-living Europe.

The following 11 study cases have been the result of field visits and interviews with several agents across Europe. The result is a diverse practical perspective with the key points of innovation regarding ageing in place, care services, urban developments and housing solutions as an inspirational source of knowledge and experiences for finding new opportunities for a longer-living Europe. The study cases are grouped by country, and they include the main points of interest.

Belgium

1. Foton in Brugge
2. Woonzorgcentrum De Weister in Aalbeke
3. Woonzorgcentrum De Zon in Bellegem
4. Huis Perrekes in Osterloo

United Kingdom

5. Dementia-Friendly Newcastle upon Tyne
6. Great Sankey Neighbourhood Hub

E. Pozo Menéndez (✉)
ABIO Research Group, Universidad Politécnica de Madrid, Madrid, Spain
e-mail: e.pozo@upm.es

E. Pozo Menéndez, E. Higueras García (eds.), *Urban Design and Planning for Age-Friendly Environments Across Europe: North and South*, Future City 19,
https://doi.org/10.1007/978-3-030-93875-8_19

7. Harmonia village, within CASCADE project: community areas of sustainable care and dementia excellence in Europe

The Netherlands

8. De Hogeweyk in Weesp
9. Kwiek beweegroute

Spain

10. Mixed-use building with collective housing for elderly people and health public equipment in Barcelona

Finland

11. Kotisatama in Helsinki

19.1 Belgium

1. **Foton in Bruges**

Type of project Community house and information point for dementia-friendly activities and support total approach in care and into the city	**Location** Bruges (Flanders)
Date Foton founded in 1995, FotonHouse opened in 2005	**Population** 118,000 inhabitants 25,000 are 65 years or older (21%)
Local stakeholders Multiple partnerships at the regional and local scale, such as the city council of Bruges, the SEL North West Flanders primary health care	
Persons interviewed Bart Deltour (Director Care Familiezorg West-Vlaanderen vzw/Coordinator Foton project) Hilde Delameillieure (coordinator of the Foton team)	**Awards** 2012 and 2020 EFID Award "Together for a dementia-friendly Bruges" Golden Ear for Foton Volunteers

Concept

Bruges is a Belgian city with a population of 118,000 inhabitants, of which 25,000 of them are 65 years or older. The number of people living with dementia is estimated at 2,600, with 70% living at home. Foton is the Centre of Expertise for Dementia in Bruges, an initiative founded by the *Familiezorg West-Vlaanderen*, an autonomous and integrated non-profit home care organization in West Flanders that offers specialized care and support at home for people living with dementia. Foton leads and coordinates the project "*Working together towards a dementia-friendly Bruges*" in a multi-scale perspective to raise awareness in all the cities.

Since 2005, the meeting point is FotonHouse, in the city centre of Brugge; it is a low-threshold walk-in house open to the community where people with dementia, caregivers and relatives are invited to go in. The centre offers what they call a *Kopje Troost* ("Cup of Comfort") to have assessment and support for the process of dementia, to better know and understand the person's needs and desires and how to provide better care. The centre also has a library and documentation centre that is open and available for people who want to learn more about dementia. They organize music afternoons, cognitive stimulation, literary evenings and testimonies for older people and caregivers, to raise awareness towards dementia. Seasonal celebrations such as Easter and Christmas are important as well to organize different cultural activities in collaboration with other volunteers and social collectives.

Foton offers advice and support for people living with dementia, working with specialists on dementia care expertise, volunteers, occupational therapy at home, dementia counsellors at home and psycho-education programme for families.

Foton activity works also at the city level to create an age-friendly and dementia-friendly community in Bruges. Activities and programmes have been delivered in multiple scales, where there is a strong commitment with different entities and organizations in order to promote activity and autonomy for residents with dementia, preventing social isolation and loneliness. Activities with the cultural sector include developing artistic projects, museum visits and city guides for encouraging people with dementia and their caregivers to join the community and keep active with meaningful activities. Also, the local economy has been activated generating a high level of awareness on dementia in local commerces and shops, so people with dementia can feel comfort and safe going around the city, being aware that the community is also taking care of them.

Best Practice

Thanks to the driving leadership, passion and enthusiasm of Foton, Bruges has been recognized as a leading community project of dementia-friendly city in Europe. The city is pioneering innovative initiatives to actively engage people with dementia in multiple activities to reduce isolation and improve well-being and quality of life. There are different projects involving cultural activities, such as dementia choirs, but also local commerces have been trained to understand the specific needs of person living with dementia, and they have installed dementia-friendly symbols in their shopfronts, an iconic and recognisable symbol that can be identified as a safe place for those people with an early-stage diagnosis of the disease. Local police have also been trained to change attitudes towards dementia in this part of Flanders, and databases have been set up to track down lost people with dementia who are frequently found confused and far away from their homes.

	FotonHouse offers a quiet space for informal and quiet meetings for people with dementia, families and caregivers. Source: Foton

	FotonHouse also provides a community space for socializing. Foton welcomes and organizes social and community activities for raising awareness about dementia throughout cultural activities along the year. Source: Foton/Lander Loeckx

	Foton action in Bruges has raised awareness about dementia in many sectors of the community, innovating in the social approach to keep people with dementia engaged to the community through multiple actions that include cultural and healthy activities. Source: Foton

Additional References

Dementia-friendly city of Bruges. https://www.dementievriendelijkbrugge.be/
European Foundations' Initiative on Dementia. https://efid.info/casestudy/38-together-for-a-dementia-friendly-bruges-belgium/
Transform Integrated Community Care. https://transform-integratedcommunitycare.com/casestudy/foton-together-for-a-dementia-friendly-bruges/

2. **Woonzorgcentrum De Weister in Aalbeke**

Type of project Small-scale care home	**Location** Aalbeke, Kortrijk (Flanders)
Completion date 2012	**Number of residents** 46, 32 living with dementia and 14 with high level of dependency Organized in 2 units of 16 residents and 1 unit of 14 residents
Client Public Centre for Social Welfare (OCMW)	**Architect/studio** THV SAR en AAPROG architecten
Person interviewed Jan Goddaer (director)	**Awards** EFID (European Foundations' Initiative on Dementia) Awards (2014)

Concept

WZC De Weister Care Home Centre is a small-scale centre for people living with dementia. The building has been constructed in 2012 by the initiative of the public Centre for Social Welfare and the City Council of Kortrijk. The building has been conceived following the principles of the person-centred model and the small-scale normalized living. The result is a home care centre organized in three units of normalized living where the attention is given to small detail of design, enabling the residents to maintain meaningful activities along the day. All the units are organized around a main living room connected with an accessible kitchen, which offers the possibility to engage in the daily activities. All the rooms have outdoor views to common terraces and gardens, where activities with nature are provided. The rooms are separated of this common space by a corridor which progressively offers a quieter space for calm. The rooms can be personalized with the furniture and preferences of the person. An accessible toilet is included in all the rooms to provide autonomy. All the rooms are individuals and have also outdoor views to a green space. There are memory shelves integrated in the entries, so the residents can propose their own pictures and objects for wayfinding. Also, pets and decoration answer to the desires of the residents in each unit.

There are a total of 45 workers in the centre, including nurses, caregivers and management staff. The team is divided into convivial units, and they function as a core team in the daily life of each unit. They have a specific room for management activities directly connected with the living room. There is also task integration with the residents, which means that caregivers do all kinds of tasks integrating the competencies related to their education/experience, so residents can take part of the daily routines, such as preparing the meals or shopping.

The connectivity and inclusion with the local community is an essential part of the daily life in De Weister. Families have an active role in the management and activities of the centre, as well as the local community. A user board is organized at least one per trimester, where the resident, the family members and the informal caregivers can take part and propose changes, activities and suggestions. They also work with volunteers and associations that usually come along to organize activities. In this sense, the centre functions also as a local equipment for different purposes. There is an outdoor green area in front of the main access to the centre with a park and facilities that work both for children and older people, such as music instruments. That is also the start point of the four reminiscence itineraries developed by the nursing team. These walks welcome anyone to take part, so during the walks, there is also an opportunity for education and raising awareness about dementia. The itineraries follow four slow roads with some spots to rest and learn about heritage and historic moments of the city, so people with dementia can also interact and tell stories from their past.

Best Practice

WZC De Weister is part of the local life of Aalbeke. The reminiscence promenade has been awarded as European Foundations' Initiative on Dementia in 2014. The four principles which this activity relies on are intergenerational relationships, reminiscence and local identity, improving mobility and accessibility and integration into the village life.

Also, the signs for identifying the promenade in Aalbeke were made by local unemployed people, so they could develop new skills and contribute to the feeling of belonging to the community.

	WZC De Weister is based on the person-centred care approach. There are no boundaries around the building, and pedestrian- and bike-friendly environments offer easier connectivity with the rest of the village. Nature is present in all the surroundings. Source: WZC De Weister

	The reminiscence promenades in Kortrijk start from the front door of WZC De Weister, in a public green area. The routes consider the main elements of the village heritage and an accessible itinerary. Source: Pozo Menéndez, 2020

	The design of the spaces and programmes of the building address the needs for creating homelike environments but also facilitate spaces for interaction with the community, such as the main cafeteria/ restaurant. Source: Pozo Menéndez, 2020

Additional References

European Foundations' Initiative on Dementia. https://efid.info/casestudy/31-a-reminiscence-promenade-in-aalbeke-belgium/
Kortrijk. https://www.kortrijk.be/zorg/adressen/woonzorgcentrum-de-weister
M4Architecten. https://m4architecten.be/projects/wzc-de-weister/

3. Woonzorgcentrum De Zon in Bellegem

Type of project Small-scale care home	**Location** Bellegem, Kortrijk (Flanders)
Completion date 2019	**Number of residents** 96, 64 living with dementia and 32 with high level of dependency Organized in 6 units of 16 residents
Client Public Centre for Social Welfare (OCMW)	**Architect/studio** B2Ai architects
Person interviewed Jan Goddaer (director)	**Awards** –

Concept

The urban planning design has been developed by WZC De Zon in Bellegem village, in Kortrijk. This new development of the village consisted of a new care home centre with 96 housing units, 24 social flats and a community centre. The project is articulated around public spaces for pedestrians and cyclists, connecting the existing nearby school with the rest of the village. The *Zomerplein* is the central core of the place between the care home and the social buildings. This square is a pedestrian-friendly space with natural areas, a park for children as well as activities for older people. The connectivity and security of these spaces generate a safe environment for all, so naturalized interaction between people is achieved. The whole operation has been finished in 2019 by the initiative of the Public Centre for Social Welfare of Kortrijk. The care home has been conceived following the principles of the person-centred model and the small-scale normalized living. The centre is organized in 6 units of normalized living with 16 residents each.

The whole project has been developed following the principles of sustainability and energy efficiency, as well as healthy parameters. For instance, all the delivery services and ambulances to the home care centre are integrated in a basement level that resolves the slope difference of the site. In this sense, circulation from cars and pollution and noise have been completely removed from the residential area, developing specific car parking in two perimetral points, offering a safe and friendly accessible space for children, older people and everyone.

The three-floor care centre is oriented following a bioclimatic design, taking profit of the south orientation for liveable spaces. In this sense, all the common terraces for outdoor activities have direct connection with the outdoor park, creating a permanent link with the community life. On the other side, the social housing blocks have been conceived from the "lifelong living" approach, and so, all the flats are completely accessible. A community centre is located in the southwest ground floor of the social apartments. Also, the cafeteria and restaurant from the home care centre are open to the whole village as an open service.

The daily life of the centre is organized in the six housing units, creating a familiar environment for residents and following the principles of the person-centred model. The management is like De Weister's one, with caregivers and nurses that work in teams in each of the convivial units. Each of the units organizes around a living room and a kitchen, which are connected with the terraces mentioned above.

Best Practice

WZC De Zon and the social housing is a new urban development in the city centre of Bellingen village, creating a new connectivity within the existing urban tissue. The principles of proximity, compacity and intergenerational inclusion have been applied when planning this urban development.

The whole project has been developed considering the principles of energy efficiency, with a compact design and high insulation. All the buildings have solar panels on the roofs.

	WZC De Zon share a pedestrian- and bike-friendly environment in the closer surroundings, interacting with the social housing developments and also with the rest of the community of Bellegem. Source: WZC De Zon

	Residents participate in the daily activities of the Centre. Each convivial unity works as a team or family establishing a close relationship of care and trust. Source: WZC De Zon

	Little details are included in the building. The feeling of belonging is constructed through the process of appropriation of the space. Source: Pozo Menéndez, 2020

Additional References

B2Ai architects. https://b2ai.com/en/projects/woonzorgproject-in-bellegem/
Kortrijk. https://www.kortrijk.be/zorg/adressen/woonzorgcentrum-de-zon
WZC De Zon. https://onshartkloptvooru.be/voorziening/de-zon/

4. Huis Perrekes in Osterloo

Type of project Small-scale care home	**Location** Osterloo, Geel (Antwerp)
Completion date 1986 2017 (extension "Villa")	**Number of residents** 64, all with dementia in different stages Organized in 1 unit of 10 day care visitors and 4 units of 16 residents
Client Huis Perrekes	**Architect/studio** Gerard Cools NU architectuuratelier Ester Goris
Local stakeholders Relatives, local home care, primary school, local community, cultural sector, local nature organizations	
Person interviewed Karen Lambrechts (worker)	**Awards** Published in *Flanders Architectural Review* N°14. "When Attitudes Take Form"

Concept

Huis Perrekes is a small-scale project for good living and well-being with dementia in Osterloo, a rural village in Belgium near Antwerp. The centre is part of the village itself, with a total of 4 small-scale normalized living units. It has been created in 1986 with the philosophy of innovation and adaptation to meet the changing needs and desires of people living with dementia.

Huis Perrekes concept offers all the different alternatives that can be necessary for living well with dementia. Apart from the care home service, it also provides care services at home for those people who want to remain in their own environment as long as possible, they offer a daycare service for people who want to spend the day in company and it is also possible to stay for a temporary hospice.

The centre is divided into normalized living environments, in which each house counts with a common living room, a kitchen and 14–16 rooms, depending on each unit. All the buildings are integrated into the surroundings and have a garden, as well as the rest houses of the village. In each building, people with different stages of dementia live together, giving support to each other. The homes are completely normalized, considering stairs as a physical challenge to promote autonomy. Rooms are equipped with a homemade cradle to relax in a calm way when needed. Contrary to many other care home projects, toilets are kept outside the rooms, to make rooms more homelike.

Common spaces such as living rooms and kitchens are designed to promote task integration, with wide spaces to let residents participate in all the daily routines. Multiple activities are embedded in daily life and also in a common workshop where children and visitors are welcomed. These activities promote creativity and working with sensory stimulation. Well-being is based on the concept of a healthy and meaningful lifestyle. For instance, vegetables are grown locally, and bread is homemade as part of the daily routines. The direct contact with nature also provides a healthy environment both for physical and mental care.

Reading workshops with children and schools are regularly celebrated. During summertime, children stay in Huis Perrekes during the day, being involved in the daily activities with the residents and generating community solid bounds.

Near the northern houses, it can be found a big green space for summer events and festivals, as well as a little farm with comfortable areas to sit down and rest. The southern house has been recently acquired and renovated, with the ampliation project by NU architectuuratelier. A green community garden is open to anyone, with swings for children to play. Opposite the main house, a pavilion for cultural activities, recitals, exhibitions, symposia, education or concerts has been developed, creating a vibrant meeting point for all the communities.

Best Practice
Huis Perrekes is an inspiration for care in many senses. All the theories about the health benefits of nature, culture and art, intergenerational meetings, healthy diet, physical activity and contact with animals are all together combined in a normalized living environment, where older people can remain part of the community. It is a project that promotes health for the whole community, not only for the person with dementia. They embrace the well-being of different generations, creating spaces for exchanging, and grow together. The built space accompanies this philosophy: the indoor rooms connect with outdoor spaces and generate common places for intergenerational activities. The little details in design also assure a good quality environment with natural and healthy materials such as wood, assuring good thermal comfort throughout the whole year and natural light for indoor spaces.

Huis Perrekes offers community normalized spaces for cooking, gardening or workshops. Living in a community in a normalized environment is the core of the philosophy. Source: Copyright Stijn Bollaert, NU architectuuratelier

	Green spaces to stay and enjoy nature surround the buildings. They also welcome children to come in and play. Source: Pozo Menéndez, 2020

	Cultural activities are part of the routines at Huis Perrekes. Cultural events, such as concerts, conferences, choirs, training courses and workshops, are celebrated in the Centre in different buildings they have for this purpose. Source: Copyright Stijn Bollaert, NU architectuuratelier

Additional References

Huis Perrekes. https://perrekes.be
NU architectuuratelier. https://www.nuarchitectuuratelier.com/en/project/huis-perrekes/
Stijn Bollaert (photographer). https://www.stijnbollaert.com/

19.2 United Kingdom

5. Dementia-friendly Newcastle upon Tyne

Type of project Dementia-friendly city	**Location** Newcastle upon Tyne (United Kingdom)
Date 2019	**Population** 279,100 hab., 2,469.9 hab/km^2
Person interviewed Lynne Livsey (Newcastle's city council) Joanne Charlton (discovery museum) Catherine Butcher (EIT/VOICE (valuing our intellectual capital and experience) project manager at NICA) Barbara Douglas (member of VOICE and elders Council of Newcastle) Rose Gilroy (co-founder of future homes project and professor at Newcastle University)	**Local stakeholders** City council (promoter), non-profit organizations, NHS, local agents and communities, charities, Tyneside cinema, museums, religious centres, business and shops, municipal services, universities, etc.

Concept

The City Council of Newcastle upon Tyne engaged a strong commitment for dementia-friendly awareness since 2019. In this year, the historical Grainger market, located in the core of the city centre, opened one of the stands with workshops and activities for raising awareness about dementia.

This action plan has multiple strategies in order to make people with dementia remain as long as possible in their homes, with care necessities met, creating a sense of belonging to a place, reinforcing community social cohesion, preventing and promoting both physical and mental health at the local scale and innovating in dementia-friendly initiatives. Raising awareness in order to understand, respect and support all the stakeholders is important in order to create enabling environments for people living with dementia, so they can feel confident and can continue to contribute to their community.

Local stakeholders support this initiative in multiple forms. Some of the most innovative ones are the following:

– VOICE (Valuing Our Intellectual Capital and Experience) at Newcastle University is a community of members of the public where public and carers meet with the common objective of contributing with their experiences to improve research and innovation in collaboration with research centres. Older people and people with dementia collaborate actively in this network. VOICE is part of the UK National Innovation Centre for Ageing (NICA), a world's leading organization on innovation and health, developed in partnership with Newcastle University and the UK Government. The new urban innovation hub where the building The Catalyst is located provides a unique centre for innovation in health and technology with smart solutions. The building has been audited by the Dementia Services Development Centre in order to be a dementia-friendly inclusive building.

- Newcastle University provides training and awareness about dementia for carers.
- Tyneside cinema was founded in 1937 as a newsreel theatre. The current cinema is the evolution of this cultural equipment in the city. The cinema has a screening programme dedicated for people with dementia, as well as their families and carers. The lighting and sound are adapted in order to create a friendlier atmosphere, and classic films are often displayed for reminiscence activities.
- Discovery museums (Tyne and Wear Archives and Museums) have specific programmes for people with dementia, so they can visit the museum in specific times where it is calmer, and a guide can explain all the information in a more comprehensive way. They also organize workshops outside the museum (i.e. in care homes) with memory boxes and historic objects for cognitive stimulation. This kind of activities reinforces the links with the community and makes people with dementia feel more confident about visiting museums and continue to take part in the active life of the city.
- Slow Shopping is an initiative for creating a friendlier environment in local shops for people with dementia or other disabilities. There are dedicated times within the week when the staff is aware of the needs of the shoppers with any cognitive problem. Also, seats to rest and help points are set to facilitate wayfinding or supporting specific needs. Music and light are also regulated to create a quieter environment.
- Dementia-friendly community in Jesmond. This community is committed to becoming dementia-friendly, providing opportunities to be at home with adequate care and services but also engaging all the community (faith centres, business, local transport, administrative services, health care and caregivers, etc.) to be aware of dementia. The Jesmond Dementia Action Alliance organizes sessions to develop knowledge and awareness about dementia.

Best Practice

The commitment of the whole city to become dementia-friendly is a common objective shared by a public general awareness. Both public and private sectors, including businesses and non-profit organizations, are willing to participate in this common public health challenge. Innovation in terms of technology and new services are intertwined with social innovation, from the individual perspective as well as community engagement. The National Innovation Centre of Ageing and new building development are positioning Newcastle upon Tyne as one of the world's leader cities in innovation in ageing.

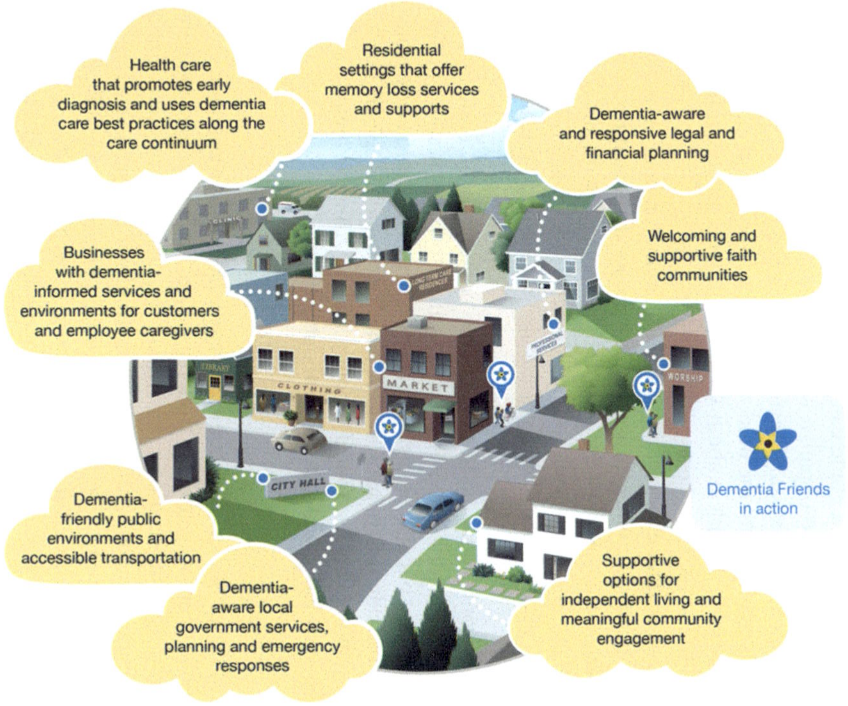

A dementia-friendly neighbourhood and city requires both a built infrastructure and social network. It is important to have an enabling environment with opportunities for housing and services in the proximity and social cohesion for support.
Source: Dementia friends in action

	The National Innovation Centre for Ageing, set in The Catalyst building, is part of a new development of the city of Newcastle, a new urban development integrated in the city, and generating an inclusive environment for innovation and research. Source: Copyright Anthony Foster

	The Grainger market is a central market in the city centre of Newcastle. It is part of the heritage of the city, and nowadays, it has been refurbished, integrating a dementia-friendly information point in order to raise awareness and organize social activities. Source: Brooke-Battersby in Newcastle City Council

Additional References

Alzheimer's society. https://www.alzheimers.org.uk/support-services/Newcastle%2BLocal%2BService%2BOffice/Dementia%2BSupport%2BNewcastle

InformationNow. https://www.informationnow.org.uk/article/dementia-friendly-newcastle/

Newcastle City Council. https://www.newcastle.gov.uk/citylife-news/community/newcastle-sets-out-journey-become-dementia-friendly-city

6. Great Sankey neighbourhood hub

Type of project Multifunction community hub	**Location** Great Sankey, Warrington (United Kingdom)
Completion date 2019	**Population** 203,700 inhabitants (Warrington)
Client Warrington Borough Council LiveWire	**Architect/studio** Walker Simpson architects
Budget £16 million	**Local stakeholders** Warrington Borough Council, NHS England, sports England, Football Foundation, Lawn Tennis Association, The National Lottery, Housing federation: Dementia and housing working group member, LiveWire (operation manager)
Person interviewed Locklynne Hall (LiveWire) John Walker (Walker Simpson architects) (Guest speakers in dementia-friendly design course at Stirling's University (2020))	**Awards** DSDC (Dementia Services Development Centre) – University of Stirling's Gold Award for dementia-friendly design

Concept

The great Sankey neighbourhood hub is a multifunctional building in Warrington, a town of 203,700 inhabitants. The neighbourhood hub is a redevelopment project for an old 1970s leisure centre located in the west side of the town, in a zoning area near residential ones. This building includes a library, quiet rooms and studios, a conference hall and fitness centre, wellness spa and swimming pools. Outdoors also include two football pitches and four tennis courts.

The promoters foresee the development of the hub as a community facility to strengthen social cohesion and feeling of belonging creating an inclusive environment for all ages through sport and cultural activities. The result is a community centre to promote physical activity, health and well-being among all residents. High technology is also integrated in the centre, as part of the services for each programme. The design of the building has met the dementia-friendly standards becoming the first public building in the world to receive the DSDC's gold award for dementia-friendly design, which means a universal accessibility approach for all the wayfinding and signage, as well as accessibility and creating comfortable and legible spaces to feel welcomed. Other dementia-friendly specific features are natural and artificial lightning to avoid glares or high contrast shadowing, accessible toilets, floor finishes and furniture, and a "quiet room."

The staff has also been trained to support families and friends to raise awareness of dementia. Programmed activities in the centre include dementia-friendly fitness classes, quiet hours in the gym and swimming sessions. The wellness spa also offers specialist treatments and products that are suitable for people with dementia as well as people undergoing cancer treatment, such as touch therapy and cancer rehabilitation programme (helping to move, eat and get better quicker after treatments, plus palliative services).

Best Practice

The commitment of the stakeholders to create a community hub that promotes physical activity and well-being has been recognized as the first public building to have a gold award on dementia-friendly design. This recognition raises awareness about dementia and creates a community intergenerational hub to be accessible and inclusive for residents. Both the design and the activities programmed in the centre are adapted to people living with dementia and their caregivers and relatives. The aim is to promote physical activity as a factor to promote healthy lifestyles and prevent many risk factors for non-communicable diseases.

	Great Sankey community hub is a health and well-being multifunctional equipment. Source: LiveWire

	The design offers an inclusive environment for users of all ages. Dementia-friendly programmes are scheduled regularly. Source: LiveWire

	Small quiet areas to relax and meet are important to create more friendly environment. Colours and contrasts have been specially selected under the dementia-friendly standards. Source: LiveWire

Additional References

Dortech architectural systems. https://www.dortech.co.uk/great-sankey-neighbourhood-hub-completes/

DSDC Dementia Services Development Centre. https://dementia.stir.ac.uk/blogs/dementia-centred/2018-03-14/designing-future-primary-healthcare

LiveWire Warrington. https://livewirewarrington.co.uk/194-great-sankey-neighbourhood-hub

Mdg design. https://www.mdg.design/great-sankey-neighbourhood-hub

The design concept. https://thedesignconcept.co.uk/media/25506/uk_case_study_great_sankey_neighbourhood_hub.pdf

The National Lottery. https://www.lotterygoodcauses.org.uk/projects/view/great-sankey-neighbourhood-hub

Warrington Borough Council. https://www.warrington.gov.uk/news/community-hub-becomes-worlds-first-public-building-secure-gold-dementia-friendly-standard

We are library people. https://wearelibrarypeople.com/project/united-kingdom/great-sankey/great-sankey-neighbourhood-hub-united-kingdom/pr/25297

7. **Harmonia village, within CASCADE project: community areas of sustainable care and dementia excellence in Europe**

Type of project Innovation Project, European Interreg 2 Seas Meers Zeeën Programme	**Location** Dover (United Kingdom)
Dates 2017–2023	**Number of residents** Six houses with five rooms each
Client East Kent hospitals university NHS foundation trust (Dover)	**Architect/studio** HMY architects
Budget £2.5 million (Harmonia Village in Dover) 9,324,211 € CASCADE project	**Local stakeholders** East Kent Hospitals University NHS Foundation Trust, Dover City Council, Dementia-Friendly Dover
Person interviewed Henry Quinn (Harmonia Village in Dover, project Lead CASCADE and founder of Harmonia Village, NHS) Samantha Robinson (Harmony House in Kent, project Lead CASCADE and Harmony House Medway community healthcare)	**Partners** East Kent Hospitals University NHS Foundation Trust (UK), Medway Community Healthcare (UK), Canterbury Christ Church University (UK), ZorgSaam (NTH), Emmaus (BE), Expertisecentrum Dementie Vlaanderen (BE), Woon en Zorg H. Hart vzw (BE), HZ University of Applied Sciences (NTH), Université de Lille (FR)

Concept

The Harmonia Village in Dover is a regeneration project in ancient hospital properties, near the Buckland hospital. The aim of the project is to create a new dementia village care home, inspired by the Dutch model, integrating smart sensors and monitoring to better support care in daily life. The objective is to keep people living with dementia integrated in the local communities and so improve quality of life and well-being. The development is in a cul-de-sac street, in continuity with a residential community and in the surrounding area of the Dover's Buckland hospital.

There is an entry gate to assure a safe environment for people that live there and a welcoming "hub", providing communal facilities for all residents and the staff and where it is possible to organize workshops and social activities with the community and neighbourhood. The hub has also a café which serves a daily menu and home-cooked food as well as dining both for residents, visitors and guests, and recreational space to provide respite care for non-residents as well as staff and caregivers.

The rest of the development consists of 6 detached buildings that have been renovated from 12 ancient semi-detached houses. Each one has 5 single rooms for the residents, with common areas such as kitchen and living room. Accessibility is totally guaranteed thanks to a platform life, and the interior design follows the dementia-friendly standards.

The surrounding landscape and garden include spaces for outdoor activities and an area for farming to be in contact with nature and animals.

Best Practice

The CASCADE project is developing a new and innovative model for dementia care facilities and services. The project has two main objectives. The first is to create new facilities for older people and people living with dementia, from which Harmonia Village in Dover has been described. The second is to test a set of telemedicine and IT technology solutions in an innovative way, integrating care in the housing and daily life.

The challenge is to provide sustainable person-centred programmes of care support in local communities. The CASCADE model promotes a holistic, person-centred approach for people living with dementia in order to support autonomy and quality of life. The five key concepts that are shared at the core of the partnership are (i) focusing on individual abilities and desires of the person, (ii) personalizing holistic care, (iii) providing opportunities to have a meaningful life, (iv) encouraging a positive public perception about dementia and (v) creating a safe environment for independent living integrated in the community.

	The Harmonia Village consists of six houses and a community centre for social activities, workshops and cafeteria service both for guests, caregivers and staff. Source: Pozo Menéndez, 2020

	The main gate separates the Harmony Village from the rest of the neighbourhood. It has a cul-de-sac street in a residential and industrial area of Dover. The surroundings are all covered by natural spaces. Source: Pozo Menéndez, 2020

	The common spaces are open to the community, and the design connects indoor and outdoor environments. Source: Pozo Menéndez, 2020

Additional References

Interreg2seas. https://www.interreg2seas.eu/en/cascade
Harmonia Village. https://www.harmoniavillage.org.uk/village-life/

19.3 The Netherlands

8. **De Hogeweyk in Weesp**

Type of project Dementia care village	**Location** Weesp (Amsterdam)
Completion date 2009	**Number of residents** 169 residents with dementia Organized in 27 apartments (small-scale households)
Client Vivium Zorggroep, Huizen (private foundation)	**Architect/studio** MBvD-Architecten Buro Kade
Budget 19.3 million €	**Local stakeholders** -
Person interviewed Eloy van Hal (founder and advisor at be advice – The Hogeweyk® care concept) (Inspirational visit in 2020)	**Awards** Hospitality Care Award 2010 Nominated for the Hedy d'Ancona Award 2010, Excellence in Health care architecture Golden label for Quality by Perspect 2006 Project of the World, Expo 2000 Hannover International Hospital Federation Award 1995, for innovative management Dien Cornelissenprijs 1993 for privacy in nursing homes

Concept

De Hogeweyk was born in 2001 from the idea of excellence in care, creating a homelike environment for long-term care for people with dementia. The concept is a home care centre that integrates many elements of the normal life: a supermarket, a theatre, a restaurant, a cafeteria, shops and services. All these facilities are connected with the residential programme, which is organized in 27 households of 6–7 residents, through accessible and comfortable outdoor spaces that recreate a normal environment that we could find in any other street of Weesp. The result is a safe environment for having a high-quality lifestyle maintaining the normality as much as possible, despite the loss of memory and progressive impairment and disability. The households have been designed with a singular identity grouped in four main themes, in order to respect the lifestyle of the person and create a familiar environment. This assures a continuity with the person's routines, likes and dislikes. The apartments have common spaces such as living room and kitchen and then the private areas for dormitories.

Outdoor spaces include nature and urban gardening facilities, comfortable spaces to seat and rest and artistic and iconic elements that support wayfinding and orientation when wandering around. Urban features to stimulate interaction and enabling environments to enjoy outdoors are located along the whole centre.

All the households are open to the outdoor private space, assuring the safety and security of the residents. Visuals connect indoor and outdoor environments, creating a contact with natural environments from indoor spaces. From the main entry to the apartments' ones, there are different filters of spaces, concentrating the most public activities in the surroundings of the main gate and progressively becoming more private and quieter spaces when approaching the entry of the apartments. Access for services is separated, and this offers a friendlier access in the main gate both for pedestrians and cyclists.

The centre welcomes events and activities in the theatre, so it also creates a connection with the community. The restaurant is also open for visitors, as well as relatives and caregivers. De Hogeweyk is surrounded by a residential neighbourhood, with both open blocks and terraced houses, with big sport courts nearby, a primary school and a religious centre. The west area is developed with tertiary activities, including many different services. When the dementia care village was created in 2009 in the area called Hogeweyk, Weesp was a small satellite town in the periphery area of Amsterdam, and the centre was in the limits of the village, far from the historical urban centre. Nowadays, recent urban planning strategies in Weesp have planned to create more links and connections between the centre and the community, as a unique equipment in the town.

Best Practice

The innovation of De Hogeweyk was to question the long-term care model from an integral perspective. The result has received many awards and recognitions because of the innovation in management and care. Ten years after the inauguration, De Hogeweyk is an inspirational model for many other long-term developments across Europe, for instance, in Ireland or France. From the point of view of urban planning, the most interesting question is appearing in the recent years, when the council of Weesp has included the home care centre in the strategic plan of the city for creating more social cohesion within the community and intergenerational activities.

	Outdoor public spaces are designed with features and elements that promote social interaction and stimulation. Source: Pozo Menéndez, 2020

	The fountain offers more comfort during the summertime, and it attracts ducks and other birds. Source: Copyright Vivium

	Comfort spaces for seating and rest are located around the whole centre to offer a friendly environment to wander around. Source: Pozo Menéndez, 2020

Additional References

Be advice. www.bethecareconcept.com
DVA and be advice. https://hogeweyk.dementiavillage.com/
Weesp City Council. http://www.welkominweesp.nl/
Vivium https://www.vivium.nl/verpleeghuis-dementie-hogewey-de-hogeweyk-weesp

9. **Kwiek beweegroute in The Netherlands**

Type of project Urban exercise route in the neighbourhood	**Location** Nuenen, Eindhoven (visited) Belgium and the Netherlands
Completion date 2013 onwards	**Architect/studio** Denovo design, Janne van Kollenburg and STUDIO LVWP
Client Own initiative	**Local stakeholders** City councils, care homes, community centres, hospitals
Person interviewed Manuel Wijffels (Denovo design team)	**Awards** UKON Uniek Prijs 2014 – first prize Your Street Design Indaba 2012 – third prize

Concept

The challenges of demographic transformation in the Netherlands arose the question to designers, in order to innovate on urban design for older people. The Kwiek route was awarded for a first pilot, and now it is a concept that has been implemented in more than 50 cities in northern Europe in The Netherlands and Belgium. It also offers an indoor version adapted to hospitals. Kwiek route promotes active ageing by designing a promenade for exercising outdoors, using the existing opportunities of the built environment and urban furniture.

Kwiek is a complete process for engaging older citizens, community centres, gyms or care homes in the appropriation of the outdoor surrounding spaces. The design team studies in collaboration with the local stakeholders the best places to integrate an accessible and friendly route for older people with different spots to train balance, flexibility, endurance and resistance.

The route supports the physical activities delivered by a physiotherapist or fitness coach, but it also creates an opportunity to meet the people in the community and to create more cohesion within the neighbourhood. The design team identifies the best spots in the surrounding areas where it is possible to create a safe environment to exercise outdoors. Benches, street lighting posts, sidewalks, fences, etc. are some of the urban furniture that are creatively used for this purpose.

The design is a set of elements that are integrated in the urban environment to create the itinerary.

Best Practice

The originality relies on the cost-efficiency of the design and resources and the social positive impact of the whole process. It consists in a small intervention with a high impact. Tiles with the exercise instructions are installed in the pavement, as well as some landmarks that are additionally installed in order to make it easier to read the itinerary.

	Benches can offer support to do stretching exercises in group. Source: Kwiek beweegroute

	Different spots are selected to train different skills for being fit. Source: Kwiek beweegroute

	Public spaces are used by all the community, raising awareness about the need for walkable and accessible environments. Source: Kwiek beweegroute

Additional References

Kwiek beweegroute. http://kwiekbeweegroute.nl/

19.4 Spain

10. **Mixed-use building with collective housing for elderly people and health public equipment in Barcelona**

Type of project Collective housing for elderly people and civic and health centre	**Location** 22@ District, Barcelona
Completion date 2016	**Number of residents** 105 senior citizens housing, organized in 3 blocks The base is a double height plinth with a health centre and a senior community centre
Client Patronat municipal de l'Habitatge de Barcelona	**Architect/studio** Peris+Toral Arquitectes; Bonell i Gil
Budget 20,227,969 €	**Local stakeholders** Barcelona's council, regional government, municipal services of Barcelona
Person interviewed –	**Awards** European Union Prize for Contemporary Architecture – Mies Van der Rohe Award 2017

Concept

Barcelona is a very dense and compact city. This project is one of the social housing projects for seniors in Barcelona. They are specific housing apartments for older adults promoted by the Barcelona City Council, in order to create housing solutions for active ageing and self-autonomy.

This project is in the urban renewal project of Barcelona 22@ in Poblenou, near Glories' square. The programme's challenge was to develop a mixed-use programme combining different real estate developments: 105 social housing for the city council, a health centre for the regional government, a senior community centre for the district and a parking area for municipal services of Barcelona. The whole district has been activated in a complex urban regeneration project to generate a diverse city tissue, with high environmental quality and good quality of life.

The building generates a continuity between the different interfaces of the Barcelona's Eixample and the Diagonal street limits. The residential buildings have common facilities in each block, such as a communal laundry in each top floor, a covered porch with clothes lines and a terrace for urban gardening. Also, there is a polyvalent room which is connected with an outdoor space. This space is an accessible terrace that is usually used by the social workers in order to organize workshops and activities. The apartments always include a terrace and are organized around a central core for services.

The whole development is connected to a district heating system, a more efficient service for sanitary hot water and central heating.

Best Practice

There are more than 1,300 housing solutions like this one in Barcelona, being a reference for social housing in Europe. There is a very interesting contrast between the high density of the city and intense dynamics and the community homelike environment created in the interior of the buildings. The management usually integrates a maintenance of the housing and also the community, which is normally entrusted to cooperatives who submit their proposals about the services and programmes to activate the common spaces.

	22@ smart district is an innovative urban renovation in Barcelona near Glories Square and Poblenou Eixample. Source: Copyright José Hevia, Bonell i Gil, Peris+Toral

The social housing has common outdoor spaces for workshops and activities for older residents.
Source: Copyright José Hevia, Bonell i Gil, Peris+Toral

	Accessibility is the key factor for creating an enabling environment to promote physical activity and healthy ageing. Source: Copyright José Hevia, Bonell i Gil, Peris+Toral

Additional References

Afasia. https://afasiaarchzine.com/2017/07/bonell-i-gil-peristoral/
Arquitectura Viva. https://arquitecturaviva.com/works/105-viviendas-para-mayores-9
European Union prize for contemporary architecture Mies van der Rohe Award. https://miesarch.com/work/3582
Peris + Toral Arquitectes. http://peristoral.com/proyectos/lugares-colectivos

19.5 Finland

11. **Kotisatama in Helsinki**

Type of project Co-housing for seniors	**Location** Kalasatama, Helsinki (Finland)
Completion date 2015	**Number of residents** 141 residents Different typologies include 22 double room apartments, 34 larger apartments and 63 studios
Client Aktiiviset Seniorit ry (active seniors association)	**Architect/studio** Kirsti Sivén and Asko Takala architects Project architect: Milja Nykänen
Budget 16,135,000 €	**Local stakeholders** Senior association
Person interviewed Helsinki city planning department, guided visit	**Awards** -

Concept

A new smart district in Helsinki, Kalasatama, is being developed in the surroundings of an old harbour and industrial area. It consists of a 175 ha area which is expected to host 25,000 residents and 10,000 jobs by 2030. The development of this area of the city was based on a high-efficient and clean urban infrastructure for energy and smart grid both for cooling and district heating, as well as a high-quality district for healthy lifestyles and smart services. Technology innovations and start-ups have been incorporated as part of the project. Waste management is solved by a pneumatic system integrated in the building blocks.

Public spaces of Kalasatama include public art and sculpture landmarks, to create identity and support wayfinding, as well as to promote mental health. Streets are accessible and priority is given to pedestrians and cyclists.

One of the housing developments is the Kotisatama co-housing for senior citizens. This housing development is located near a bike lane and the new metro station, also close to health and well-being centres, offering a wide range of health care and social support. It also connects by a pedestrian bridge to the island of

Mustikkamaa, a green open space for walking. The apartments are mixed, to welcome multiple profiles of older adults although all of them need to be completed independently. The building is free of architectonic barriers, and both indoor and outdoor common spaces acquire a relevant interest as they are the main spaces for common activities and services, such as workshop rooms, a canteen, two saunas, a small gym, terraces, offices, a library or a laundry room. There is also an apartment for guests and many gadgets and facilities to enable a more sustainable lifestyle, for example, a little shower in the front door to clean the pets or the bikes, two community bike shelters as well as screens in common areas in order to have all the real-time information, such as the forecast or the public transport schedules.

The active participation of the future residents (investors of the project and members of the cooperative) during the project design phase was key to identifying needs and desires.

On the other side of the block, a social housing building was also constructed by the Department of Housing of Helsinki.

Best Practice

This co-housing for senior adults is a high-quality environmentally performance building integrated in a smart and healthy district that is still under construction. The context in which it is located is a great urban renovation project that brings all the best practices for energy efficiency and reduction of the emissions of carbon dioxide. The motto of Kalasatama district is that all "*new residents will save an hour per day*" thanks to the good connectivity and frequency of the public transport. The co-housing also has invested in a good quality habitat with multiple common spaces and services. The result is a building block with three different types of apartments, in a pedestrian- and bicycle-friendly environment with green spaces and public art in the urban landscape.

	Outdoor common spaces have been designed to promote social meeting and outdoor activities. Source: Copyright Tuomas Uusheimo

	Landscape design and sustainable management of the water and green spaces are important elements that have been considered in the design. Source: Copyright Tuomas Uusheimo

	The building's façade is all in brick. Waste management is integrated in the building for a pneumatic collection system. Source: Copyright Tuomas Uusheimo

Additional References

Fiksu Kalasatama. https://fiksukalasatama.fi/kotisatama-housing-community-for-elderly-people/
Forum Virium Helsinki. https://forumvirium.fi/en/kotisatama-housing-community-for-elderly-people/
This is Finland. https://finland.fi/life-society/enabling-active-ageing-in-finland/
Sivén & Takala. https://arksi.fi/en/kotisatama-community-leonsatama-helsinki/

The eleven study cases illustrate the different tendencies that are emerging in different countries across Europe. The selection tried to illustrate best practices and innovative projects. Table 19.1 shows a general overview of these cases. There are all projects from the last two decades; the oldest ones are the Foton community centre in Brugge (Belgium) and De Hogeweyk in Weesp (Amsterdam). Many of them are located in urban or suburban contexts, usually with a good connectivity by public transport (bus or tram) and with a bike parking in the immediate surroundings, or integrated in a sheltered space. The main entry and the surrounding spaces are pedestrian-friendly, with green areas around or trees. The buildings are also designed respecting the views to outdoor spaces from windows.

The selected examples are goodly connected with natural spaces, sometimes they are part of the building development; othertimes the project is near a public space with green areas. Public outdoor spaces and contact with nature are the two main characteristics that are usually found in the proximity. It is interesting to note different filters of intimacy that are applied both at the urban scale (streets, squares, courtyards) and the architectural project (terraces, shared spaces, common rooms, living rooms and kitchens, etc.). In turn, all of the building projects include common spaces for meeting and shared services. Depending on the level of dependency, the facilities and equipment differ, as well as the necessity of specific needs and additional spaces for caregivers. For instance, the selected care homes integrate activities and spaces that welcome visits and intergenerational activities from local communities. Senior homes also have common spaces for social activities.

The heterogeneity of the projects makes it difficult to establish any tendency or comparative element in terms of design parameters, but it is clear that age-friendly spaces are becoming more important in urban design and planning. We can recognize a continuity in the relation of the building with the green outdoor spaces, the accomplishment of the accessibility standards, the need to provide small-scale spaces and an integration with the built environment local landscape. The importance of design and planning with an age-friendly perspective is key to creating comfortable and attractive spaces for older adults but also to provide opportunities to intergenerational activities. The integration of care services and home care centres in urban planning and design can develop other compatible uses that are aligned with the principles of a healthy and sustainable territory. Multi-scale alliances are required in many of the cases to find common and complementary interests and develop new urban programmes that optimize resources as well as they promote wellbeing and social cohesion. On the other hand, multidisciplinary teams are also necessary in order to understand the current challenges with a global perspective

Table 19.1 Synthesis of the main characteristics of the 11 study cases across Europe

		Identification and main characteristics of the centre					Connectivity and accessibility			Relation with other activities and surroundings		Relation with nature	
		Project	Location (U= urban, S= suburban, R= rural)	Type of project	N° residents	Completion date	Parking	Public transport (B=bus, T=tram, t= taxi)	Connectivity/ accessibility (b=bike, f= low quality, F= high quality)	Other activities r<400 m	Perimetral limits	Green areas	Accessibility/ connectivity with surrounding green areas
Belgium	1	Foton in Brugge Biskajersplein 3, 8000	U	Community house and information point for DF-activities and support	-	Fotonhouse open on 2005	Restricted area, there are reserved parking places for handicap and logistics	(B) 80 m	(b) bike parking (F) pedestrian-friendly area	residential, primary school, health care services, shops, cafés, hotels, museums, art galleries, touristic attractions	windows are oriented both to the public square and the indoor court-yard	No	The entry is a public square with big trees (5) and benches and places to rest
	2	WZC De Weister in Aalbeke Lijsterstraat 2, 8511	S	Small-scale care home	46 p.	2012	Restricted parking for the service and logistics and available places nearby	(B) 190 m	(b) sheltered bike parking (F) pedestrian-friendly area	residential, urban centre, shops, groceries, cultural centre, restaurants, cafés, banks, primary school, library, industries	windows are always connected with green spaces, terraces and gardens from the centre and public space	Yes	Main entry is a green space and park. The centre also has green terraces and gardens for outdoor activities
	3	WZC De Zon in Bellegem Zomerplein 4, 8510	S	Small-scale care home	96 p.	2019	Restricted underground parking for the service, residents and logistics	(B) 290 m	(b) sheltered bike parking (F) pedestrian-friendly area	residential, school, social housing, urban centre, shops, groceries, restaurants, cafés, services	windows are always oriented to green spaces and open public space	Yes	Main entry is a public square with green areas and equipment for children and older adults
	4	Huis Perrekes in Osterloo Oosterloseweg 92, 2440	R	Small-scale care home	64 p.	1986 2017 (extension "Villa")	Public parking in the immediate surroundings	90 m	(b) bike parking (F) pedestrian-friendly area	residential, town centre, school, pharmacy, hospital, restaurants, groceries, church, health services	windows are always oriented to green gardens and outdoor public space	Yes	Each house has its own garden, plus the farm, plus the community garden for children and surrounding nature
		Identification and main characteristics of the centre					Connectivity and accessibility			Relation with other activities and surroundings		Relation with nature	
		Project	Location (U= urban, S= suburban, R= rural)	Type of project	N° residents	Completion date	Parking	Public transport (B=bus, T=tram, t= taxi)	Connectivity/ accessibility (b=bike, f= low quality, F= high quality)	Other activities r<400 m	Perimetral limits	Green areas	Accessibility/ connectivity with surrounding green areas
United Kingdom	5	Dementia-friendly Newcastle-upon-Tyne	U	Dementia-Friendly City	-	2019	-	DF-campaigns for bus drivers open labs with VOICE to design universal accesible new trains	the city is not yet pedestrian-friendly nor bike-friendly, although there are progressively urban transformations, such as The Helix	Dementia-friendly communities and initiatives are spread around the city, such as the cinema, museums, churches, services	-	-	the city is working towards introducing more nature-based solutions and green infrastructure
	6	Great Sankey Neighbourhood Hub Barrow Hall Lane, Great Sankey, Warrington, WA5 3AA	S	Multi-function Community Hub	-	2019	Big parking areas are located around the hub	(B) 110 m	(b) bike parking	residental (low density), college, school, nursery, services, industries	the Hub is open to the neighbourhood, with transparent facades	Yes	there is a lack of designing with nature in the immediate surroundings, although there are multiple green areas in the surroundings
	7	Harmonia Village, within CASCADE project Randolph Rd, CT17, Dover	S	Innovation Project, care home + hub	30 p.	2020	Parking area in front of the main entry	(B) 190 m	(b) bike parking (f) pedestrian-friendly area	residental (low density), hospital, school, home care, industries, shops	all the houses' windows have views to green areas or public common spaces	Yes	each house is surrounded by green areas, a natural space is in the immediate limits of the village care centre
The Netherlands	8	De Hogeweyk in Weesp	S	Dementia Care Village	150 p.	2009	Parking areas in some of the surrounding streets	-	(b) bike parking (F) pedestrian-friendly area	residental, primary school, moskee, shops, services, sports centre, industries,	all the windows have views to the green public courts or surrounding streets with trees	Yes	all the entries are through a green quiet space, the main gate is also in a pedestrian-friendly street
	9	Kwiek beweegroute	all scales	Urban exercise route in the	-	2013 onwards	-	-	(F) pedestrian-friendly area	it connects day care or care home centre with other spaces across the city/town	-	Yes	the routes are designed taking into account the pedentrian-friendliness, comfort and connection to green areas and public spots
		Identification and main characteristics of the centre					Connectivity and accessibility			Relation with other activities and surroundings		Relation with nature	
		Project	Location (U= urban, S= suburban, R= rural)	Type of project	N° residents	Completion date	Parking	Public transport (B=bus, T=tram, t= taxi)	Connectivity/ accessibility (b=bike, f= low quality, F= high quality)	Other activities r<400 m	Perimetral limits	Green areas	Accessibility/ connectivity with surrounding green areas
Spain	10	Mixed-use building with collective housing for elderly people in Barcelona Carrer de Bolivia, 45X, 08018	U	Collective housing for elderly people and civic and health centre	105 p.	2016	Parking integrated in the block, public parking in the streets	(B) 100 m (T) 350 m (t) 100 m	(b) bike parking (F) pedestrian-friendly area	mixed area, city centre, residential, offices, health and social centres, cultural centres, shops, groceries, restaurants, hotels, services, administration	windows are open to outdoor spaces (streets, public areas) and common spaces of the blocks, the blocks are integrated in the city	Yes	the housing blocks are located near a pedestrian area with green areas and trees, streets have tree aligments and there are public parks in the surroundings
Finland	11	Kotisatama in Helsinki Arielinkatu 8 B, 00540	U	Cohousing for seniors	141 p.	2015	Parking integrated in the block	(B) 100 m (T) 290 m	(b) bike parking (F) pedestrian-friendly area	mixed area, residential, nursery school, school, gym, cultural centres, shops, groceries, restaurants, services	windows are open to outdoor spaces (streets, public areas) and common spaces of the courtyard	Yes	there is green areas in the public spaces and there are public parks in the surroundings

Source: Author's own

and to bring innovation adapted to the local context. New developments include good quality performance certificates and designs, both for the building and for the urban scale, but also new technologies, services and IT devices. Proximity, mixed uses, sustainable mobility, urban gardens and food production, biodiversity and water management, as well as heritage and urban landscape are elements that need to be intertwined and integrated under a common strategy. Age-friendliness requires a pedestrian-friendly urban space, in order to be accessible and safe, but also to be bike-friendly, as it is important to have separate spaces both for pedestrians and bikes, and promoting cycling reduces the place of the car as well as air and noise pollution associated with traffic. Also, green spaces with multiple functions are required to create intergenerational places with different elements both for calm and relax and passive activities, as well as programmes and elements for exercising and engaging in multiple activities, such as gardening, board games, reminiscence activities or doing exercise. Finally, both indoor and outdoor spaces for intergenerational meetings require a welcoming design for all ages with multiple choices for interaction and stimulation.

Chapter 20
Learning and General Principles for Healthy and Inclusive Design Considering Regional Context

Elisa Pozo Menéndez and Ester Higueras García

Abstract The rapid growth of the population and densification in cities is creating new challenges in a transforming scenario, where care systems' frailty, climate change and unhealthy lifestyles triggers well-being. Taking into consideration the demographic trends and the need to reduce health inequities the need for a common objective for improving quality of life and reducing inequities is key to generate prosperity and wealth. "*Health in all policies*" presents a framework that integrate a multisectoral perspective where health and well-being is intertwined with sustainability. Each contributor has been asked to answer the three questions proposed at the beginning of the book. In this final chapter their perspectives are discussed in order to better understand the challenges and learnings from the different contexts across Europe.

Which is the main challenge for addressing urban design from the perspective of ageing? How can societies transform into value having inclusive environments for all ages? What are the main strategies in cities that should be implemented in urban planning and design to encourage ageing in place? Urban planning and design require a major transformation in order to provide healthy and inclusive environments for all, as well as providing alternatives for services, housing and communitary support. The three questions proposed at the beginning of the book have been discussed by each contributor in order to present a multidisciplinary and international vision across Europe. The urban phenomena and the built environment are motivating multiple research from different perspectives and disciplines all over the world. The demographic trends and the objective of reduce health inequities are issues that are closely related with the challenges of economic and environmental sustainability, ecology and social well-being. They are indeed common challenges

E. Pozo Menéndez (✉) · E. Higueras García
ABIO Research Group, Universidad Politécnica de Madrid, Madrid, Spain
e-mail: e.pozo@upm.es

E. Pozo Menéndez, E. Higueras García (eds.), *Urban Design and Planning for Age-Friendly Environments Across Europe: North and South*, Future City 19,
https://doi.org/10.1007/978-3-030-93875-8_20

that are requiring innovation and a more flexible and multidisciplinary approach, in order to better understand the present to create evidence-based decision making and proposals that help to optimize resources and to develop solutions with higher positive impact.

20.1 Main Challenges for Addressing Urban Design from the Perspective of Healthy Ageing

The main challenges addressing a health equity-oriented urban planning are to redesign or adapt the urban infrastructure for a life course approach that is capable to answer the evolving needs of the population. Planning the cities with an ageing-inplace perspective requires to think about all groups of ages' needs and consider the multiple opportunities to build a safe and healthy environment. Freitas and Santana recalled the impact of urban environment on health, stating that urban planning should recognise the three main components for a successful ageing process: avoidance of disease, maintenance of functionality and active engagement in public life.

Román López, Córdoba Hernández and Eiroa Escalada established the common challenges that a sustainable and healthy city should answer. Baquero Larriva pointed out the physiological requirements of older people for designing urban environments. She argues that climate change and air and noise pollution will affect more to older adults. In this sense, addressing urban planning and the design of urban spaces from this perspective of vulnerability will enable friendlier and healthier environments for children, for people with different abilities and for all the citizens. Picard also reinforces the need for social inclusion in the process of urban design. Cities are often built from the perspective of younger professionals, imposing an ideal that is not always answering the needs of everyone. In this sense, the author Pozo Menéndez also drove attention to mental health, which is one of the major challenges nowadays and affecting all ages and sectors of population. Designs of public spaces – streets, squares and parks – have an important role to also promote mental health, not only with strategies like citizen participation in part of the design process, but also create specific elements that can support the future development of memories and well-being. For example, spaces with public art, regulation for urban landscape, biodiversity and thematic activities or itineraries. Baert, Cornelis and De Vriendt are aligned with the previous considerations, and they suggested that this is also a way to warrant better proposals, as designers can consider the wishes and dislikes of persons with dementia and the carers, deriving in fewer problems and unwanted features afterwards. Douglas completed these perceptions with the need to mediate and establish ways to look for a consensus in the process of development and design, establishing a balance between different stakeholders.

In this sense, Calvo and Ondiviela support this need for inclusion addressing a wider perspective to combat ageism in general; considering the professional sphere, gerontophobia and improvement of intergenerational relationships; changing the pension model; achieving healthy ageing, so it reduces the social assistance needed; and adapting homes and cities (public spaces, transport, shops and services). Reducing ageism, stereotyping and discrimination is also acknowledged as a challenge by García-Soler et al., who also reinforced the importance of considering ageing as a life course itinerary, and in this sense, prevention and promotion of healthy habits to exercise the body and mind, good nutrition, control over stress and care of social networks are encouraged to reduce the risk of developing chronic diseases and avoiding frailty situations.

In terms of innovation, Castro Gonçalves reminded the need to work from the innovation sector with the network of associations, which represents different categories of older adults and supports the exchange between the innovation labs in territories and the accessibility to the final users that require a closer approach. Integrating older adults into governance of living labs, for instance, could be also an inclusive measure to put their voice at the centre of the process of co-designing innovation, housing models, services or urban spaces. The multiscale approach is again highlighted, not only for the design itself but also for the processes that occur in the city. This perspective is included in Hamza's methodology to transfer this knowledge to younger generations or architects and innovators that will have to face the greater impact of the current challenges.

However, this is directly related with the characteristics of the place where older people live, as this impacts their health status and quality of life. For Rodríguez-Blázquez and Forjaz, building age-friendly environments that guarantee all the specific needs for older people should be considered when designing living facilities. Taking into account that ageing in place is the preferred option for most persons as they age due to the special meaning that places have for people and the sense of control over their own life, it is important to include regeneration and renovation programmes as well as alternatives at the scale of the neighbourhood to enable accessible and adapted spaces with services for assuring the social conditions and elements to allow people to age in place.

This perspective is also illustrated by Gallerani, who questioned the processes of gentrification in cities wondering the mechanisms to reimagine our social places and how to make people remain in the place they chose. Aligned with this question, Padeiro et al. put the emphasis on the reduction on the social risks that are usually related with the modalities of urban management and development, with a domination of the entrepreneurial managerialism, the financialisation of the land and real estate market and an overall acceptance of the growing of the touristic sector in the most attractive neighbourhoods of the city.

Fernández-Mayoralas et al. complete this vision about living environment showing evidence on how much it contributes to social interaction and participation and how this is related with the self-image of the process of ageing depending on the type of residential environment and the relationships that are developed between the inhabitants, either friends, co-housers or mates or support and service staff.

The analysis and diagnosis of the situation of healthy ageing are common frameworks in many countries. This challenge is also identified by Gilroy, specifically on the housing developers in the UK, to continue to promote segregated spaces, while integration of communal spaces and facilities is emerging as a social need. Lantarón, Sancho and Azurmendi accompany this vision from the learnings from Denmark and the application in the Spanish context. They also envisaged the necessity of flexibility of the new housing and accommodation models, in order to adapt to constant changing and evolving process, and also the need of variety, to satisfy the heterogeneity of an important group of the population. Rojo-Pérez et al. also recognised the same challenge from the co-housing initiative perspective, as the location and construction features of these kinds of residential facilities are directly determined by financial and availability of land factors, as well as cultural and geographical ones. However, in these initiatives, there is already a common vision of the additional value of shared spaces and services for community engagement.

20.2 Inclusive Environments for All Ages to Healthier and Better Quality Cities

The word 'inclusivity' indicates an invitation to the people of the neighbourhood to join in an ever larger and more solid community. For Gallerani, inclusivity should be difference of integration, putting the emphasis on the inclusion concept, that embraces the uniqueness of the individuals and minorities as a value, instead of the concept of integration, which sometimes focuses on the "homologation" of a minor group to the rules of the more deeply rooted community. The "city of proximity" urban model has recently emerged as the post-pandemic city paradigm for enabling healthier and more inclusive environment with an accessible distribution of urban services, at the same time it promotes active mobility as distances are shorter. This city of proximity also interacts with the ideal city for elderly – Silver City – presented by Calvo and Ondiviela, a combination of the Attractive City for talented citizen drivers (Ondiviela, 2020). Thus, the best Silver City would be a top Attractive City for talent, with a strong identity and special emphasis on opportunities for a healthy lifestyle. But it would also require a greater heterogeneity and mix of uses in a smaller scale.

Baquero Larriva also corroborated this perspective with the results of her specific study cases. She found that the spaces offering the greatest variety of activities and users were those who attracted the greatest number of older adults. It has been proved that intergenerational activities are very important for the psychological and cognitive development of all citizens, it is part of social participation and it helps older people to feel useful and less lonely in big cities. This intergenerational perspective and need for urban innovation are shared also by Freitas and Santana, addressing the multiple perspectives to adequate community services (integrated care), housing (adaptation, accessibility, maintenance) and efficiency of the public transport and embracing the way people move and improve the service to the interests and needs of an ageing population. Pozo Menéndez also argues that the feeling of belonging with a place and a neighbourhood contributes to respect the public spaces; to improve the social cohesion and the trust, which is directly related with reduction of mortality, as well as cognitive impairment; and to reduce costs of the health systems.

Moreover, as Soja (2010) argues in "*seeking spatial justice*," justice has a geography, and so the equitable distribution of resources and services and access to them are basic human rights. When a city is creating environments where all people are respected and have opportunities to participate and contribute, it is also planning for health equity, for all ages, genders, social positions, health status or disabilities. The vision from active ageing strategies in Japan as one of the most longevity places in the world is compared with France, as European example, by Picard. Her work helps to better understand social innovation in both countries to provide new solutions to identified needs that have not been met yet.

When inspiring other innovative projects, it is important to understand who are the principal agents that initiate the innovations. Depending on the status of the projects' instigators, social innovation emerges in coherence with or in disruption with the pre-existing systems. The unprecedented socio-demographic developments and the issues they create are calling for innovation and unprecedented solutions. This is also related with the sociocultural differences, and so this sensitivity needs also to be applied between the different countries in Europe, considering the local contexts to adapt the innovations. Douglas suggested to create opportunities for citizens to engage in negotiating what makes an inclusive environment for all. This can help to build understanding amongst citizens as well as between citizens and professionals of what creates an inclusive environment and gives everyone a stake in making places the best they can be.

Castro Gonçalves discussed also the challenges concerning the convergence of economical and inclusive approaches, the development of a broader understanding of the smart city that goes beyond the digital approach to include more of the social dimension and finally the strengthening of the participation of the elderly in the co-design activities and decision-making process.

García-Soler et al. stated that the design of support for the care of people during their life course is highly complex given the different areas involved, the levels of analysis and the agents and beneficiaries. Thus, the challenge is to coordinate the design itself in a common and holistic approach, instead of setting objectives for specific areas and attending to variables at a single level of analysis. Again, considering the necessities of the beneficiaries and looking with them at the solutions from a co-creative approach, in many cases, will show the requirements for coordination of care and so incorporating them into the design objectives.

Regarding innovation, new technology and services are addressed by Blázquez-Rodriguez and Forjaz. They recalled that recent developments such as assisted living technologies and new housing models that have been presented in different study cases can help older people to remain at a comfortable and functional accommodation that covers their needs. Smart homes and monitoring systems can provide additional care, support and safety for older people to remain at home. Also, when ageing in place is no longer possible, long-term care facilities can provide a supportive environment, compensating the loss of functional capacity, independence and social network some older people may suffer. In this sense, residential characteristics, highly skilled staff and the implementation of psychosocial interventions are required for creating an environment that promotes healthy and active ageing and improves older persons' quality of life.

This improvement of the quality of life and renovation of the built environment, combined with technologies and services, is a desired itinerary, but it requires to develop an integrative process co-constructed with older people themselves. Padeiro et al. recall that in this moment, when the shape of cities is called to be redesigned, as well as the proximity of services, and inclusive design, it is important to consider that the established model for urban production is mainly favourable to the private sector. Building more inclusive environments through a process of improving urban design can improve the conditions of space, but not necessarily the conditions of the people. It is important to reconcile urban planning imperatives with local social needs, which implies a revaluation of the public competencies and its collective service mission. Fernandez-Mayoralas et al. highlight the need of public authorities' commitment to further develop and promote three specific areas amongst the WHO's Age-Friendly Cities and Communities framework that might influence a positive and non-stereotyped self-image amongst older adults, which are social participation and inclusion, physical environment and civic participation.

For Gilroy, the COVID-19 lockdown response is seen as an opportunity to recuperate, both as individuals and as society, and move to a renewal paradigm. Part of this process is the re-evaluation of the neighbourhoods and the home, understanding the importance of the community support that may have provided a new set of aspirations centred on well-being and nurture and a broader understanding of sustainability that embraces both social and environmental commitments.

In this sense, Lantarón, Sancho and Azurmendi highlighted the potential that older people as a group represent for the society and the great opportunity that arises for professionals from this challenge. Inclusive environments to participate are needed for the definition of a housing model that can be adapted for multiple stages of life and thus care needs but also consider sustainability and the local context in which they are located. In addition, Rojo-Pérez et al. drove attention on the necessity of including the gender perspective. In the context of new residential and care models, there is a higher demand of better understanding and collaborating amongst all the fields and sectors involved, including researchers, policymakers, service providers and older people themselves. Baert, Cornelis and De Vriendt added the need for citizens to understand the challenges of ageing and dementia, before it is possible to take age-friendly or dementia-friendly actions. In this sense, it is also important to ensure that the level of information is appropriate to the audience, as raising awareness is different to attending an accredited training course. Also, local governments can share good practices or form learning communities.

20.3 Main Strategies to Be Implemented in Urban Planning and Design to Encourage Ageing in Place

Ageing in place requires a collaborative approach to health and urban planning. From the research of Freitas and Santana, it is highlighted that the local governments should also work together with the public health leaders and promote citizen's participation in order to be more sensitive to needs and demands for a healthier lifestyle. Innovation is required for social interventions and methodologies for participation and co-creation processes, in order to overcome existent barriers to improve health and well-being, considering all population groups and the scale of the neighbourhood, but also for regulations and administrative processes.

The study case by Picard from Japan and France offers this vision of opportunity to develop social innovation by the elderly people themselves. But cities need both from the built environment and the social structure. Indeed, self-perception differs depending on the residential environment in which the person is living. Fernandez-Mayoralas et al. have shown how personal, social and health-related determinants of active ageing are related with the residential environment, and so it might play an important role in whether the person takes part in activities and initiatives in their environment.

In this sense, fostering social innovations as strategy in cities and neighbourhoods can encourage inclusivity and ageing in place. First, the initiatives and the process enable elderly people to remain independent in their living environment. Second, the innovations are developed for and by the elderly. And this has also to consider both the public and private innovations, as a complete ecosystem of research and improvement. Blázquez-Rodriguez and Forjaz agree in this sense, considering mandatory to commit older adults in the design of age-friendly environments as well as to promote research that evaluates the effect of interventions on older adults' quality of life. Moreover, García et al. completed this vision with the additional interest to include through the participatory process a research approach in order to document developments in a descriptive way but also to establish objectives, indicators and metrics that help to monitor the development of the project as well as the main outputs. They also highlight the need of considering long-term strategies to make it possible to accompany the project and consolidate the new behaviours and practices into a new culture of care.

Padeiro et al. also recalled the need to incorporate global, intergenerational and socially inclusive dimensions in the post-COVID planning measures. The COVID-19 crisis has highlighted the need to consider management and services in urban planning and drive attention to the frailty group of the older people. Gallerani also recognised the need to build a strong solidarity between the inhabitants of a neighbourhood as the only way to undermine the individualism and loneliness that are contributing to the territorial and urban inequities. From the design point of view, local public space has also to be redefined considering all the ages and social categories, but reducing the place of the automobile is the first step in this sense with important healthy outcomes from this measure.

Baquero Larriva's research identified some environmental factors, subjective perception of comfort and design elements that influence the use of public spaces by older adults. These factors are important to design more inclusive, accessible and healthy urban spaces. Some of those strategies, which have been also identified by Román López, Córdoba Hernández and Eiroa Escalada, should be greener public spaces protected from wind, noise and sun or shade according to the season, with sufficient places to sit, public toilets and equipment to facilitate exercise for the elderly and ensure that they are located close to the places of residence of older people and are easily accessible. Besides, quiet areas with pleasant soundscapes can be designed to generate a mental rest for the busy life of the city. Pozo Menéndez also acknowledges the role both of public and private sectors to create intermediate spaces that can be dedicated for conviviality activities and contribute to the creation of spaces that can build a feeling of belonging in a community.

One of the strategies highlighted by Gilroy is to implement urban rooms in cities (already set up in the UK, including Newcastle upon Tyne). These rooms are places where both *experts by experience* and *experts by knowledge* can consider together the future of their place, an initiative that should be the start point for changing the planning, promoting an inclusive debate to make older voices count. A similar concept, the living labs in France presented by Castro Gonçalves, showed the interest on going to a more detailed analysis of innovation projects dedicated to cities and urban spaces in order to explore closely the technological, social and political challenges that remain to be met to inclusive co-design environments.

This idea is also present in the contribution by Lantarón, Sancho and Azurmendi, who realised this need of interaction of professionals and older people, in order to define local and adapted solutions that answer the requirements and desires from society dynamics. Although design is a determining factor in the improvement of living conditions and so the integration of long-term care, a new culture of the care model and services is also needed.

This new concept requires also new frameworks for legal, administrative and financial alternatives, to facilitate its implementation. For instance, in the co-housing experience explained by Rojo-Pérez et al., there is still no legal framework for this shared residential model in Spain, which results on problems with financing, lack of management experience and sometimes disagreements between participants during the process that might even lead to the failure of the project development. In this sense, providing evidence to strengthen the knowledge base for promoting this kind of initiatives is important to support the decision-making, both for individuals who seek alternatives for their ageing process and for administrations and services, to better understand the desires and needs of older people and progressively advance in healthier and more inclusive models.

Douglas resumed the process to make ageing *in* place possible in two main elements: first, the design of the physical environment (our homes and neighbourhoods), and second, the related services that need to be accommodated to answer to our changing needs across the life course. This challenges urban planners and designers to develop their skills in facilitation of creative methods of walking alongside citizens to understand their lived experiences and to draw on that knowledge in their work but also has significant resource implications for the planning and design process. The loss of resource because of the scarcity we are all facing in Europe, coupled with the failure to prioritise age issues, is driven partly by competing high priorities and ageism and fuels a lack of understanding of the social and economic contribution of older people to a city and the value in creating ownership and pride in the places in which we live.

Probably, one of the most recent policy framework that shows a clear commitment to address these challenges is in Belgium, where the development of the dementia-friendly communities' initiatives goes hand in hand with the tendency of socialisation of care and community-based care. Baert, Cornelis and De Vriendt explained how Belgian municipalities have important tools to enhance dementia-friendly communities, and in the last months (2021), the Flemish government announced to invest in "caring neighbourhoods" in the coming years. So again, the scale of the neighbourhood is resulting in the alignment of more sustainable and healthier urban lifestyles with the concept of the caring neighbourhoods, as the future model for the organisation of support. This model offers opportunities to increase the quality of life and reduce the costs of care and support.

References

Soja, E. (2010). *Seeking spatial justice* (6th Edition). University of Minnesota Press.

Ondiviela, J. A. (2020). Beyond SmartCities: How to create an Attractive City for Talented Citizens. *Kult-Ur, 7*(13), 205–232. https://doi.org/10.6035/Kult-ur.2020.7.13.8

Chapter 21
Conclusion and Recommendations for Further Work

Elisa Pozo Menéndez and Ester Higueras García

This book contributes to the current dialogue of the post-pandemic cities. Although healthy urban environments were already acknowledging the need for an intertwined perspective addressing both health and urban planning, it has been the COVID-19 outbreak crisis that has accelerated this process. The pandemic has supposed the highlight of many vulnerabilities that were already present, where older people have been the most affected for many reasons. Ageism, loneliness and unhealthy living conditions are still consequences that we can find in our cities. Care and health equity are core principles that shall be integrated in all policies, specially in urban planning and design as the built environment play a key role to provide and facilitate health and wellbeing.

After the COVID-19 pandemic, the "15-minute city" has emerged as a new paradigm for urban renovation and economic recovery. This idea has been consolidated by Moreno in Paris (Moreno et al., 2021), but it is indeed a model that has been already developed in other cities. Cities such as Barcelona, Copenhagen, Helsinki and also those outside Europe, like Vancouver, Melbourne or Tokyo, have been working towards a city of proximity that enables healthier and more sustainable environments to live.

According to the World Health Organization (1946), "*health is a state of complete physical, mental and social well-being and not merely the absence of disease or infirmity*". But remaining in good health in the later stages of life is still a challenge for societies and here is where the physical environment has an important role to promote healthy lifestyles and social cohesion. Cities are rapidly growing and evolving, but the reduction of health inequities and inclusion of all the groups of the population is a necessity that is not yet met.

E. Pozo Menéndez (✉) · E. Higueras García
ABIO Research Group, Universidad Politécnica de Madrid, Madrid, Spain
e-mail: e.pozo@upm.es; ester.higueras@upm.es

E. Pozo Menéndez, E. Higueras García (eds.), *Urban Design and Planning for Age-Friendly Environments Across Europe: North and South*, Future City 19,
https://doi.org/10.1007/978-3-030-93875-8_21

There is a need to align public and private interests and find a balanced development of our territories, in order to have the possibility to choose where to live when getting older and have access to high quality of services and meaningful relationships. Innovation is needed from the multiscale perspective and opening new frameworks to transform cities and territories to achieve healthier and more sustainable spaces to live. In turn, longitudinal studies and assessments are necessary in order to understand the health impact of the built environment on our health, both physical and mental. Also, further research is required in order to understand the ageing process in urban contexts of different social and cultural groups. The quality of life and the feeling of belonging are related to a place and with a community, and both are needed to have more opportunities to become older in a good health.

This book offers a wide vision from different countries and contexts in Europe. Research, study cases and best practices are illustrating the new paradigm of a healthy and age-friendly city, structured in communities and neighbourhoods, answering to the paradigm of the 15-minute city. Encouraging the creation of spaces that can build a feeling of belonging in a community, promoting intergenerational and intercultural meetings and activities naturally, is the current challenge of urban planning and design.

Index

E. Pozo Menéndez, E. Higueras García (eds.), *Urban Design and Planning for Age-Friendly Environments Across Europe: North and South*, Future City 19,
https://doi.org/10.1007/978-3-030-93875-8

The manufacturer's authorised representative in the EU is Springer Nature Customer Service Centre GmbH, Europaplatz 3, 69115 Heidelberg, Germany. If you have any concerns regarding our products, please contact ProductSafety@springernature.com

Printed and bound by CPI Group (UK) Ltd, Croydon, CR0 4YY
30/09/2026
02225493-0018